Mahendra Rai
Donatella Mares
Editors

Plan[...] Anti[...]
Curr[...] and Futu[...] Prospects

Pre-publication
REVIEW . . .

"There is compelling logic in the expectation that the billion-year evolutionary war between plants and fungi will have thrown up some really effective defenses against fungal infection that we could exploit. True, there are no exciting examples to show yet. Nothing like the strobilurins (which are fungus-derived antimycotics), though these provide the object lesson of the promise at the end of the rainbow. The work brought together for the first time in this book sets the scene extremely well and gives a solid account of the experimental successes and failures, challenges and opportunities.

The first chapter gives a succinct and useful introduction to the mode of action of antimycotics and methodology used in this research. It is followed by twenty-three chapters that range widely across different chemical species, different plant species, and different geographical and environmental regions. The result is a comprehensive and extensive account of the detection, evaluation, and nature of plant antimycotics. Several novel procedures for evaluating antifungal activity are described, and I found particular interest in the discussion of light-activated antimycotics. However, the coverage is sufficiently comprehensive for the book to be an essential starting point for anyone (or any company) with ambitions to find that Holy Grail of a new, safe, selective fungicide.

At a time when there is wide recognition of the need in clinical medicine for more effective and greater variety of antifungal agents, this is an extremely valuable overview of the current state of knowledge. All aspects are adequately covered and the overall quality is very high."

Dr. David Moore
Reader in Genetics,
School of Biological Sciences,
University of Manchester,
United Kingdom

NOTES FOR PROFESSIONAL LIBRARIANS AND LIBRARY USERS

This is an original book title published by Food Products Press®, an imprint of The Haworth Press, Inc. Unless otherwise noted in specific chapters with attribution, materials in this book have not been previously published elsewhere in any format or language.

CONSERVATION AND PRESERVATION NOTES

All books published by The Haworth Press, Inc. and its imprints are printed on certified pH neutral, acid free book grade paper. This paper meets the minimum requirements of American National Standard for Information Sciences-Permanence of Paper for Printed Material, ANSI Z39.48-1984.

Plant-Derived Antimycotics
Current Trends and Future Prospects

FOOD PRODUCTS PRESS®
Crop Science
Amarjit S. Basra, PhD
Senior Editor

New, Recent, and Forthcoming Titles of Related Interest:

Conservation Tillage in U.S. Agriculture: Environmental, Economic, and Policy Issues by Noel D. Uri

Cotton Fibers: Developmental Biology, Quality Improvement, and Textile Processing edited by Amarjit S. Basra

Heterosis and Hybrid Seed Production in Agronomic Crops edited by Amarjit S. Basra

Intensive Cropping: Efficient Use of Water, Nutrients, and Tillage by S. S. Prihar, P. R. Gajri, D. K. Benbi, and V. K. Arora

Physiological Bases for Maize Improvement edited by María E. Otegui and Gustavo A. Slafer

Plant Growth Regulators in Agriculture and Horticulture: Their Role and Commercial Uses edited by Amarjit S. Basra

Crop Responses and Adaptations to Temperature Stress edited by Amarjit S. Basra

Plant Viruses As Molecular Pathogens by Jawaid A. Khan and Jeanne Dijkstra

In Vitro Plant Breeding by Acram Taji, Prakash P. Kumar, and Prakash Lakshmanan

Crop Improvement: Challenges in the Twenty-First Century edited by Manjit S. Kang

Barley Science: Recent Advances from Molecular Biology to Agronomy of Yield and Quality edited by Gustavo A. Slafer, José Luis Molina-Cano, Roxana Savin, José Luis Araus, and Ignacio Romagosa

Tillage for Sustainable Cropping by P. R. Gajri, V. K. Arora, and S. S. Prihar

Bacterial Disease Resistance in Plants: Molecular Biology and Biotechnological Applications by P. Vidhyasekaran

Handbook of Formulas and Software for Plant Geneticists and Breeders edited by Manjit S. Kang

Postharvest Oxidative Stress in Horticultural Crops edited by D. M. Hodges

Encyclopedic Dictionary of Plant Breeding and Related Subjects by Rolf H. G. Schlegel

Handbook of Processes and Modeling in the Soil-Plant System edited by D. K. Benbi and R. Nieder

The Lowland Maya Area: Three Millennia at the Human-Wildland Interface edited by A. Gómez-Pompa, M. F. Allen, S. Fedick, and J. J. Jiménez-Osornio

Biodiversity and Pest Management in Agroecosystems, Second Edition by Miguel A. Altieri and Clara I. Nicholls

Plant-Derived Antimycotics: Current Trends and Future Prospects edited by Mahendra Rai and Donatella Mares

Concise Encyclopedia of Temperate Tree Fruit edited by Tara Auxt Baugher and Suman Singha

Plant-Derived Antimycotics
Current Trends and Future Prospects

Mahendra Rai
Donatella Mares
Editors

Food Products Press®
An Imprint of The Haworth Press, Inc.
New York • London • Oxford

Published by

Food Products Press®, an imprint of The Haworth Press, Inc., 10 Alice Street, Binghamton, NY 13904-1580.

Cover design by Marylouise E. Doyle.

Library of Congress Cataloging-in-Publication Data

Plant-derived antimycotics : current trends and future prospects / Mahendra Rai, Donatella Mares, editors.
p. cm.
Includes bibliographical references and index.
ISBN 1-56022-926-8 (alk. paper)—ISBN 1-56022-927-6
1. Antifungal agents. 2. Pharmacognosy. 3. Materia medica, Vegetable. I. Mahendra, Rai. II. Mares, Donatella.

RS431.A62 P56 2003
616.9'69061—dc21

2002027165

CONTENTS

ABOUT THE EDITORS

Dr. Mahendra Rai is Professor and Head of the Department of Biotechnology at Amravati University in Maharashta, India. He has published more than 125 research papers in Indian and foreign journals and five books. He is a member of several scientific societies and has been a National Scholar for five years. He has received several prestigious awards, including the Father T.A. Mathias award (1989) from the All India Association for Christian Higher Education, and the Medini Award (1999) from the Department of Environ and Forest, Government of India. He has approximately two decades of teaching and research experience.

Donatella Mares is Associate Professor, Department of Biology, Botany Section (faculty of pharmacy) at the University of Ferrara in Italy. She teaches advanced courses in plant biology and plant ultrastructures. She has authored numerous articles related to algae and fungi ultrastructure and the mechanism of fungicidal action within many substances, both synthetic and natural. She is a member of Società Botanica Italiana and also of the Italian Society for Electron Microscopy.

CONTRIBUTORS

Acham Abraham, Tropical Botanic Garden and Research Institute, Palode, Trivandrum, Kerala State, India.

Deepak Acharya, Microbiology Research Lab, Danielson College, Chhindwara, M.P. State, India; e-mail <deepak@chhindwaraonline. com>.

M. S. Ali-Shtayeh, Department of Biological Sciences, An-Najah University, Nablus, Palestine.

S. Brocklehurst, Biomathematics and Statistics Scotland, Hannah Research Institute, Ayr, Scotland, United Kingdom.

María Cecilia Carpinella, CEPROCOR, Alvarez de Arenales, Barrio Juniors, Córdoba, Argentina; e-mail: <ccarpine@cordoba.com. ar>.

Jean-Pierre Chaumont, Member of the French National Academy of Pharmacy, Laboratoire de Botanique et de Cryptogamie, Faculté de Médecine et de Pharmacie, Besançon Cedex, France.

Pier Luigi Cioni, Dipartimento di Chimica Bioorganica e Biofarmacia, Via Bonanno, Pisa, Italy.

Aijun Dong, Institute of Research and Development of Drugs, Northeast Pharmaceutical Corporation, Zhonggong North Avenue, Shenyang, Liaoning, China.

R. Daniel Enriz, Química General, Facultad de Química, Bioquímica y Farmacia, Universidad Nacional de San Luis, Chacabuco y Pedernera, San Luis.

Carlos Guillermo Ferrayoli, CEPROCOR, Alvarez de Arenales, Barrio Juniors, Córdoba, Argentina.

Valdir Cechinel Filho, Núcleo de Investigações Químico-Farmacêuticas (NIQFAR)—Universidades do Vale do Itajaí, Itajaí SC, Brasil.

Guido Flamini, Dipartimento di Chimica Bioorganica e Biofarmacia, Via Bonanno, Pisa, Italy; e-mail: <flamini@farm.unipi.it>.

Christina Gip, Golfbanevagen, Hovas, Sweden.

Lennart Gip, Golfbanevagen, Hovas, Sweden.

Kazimierz Glowniak, Department of Vegetable and Medical Herbs, Agriculture University, Lublin, Poland; e-mail: <katwarz@consus.ar.lublin.pl>.

J. B. Hampson, Plant Biology Department, SAC Auchincruive, Ayr, Scotland, United Kingdom.

Ke Hu, Department of Pharmacology and Therapeutics, Grace Cancer Drug Center, Roswell Park Cancer Institute, Elm and Carlton Streets, Buffalo, New York; e-mail: <huke98@hotmail.com>.

Shigeo Iwasaki, Research Center for Biological Functions, The Kitasato Institute, Shirokane, Minato-ku, Tokyo, Japan.

Rana M. F. Jamous, Department of Biological Sciences, An-Najah University, Nablus, Palestine.

Zdenka Kalodera, Department of Microbiology and Department of Pharmacognosy, Faculty of Pharmacy and Biochemistry, University of Zagreb, A. Kovacica, Zagreb, Croatia.

Süheyla Kirmizigül, Ege University, Faculty of Science, Chemistry Department, Organic Chemistry Section, Bornova-Izmir, Turkey; e-mail: <kgul@sci.ege.edu.tr>.

Hisayoshi Kobayashi, Institute of Molecular and Cellular Biosciences, The University of Tokyo, Yayoi, Bunkyo-ku, Tokyo, Japan.

Jinichiro Koga, Bio Science Laboratories, Meiji Seika Kaisha, Ltd., Chiyoda, Sakado-shi, Saitama, Japan; e-mail: <jinichiro_koga@meiji.co.jp>.

Ivan Kosalec, Department of Microbiology and Department of Pharmacognosy, Faculty of Pharmacy and Biochemistry, University of Zagreb, A. Kovacica, Zagreb, Croatia.

Vladimir Kouznetsov, Laboratorio de Química Orgánica Fina, Facultad de Química, Universidad Industrial de Santander A.A., Bucaramanga, Colombia.

Danica Kuštrak, Department of Microbiology and Department of Pharmacognosy, Faculty of Pharmacy and Biochemistry, University of Zagreb, A. Kovacica, Zagreb, Croatia.

S. K. Kyle, Plant Biology Department, SAC Auchincruive, Ayr, Scotland, United Kingdom.

P. G. Latha, Tropical Botanic Garden and Research Institute, Palode, Trivandrum, Kerala State, India.

Narayan Prasad Manandhar, Ka, Naya Baneshwar, Kathmandu, Nepal; e-mail: <valley@mos.com.np>.

Olívia de Matos, Estação Agronómica Nacional—Departamento Fisiologia Vegetal, Oeiras, Portugal; e-mail: <papoila@mail. telepac. pt>.

Elena Mongelli, Cátedra de Farmacognosia, IQUIMEFA-CONICET, Facultad de Farmacia y Bioquímica, Universidad de Buenos Aires, Junín, Buenos Aires, Argentina; e-mail: <emongelli@huemul. ffyb.uba.ar>.

Kyoko Nakashima, Laboratory of Food Material and Engineering, Faculty of Agriculture, Gifu University, Yanagido, Gifu, Japan.

Shigeki Nakashima, Laboratory of Food Material and Engineering, Faculty of Agriculture, Gifu University, Yanagido, Gifu, Japan.

Kazuyoshi Okubo, Laboratory of Food Material and Engineering, Faculty of Agriculture, Gifu University, Yanagido, Gifu, Japan.

María Sara Palacios, CEPROCOR, Alvarez de Arenales, Barrio Juniors, Córdoba, Argentina.

A. K. Pandey, Centre for Forestry Research and Human Resource Development, Poama, Chhindwara, M.P. State, India.

Stjepan Pepeljnjak, Department of Microbiology and Pharmacognosy, Faculty of Pharmacy and Biochemistry, University of Zagreb, A. Kovacica, Zagreb, Croatia; e-mail: <peplj@nana.pharma. hr>.

Cândido P. Pinto Ricardo, Instituto de Tecnologia Química e Biológica, Oeiras, Portugal; e-mail: <ricardo@itqb.unl.pt>. Instituto Superior de Agronomia, Lisboa, Portugal; e-mail: <cdisa@isa.utl. pt>.

S. Rajasekharan, Tropical Botanic Garden and Research Institute, Palode, Trivandrum, Kerala State, India.

Juan C. Ribas, Instituto de Microbiología Bioquímica, Campus "Miguel de Unamuno," Edificio Deapartamental, Universidad de Salamanca, Salamanca, Spain.

Carlo Romagnoli, Department of Animal Biology, Section Botanical Garden, University of Modena and Reggio Emilia, V.le Caduti in Guerra, Modena, Italy; e-mail: <rmc@unife.it>.

Gianni Ruzickova, Plant Biology Department, SAC Auchincruive, Ayr, Scotland, United Kingdom.

Gianni Sacchetti, Department of Biology—Pharmaceutical Biology Lab., University of Ferrara, C.so Porta Mare, Ferrara, Italy.

Makoto Shimoyamada, Laboratory of Food Material and Engineering, Faculty of Agriculture, Gifu University, Yanagido, Gifu, Japan; e-mail: <saponin@cc.gifu-u.ac.jp>.

Subramaniam Sotheeswaran, School of Pure and Applied Sciences, The University of the South Pacific, Suva, Fiji Islands; e-mail: <sotheeswaran@usp.ac.fj>.

Kannoth Panicker Sreekumar, School of Pure and Applied Sciences, The University of the South Pacific, Suva, Fiji Islands.

S. R. Suja, Tropical Botanic Garden and Research Institute, Palode, Trivandrum, Kerala State, India.

K. P. Svoboda, Plant Biology Department, SAC Auchincruive, Ayr, Scotland, United Kingdom; e-mail: <k.svoboda@au.sac.ac.uk>.

Saradha Vasanth, Central Research Institute for Siddha, Arumbakkam, Chennai, India.

Prasad Wadegaonkar, Department of Biotechnology, Amravati University, Amravati, Maharastra, India.

Kenji Watanabe, Laboratory of Food Material and Engineering, Faculty of Agriculture, Gifu University, Yanagido, Gifu, Japan.

Tadeusz Wolski, Department of Pharmacognosy, Medical University, Lublin, Poland; e-mail: <xlow@hipokrates.am.lublin.pl>.

Xinsheng Yao, Department of Natural Products Chemistry, Shenyang Pharmaceutical University, Wenhua Road, Shenyang, Liaoning, China.

Rosendo A. Yunes, Departamento de Química, Universidad Federal de Santa Catarina (UFSC), Brasil.

Susana A. Zacchino, Farmacognosia, Facultad de Ciencias Bioquímicas y Farmacéuticas, Universidad Nacional de Rosario, Suipacha, Rosario, Argentina; e-mail: <szaabgil@citynet.net.ar>.

Rabee A. G. Zayed, Department of Biological Sciences, An-Najah University, Nalbus, Palestine.

Foreword

Among the different creatures living on Earth, the fungi constitute a separate kingdom, being characterized by a peculiar metabolism. Nevertheless, structurally they are eukaryotic organisms. This fact constitutes the main problem in studying and setting forth new antifungal drugs. In fact, because of their low specificity, antifungal drugs are not only active against fungal cells but often cause severe problems in host cells.

In researching new active antimycotic compounds, it is important to focus the interest on drugs with a selective activity, i.e., molecules that can target the fungus cells and not the host. For this reason, new antifungal compounds that act selectively on the main peculiar fungal structures are of interest: (1) compounds that act on the cell wall, which is absent in animal cells and is different in chemical composition from the wall of plant cells, and (2) compounds that affect the fungal plasmalemma, which has a peculiar chemical composition.

Among the synthetic substances used as antifungals, azoles, polyenes such as nystatin and amphotericin, and griseofulvin are mainly employed in the treatment of fungal diseases in animals and humans, but they can provoke important toxicological effects. In the same way, chemical fungicides targeted at plant pathogens in many cases lead to the development of fungicide resistance, and many fungicides are toxic to the environment or produce toxic residues and therefore may be detrimental to beneficial microorganisms.

The development of biological control systems for fungal pathogens is a possible way forward, but these systems are of lower efficacy unless they are integrated using natural organisms and their products.

Phytochemicals from medicinal plants are receiving ever greater attention in the scientific literature, in medicine, and in the world economy in general. The field also has benefited greatly in recent years from the interaction of the study of traditional ethnobotanical knowledge, the application of modern phytochemical analyses, and biological activity studies on medicinal plants. The isolation of natural products has been facilitated during the past twenty years by the advances which have been made in all branches of chemical techniques (chromatography; electrophoresis; mass, H, and NMR spectroscopy, etc.). Increasingly sophisticated methods of analysis are opening new opportunities for understanding the nature and functions of plant constituents and for manipulating them to prepare new phytomedicines.

The chapters in this book represent the emerging scope of the field of antimycotics of plant origin. The book provides a series of reviews about plant secondary products.

The purposes of this book are (1) to provide an essential context to plant-derived antifungal developments, directing the reader to specialized information which cannot be found elsewhere; and (2) to better identify and characterize novel biologically active phytochemicals with pharmacological and/or antifungal properties.

This book contains a multidisciplinary approach for the review and evaluation of the progressing and expanding knowledge in the field of antifungal substances of plant origin.

Alessandro Bruni
Dean of the Faculty of Pharmacy
Department of Biology
Pharmaceutical Laboratory
Ferrara, Italy

Preface

Since the inception of civilization humans have been using herbs against various mycotic infections in plants as well as in human beings. Age-old tropical plant drugs, such as *Allium sativum* and *Azadirachta indica,* are gaining ground as antifungal drugs due to their efficacy in extensive laboratory and animal trials. In the recent past, several antimycotic agents have emerged on the market due to their rapid curative properties, but there has always been a quest for newer antifungal agents of *fungicidal,* rather than *fungistatic* nature. Furthermore, there has been a dramatic increase in the new spectrum of fungal infections, also referred to as opportunistic fungal pathogens. The existing antimycotics, such as azoles, sometimes do not respond well to the newer generation of fungal pathogens, which are associated with secondary infections in AIDS and cancer patients due to the immunocompromised position of the host. Moreover, these antimycotics may cause allergic reactions or other side effects. The medical community is aware of and has developed an estrangement from synthetic antifungals. Consequently, plant-derived antimycotics are attracting the attention of botanists and mycologists because they are natural, cheaper, safer, ecofriendly, and within the reach of the current medical community. For this reason, plant extracts are preferred in the cure of fungal infections. This fact is based on the knowledge that plants have their own defense system against microbial infections in general, and mycotic infections in particular. It is assumed that in the twenty-first century plant-derived antimycotics will create a revolution in the field of a new generation of ecofriendly fungicides for the control of plant as well as human mycotic diseases. With this aim, pharmacologists and microbiologists are delving into the field of plant-derived antimicrobials. There seems to be revitalization of use of traditional antimycotic extracts, essential oils, and herbal drugs, which form the basis of the ancient science of life, i.e., Ayurveda.

The aim of this book is to gather the reviews concerning plant-derived antimycotics to reflect what is being accomplished all over the world. The response from authors was overwhelming. This book is the first of its kind and covers new techniques and plant-based antifungals for humans and plants infected by tiny but mighty fungal pathogens.

We are grateful to Dr. Sudhir Patil, Honorable Vice Chancellor, Amravati University, Amravati, Maharastra, India; Professor S. A. Brown, Principal, Danielson College, Chhindwara, M.P. State, India; and Dr. Katerina Svoboda, Plant Biology Department, SAC Auchincruive, Scotland, for constant encouragement. Thanks are also due to Dr. Mani Jacob, General Secretary, All-India Association for Christian Higher Education, New Delhi, for providing partial financial assistance for this work. We thank Jeemi Malviya and Ravindra Nafde for help in typesetting.

Sincere thanks are due to Dr. Amarjit Basra and Bill Palmer at The Haworth Press for their encouragement and also for their cooperation in bringing this book to publication.

Chapter 1

The Need for New Antifungal Drugs: Screening for Antifungal Compounds with a Selective Mode of Action with Emphasis on the Inhibitors of the Fungal Cell Wall

Susana A. Zacchino
Rosendo A. Yunes
Valdir Cechinel Filho
R. Daniel Enriz
Vladimir Kouznetsov
Juan C. Ribas

INTRODUCTION

Fungal infections have increased in frequency in the past two decades, especially involving immunocompromised patients (Walsh, 1992; Selitrennikoff, 1995; Jackson et al., 1995). Invasive fungal infections as well as dermatomycoses produced by fungal organisms with even low virulence

---

The authors wish to thank Professor Mahendra Rai for his kind invitation to participate in this text, and Professor A. Durán (CSIC/Univ. de Salamanca, Spain) for helping and supporting part of this work. Grants from the Agencia Nacional de Promociones Cientificas y Tecnológicas de la Argentina to S.A.Z. (grant # 06-06454), from Conselho Nacional de Desenvolvimento Cientifico e Technológico/CNPq (Brazil) to R.A.Y., from Universidad do Vale do Itajai, (SC Brazil) to V.C.F. and Colciencias N°.-1 15-05-353-96 (Colombia) to V.K. are gratefully acknowledged. This work is part of the Iberoamerican project PIBEAFUN (search and development of new antifungal agents) from RIPRONAMED (Iberoamerican Network on Medicinal Natural Products) which belongs to the Iberoamerican Program of Science and Technology for the Development (CYTED). S.A.Z. wishes to thank collaboration from RIIDDMED (Iberoamerican Network on Investigation, Design, and Development of Medicines) of CYTED and particularly to Professor Mahabir Gupta (CIFLORPAN; University of Panama, CYTED Sub-program X), Professor Arturo San Feliciano (University of Salamanca, Spain, RIPRONAMED) and Professor Antonio Monge (University of Navarra, Spain, RIIDDMED). Finally, we thank the many collaborators whose names appear in the reference list.

can be life-threatening (Georgopapadakou and Tkacz, 1995; Nagiec et al., 1997) for patients such as neonates, cancer patients receiving chemotherapy, organ transplant patients, and burn patients, apart from those with acquired immunodeficiency syndrome (AIDS). Other risk factors include corticosteroid and antibiotic treatments, diabetes, lesions of epidermis and dermis, malnutrition, neutropenia, and surgery (Selitrennikoff, 1992). In addition, an increasing number of normal individuals, including children in third-world nations (Freixa et al., 1998; Ablordeppey et al., 1999) that suffer deficient sanitation and education, have fungal infections—especially those involving the skin and mucosal surfaces.

Although it appears that many drugs are available for the treatment of systemic and superficial mycoses, there are in fact a limited number of efficacious antifungal drugs (Timberlake et al., 1994; Selitrennikoff, 1995). Many of the drugs currently available have undesirable effects or are very toxic (amphotericin B), are fungistatic and not fungicidal (azoles), or lead to the development of resistance, as in flucytosine (also called 5-fluorocytosine or 5-FC) (White et al., 1998). According to Polak (1999) ideal drugs to cure fungal infections have not been discovered yet. In the meantime, resistance to currently available antifungal agents continues to grow (Bartroli et al., 1998a). Although combination therapy has emerged as a good alternative to bypass these disadvantages (Beninger et al., 1994; Bartroli et al., 1998; Polak, 1999), there is a real need for a next generation of safer and more potent antifungal agents (Selitrennikoff, 1992; Bartroli et al., 1998a).

Certain conditions are required for a compound to be a good antifungal agent: it must be fungicidal rather than fungistatic and have a broad spectrum of activity, a minimum emergence of resistant strains, and a selective mechanism of action. In addition, the agent should have minor toxic side effects and good availability (Ablordeppey et al., 1999; Polak, 1999).

This chapter reviews the most common drugs currently used in antifungal therapy, and the benefits and drawbacks derived from their mechanisms of action. Other biochemical targets useful for the discovery of selective antifungal drugs are also covered. The methodology to find compounds with inhibitory activities against the fungal cell wall, together with some efforts made by our research group to discover new molecules with inhibitory properties of the synthesis or assembly of the cell wall polymers, is reported at the end of this review.

DRUGS OF CLINICAL IMPORTANCE: BENEFITS AND DRAWBACKS

The most common drugs currently in use for treating systemic fungal infections are amphotericin B, 5-flucytosine, fluconazole, ketoconazole, and itraconazole. Naftifine, in turn, is in clinical use as a topical antimycotic. Table 1.1 shows the chemical characteristics for, the mode of action of, and the type of fungal infections treated with each drug (Selitrennikoff, 1992, 1995).

Drugs That Form Complexes with Ergosterol

Amphotericin B is a drug of the polyene class of antifungal agents, which has long been considered the most effective of the systematically administered antifungal agents (White et al., 1998). It has the broadest spectrum of action among all antifungal agents and acts as a fungicide by binding to the sterols of the fungal cell membrane causing its disruption and causing leakage of intracellular contents (Hammond et al., 1974; Selitrennikoff, 1992).

Unfortunately, the frequent occurrence of nephrotoxicity and side effects including fever, anemia, nausea, and vomiting (Selitrennikoff, 1995) have limited the utility of this drug. In addition it is associated with water insolu-

TABLE 1.1. Antifungal Drugs Currently in Use

Antifungal Drug	Type of Chemical Compound	Fungal Infection	Mode of Action
Amphotericin B	Polyene	Aspergillosis, Blastomycoses, Candidasis, Cryptococcosis, Histoplasmosis, Paracoccidiodomycoses	Complex with ergosterol in fungal plasma membrane
5-Fluorocytosine	Fluor substituted pyrimidine	Candidasis or in combination therapy	Inhibits DNA and RNA synthesis
Terbinafine Naftifine	Allylamines	Dermatomycoses or mucosal infections	Inhibits ergosterol biosynthesis by blocking squalene epoxidase
Ketoconazole Miconazole Itraconazole Fluconazole	Imidazol Triazoles	Blastomycoses Coccidiodomycoses Histoplasmosis Paracoccidiodomycoses	Inhibits ergosterol biosynthesis by blocking 14-demethylation of lanosterol

bility (Ablordeppey, 1999). Resistance to amphotericin B is rare and polyene resistance has not been a major clinical problem to date; some combinations of amphotericin with other drugs are in clinical use in attempts to reduce toxicity, for example, amphotericin B and 5-FC, or amphotericin plus azoles (Polak, 1999).

Drugs That Inhibit the RNA/DNA Synthesis

5-Fluorocytosine acts by disrupting pyrimidine metabolism, avoiding the synthesis of RNA/DNA, and consequently the proteins of the fungal cell. The active compound is its metabolite, 5-fluorouracil, which is obtained through desamination by the enzyme cytosine deaminase inside the fungal cell. 5-Fluorocytosine has low toxicity in mammalian cells due to the absence of this enzyme, thus preventing the presence of the cytotoxic 5-fluorouracil (Selitrennikoff, 1995).

5-Fluorocytosine has limited clinical utility owing to its narrow spectrum of action, the presence of hematological toxicity, and the rapid development of resistance, which reduces its efficacy in combination therapy (Polak, 1999). As stated in the previous section, 5-FC is used in combination with amphotericin B due to their synergistic effect (Medoff et al., 1971), and with azoles (White et al., 1998).

Drugs That Act by Inhibiting the Biosynthesis of Ergosterol

Sterols are hydrophobic, passive components of the fungal cell membranes whose primary role is to maintain membrane fluidity and integrity. Sterols play vital hormonal, regulatory, and architectural roles in fungi (Oehlschlager and Czyzewska, 1992; Parks et al., 1995).

The principal sterol of yeast and many other fungi is ergosterol. It appears to fulfill the multiple roles attributed to sterols. If ergosterol is absent, critical physiological functions needed for the survival of the fungal cells may not be accomplished (Parks et al., 1992). Since the regulation of sterol biosynthesis varies among animals and fungi, the inhibition of some of its steps seems to offer significant prospects in the rational design of therapeutic agents (Barrett-Bee and Ryder, 1992), and the inhibition of the synthesis of ergosterol has been used as a target for detecting antifungal agents. However, the biosynthetic steps prior to squalene oxide are essentially the same in both animals and fungi, and to date all antifungal agents in use inhibit some of those common steps of the biosynthesis of ergosterol in fungi and cholesterol in mammals. Figure 1.1 shows the ergosterol and cholesterol biosynthesis pathways and the steps that are inhibited by some of the differ-

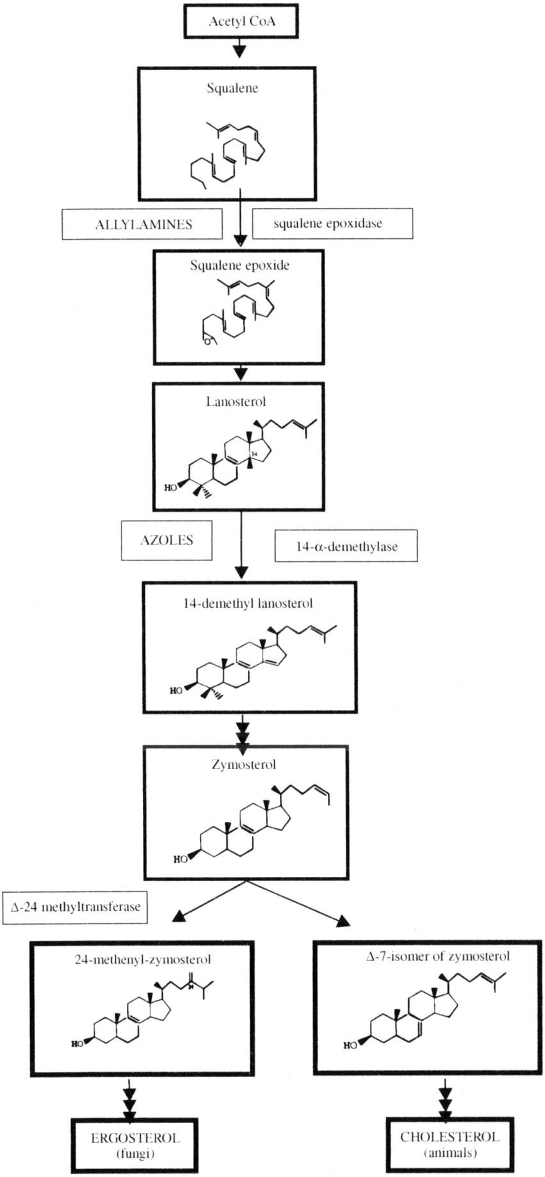

FIGURE 1.1. Biosynthesis of ergosterol and cholesterol. Some of the involved enzymes (to the right of the arrows) and their known inhibitors in clinical use (to the left of the arrows).

ent antifungal agents currently in use. Azoles inhibit 14-α-demethylase, and allylamines inhibit the enzyme squalene epoxidase. There clearly is a potential for mammalian toxicity from these agents as most inhibitors affect the bulk sterol composition in both invading and host organisms (Oehlschlager and Czyzewska, 1992). This toxicity was particularly observed on enzyme pathways that metabolize sterols to form hormones—especially sex hormones (Santen et al., 1983).

The most promising antifungal agent dealing with inhibition of the ergosterol biosynthesis would be one inhibiting the step of 24-methenylation, which is unique to the ergosterol pathway. Although some efforts have been made in searching for inhibitors of this step (Rahier et al., 1980, 1984; Narula et al., 1981; Chung et al., 1998), an antifungal agent with this mechanism of action has not yet been discovered.

Inhibition of Fungal Squalene Epoxidase

Epoxidation at the 2,3 position is the first step in the conversion of the 30-carbon-chain squalene to the tetracyclic sterol skeleton, which is catalyzed by the enzyme squalene epoxidase. It is a key enzyme in the biosynthetic sequence from acetyl CoA to ergosterol in fungi and to cholesterol in mammals (Ryder, 1987; Ryder et al., 1994). Squalene epoxidase has emerged as a major target for antifungal drugs because it is not a member of the P-450 family, it can be inhibited selectively, and its inhibition results in fungicidal action. Although fungal and mammalian enzymes appear to be quite similar in properties, the rat liver enzyme is several orders of magnitude less sensitive than the *Candida albicans* enzyme to inhibition by allylamines (Barrett-Bee and Ryder, 1992).

Naftifine, a fungicide allylamine with a broad spectrum of action, was the first specific inhibitor of squalene epoxidase to be reported (Rahier et al., 1980), and it is clinically used as topical antifungal agent. Its analog, terbinafine, is also in clinical use as it shows both oral and topical efficacy (Ryder et al., 1984; Ryder, 1992). Nevertheless, considering that the fungicidal action of inhibitors of squalene epoxidase (Petranyl et al., 1984; Ryder, 1985, 1990; Ryder and Dupont, 1985; Nussbaumer et al., 1991, 1994; Rashid et al., 1995) is due to the accumulation of squalene, which is toxic to cells, Georgopapadakou et al. suggested (1987) that an important point to be determined is whether squalene epoxidase could be a useful target for antifungal agents or for modulators of cholesterol biosynthesis. In addition, it was reported that treatment of fungi-infected plants with allylamines also resulted in the modification of the whole sterol composition in the host plant (Barrett-Bee and Ryder, 1992).

Thiocarbamates such as tolnaftate are also antifungal agents that act by inhibiting squalene epoxidase (White et al., 1998; Oehlschlager and Czyzewska, 1992; Barrett-Bee and Ryder, 1992).

Inhibitors of 14-Demethylation of Lanosterol

Once lanosterol has been formed by the ciclization of squalene epoxide, it undergoes several sequential transformations to form ergosterol (Barrett-Bee and Ryder, 1992), which, upon removal of the 14-α-methyl, leads to 14-demethyl lanosterol. The demethylation is catalyzed by the 14-α-methyldemethylase, a membrane enzyme bound to oxidase (Oehlschlager and Czyzewska, 1992).

The major antifungal drugs, the azoles, inhibit this demethylation step (Kelly et al., 1995), producing an accumulation of trimethylated sterols that are bulkier than ergosterol, making fungal membranes less efficient (Barrett-Bee and Ryder, 1992; Bartroli et al., 1998b). This type of compound possesses one or more five-membered rings containing either two (imidazoles) or three (triazoles) nitrogens (Selitrinnikoff, 1995). Though they have a broad spectrum of action, they are fungistatic but not fungicidal, thus producing recurrence. Among azoles, ketoconazole, itraconazole, and fluconazole are important therapeutic agents in clinical use.

Ketoconazole was the first oral water-soluble therapeutic antifungal azole and is used currently in nonimmunosuppressed patients. Nevertheless, it possesses many adverse effects such as hepatotoxicity, nausea, vomiting, and anorexia, and its toxicity is also associated with the inhibition of testosterone biosynthesis (Santen et al., 1983; Selitrinnikoff, 1995).

Fluconazole and itraconazole, azoles derived from ketoconazole, are the azoles of choice for the treatment of deep mycoses. They possess a broader spectrum of action than does ketoconazole and a less toxic profile, but they have not fulfilled all hopes (Polak, 1999). Itraconazole is, together with amphotericin B, the only available drug to treat infections produced by *Aspergillus fumigatus*. However, it showed only moderate response rates (Bartroli et al., 1998a), and side effects such as allergic rash and gastrointestinal disorders have been reported with its use (Selitrinnikoff, 1995). Fluconazole, in turn, which showed to be active against *C. albicans* in vitro and in vivo, is the preferred azole due to its broad spectrum of action, high oral bioavailability, and low occurence of secondary effects (White et al., 1998; Bartroli et al., 1998a). Nevertheless, resistance has been reported to occur in response to cumulative doses (Sangeorzan et al., 1994; Sandven et al., 1997; White et al., 1998).

Many other azole derivatives and their structure-activity relationships have been described in the literature (Bartroli et al., 1998a,b) and some of them are still in clinical development (Ablordeppey et al., 1999).

Resistance to all azoles in clinical use is emerging due to the prolonged theraphies in immunocompromised patients (Kelly et al., 1995; Sanglard et al., 1995; Onishi et al., 2000). Azoles, for example, should probably not be used prophylactically for oral or vaginal candidiasis since they may predispose to resistance (White et al., 1998). The drawbacks of the antifungal agents in clinical use emphasize the need for new, potent, and safe antifungal drugs.

TRENDS IN THE SEARCH FOR NEW ANTIFUNGAL AGENTS

New Compounds from Natural Sources

One of the trends in the search for new antifungal compounds is to look for new structures from natural sources, preferably based on an ethnobotanical approach, screening plants that have been used in traditional medicine for the treatment of various infectious diseases. Shaman Pharmaceuticals' program follows this approach and, according to Hector et al.'s report (1994), it has found a high rate of activity, particularly against dermatophytic fungi.

Many researchers, particularly from countries with a rich biodiversity, have contributed to the detection of new antifungal compounds from medicinal plants. In most cases, active extracts are detected by using agar diffusion (Carrillo-Muñoz et al., 1994), bioautography (Rahalison et al., 1991), or agar (Mitscher et al., 1972) or broth (Vanden Berghe and Vlietink, 1991) dilution assays, the latter ones seeming to be the most convenient methods for routine testing of complex samples such as plant extracts. In these methods, dilution series of samples are mixed with a suitable culture medium (solid or liquid) and then inoculated with a quantified fungal inoculum. After incubation, the lowest compound concentration showing no visible fungal growth after the incubation time (the minimum inhibitory concentration, or MIC) is recorded (Figure 1.2). By plating out samples of the completely inhibited cultures to a medium without the test compound, or any other antifungal agent, the minimum fungicidal concentration (MFC) can also be determined.

In recent years, diffusion (Sur-Altiner, Gurkan, Sarioglu, and Tuzlagi, 1997; Sur-Altiner, Gurkan, Sarioglu, Tuzlagi, and Büyükbaba, 1997; Mehta et al., 1997; Ohiri and Okelu, 1997; Freixa et al., 1998; Thopp et al., 1998;

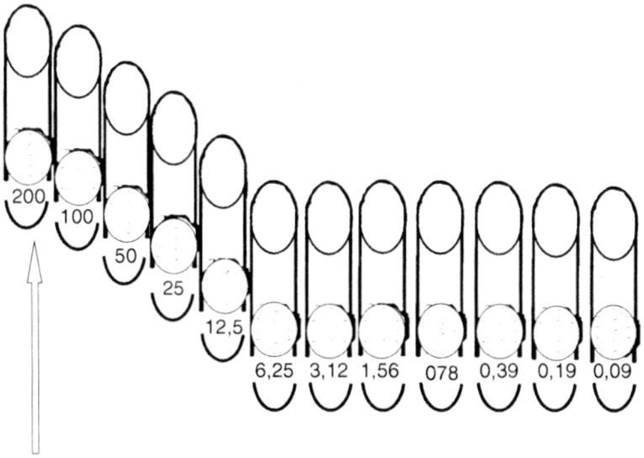

MIC: Minimum Inhibitory Concentration

FIGURE 1.2. Determination of MIC in broth dilution assays

Rajbhandari and Schöpke, 1999; Mahasneh and El-Oqlach, 1999), dilution (Salie et al., 1996; Larhsini et al., 1997; Hammer et al., 1998; Elmahi et al., 1997; Zacchino et al., 1997; Zacchino, Rodriguez, et al., 1998; Zacchino, Santecchia, et al., 1998; Zacchino et al., 1999; Lentz et al., 1998; Esimone and Adikwu, 1999; Saxena and Sharma, 1999; Lee et al., 1999), and bioautography techniques (Rahalison et al., 1995; Chandravadana et al., 1997; Nidiry, 1999; Andrianaivoravelona et al., 1999) have proven very useful not only in detecting natural extracts with antifungal properties but also in performing bioassay-guided fractionation and in discovering a great number of interesting and varied antifungal structures. For example, these techniques have been used for the isolation of antifungal neolignans and phenylpropanoids from *Myristica fragrans* (Miyazawa et al., 1996; Zacchino et al., 1997; Zacchino et al., 1999), polyacetylenes from *Bellis perennis* (Avato et al., 1997), bisphosphocholines from *Irlbachia alata* (Lu et al., 1999), triterpenoids from *Terminalia bellerica* (Nandy et al., 1997) and *Lantana indica* (Verma et al., 1998), anthraquinones from *Morinda lucida* (Rath et al., 1995), sesquiterpenelactones from *Inula viscosa* (Maoz et al., 1999), biflavones from *Cupressocyparis leylandii* (Krauze-Baranowska et al., 1999), and many other compounds which could contribute to the development of new antifungal agents (Iwasa et al., 1997; Al Magboul and Bashir, 1997; Smirnov et al., 1998). Nevertheless, these methods must be used as a preliminary screening device since they do not give information about the mechanism of action and, therefore, about the selectivity or the antifungal properties needed to bypass the drawbacks of the drugs in clinical use.

Selective Targets

Another trend in the search for new antifungal agents is to look for antifungal compounds with an entirely different and selective mode of action. According to Selitrennikoff (1992), new antifungal agents must inhibit a pathogen's molecular process which is either absent or sufficiently different in the host, so that the host metabolism will be minimally affected. Although fungi and humans are both eukaryotic, the extensive differences between them are yet to be exploited in the development of new antifungal therapies (Timberlake et al., 1994).

In fact, some other antifungal targets have been used apart from complexation with ergosterol, ergosterol biosynthetic pathway, and RNA/DNA synthesis, which were discussed previously in this review. Microtubules, protein N-myristoyl transferase (Devadas et al., 1995), sphingolipid biosynthesis (Nagiec et al., 1997), dihydrofolate reductase (Chan et al., 1995), or sulfate assimilation pathway (Aoki et al., 1995) have been used as targets for mechanistic-based studies on antifungal compounds. Because of this, the search for inhibitors of fungal topoisomerase enzymes, as well as inhibitors of fungal cell wall synthesis, has recently attracted considerable attention. In the following sections the main advantages of both targets are discussed, with particular focus on the methodology used to discover fungal cell wall inhibitors.

Topoisomerases As Targets for Selective Antifungal Drugs

DNA topoisomerases are a group of ubiquitous enzymes that catalyze the interconversion between both topological forms (relaxed and supercoiled) of the covalently closed circular DNA. Many important cellular processes are sensitive to the DNA shape, including DNA replication, repair, transcription, and recombination (Shen et al., 1989; Jackson et al., 1992). Based on their mechanism of action, DNA topoisomerases are classified into two types: type I enzymes can break and rejoin one DNA strand at a time; type II enzymes work by making concerted breaks in both strands (Jiang et al., 1997).

DNA topoisomerases have been demonstrated to be the target of a number of antineoplasic (Liu, 1989) and antibacterial (Shen, 1993) agents whose effects arise from the inhibition of the religation step of the topoisomerase reaction, resulting in the stabilization of a catalytic intermediate complex enzyme-DNA drug. This complex would act as a cellular poison, leading to cell death (Pilch et al., 1997).

In turn, DNA topoisomerase I seems to be a promising target for antifungal drugs since its fungal enzymes have proven to be quite different in sensitivity to some drugs from their human counterpart, thus opening the possibility of using them for detecting selective antifungal agents. According to this, Fostel and colleagues (1996) showed that 5-OH-1H-indolacetic acid (5HIAA) and the compound A3253 possess a different capacity of inhibition for human and *Candida* topoisomerases I. In turn, Goldman et al. (1997) showed that, unlike human topoisomerase, *Aspergillus* topoisomerase I is completely resistant to monobenzimidazoles, protoberberines, and nitidine. In addition, studies with *Cryptococcus neoformans* and with *Candida albicans* showed that fungal and mammalian topoisomerases I are structurally different (Taylor et al., 1996; Jiang et al., 1997; Del Poeta et al., 1999). Because there are identifiable differences in the topoisomerase enzymes of fungi and mammals, the goal is to look for inhibitors with differential activity against the fungal enzyme that have a favorable toxicological profile (Hector et al., 1994).

Fungal Cell Wall As a Target for Selective Antifungal Drugs

The cell wall is a multilayer dynamic structure, essential to the integrity and shape of the fungal cell, whose function is to counteract the osmotic forces that could otherwise produce the lysis of the fungal cell. The fungal cell wall is unique to lower eukaryotes, therefore representing a useful target in discovering selective drugs for the treatment of fungal infections without producing toxicity in the host (Tkacz, 1992).

Although fungi exhibit a considerable diversity in their cell wall structure, the major components are (1,3)-β- and (1,6)-β-glucans, chitin, and mannoproteins. Polysaccharides appear to have a structural function, whereas mannoproteins may play different roles in the maintenance of the overall architecture of the walls (Sentandreu et al., 1994; Kollar et al., 1997).

Inhibition of (1,3)-β-Glucan Synthase Activity

(1,3)-β-Glucan is essential for the normal growth and development of fungi. Its polymerization is catalyzed by the (1,3)-β-glucan synthase enzyme, using UDP-Glc as a substrate. Since mammals do not have any of the enzymatic machinery to synthesize this polymer, (1,3)-β-glucan synthase is an attractive target for antifungal drugs (Selitrennikoff, 1995).

The most important known inhibitors of (1,3)-β-glucan synthase enzyme are the glycolipids papulacandins and the lipopeptides echinochandins and pneumocandins. Recently, a group at Merck Research Laboratories has reported the identification of natural acidic terpenoids that selectively inhibit the (1,3)-β-glucan synthase enzyme (Onishi et al., 2000). In addition, sugars, uridine dinucleotides, and dyes were found to inhibit (1,3)-β-glucan synthesis (Selitrennikoff, 1995).

Papulacandins are a family of closely related glycolipids composed of a spirocyclic diglycoside (galactose and glucose), with each sugar molecule linked by ester bonds to an insaturated fatty acid. The differences among them are the type of fatty acid side chains. Figure 1.3 shows the structure of papulacandin B, isolated from *Papularia sphaerosperma,* by Ciba-Geigy in 1975. Papulacandins showed potent growth-inhibitory effects against *Candida albicans,* caused morphological aberrations, and specifically inhibited the (1,3)-β-glucan synthase (Varona et al., 1983.; Kopecka, 1984; Davila et al., 1986; Jackson et al., 1995). Papulacandins appear to be very low in toxicity and show a potent inhibitory activity, both in vitro and in vivo. However, their effectiveness in animal models is poor and their low solubility in water make them orally inactive.

Lipopeptides are compounds consisting of a linear or cyclic peptide of five to fifteen amino acids and one or more fatty acid groups. Echinocandin (Figure 1.4), the first reported antifungal lipopeptide, was discovered in *Aspergillus fumigatus* var. *echinulatus* by Ciba-Geigy in 1974 (Sawistowska-Schroder et al., 1984).

However, since echinocandin B showed moderate toxicity in vivo, poor solubility in water, and a narrow spectrum of antifungal activity, it is not

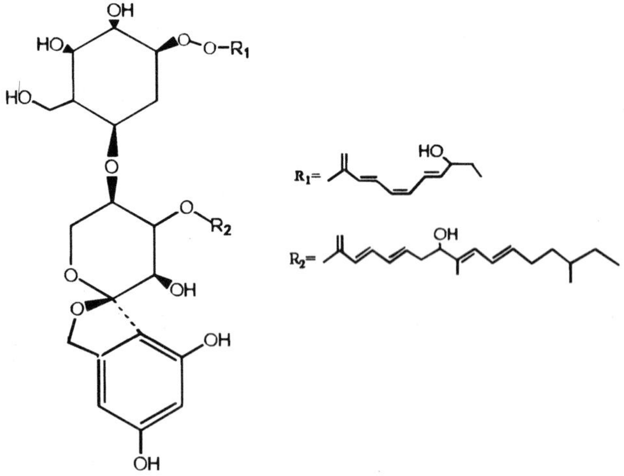

FIGURE 1.3. Structure of Papulacandin B

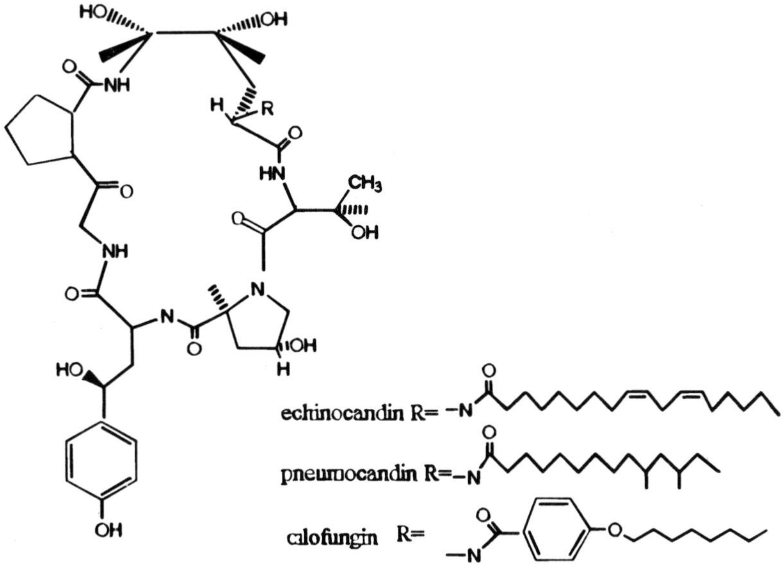

FIGURE 1.4. Different lipopeptide inhibitors of the fungal (1,3)-β-glucan synthase

useful as a therapeutic agent. A structurally related natural compound, aculeacin A, was discovered by Toyo Jozo, but it possesses the same drawbacks as echinochandin B (Mizoguchi et al., 1977). Semisynthetic derivatives of echinochandin B were assayed, leading to the discovery of cilofungin (*N*-*p*-octyloxybenzoyl echinochandin B, Figure 1.4), which was one the best studied echinochandin-type antifungal agents by Lilly Laboratories and the first echinochandin-type compound tested in clinical trials. It is as effective as amphotericin B in treating fungal infections. But contrarily to amphotericin B, it is not toxic to laboratory animals. Nevertheless, problems in formulation stopped the efforts of Eli Lilly to develop it as an antifungal drug (Debono et al., 1989; Denning and Stevens, 1991).

Other natural echinocandin-type (1,3)-β-glucan synthase inhibitors, named pneumocandins (Figure 1.4), were isolated from extracts of *Zalerion arboricola* by Merck in 1989, but unfortunately they were water insoluble. The semisynthesis of many analogs was induced in order to improve the efficacy and the water solubility of this type of compound, and new derivatives, such as MK 0991, reported by Merck (Abbruzzo et al., 1997), and LY 303366, developed by Eli Lilly (Petraitiene et al., 1999), were found to be potent inhibitors of (1,3)-β-glucan synthase in vitro and to display a high efficacy in vivo. These derivatives have actually been tested in clinical trials.

Inhibition of Chitin Synthase Activity

Chitin, a β-1,4-linked polymer of *N*-acetylglucosamine, is one of the essential components of the fungal cell wall and is synthesized by the enzyme chitin synthase (Cabib et al., 1987).

Three chitin synthases have been described in *Saccharomyces cerevisiae* (Chs1p, Chs2p and Chs3p) which possess functional analogies with CaChs1p, CaChs2p, and CaChs3p reported for *Candida albicans,* respectively (Cabib et al., 1987; Shaw et al., 1991; Sudoh et al., 2000). Chs1p repairs damaged chitin during cell separation; Chs2p is required for primary septum formation; and Chs3p is reported to synthesize the chitin ring between mother and daughter cells and to be necessary for chitin synthesis around the cell during vegetative growth, in the shmoos of the cells during mating, and in the sporulation process (Nagahashi et al., 1995; Sudoh et al., 2000). In *Saccharomyces cerevisiae,* the simultaneous disruption of *CHS2* and *CHS3* genes is lethal, while in *Candida albicans* it appears that inhibition of CaChs1p causes cell growth arrest, and simultaneous disruption of *CaCHS1* and *CaCHS2* genes is lethal.

In vitro inhibitors of chitin synthase have not always been shown to be effective antifungal agents, mainly due to their differential inhibition of the different chitin synthases (Georgopapadakou, 1992). The known competitive inhibitors of chitin synthases polyoxin D, nikkomicin X, and nikkomicin Z, substrate analogs of the UDP-*N*-acetylglucosamine (Figure 1.5), showed differential inhibition of chitin synthetases 1 and 2 from *S. cerevisiae* (Cabib, 1991), and more recently nikkomicin Z showed to be a more specific in vitro and in vivo inhibitor of Chs3p from *S. cerevisiae* (Gaughran et al., 1994). In addition, (±)-cathechin and (–) epicathechin, isolated from *Taxus cuspidata* (Kim et al., 1999), have been shown to be inhibitors of chitin synthase 2 but not of Chs3p, and ursolic acid isolated from *Crataegus pinnatifida* (Jeong et al., 1999) selectively inhibits chitin synthase 2 from *S. cerevisiae.*

Although some inhibitors of the fungal cell wall are now in clinical trials, new structures with selective modes of action, including the inhibition of the fungal cell wall, are still needed and are actively being sought at many laboratories in order to develop useful antifungal drugs. The next section deals with the current available methods in the search for new antifungal drugs with inhibitory activities against fungal cell wall synthesis.

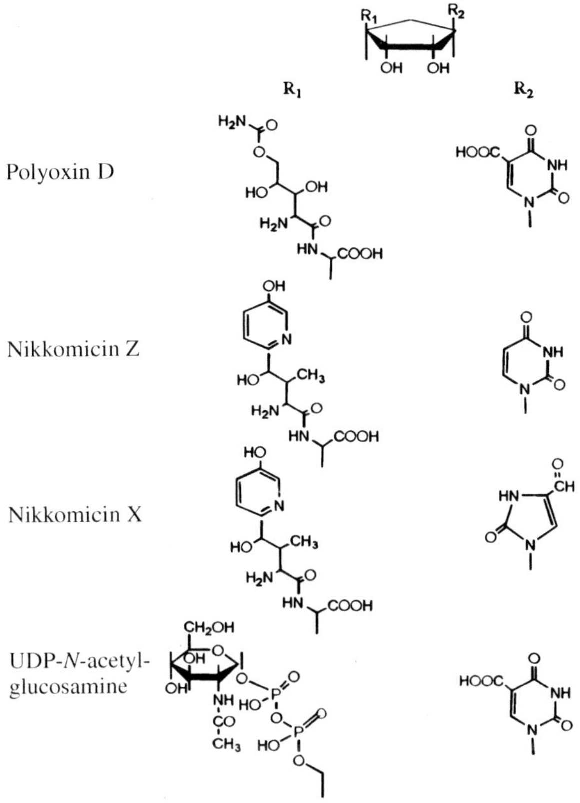

FIGURE 1.5. Structure of chitin synthase inhibitors and the substrate of chitin biosynthesis. UDP-*N*-acetyglucosamine

CURRENT METHODS FOR THE DETECTION OF FUNGAL CELL WALL INHIBITORS

In the screening for compounds that could act by inhibiting the fungal cell wall polymer biosynthesis or assembly, the following methodology is currently in use:

Whole Cell Assays

Whole cell assays allow the detection of inhibitors of fungal cell wall polymer biosynthesis or assembly, but they do not give information about the inhibition step. They are based on the following premises:

- Susceptible organisms undergo morphological changes upon treatment with fungal cell wall inhibitors (Gunji et al., 1983; Fukushima et al., 1993).
- An agent that disrupts cell wall integrity will cause lysis in the absence of an osmotic support. Nevertheless, in its presence, fungal cells can grow to a much higher concentration of the cell wall inhibitor (Frost et al., 1995).

The following assays can be used in a first instance of screening.

Hyphal Malformation Inducer Assay

Screening to detect activities inducing morphological abnormalities in fungal cells has demonstrated to be useful to detect specific inhibitors of the fungal cell wall. Aculeacin and polyoxin, for example, known inhibitors of glucan and chitin synthesis respectively, caused swelling of yeasts or filamentous fungi, among other morphological changes (Selitrennikoff, 1995).

Nevertheless, this method must be used with cautious since morphological changes of fungal cells caused by antibiotics are considerably affected by physiological conditions such as medium composition, growth phase, or concentration of the compound. In addition, when testing complex substances such as plant extracts, sometimes the presence of other compounds causes antagonic suppression of the antifungal effect (Gunji et al., 1983).

Sorbitol Protection Assay

Although fungal morphology has been used as a primary screening method for novel antifungal agents, it is laborious. As an alternative, Frost et al. (1995) developed the Sorbitol Protection And Morphology assay (SPAM). It is a whole-cell screening method based on the fact that sorbitol-protected cells can grow in the presence of fungal cell wall inhibitors, while at these concentrations the cells will lyse in the absence of an osmoprotector. When paired dilution assays are conducted with and without sorbitol, in the presence of a fungal cell-wall inhibitor, a great MIC difference is observed after incubation. The scheme in Figure 1.6 clearly shows this osmotic-protection effect. The assay can be used with extracts as well as with pure compounds.

Neurospora crassa *Assay*

Neurospora crassa assay is an agar diffusion method whose advantage is to allow a macroscopical detection of fungal cell wall inhibitors (Fukuda

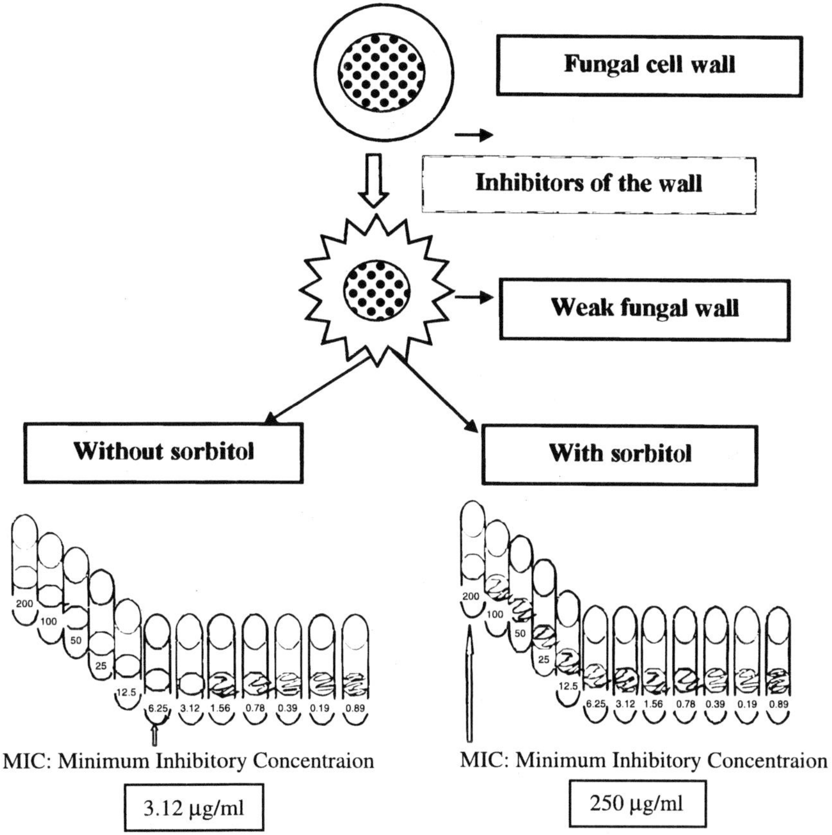

FIGURE 1.6. Sorbitol protection assay: Results that can be obtained, in the presence or in the absence of sorbitol, by a compound that acts by inhibiting synthesis or assembly of fungal cell wall polymers.

et al., 1991). Test compounds are deposited in paper disks over an *N. crassa*-inoculated agar plate. After 24 hours incubation, the types of inhibition halos around the disk are observed. Fungal cell wall inhibitors macroscopically produce mottled inhibition zones due to the presence of surviving protoplasts, whereas antifungal agents that do not disrupt the cell wall, such as ketoconazole, produce clear inhibition zones (Figure 1.7). The microscopic observation of those hazy inhibitor-produced zones shows shortening and branching of the *N. crassa* hyphae (Figure 1.8B), which in normal conditions grow in a branched diffuse way (Figure 1.8A). This morphological growth is due to a sublethal inhibitor concentration in the area.

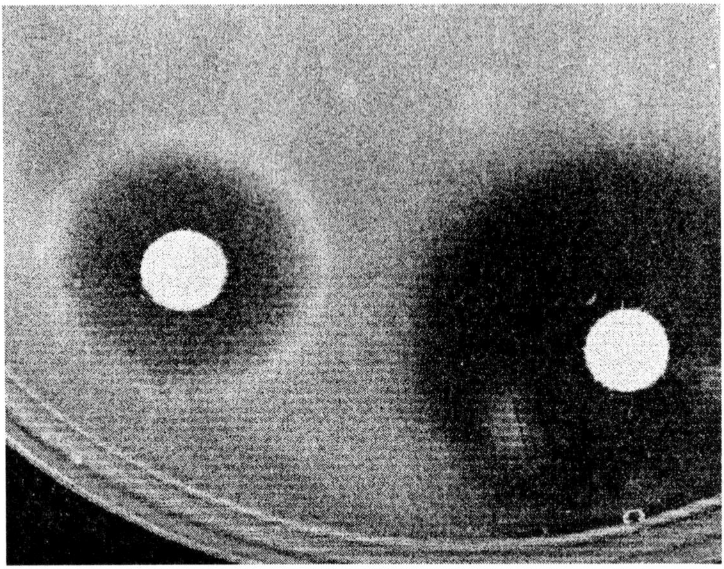

FIGURE 1.7. Positive result of the *Neurospora crassa* assay. At left, the hazy halo produced by an antifungal compound that acts by inhibiting fungal cell wall polymer synthesis or assembly. At right, clear inhibition halo produced by ketoconazole.

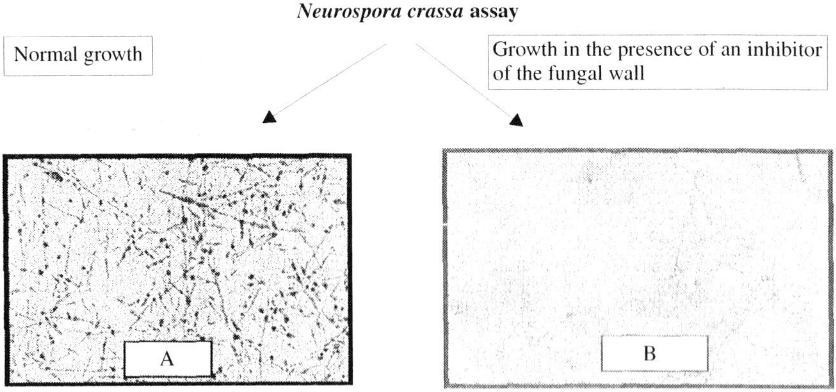

FIGURE 1.8. Microscopic appearance of *Neurospora crassa* hyphae in the presence (B) or absence (A) of a fungal cell wall inhibitor.

Enzymatic Assays

Inhibition of (1,3)-β-D-Glucan Synthase Enzyme

This assay measures the incorporation of glucose from soluble [^{14}C] UDP-glucose into insoluble linear [^{14}C] (1,3)-β-D-glucan by the (1,3)-β-D-glucan synthase enzyme, in the presence and in the absence of inhibitors (Figure 1.9). After incubation, insoluble labeled glucan is measured in a β-emission counter as previously described (Cabib et al., 1987; Frost et al., 1994) and the percent of inhibition is calculated: (1 – [cpm obtained with the inhibitor (*b*)/cpm obtained in reaction control without inhibitor (*a*)] × 100). For example see reported paper (Urbina et al., 2000).

Chitin Synthase Enzyme Assay

This assay measures the incorporation of *N*-acetylglucosamine from [^{14}C]-UDP-*N*-acetylglucosamine into insoluble [^{14}C]-chitin by the chitin synthase enzyme, in the presence or in the absence of inhibitors, according to the reactions described in (*c*) and (*d*) (Figure 1.10). The percent of inhibi-

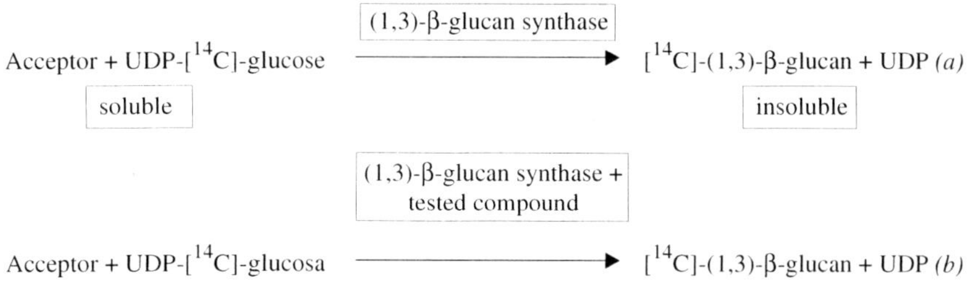

FIGURE 1.9. β-D-Glucan synthase assay

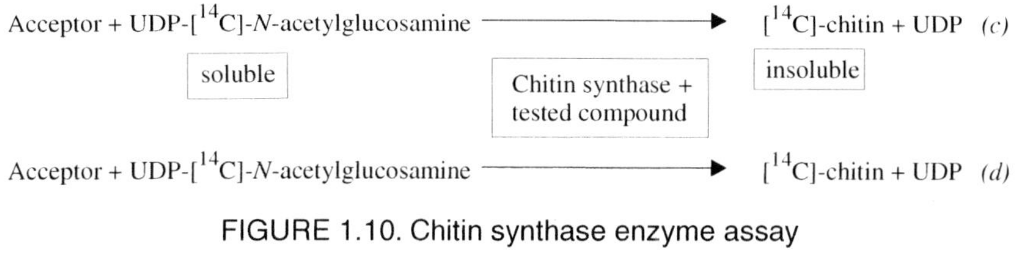

FIGURE 1.10. Chitin synthase enzyme assay

tion is calculated similarly to the (1,3)-β-D-glucan synthase assay described previously. Conventional assays measure the predominant chitin synthase 1 activity from *Saccharomyces cerevisiae* membrane extracts (Cabib et al., 1987). Nevertheless, since differential activity from the different chitin synthases exists (Choi and Cabib, 1994), it is important to take into account the enzyme and assay conditions when chitin synthesizing enzymes are used in screens for possible antifungal agents.

DETECTION OF NEW ANTIFUNGAL COMPOUNDS WITH INHIBITORY CAPACITY OF FUNGAL CELL WALL SYNTHESIS

As part of a collaborative research project among colleagues of Universities of Santa Catarina (Brazil), do Vale do Itajai (SC, Brazil), Santander (Bucaramanga, Colombia); Instituto de Microbiologia Bioquimica, (Salamanca, Spain), and University of San Luis and University of Rosario (Argentina), a variety of natural plant extracts and series of synthetic products were screened initially for their antifungal properties with dilution assays, and then for their capacity of inhibiting the fungal cell wall.

Evaluation of Plant Extracts

Thirty extracts belonging to eight species of Argentinian flora, reported as antifungal in folk medicine, were assayed for antifugal properties using the agar dilution method against a panel of yeasts, filamentous fungi, and dermatophytes. Among the tested species, *Xanthium spinosum* is a widely distributed bush in northwest Argentina. Decoctions of its aerial parts are used in folk medicine for treating skin pruritus and healing infected wounds (Pei-Gen, 1989). Crude extracts and fractions from this species have demonstrated in vitro antifungal properties against a panel of phytopathogenic fungi (Choi and Cabib, 1994).

Regarding species of *Polygonum,* they are perennial herbs found in the center, north, and northwest of Argentina, as well as south of Brazil. Extracts from ground leaves of *Polygonum acuminatum* (vernacular name *sanguinaria del agua, catay,* or *catay grande*), *P. punctatum* (*yerba del bicho* or *yerba picante*), *P. stelligerum,* and *P. ferrugineum* were reported in traditional medicine as useful in healing infected wounds (Hyeronimus, 1882; Rodríguez, 1915; Martinez Crovetto, 1964; Gonzálaz Torres, 1981).

Peschiera australis (syn. *Tabernaemontana australis* Muell. Arg.), of the family Apocynaceae, is exclusively an American tree, popularly known in

Argentina as *palo vibora*. It is found in Argentina, Uruguay, Paraguay, and southern Brazil. The plant is used in folk medicine to relieve toothaches and for healing infected wounds (Hyeronimus, 1929).

Inga urugüensis (Ingá colorado) is the only species among the 250 American species of *Inga* that can be found in Argentina, specifically in the state of Entre Rios—this region being the austral limit for the genus. Decoctions of its leaves are used externally in Argentina to treat infected wounds or grains.

Luehea divaricata (azota caballos) is a tree that grows in northest Argentina and in the riverside forests of south Brazil and Paraguay. Its bark and aerial parts are used in traditional medicine for healing skin wounds, for cleaning grains, and for vaginal washes (Rojas Acosta, 1907; Jozami and Muñoz, 1984).

Among the tested extracts, nine of them belonging to six species, exhibited a broad spectrum of activities against *Microsporum canis, Microsporum gypseum, Tricophyton mentagrophytes, Tricophyton rubrum,* and *Epidermophyton floccosum,* with MICs ranging from 25 to 900 μg/ml. Dichloromethanic extract of *Polygonum ferrugineum* was the most active extract with MICs of 25-50 μg/ml (Table 1.2).

Results obtained in the *N. crassa* assays for active extracts are shown in the far-right column of Table 1.2. The fact that dichloromethane extracts from aerial parts of *Polygonum punctatum* and *P. ferrugineum,* bark of *Luehea divaricata,* and methanolic and nonalkaloidic fractions of *Pescheria australis* showed a blotchy zone around the paper disk suggests that the mode of action of these extracts against dermatophytes could be associated with an inhibition of cell wall polymer synthesis or assembly. The clear inhibition zone shown by the dichloromethane extract of *Xanthium spinosum* could be ascribed to the fact that this extract has another effect on fungal cells, in addition to the inhibition of their cell wall synthesis (Zacchino, Santecchia, et al., 1998).

Triterpenoid Saponins from **Phytolacca tetramera**

Among plants of the Phytolaccaceae family, *Phytolacca tetramera* (Hauman) is an endemic species of Argentina that has not been previously studied. It has been submitted to a strong anthropic impact, which has caused its numerical population retraction, having been appointed to the category of species in critic risk (CR) (Hernández et al., 1998). Phytochemical analyses on several species of the genus *Phytolacca* showed that they are rich in saponins (Woo et al., 1977; Woo and Kang, 1975, 1976; Suga et al., 1978; Harkar et al., 1984; Kang and Woo, 1987; Hua, 1990,

TABLE 1.2. In vitro evaluation of antifungal properties of extracts from selected species of Argentinian flora with agar dilution assays and studies on their mode of action with *Neurospora crassa* assay

Plant species Family Voucher specimen Common name	Plant part[a]	Extract Type[b]	MIC (µg/ml) M.c.[d]	M.g.[e]	T.m.[f]	T.r.[g]	E. f.[h]	*N. crassa* assay[i]
Xanthium spinosum L.	Ae	Aq	>1000	>1000	>1000	>1000	>1000	–
Asteraceae		Dichl	250	250	600	500	100	clear
UNSL 154		MeOH	200	300	250	600	80	clear
Polygonum stelligerum	Ae	Aq	>1000	>1000	>1000	>1000	>1000	–
Cham.		Dichl	600	>1000	>1000	>1000	500	–
Polygonaceae		MeOH	>1000	>1000	>1000	>1000	>1000	–
UNR 212								
Polygonum punctatum	Ae	Aq	>1000	>1000	>1000	>1000	>1000	–
Elliot.		Dichl	250	250	250	50	50	hazy
Polygonaceae		MeOH	>1000	>1000	>1000	>1000	>1000	–
UNR 97								
yerba del bicho								
Polygonum ferrugineum	Ae	Aq	>1000	>1000	>1000	>1000	>1000	–
Wedd.		Dichl	25	50	50	25	25	hazy
Polygonaceae		MeOH	900	500	900	700	700	–
UNR 99	Ae							
Pescheria australis	Bk	MeOh	>1000	>1000	>1000	650	500	hazy
Miers.		A	>1000	>1000	>1000	250	200	hazy
Apocynaceae		B	>1000	>1000	>1000	>1000	>1000	–
UNR		MeOH	>1000	>1000	>1000	>1000	>1000	–
palo víbora	Ae							
Luehea divaricata	Bk	Aq	>1000	>1000	>1000	>1000	>1000	–
Mart.		Dichl	250	250	700	900	500	hazy
Tiliaceae		MeOH	>1000	>1000	>1000	>1000	>1000	–
ERA 4307	Ae	Aq	>1000	>1000	>1000	>1000	>1000	–
Azota caballos		Dichl	>1000	>1000	>1000	>1000	>1000	–
		MeOH	>1000	>1000	>1000	>1000	>1000	–
Amphotericin			>50	6.25	6.25	25	0.3	–
Ketoconazole			15	6.25	12.5	15	25	clear
Cilofungin								hazy

[a]Ae: Arial parts, Bk: bark; [b]Aq: Aqueous, Dichl: Dichlorometanic, MeOH: Methanolic, A: Nonalkaloidic fraction, B: Alkaloidic fraction; [c]g of plant extract/100 g total plant; [d]*Microsporum canis* C 112; [e]*Microsporum gypseum* C 115; [f]*Trichophyton mentagrophytes* ATCC 9972; [g]*Trichophyton rubrum* C 113; [h]*Epidermophyton floccosum* C 114: Type of inhibition zone.

1991,1992; Spengel et al., 1995; Strauss et al., 1995; Spengel and Schaffner, 1993; Nielsen et al., 1995), which were described as displaying important biological activities (Parkhurst et al., 1973; Hua and Chu-Lu, 1989; Moreno and Rodríguez, 1981; Favel et al., 1994; Rao and Sung, 1995; Kobayashi et al., 1995).

From the berries of *Phytolacca tetramera* our team isolated pure phytolaccoside B [3-*O*-β-xylopiranosyl-phytolaccagenin (Woo et al., 1977; Suga et al., 1978; Kobayashi et al., 1995) [=phytolaccasaponin G (Spengel et al., 1995) (=esculentoside B (Spengel and Schaffner, 1993), phytolaccoside E [3-*O*-β-D-glucopyranosy l-(l→ 4)-β-D-xylopiranosyl-phytolaccagenin] (Woo and Kang, 1975) [=phytolaccasaponin E (Kobayashi et al., 1995) =esculentoside A (Spengel and Schaffner, 1993), and phytolaccoside F (Kang and Woo, 1987; Strauss et al., 1995) (Figure 1.11) at 0.51, 0.35, and 0.1 percent, respectively. All of them are monodesmosides derived of oleic triterpenes of β-amyrin series, possessing 28,30 dicarboxylic groups and a double bond on C-12.

The antifungal evaluation of the three isolated saponins was initially performed with agar dilution assays, at concentrations up to 250 μg/ml, according to reported procedures (Zacchino et al., 1997, 1999). Compounds producing no inhibition at that level were considered inactive. Results showed that phytolaccoside B possesses a broad spectrum of activity. It inhibited yeasts, hyalohyphomycetes, and dermatophytes, with MICs between 25 and 125 μg/ml, and *Trichophyton mentagrophytes* being the most susceptible species. Its analog with a longer glycoside chain, phytolaccoside E, also possesses antifungal activity, although showing a lower and narrower spectrum of activity than phytolaccoside B. It inhibited only dermatophyte spe-

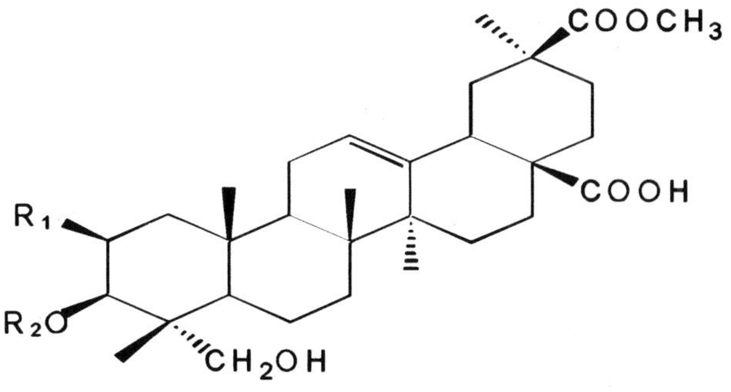

R_1	R_2	*Compound*
OH	xyl	Phytolaccoside B
OH	xyl $\xrightarrow{4}$ glc	Phytolaccoside E
H	xyl $\xrightarrow{2}$ glc $\xrightarrow{2}$ rha	Phytolaccoside F

xyl: *β*-D-xylopyranosyl; glc: *β*-D-glucopyranosyl; rha: *α*-L-rhamnopyranosyl

FIGURE 1.11. Phytolaccosides isolated from *Phytolacca tetramera* cies, with MICs between 125 and 250 μg/ml. In turn, phytolaccoside F was devoid of activity against all of the tested fungi.

Regarding the mode of action, the fact that phytolaccosides B and E showed a blotchy zone around the paper disk suggests that the mode of action of these saponins could be associated with an inhibition of cell wall polymer synthesis or assembly. The microscopic observation of the hazy zones shows morphological changes that result in a shortening of the hyphae (Zacchino et al., in press). According to these results, it is interesting to note that other terpenoid saponins with only one sugar as the glucon moiety have recently been described as specific inhibitors of the fungal cell wall (1,3)-β-D-glucan synthesis (Onishi et al., 2000).

Natural 8.0.4'-Neolignans from the Myristicaeae Family and Synthetic Analogs

Chemical investigations of species of *Virola* and related genera (Myristicaceae family) from the Amazonian region led to the hypothesis that the alleged usefulness of plasters made from their leaves or bark resin in the treatment of skin fungal infections may be due to the fungistatic or fungitoxic activity of neolignans (Schultes and Holmstedt, 1971) since different ketones and alcohols of *threo* and *erythro* relative configuration (Herrera Braga et al., 1984) have been found only in the Myristicaceae family from *Virola surinamensis* (Barata et al., 1978), *V. carinata* (Cavalcante et al., 1985), *V. pavonis* (Ferri and Barata, 1992), and *Myristica fragrans* (Forrest et al., 1974; Hattori et al., 1987; Hada et al., 1988). We carried out an in vitro antifungal evaluation of this type of compound in order to establish whether the observed activity could be ascribed to 8.O.4'-neolignans (Figure 1.12).

In preliminary studies made with agar-dilution assays, we reported that synthetic alcohols 7 through 18 (Calixto et al., 1990; Zacchino and Badano, 1991; Cechinel Filho, 2000) but no ketones 1 through 6 possess a significant antifungal activity (MICs 5-250 μg/ml) against dermatophytes. This activity showed to be dependent upon relative stereochemistry *(erythro* up to three times more active than *threo* alcohols) and upon substitution patterns at rings A and B. In addition, *Epidermophyton floccosum* was the most susceptible species (Zacchino et al., 1997).

Compounds 7 through 18 were assayed for their capacity to inhibit fungal cell wall synthesis. The first bioassay used was the *Neurospora crassa* hyphal growth inhibition. Then, inhibition of (1,3)-β-glucan synthase activity, the enzyme that catalyzes polymerization of glucose into (1,3)-β-glucan, was measured according to reported procedures. In addition, ketones 1 through 6 were assayed in this enzymatic assay.

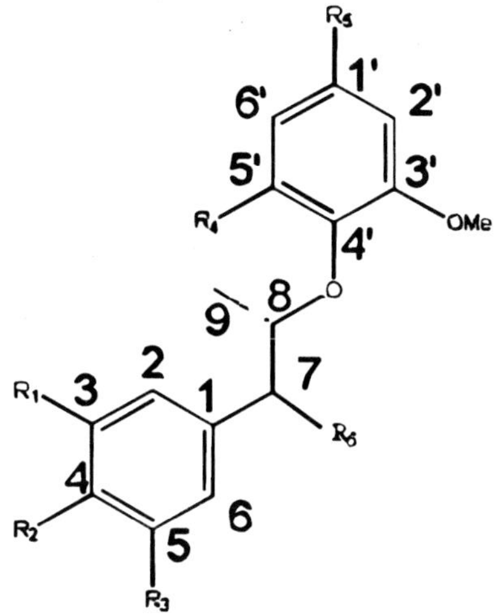

Comp	R_1	R_2	R_4	R_5	R_6	Confi	
1	OMe	OM	H	OM	allyl	=O	–
2	OMe	OM	OM	OM	allyl	=O	–
3	-O-CH_2-O-		H	OM	allyl	=O	–
4	OMe	OM	H	H	allyl	=O	–
5	OMe	OM	H	H	trans-propenyl	=O	–
6	OMe	OM	OM	H	trans-propenyl	=O	–
7	OMe	OM	H	OM	allyl	OH	erythro
8	OMe	OM	H	OM	allyl	OH	threo
9	OMe	OM	OM	OM	allyl	OH	erythro
10	OMe	OM	OM	OM	allyl	OH	threo
11	-O-CH_2-O-		H	OM	allyl	OH	erythro
12	-O-CH_2-O-		H	OM	allyl	OH	threo
13	OMe	OM	H	H	allyl	OH	erythro
14	OMe	OM	H	H	allyl	OH	threo
15	OMe	OM	H	H	trans-propenyl	OH	erythro
16	OMe	OM	OM	H	trans-propenyl	OH	threo
17	OMe	OM	OM	H	trans-propenyl	OH	erythro
18	OMe	OM	OM	H	trans-propenyl		threo

FIGURE 1.12. Structure of 8.O.4' neolignans tested for antifungal properties

Most 8.O.4' neolignans tested (compounds 7, 9-11, 13-15, and 18) showed a hazy zone around the paper disk (Table 1.3, column 3), suggesting that the mode of action of these compounds acting against dermatophytes could be associated with inhibition of cell wall polymer synthesis or assembly. The fact that compounds 8, 16, and 17, showed clear inhibition zones could be ascribed to the fact that these compounds may have another effect on fungal cells in addition to the inhibition of fungal cell wall synthesis.

Regarding the (1,3)-β-glucan synthase assay, results showed that compounds 7 through 18 inhibited (1,3)-β-glucan synthase in a percent between 2 and 71.6 at 250 μg/ml (Table 1.3, column 4). Although ketones 12 through 18 did not display any activity in agar dilution assays, they inhibited (1,3)-β-glucan synthase at 3 to 56 percent at 250 μg/ml.

Figure 1.13 shows the diminution of incorporation of [^{14}C]-glucose into insoluble labeled glucan produced by an increased concentration of 8.O.4' neolignans. The IC_{50} for compounds whose inhibition were > 50 percent is shown in column 5 (Urbina et al., 2000).

As a conclusion, we state that one of the modes of action of antifungal 8.O.4' neolignans, could be the inhibition of fungal cell wall synthesis. Their inhibitory activities of (1,3)-β-glucan synthase, the enzyme that catalyzes the polymerization of the most abundant cell wall polymer, allowed us to infer that these compounds are moderate inhibitors of this enzyme, although this may not be its primary mode of action since we could not find a strict correlation between whole cells and enzymatic assays. Nevertheless, considering that the crude enzyme was prepared from *Saccharomyces cerevisiae*, the selective and specific species/genus responses could account for the absence of correlation between results in these assays and those obtained in agar dilution.

d-Xanthoxyline from Sebastiania schottiana *and Synthetic Analogs*

Sebastiania schottiana Muell. Arg. (Euphorbiaceae) is popularly known as *sarandi negro, quebra pedra,* or *branquicho* and is found in the riverside of Itajai-açú river in Santa Catarina, Brazil. It is used in traditional medicine to treat renal diseases and different type of infections (Cechinel Filho, 2000).

From the hexanic extract of stems and leaves of *S. schottiana,* 2-hydroxy-4,6-dimethoxyacetophenone (compound 19, see Table 1.4) (xanthoxyline) was isolated in ~0.25 percent (Calixto et al., 1990). This compound has also been reported in *Phyllantus sellowianus* (Cechinel Filho, Lima, et al., 1996; Pinheiro et al., 1999) and *Citrus limon* (Hertmann and Nienhaus, 1974).

TABLE 1.3. In vitro evaluation of antifungal properties of 8.O.4' neolignans alcohols in their *erythro* and *threo* forms and ketones, with different types of assays

Compound	Agar dilution assay against *E. floccosum* MIC (μg/ml)	*N. crassa* assay[a]	(1,3)-β-Glucan synthase assay[b]	IC_{50} (μg/ml)
1	>250	n.t.	20	–
2	>250	n.t.	3	–
3	>250	n.t.	17	–
4	>250	n.t.	56	205
5	>250	n.t.	41	–
6	>250	n.t.	43	–
7	40	hazy	47	–
8	60	clear	21	–
9	50	hazy	41	–
10	60	hazy	n.t.[d]	–
11	5	hazy	46	–
12	15	–	2	–
13	15	hazy	51	244
14	25	hazy	52	237
15	70	hazy	43	–
16	100	clear	31	–
17	25	clear	72	175
18	50	hazy	34	–
Papulacandin A	n.t.	hazy	–	100
Ketoconazole	25	clear	–	–

[a]Type of inhibition zone; [b]Type of inhibition halo; [c]Percent of inhibition at 250 μg/ml; [d]n.t.: Not tested

Previous studies with agar diffusion assays show that compound 19 possesses antifungal properties (Lima et al., 1995).

Our results in agar dilution assays showed that xanthoxyline selectively inhibits the dermatophytes *Tricophyton mentagrophytes* and *T. rubrum,* with MICs between 75 and 100 μg/ml (see Table 1.4). *Neurospora crassa* assay showed that xanthoxyline produced a blotchy appearance, thus suggesting it could be a inhibitor of fungal cell wall polymer synthesis or assembly (Pinheiro et al., 1999). This finding led us to synthesize a series of

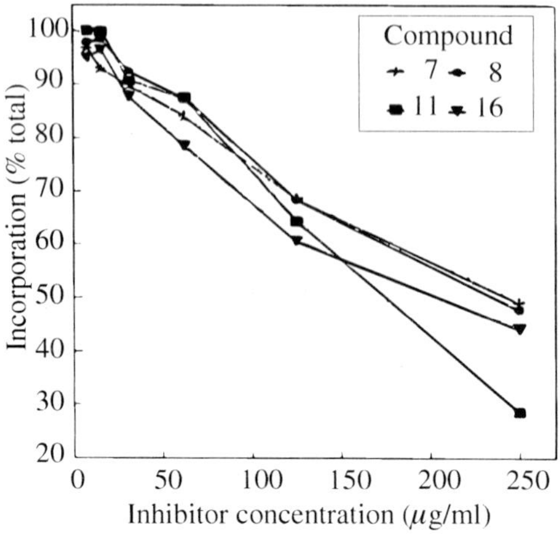

FIGURE 1.13. Effect of 8.O.4' neolignans on incorporation of [^{14}C]-glucose into insoluble labeled glucan. Points are the means of duplicate incubations.

analogs of xanthoxyline (Cechinel Filho, 1995; Cechinel Filho et al., 1995; Cechinel Filho, Vaz, et al., 1996) compounds 20 through 28, which were tested for antifungal activities with agar dilution assays and then with the *N. crassa* assay (Pinheiro et al., 1999) (see Figure 1.14).

Results in Table 1.4 show that, among the tested xanthoxyline derivatives, benzofurane analog (compound 25) possesses the broadest spectrum of action, being moderately active (MICs 20-75 μg/ml) against all of the tested dermatophytes. Compound 28, although inhibitor of only three dermatophytes, was the most active derivative with MICs between 6.25 and 40 μg/ml. It inhibited *Epidermophyton floccosum* with MICs as low as 6.25 μg/ml and it displayed the same activity as xanthoxyline against *T. mentagrophytes,* and was 12 times higher against *T. rubrum.*

Active compounds 23, 25, and 28 were tested with the *N. crassa* assay for their capacity to inhibit fungal cell wall synthesis. All of them showed a hazy zone around the paper disk, meaning these compounds could be interesting structures for the development of new and more potent antifungal agents.

Synthetic Homoallylamines, Tetrahydroquinolines, and Quinolines

In an attempt to generate leading structures that might have selective antifungal activities, and considering that certain homoallylamines had shown in vitro antifungal properties against a panel of phytopathogenic fungi

TABLE 1.4. MICs values (μg/ml) of xanthoxyline and its derivatives against dermatophytes, obtained with the agar dilution method (columns 2-6). Column 7 shows the results of the *Neurospora crassa* assay.

Compound	[a]*M.c.*	[b]*M.g.*	[c]*T.m.*	[d]*T.r.*	[e]*E.f.*	*N. crassa* assay[f]
19	>100	>100	50	75	>100	hazy
20	>100	>100	>100	>100	>100	n.t.
21	>100	>100	>100	>100	25	hazy
22	>100	>100	>100	>100	25	hazy
23	>100	>100	40	40	25	–[g]
24	>100	>100	>100	>100	>100	n.t.
25	75	50	20	70	75	hazy
26	>100	>100	>100	>100	>100	n.t.
27	>100	>100	>100	>100	>100	n.t.
28	>100	>100	40	6.25	12.5	hazy
Ketoconazole	15	6.25	12.5	15	15	clear
Amphotericin	30	6.25	6.25	25	0.3	clear
Cilofungin	–	–	–	–	–	hazy

[a]*Microsporum canis* C 112; [b]*Microsporum gypseum* C 115; [c]*Trichophyton mentagrophytes* ATCC 9972; [d]*Trichophyton rubrum* C 113; [e]*Epidermophyton floccosum* C 114; [f]Type of inhibition halo; [g]No inhibition halo is produced; n.t.: Not tested

(Kouznetsov et al., 1996), we synthesized a series of aryl and alkyl homoallylamines 40-50 from aldimines 29-37 (Figure 1.15) (Kouznetsov et al., 1992, 1994, 1998). Then, we transformed some of them into 4-methylquinolines (compounds 53 and 54) via the 4-methyl-1,2,3,4-tetrahydroquinolines (compounds 51 and 52) by using a novel synthetic pathway, showing that homoallylamines could serve as useful precursors for this type of biologically important compound.

Our work with agar dilution assays using human pathogenic fungi showed that structures 40-42, 46, and 48 possessed potent antifungal properties, in particular against *E. floccosum* (Table 1.5). Tetrahydroquinolines 51 and 52 and quinolines 53 and 54 also displayed antifungal activities. Compound 54 was particularly active against *M. canis* (Urbina et al., 2000).

It is interesting to note that 4-aryl homoallylamines and their derivatives display similar or stronger antifungal activities than amphotericin B or ketoconazole on the tested species (Urbina et al., 2000).

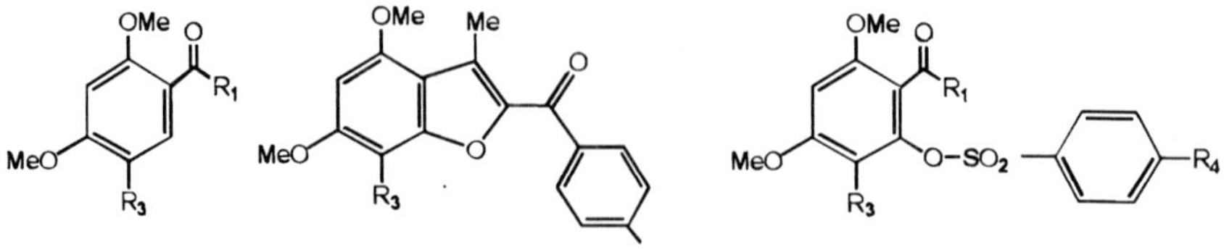

A B C

Compound	Type	R_1	R_2	R_3	R_4
19	A	$-CH_3$	-OH	-H	–
20	A	$-CH_3$	-OH	-Br	–
21	A	$-CH_3$	$-O-CO-CH_3$	-Br	–
22	A	$-CH_3$	-H	-H	–
23	A	$-CH_3$	$-O-CO-CH_3$	-H	–
24	A	-CH=CH-Ph	-OH	-Br	–
25	B	–	–	–	–
26	C	–	–	–	$-OCH_3$
27	C	–	–	–	-Br
28	C	–	–	–	$-NO_2$

FIGURE 1.14. Structures of natural xanthoxyline (compound 19) and synthetic analogs 20 thorugh 28

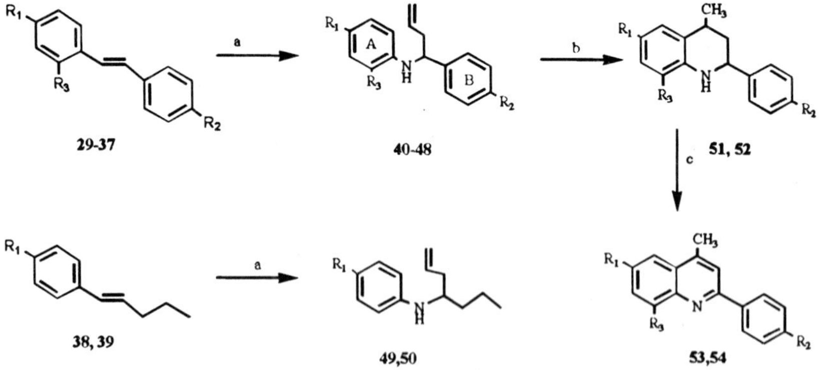

(a)Allylbromide + Mg/Et$_2$O, 10 °C; (b) H$_2$SO$_4$ 75% W/V; (c) DDQ/Bz

FIGURE 1.15. Preparation of homoallylamines, 4-methyl-1,2,3,4-tetrahydroquinolines, and 4-methylquinolines

TABLE 1.5. MIC values (μ/ml) of homoallylamines, tetrahydroquinolines, and quinolines acting against dermatophytes

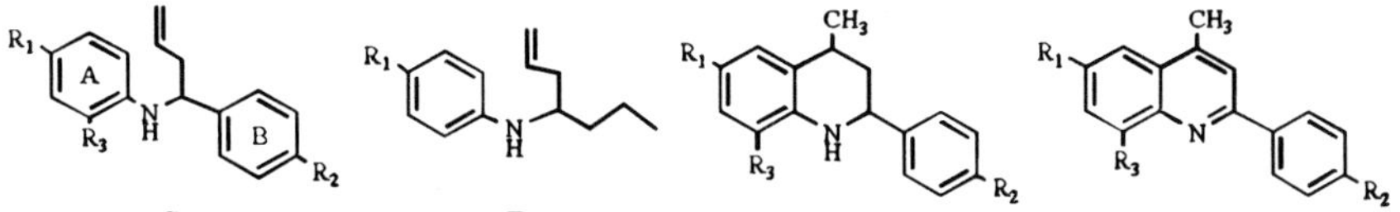

C D E F

Com.	Type	R_1	R_2	R_3	*M.c.*[a]	*M.g.*[b]	*T.m.*[c]	*T.r.*[d]	*E.f.*[e]
40	C	H	H	H	30	30	30	30	12.5
41	C	CH_3	H	H	30	>50	>50	>50	3.12
42	C	OCH_3	H	H	30	>50	30	30	3.12
43	C	F	H	H	>50	>50	>50	>50	30
44	C	Cl	H	H	>50	>50	>50	>50	30
45	C	Br	H	H	>50	>50	>50	>50	30
46	C	H	OCH_3	H	30	>50	>50	>50	3.12
47	C	Cl	$N(CH_3)_2$	H	>50	>50	>50	>50	>50
48	C	H	H	CH_3	30	>50	>50	>50	3.12
49	D	H	–	–	>50	>50	>50	>50	>50
50	D	CH_3	–	–	>50	>50	>50	>50	>50
51	E	H	H	H	50	25	25	25	12.5
52	E	CH_3	H	H	50	25	25	25	12.5
53	F	H	H	H	25	12.5	12.5	25	12.5
54	F	CH_3	H	H	0.75	12.5	25	12.5	12.5
Amp.[f]					>50	6.25	6.25	25	0.3
Ket.[g]					15	6.25	12.5	15	25

[a]*Microsporum canis* C 112; [b]*Microsporum gypseum* C 115; [c]*Trichophyton mentagrophytes* ATCC 9972; [d]*Trichophyton rubrum* C 113; [e]*Epidermophyton floccosum* C 114; [f]Amp. = amphotericin B; [g]Ket. = ketoconazole

Regarding the mode of action, active compounds 40-42, 46, 48, and 51-54 showed inhibitory activity against (1,3)-β-glucan synthase and chitin synthase 1 (from *Saccharomyces cerevisiae)* enzymes (Table 1.6). Figures 1.16 and 1.17 show the effect of different concentrations of antifungal homoallylamines, 4-methyl-1,2,3,4-tetrahydroquinolines, and 4-methy-

TABLE 1.6. Capacity of inhibiting (1,3)-β-glucan synthase and chitin synthases expressed in percent of inhibition and IC_{50} values (μg/μl)

Compound	Glucan synthase assay		Chitin synthase assay	
n[r]	%I[h]	IC[i]50	%I[j]	IC[i]50
40	60.10±4.36	0.25	87.07±1.04	0.09
41	53.35±3.27	0.32	56.36±2.57	0.19
42	25.34±5.78	>0.50	68.42±1.02	0.15
46	36.02±4.63	>0.50	75.26±0.65	0.07
48	51.67±3.94	0.33	70.01±1.77	0.17
51	53.48±4.90	0.23	73.41±0.10	0.10
52	41.64±7.16	>0.50	49.75±0.28	0.40
53	59.46±9.20	0.23	87.90±0.88	0.02
54	34.32±3.08	>0.50	84.11±0.50	0.04
Papulacandin B		0.10		
Aculeacin A		>0.50		
Nikkomicin				0.0006

[h]Percent of inhibition at 20 μg/assay (total volume: 40 μl): mean ± SEM; [i]Concentration (μg/μλ) that produces 50 percent of inhibition; [j]percent of inhibition at 20 μg/assay (total volume: 50 μL): mean ± SEM; SEM: standard error mean; n[r]: number

iquinolines on in vitro incorporation of [^{14}C]-glucose into insoluble [^{14}C] (1,3)-β-glucan and [^{14}C] *N*-acetylglucosamine into labeled chitin, expressed as residual activity of the enzyme (percent incorporation compared to the incorporation in the absence of the compound). The fact that these compounds inhibit the synthesis of both glucan and chitin at the same time is particularly attractive. We must consider that the combination of chitin synthase and glucan synthase inhibitors might be required (Hector and Braun, 1986).

The antifungal activity of homoallylamines, tetrahydroquinolines, and quinolines against some dermatophytes, combined with their selective modes of action, make these compounds attractive options for the development of more potent and safer antifungal drugs.

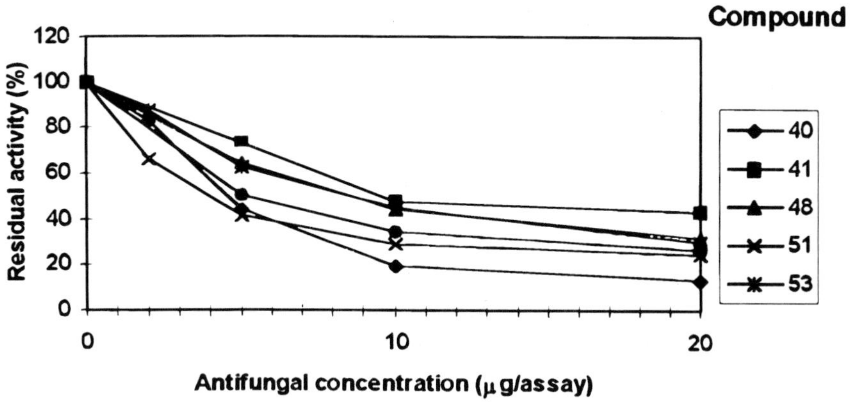

FIGURE 1.16. Effect on different concentrations of antifungal homoallylamines, 4-methyl-1,2,3,4-tetrahydroquinolines, and 4-methylquinolines on in vitro incorporation of [^{14}C]-glucose into insoluble (1-3)-β-glucan expressed as residual activity of the enzyme (percent incorporation compared to the incorporation in the absence of antifungal compounds).

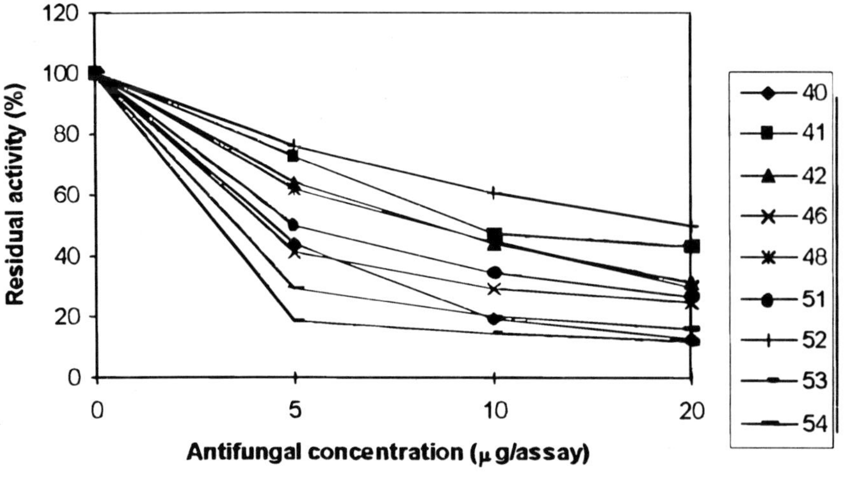

FIGURE 1.17. Effect of different concentrations of antifungal homoallylamines, 4-methyl-1,2,3,4-tetrahydroquinolines, and 4-methylquinolines on the in vitro chitin synthase 1 activity, expressed as residual activity of the enzyme (see legend of Figure 1.16 for details).

CONCLUSIONS

Despite the increasing incidence of fungal infections, especially involving immunocompromised patients, life-threatening mycoses remain very difficult to treat. There are, in fact, a very limited number of useful and efficaceous antifungal drugs, which often possess diminished therapeutic capabilities due to the rapid development of resistant fungal strains; are fungistatic and not fungicide, thus producing recurrence; or produce very toxic side effects. There is a general consensus that newer, more potent, and safer antifungal agents are needed to treat unmanageable fungal diseases.

The search is oriented toward new antifungal agents that act through an entirely different and selective mode of action, particularly by inhibiting a molecular process of the pathogen which is either absent in the host or sufficiently different so that the host metabolism will be minimally affected.

The targets that offer the best possibilities for finding safe antifungal drugs are ergosterol biosynthesis, the fungal topoisomerases, and the fungal cell wall. Among them, fungal topoisomerase-I has recently demonstrated to be sufficiently different from the human enzyme, thus becoming a new target alternative for the discovery of selective antifungal agents.

Regarding the fungal cell wall, the fact that fungal but not mammalian cells possess a cell wall, which is essential for the fungal cell's survival, makes this structure an ideal target for new antifungal drugs. New structures with a capacity for inhibiting fungal cell wall synthesis—the biosynthesis of its polymers or their assembly—are actively being sought with the aim of developing new drugs that will be more potent and safer than those in use today.

FUTURE DIRECTIONS IN THE DISCOVERY OF NEW ANTIFUNGAL AGENTS

Future strategies for the discovery of safer and more potent antifungal drugs include the identification and characterization of novel selective targets, the detection of new classes of antifungal compounds, and the discovery of methods to avoid and suppress the emergence of antifungal resistance.

One of the trends in an antifungal discovery program is an ample screening of natural products from different sources. If the source are plants, this search is preferably based on an ethnobotanical approach. Then, active natural products could be improved by semisynthesis. Combinatorial chemistry will undoubtedly be a source for new and sometimes unexpected drugs. In addition, the identification of screens based on selective targets unique to fungi may offer new opportunities for drug discovery. Although fungi and mammals are both eukaryotic, there are many differences between them, that could be used to find fungal-specific drugs. Although the fungal cell wall is currently used as a target for the discovery of selective antifungal drugs because of the uniqueness of its structure. many processes of its synthesis and assembly remain unknown. As a consequence, investigation into these processes constitutes an opportunity to detect new, useful antifungal targets.

At present, virulence factors have emerged as a new target for the discovery of new antifungal drugs. They are based in the fact that if only few fungal species are pathogens, they surely have distinctive processes that produce virulence. Although they are not yet well determinated, efforts must be done toward the examination of host/pathogen interaction in order to delineate the basis of fungal virulence.

If current trends continue, an expanded array of new drugs with a selective mode of action will find their way to development, thus improving life expectations—especially in patients with compromised immune defenses. Although the ideal drugs to prevent or cure all invasive and dermal fungal infections have yet to appear, antifungal therapy seems to be oriented toward the combined use of different types of drugs. Combination therapy of known drugs began to be considered as an alternative to the existing ones since the discovery of the excellent synergistic effect of 5-flucytosine and amphotericin B, reported by Medoff and colleagues in 1971.

BIBLIOGRAPHY

Abbruzzo, G., Flattery, A., Gill, Ch., Kong, L., Smith, J., Pikounis, V., Balkovec, J., Bouffard, A., Dropinski, J., Rosen, H., Kropp, H., Bartizal, K. (1997). Evaluation of the echinochandin antifungal MK-0991 (L.743,872): Efficacies in mouse models of disseminated aspergillosis, candidiasis and cryptococcosis. *Antimicrob Agents Chemother* 41:2333-2338.

Ablordeppey, S., Fan, P., Ablordeppey, I., Mardenborough, L. (1999). Systemic antifungal agents against AIDS-related opportunistic infections: Current status and emerging drugs in development. *Current Med Chem* 6:1151-1195.

Al Magboul, A., Bashir, A. (1997). Antimicrobial activity of vernolepin and vernoladin. *Fitoterapia* 68:83-84.

Andrianaivoravelona, J., Sahpaz, S., Terreaux, C., Hostesttmann, K., Stoeckli-Evans, H., Rasolondramanitra, J. (1999). Two 6-substituted, 5,6-dihydro-alpha-pyrones from *Ravensara anisata*. *Phytochem* 52:265-269.

Aoki, Y., Kamiyama, T., Fujii, T., Yamamoto, M., Ohwada, J., Arisawa, M. (1995). Design of antifungal methionine inhibitor not antagonized by methionine. *Biol Pharm Bull* 18:1267-1271.

Avato, P., Vitali, C., Mongelli, P., Tava, A. (1997). Antimicrobial activity of polyacetylenes from *Bellis perennis* and their synthetic derivatives. *Planta Med* 63:503-507.

Barata, L., Baker, P., Gottlieb, O., Rúveda, E. (1978). Neolignans of *Virola surinamensis*. *Phytochem* 17:783-786.

Barrett-Bee, K., Ryder, N. (1992). Biochemical aspects of ergosterol biosynthesis inhibition. In *Emerging targets in antibacterial and antifungal chemotherapy*, Sutcliffe, J. and Georgopapadakou, N. (Eds.). New York: Chapman and Hall, pp. 411-435.

Bartroli, J., Turmo, E., Algueró, M., Boncompte, E., Vericat, M., Conte, L., Ramis, J., Merlos, M., García-Rafanell, J., Forn, J. (1998a). New azole antifungals. 2. Synthesis and antifungal activity of heterocyclecarboxamide derivatives of 3-amino-2-aryl-1-azolyl-2-butanol. *J Med Chem* 41:1855-1868.

Bartroli, J., Turmo, E., Algueró, M., Boncompte, E., Vericat, M., Conte, L., Ramis, J., Merlos, M., García-Rafanell, J., Forn, J. (1998b). New azole antifungals. 3. Synthesis and antifungal activity 3-substituted-4(3H)-quinazolinones derivatives of 3-amino-2-aryl-1-azolyl-2-butanol. *J Med Chem* 41:1869-1882.

Beninger, P., Webb, D., Mc Neil, M., Gutman, S., Gitterman, S., Dixon, D., Laughon, B., Branche, W. (1994). Private, public and academic partnerships, 2nd NIAID Workshop in Medical Mycology, University of Arizona, Northern Arizona University, Flagstaff, AZ, June 8-11.

Cabib, E. (1991). Differential inhibition of chitin synthetases 1 and 2 from *Saccharomyces cerevisiae* by polyoxin D and nikkomycin. *Antimicrob Agents Chemother* 35:170-173.

Cabib E., Kang M. (1987). Fungal 1,3-beta-glucan synthase. In *Methods in enzymology* 138:637-642.

Cabib, E., Kang, M., Au-Young, J. (1987). Chitin synthase from *Saccharomyces cerevisiae*. In *Methods in enzymology*, 138:643-649.

Calixto, J., Yunes, R., Miguel, O., Rae, G. (1990). Action of 2-hydroxy-4,6-dimethoxy-acetophenone isolated from *Sebastiania schottiana*. *Planta Med* 56:31-35.

Carrillo-Muñoz, L., Abarca-Salat, G., Quindós, G. (1994). Pruebas de sensibilidad a los antifúngicos. I. Factores y variables que influyen en su realización en el laboratorio. *Rev Iberoamer Micol* 11:105-110.

Cavalcante, S., Yoshida, M., Gottlieb, O. (1985). Neolignans from *Virola carinata* fruit. *Phytochem* 24:1051-1055.

Cechinel Filho, V. (1995). Obtençao de compostos farmacologicamente ativos a partir de produtos naturais: Correlaçao estrutura-atividade. Doctoral thesis, UFSC, Florianopolis, SC, Brasil.

Cechinel Filho, V. (2000). Principais avanços e perspectives na area de produtos naturais: Estudos desenvolvidos no NIQFAR/UNIV ALI. *Quimica Nova* 23: 680-685.

Cechinel Filho, V., Lima, E., Morais, V., Gomes, S., Miguel, O.,Yunes, R. (1996). Fungicide and fungistatic effects of xanthoxyline. *J Ethnopharm* 53:171-173.

Cechinel Filho, V., Miguel, O., Nunes, R., Calixto, J., Yunes, R. (1995). Antispasmodic activity of xanthoxyline derivatives: Structure-activity relationships. *J Pharm Sci* 84:473-475.

Cechinel Filho, V., Vaz, Z., Zunino, L., Calixto, J., Yunes, R. (1996). Synthesis of xanthoxyline derivatives with antinoceptive and antiedematogenic activities. *Eur J Med Chem* 31:833-839.

Chan, J., Hong, J., Kuyper, L., Baccanari, D., Joyner, S., Tansik, R., Boytos, C., Rudolph, S. (1995). Selective inhibitors of *Candida albicans* dihydrofolate reductase: Activity and selectity of 5-(arylthio)-2,4-diaminoquinazolines. *J Med Chem* 38:1837-1840.

Chandravadana, M., Nidiry, E., Venkateshwarlu, G. (1997). Antifungal activity of momordicines from *Momordica charantia. Fitoterapia* 68:383-384.

Choi, W.J., Cabib, E. (1994). The use of divalent cations and pH for the determination of specific yeast chitin synthetases. *Anal Biochem* 219:368-372.

Chung, S., Ryoo, Ch., Yang, H., Shim, J., Kang, M., Lee, K., Kang, H. (1998). Synthesis and bioactivities of steroid derivatives as antifungal agents. *Tetrahedron* 15899-15914.

Cuñat, P., Primo, E., Sanz, I., Garcerá, M., March, C., Bowers, W., Martínez Pardo, R. (1990). Biocidal activity of some Spanish Mediterranean plants. *J Agric Food Chem* 38:497-500.

Davila, T., San Blas, G., San Blas, F. (1986). Effect of Papulacandin A on glucan synthesis in *Paracoccidioides brasiliensis. J Med Vet Mycol* 24:193-202.

Debono, M., Abbott, B., Fukuda, D. (1989). Synthesis of new analogs of echinocandin B by enzymatic deacylation and chemical reacylation of the echinocandin B peptide: Synthesis of the antifungal agent cilofungin (LY 121019). *J Antib* 42:389-397.

Del Poeta, M., Toffaletti, D., Rude, T., Dykstra, C., Heitman, J., Perfect, J. (1999). Topoisomerase I is essential in *Cryptococcus neoformans*: Role in pathobiology and as antifungal target. *Genetics* 152:167-178.

Denning, D., Stevens, D. (1991). Efficacy of cilofungin alone and in combination with amphotericin B in a murine model of disseminated aspergillosi. *Antimicrob Agents Chemother* 33:1329-1333.

Devadas, B., Zupec, M., Freeman, S., Brown, D., Nagarajan, S., Sikorski, J., McWerther, Ch., Getman, D., Gordon, G. (1995). Design and synthesis of potent and selective dipeptide inhibitors of *C. albicans* myristoyl-CoA:Protein N-myristoyltransferase. *J Med Chem* 38:1837-1840.

Elmahi, M., Essassi, E., Hmamouchi, M., Hmamouchi, J. (1997). Etude de l'action antimicrobienne et antibilharzienne du *Zizyphus vulgaris. Fitoterapia* 68:284-285.

Esimone, C., Adikwu, M. (1999). Antimicrobial activity and citotoxicity of *Ramalina farinaceae. Fitoterapia* 70:428-431.

Favel, A., Steinmetz, M.D., Vidal-Ollivier, E., and Balansard, G. (1994). In vitro antifungal activity of triterpenoid saponins. *Planta Medica* 60: 50-53.

Ferri, P., Barata, L. (1992). (-)-Di de O-methylgrandisin, a lignan from *Virola pavonis* leaves. *Phytochem* 31:1375-1377.

Forrest, J., Heacock, R., Forrest, T. (1974). Diarylpropanoids from nutmeg and mace (*Myristica fragrans* Houtt). *J Chem Soc Perkin Trans I* 205-209.

Fostel, J., Montgomery, D., Lartey, P. (1996). Comparison of responses of DNA-topoisomerase I from *Candida albicans* and human cells to four new agents which stimulate topoisomerase-dependent DNA nicking. *FEMS Microb Lett* 138:105-111.

Freixa, B., Vila, R., Vargas, L., Lozano, N., Adzet, T., Cañigueral, S. (1998). Screening for antifungal activity of nineteen Latin American plants. *Phytother Res* 12:427-430.

Frost, D., Brandt, K., Capobianco, J., Goldman, R. (1994). Characterization of 1,3-β-glucan synthase in *Candida albicans*: Microsomal assay from the yeast or mycelial morphological forms and a permeabilized whole-cell assay. *Microb* 140:2239-2246.

Frost, D., Brandt, K., Cugier, D., Goldman, R. (1995). A whole-cell *C. albicans* assay for the detection of inhibitors towards fungal cell wall synthesis and assembly. *J Antib* 48:306-310.

Fukuda, D., Nakatsukasa, W., Yao, R., Hunt, A., Gordee, R., Zeckner, D., Mynderse, J. (1991). A45507, a complex of fungal cell wall inhibitors produced by a mold, 31st ICAAC meeting, Chicago, IL, poster #211.

Fukushima, Y., Sakagami, Y., Marumo, S. (1993). β-glucan biosynthesis inhibitors isolated from fungi as hyphal malformation inducers. *Bioorg and Med Chem Lett* 3:1219-1222.

Gaughran, J., Lai, M., Kirsh, D., Silverman, S. (1994). Nikkomicin Z is a specific inhibitor of *Saccharomyces cerevisiae* chitin synthase isozyme CHS3 in vitro and in vivo. *J Bacteriol* 176:5857-5860.

Georgopapadakou, N. (1992). Chitin synthase as a chemotherapeutic agent. In *Emerging targets in antibacterial and antifungal chemotherapy,* Sutcliffe, J. and Georgopapadakou, N. (Eds.). New York: Chapman and Hall, pp. 476-493.

Georgopapadakou, N., Dix, B., Smith, S., Feundenberger, J., Funke, P. (1987). Effect of antifungal agents on lipid biosynthesis and membrane integrity in *Candida albicans. Antimicrob Agents Chemother* 31:46-51.

Georgopapadakou, N., Tkacz, J. (1995). The fungal cell wall as a drug target. *Trends Microbiol* 3:98-104.

Goldman, G., Yu, Ch.,Wu, H., Sanders, M., La Voie, E., Liu, L. (1997). Differential poisoning of human and *Aspergillus nidulans* DNA-topoisomerase I by bi and terbenzimidazoles. *Biochem* 36:6488-6494.

González Torres, D. (1981). *Catálogo de plantas medicinales (y alimenticias y útiles) usadas en el Paraguay.* Paraguay: Comuneros Press, p. 218.

Gunji, S., Arima, K., Beppu, T. (1983). Screening of antifungal antibiotics according to activities inducing morphological abnormalities. *Agric Biol Chem* 47:2061-2069.

Hada, S., Hattori, M., Tezuka, Y., Kikiuchi, T., Namba, T. (1988). New lignans and neolignans from the aryl of *Myristica fragrans. Phytochem* 27:563-568.

Hammer, K., Carson, C., Riley, T. (1998). In vitro activity of essential oils, in particular *Melaleuca alternifolia* (tea-tree) oil and tea-tree oil products, against *Candida* spp. *J Antimicrob Chemother* 42:591-595.

Hammond, S., Lambert, P., Kliger, B. (1974). The mode of action of polyene antibiotics; induced entry of hydrogen ions as a consequence of polyene action on the cell membrane of *Candida albicans. J Gen Microbiol* 81:331-336.

Harkar, S., Razdan, T.K., and Waight, E.S. (1984). Further triterpenoids and ^{13}C NMR spectra of oleananes derivatives from *Phytolacca acinosa. Phytochem* 23: 2893-2898.

Hattori, M., Hada, S., Shu, Y., Kakiuchi, N., Namba, T. (1987). New acyclic bisphenylpropanoids from the aryl of *Myristica fragrans. Chem Pharm Bull* 35:668-674.

Hector, R., Braun, P. (1986). Synergistic action of nikkomycins X and Z with papulacandin B on whole cells and regenerating protoplasts of *Candida albicans. Antimicrob Agents and Chemother* 29:389-394.

Hector, R., Lartey, P., Kurt, M. (1994). Drug Discovery, 2nd NIAID Workshop in Medical Mycology, University of Arizona, Northern Arizona University, Flagstaff, AZ, June 8-11.

Hernández, A., Delucchi, H.M. et al. (1998). Abstracts. XXVI. *Jornadas Argentinas de Botánica* November 22-27.

Herrera Braga, A., Zacchino, S., Badano, H., Gonzalez Sierra, M., Rúveda, E. (1984). ^{13}CNMR spectral and conformational analysis of 8.O.4'-neolignans. *Phytochem* 23:2026-2028.

Hertmann, G., Nienhaus, F. (1974). The isolation of xanthoxyline from the bark of *Phytophtora-* and *Hendersonula*-infected *Citrus limon* and its fungitoxic effect. *Phytopath Z* 81:97-113.

Hua, Y.Y. (1990). Esculentoside L and K: Two new saponins from *Phytolacca esculenta. Planta Medica* 56: 301-303.

Hua, Y.Y. (1991). A triterpenoid and its saponin from *Phytolacca esculenta. Phytochem* 30: 4179-4181.

Hua, Y.Y. (1992). A triterpenoid saponin from *Phytolacca esculenta. Phytochem* 31: 2552-2554.

Hua, Y.Y. and Chu-Lu, W. (1989). A new active saponin from *Phytolacca esculenta. Planta Medica* 55: 551-552.

Hyeronimus, J. (1882). In *Plantas diaphoricae flora argentinae,* Academia de Córdoba (Ed.). Boletín de la Academia 4ª entrega, tomo IV, Córdoba (Argentina), 175-178.

Hyeronimus, J. (1929). In *Plantas diaphoricae flora argentinae,* Second edition, Atlántida (Ed.). Buenos Aires (Argentina), 107.

Ishiguro, J., Saitou, A., Durán, A., Ribas, J. (1997). Cps1^{+}, a *Schizosaccharomyces pombe* gene homolog of *Saccharomyces cerevisiae* FKS genes whose mutation confers hypersensitivity to cyclosporin A and papulacandin B. *J Bacteriol* 179:7653-7662.

Iwasa, K., Kamigauchi, M., Sugiura, M., Nanba, H. (1997). Antimicrobial activity of some 13-alkyl substituted protoberberinium salts. *Planta Med* 63:196-198.

Iwasa, K., Lee, D., Kang, S., Wiegrebe, W. (1998). Antimicrobial activity of 8-alkyl and 8-phenyl-substituted berberines and their 12-bromo derivatives. *J Nat Prod* 61:1150-1153.

Jackson, D., Figgitt, D., Denyer, S. (1992). Topoisomerase II inhibitors: Prospects for new antifungal agents. In *Emerging targets in antibacterial and antifungal chemotherapy,* Sutcliffe, J. and Georgopapadakou, N. (Eds.). New York: Chapman and Hall, pp. 374-391.

Jackson, M., Frost, D., Karwowski, J., Humphrey, P., Dahod, S., Choi, W., Brandt, K., Malmberg, L., Rasmussen, R., Scerr, M., Flamm, R., Kadam, S., Mc Alpine, J. (1995). Fusacandins A and B; novel antibiotics of the papulacandin class from *Fusarium sambucinum. J Antib* 48: 608-613.

Jeong, T., Hwang, E., Lee, H., Lee, E., Kim, Y., Min, B., Bae, K., Bok, S., Kim, S. (1999). Chitin synthase II inhibitory activity of ursolic acid, isolated from *Crataegus pinnatifida. Planta Med* 65:261-263.

Jiang, W., Gerhold, D., Kmiec, E., Hauser, M., Becker, J., Koltin, Y. (1997). The topoisomerase I gene from *Candida albicans. Microb* 143:377-386.

Jozami, J., Muñoz, J. de D. (1984). In *Arboles y arbustos indigenas de la provincia de Entre Rios,* Third edition, Ipnays-CONICET-UNL (Eds.). Santa Fe, Argentina: Medios Audiovisuales y Gráficos CERIDE, pp. 115-118.

Kang, S.S. and Woo, W.S. (1987). Two new saponins from *Phytolacca Americana. Planta Medica* 338-340.

Kelly, S., Lamb, D., Corran, A., Baldwin, B., Kelly, D. (1995). Mode of action and resistance to azole antifungals associated with the formation of 14-α-methyl-ergosta-8,24(28)dien-3-β, 6-α-diol. *Biochem Biophys Res Commun* 207:910-915.

Kim, S., Hwang, E., Nam, J., Bok, S., Kim, H., Kwon, B. (1999). Inhibition of chitin synthase II by cathechins from stem bark of *Taxus cuspidata. Planta Med* 65:97-98.

Kobayashi, A., Kohtatsu, H., Kajiyama, S. (1995). Antifungal compounds induced in the dual culture with *Phytolacca americana callus* and *Botrystis fabae. Z. Naturforsh.* (Drug research) 50C: 398-402.

Kollar, R., Reinholds, B., Petrakova, E., The, H., Ashwell, G., Drgonova, J., Kapteyn, J., Klis, F., Cabib, E. (1997). Architecture of the yeast cell wall. *J Biol Chem* 272:17762-17775.

Kopecka, M. (1984). Papulacandin B: Inhibitor of biogenesis of 1,3-β-glucan fibrillar component of the cell wall of *Saccharomyces cerevisiae* protoplasts. *Folia Microbiologica* 29:115-119.

Kouznetsov, V., Aliev, A., Prostakov, N. (1992). USSR Patent 1991 169429. *Chem Abst,* 117:48.356.

Kouznetsov, V., Aliev, A., Prostakov, N. (1994). Synthesis of 2-alkyl(aryl,hetaryl)-4-methyl-1,2,3,4-tetrahydroquinolines. *Khim Geterotsikl Soedin* 1:3-78.

Kouznetsov, V., Andreeva, E., Prostakov, N. (1996). Synthesis and pesticide activity of new derivatives of 2,4-disubstituted 1,2,3,4-tetrahydroquinolines. *Khim Farm Zh* 29:61-62.

Kouznetsov, V., Öcal, N., Turgut, Z., Zubkov, F., Kaban, S., Varlamov, A. (1998). Allylation and heterocycoaddition reactions of aldimines: Furan- and quinoline-carboxaldehydes. *Mon Chem* 129:671-677.

Krauze-Baranowska, M., Cisowski, W., Wiwart, M., Madziar, B. (1999). Antifungal biflavones from *Cupressocyparis leylandii. Planta Med* 65:572-573.

Lahrsini, M., Bousaid, M., Lazrek, H., Jana, M., Amarouch, H. (1997). Evaluation of antifungal and molluscicidal properties of extracts of *Calotropis procera. Fitoterapia* 68:371-373.

Lee, S., Lee, J., Lundi, C., Kubo, I. (1999). In vitro antifungal susceptibilities of *Candida albicans* and other fungal pathogens to polygodial, a sesquiterpene-dialdehyde. *Planta Med* 65:204-208.

Lentz, D., Clark, A., Hufford, Ch., Meurer-Grimes, B., Passreiter, C., Cordero, J., Ibrahimi, O., Okunade, A. (1998). Antimicrobial properties of Honduran medicinal plants. *J Ethnopharm* 63:253-263.

Lima, E., Morais, V., Gomes, S., Cechinel Filho, V., Miguel, O., Yunes, R. (1995). Preliminary evaluation of antifungal activity of xanthoxyline. *Acta Farm Bon* 14:213-216.

Liu, L. (1989). DNA topoisomerases poisons as antitumor drugs. *Ann Rev of Biochem* 58:351-375.

Lu, Q., Ubillas, R., Zhou, Y., Dubenko, L., Dener, J., Litvak, J., Phuan, P., Flores, M., Ye, Z., Gerber, E., Truong, T., Bierer, D. (1999). Synthetic analogues of irlbacholine: A novel antifungal plant metabolite isolated from *Irlbachia alata. J Nat Prod* 62:824-828.

Mahasneh, G., El-Oqlach, A. (1999). Antimicrobial activity of herbal plants used in the traditional medicine of Jordan. *J Ethnopharm* 64:271-276.

Maoz, M., Kashman, Y., Neeman, I. (1999). Isolation and identification of a new antifungal sesquiterpene lactone from *Inula viscose. Planta Med* 65:281-282.

Martinez Crovetto, R. (1964). Estudios etnobotánicos I. Nombre de las plantas y su utilidad según los indios Tobas del este del Chaco. *Bomplandia,* 1:280-283.

Medoff, G., Kobayashi, G., Kwang, C. (1971). Potentiation of rifampicin and flucytosine as antifungal antibiotics by amphotericin B. *Proc Soc Exp Biol Med* 138:571-574.

Mehta, B., Neogi, R., Kotra, S., Mall, O. (1997). Antimicrobial activity of *Abutilon indicum. Fitoterapia* 68:273-274.

Mitscher, L., Leu, R., Bathala, M., Wu, W., Beal, J. (1972). Antimicrobial agents from higher plants. I. Introduction, rationale and methodology. *Lloydia* 35: 157-166.

Miyazawa, M., Kasahara, H., Kameoka, H. (1996). Antifungal activities of 8.O.4'-neolignans from *Myristica fragrans. Nat Prod Lett* 8:271-273.

Mizoguchi, J., Saito, T., Mizuno, K., Hayano, K. (1977). On the mode of action of a new antifungal antibiotic, aculeacin A: Inhibition of cell wall synthesis in *Saccharomyces cerevisiae. J Antib* 304:308-313.

Moreno, M. and Rodríguez, V.M. (1981). Yiamoloside B, a fungistatic saponin of *Phytolacca octandra. Phytochem* 20:1446-1447.

Nagahashi, S., Sudoh, M., Ono, N., Sawada, R., Yamaguchi, E., Uchida, Y., Mio, T., Takagi, M., Arisawa, M., Yamada-Okabe, H. (1995). Characterization of chitin *Saccharomyces cerevisiae. J Biol Chem* 270:13961-13967.

Nagiec, M., Nagiec, E., Baltisberger, J., Wells, G., Lester, R., Dickson, R. (1997). Sphingolipid synthesis as a target for antifungal drugs. *J Biol Chem* 272: 9809-9817.

Nandy, A., Chakraborty, A., Podder, G. (1997). Antimicrobial activity of *Terminalia bellerica. Fitoterapia* 68:178-180.

Narula, A., Rahier, A., Banveniste, P., Schuber, F. (1981). 24-methyl-25-azacycloartanol, an analog of a carbonium ion high-energy intermediate, is a potent inhibitor of (S)-adenosyl-L-methioninesterol C-24 methyltransferase in higher plants. *J Am Chem Soc* 103:2408-2409.

Nidiry, E. (1999). Antifungal activity of tomato seed extracts. *Fitoterapia* 70:181-183.

Nielsen, S.E., Anthoni, U., Christophersen, C., and Cornett, C. (1995). Triterpenoids saponins from *Phytolacca rivinoides* and *Phytolacca bogotensis. Phytochem* 39: 625.

Nussbaumer, P., Leitner, I., Stütz, A. (1994). Synthesis and structure-activity relationships of the novel homopropargylamine antimycotics. *J Med Chem* 37:610-615.

Nussbaumer, P., Ryder, N., Stütz, A. (1991). Allylamine antimycotics: Recent trends in structure-activity relationships and synthesis. *Pest Sci* 31:437-455.

Oehlschlager, A., Czyzewska, E. (1992). Rationally designed inhibitors of sterol biosynthesis. In *Emerging targets in antibacterial and antifungal chemotherapy*, Sutcliffe, J. and Georgopapadakou, N. (Eds.). New York: Chapman and Hall, pp. 437-475.

Ohiri, F., Okelu, V. (1997). Antimicrobial properties of *Mitracarpus villosus. Fitoterapia* 68:287-288.

Onishi, J., Meinz, M., Curotto, J., Dreikorn, S., Rosenbach, M., Douglas, C., Abruzzo, G., Flattery, A., Kong, L., Cabello, A., Vicente, F., Pelaez, F., Diez, M., Martin, I., Bills, G., Giacobbe, R., Dombrowski, A., Schwartz, R., Morris, S., Harris, G., Tsipouras, A., Wilson, K., Kurtz, M. (2000). Discovery of novel antifungal (1,3)-β-D-glucan synthase inhibitors. *Antimicrob Agents Chemother* 44:368-377.

Parkhurst, R.M., Thomas, D.W., and Skinner, W.A. (1973). Molluscicidal saponins of *Phytolacca dodecandra:* Oleanoglycotoxin-A. *Phytochem* 12: 1437-1442.

Parks, L., Smith, S., Crowley, J. (1995). Biochemical and physiological effects of sterol alterations in yeast—a review. *Lipids* 30:227-230.

Parks, L., Todd Lorenz, R., Casey, W. (1992). Functions for sterols in yeast membranes. In *Emerging targets in antibacterial and antifungal chemotherapy*, Sutcliffe, J. and Georgopapadakou, N. (Eds.). New York: Chapman and Hall, pp. 393-409.

Pei-Gen, X. (1989). Excerpts of the Chinese pharmacopeia. In Herbs, spices, and medicinal plants: Recent advances in botoany, horticulture, and pharmacology, Volume 4. Cracker, L. and Simon, J. (Eds.). Phoenix, AZ: Oryx Press, p. 105.

Petraitiene, R., Petraitis, V., Groll, A., Candelario, M., Sein, T., Bell, A., Lyman, C., Mc Millan, C., Bacher, J., Walsh, T. (1999). Antifungal activity of LY 303366, a novel echinochandin B, in experimental disseminated candidiasis in rabbits. *Antimicrob Agents Chemother* 43:2148-2155.

Petranyl, G., Ryder, N., Stütz, A. (1984). Allylamine derivatives: New class of synthetic antifungal agents inhibiting fungal squalene epoxidase. *Science* 224: 1239-1241.

Pilch, D., Yu, Ch., Makhey, D., La Voie, E., Srinivasan, AR., Olson, W., Sauers, R., Breslauer, K., Geacintov, N., Liu, L. (1997). Minor groove-directed and intercalative ligand-DNA interactions in the poisoning of human DNA topoisomerase I by protoberberine analogs. *Biochem* 36:12542-12553.

Pinheiro, T., Yunes, R., López, S., Santecchia, C., Zacchino, S., Cechinel Filho, V. (1999). In vitro antifungal evaluation and studies on mode of action of xanthoxyline derivatives. *Arzneim-Forschung/Drug Research* 49:1939-1043.

Polak, A. (1999). Past, present and future of antimycotic combination therapy. *Mycoses* 42, 355-370.

Rahalison, L., Benathan, M., Monod, M., Frenk, E., Gupta, M., Solís, P., Fuzari, N., Hostettmann, K. (1995). Antifungal principles of *Baccharis pedunculata*. *Planta Med* 61:360-362.

Rahalison, L., Hamburger, M., Hostettmann, K., Monod, M., Frenk, E. (1991). A bioautographic agar overlay method for the detection of antifungal compounds from higher plants. *Phytochem Anal* 2:199-203.

Rahier, A., Genot, J., Narula, A., Benveniste, P., Schmidt, P. (1980). 25-azacycloartanol, a potent inhibitor of S-adenosyl-L-methionine-sterol C-24 and C-28 methyltransferase in higher plants cells. *Biochem Biophys Res Commun* 92:20-25.

Rahier, A., Genot, J., Schuber, F., Benveniste, P., Narula, A. (1984). Inhibition of S-adenosyl-L-methionine-sterol C-24 methyltransferase by analogs of a carbocationic ion high-energy intermediate. Structure-activity relationships for C-25 heteroatom (N, As, S) substituted triterpenoid derivatives. *J Biol Chem* 259: 15215-15223.

Rajbhandari, M., Schöpke, T. (1999). Antimicrobial activity of some Nepalese medicinal plants. *Pharmazie* 54:232-234.

Rao, A.V. and Sung, M.K. (1995). Saponins as anticarcinogens. *American Institute of Nutrition,* 717S-724S.

Rashid, A., Scott, E., Richardson, M. (1995). Inhibitory effect of terbinafine on the invasion of nails by *Trichophyton mentagrophytes*. *J Am Acad Dermatol* 33: 718-723.

Rath, G., Ndonzao, M., Hostettmann, K. (1995). Antifungal anthraquinones from *Morinda lucida*. *Int J Pharmacogn* 33:107-114.

Rodríguez, P. (1915). *Plantas medicinales del Paraguay*. Asunción, Paraguay: La Mundial Press, pp. 52-53.

Rojas Acosta, N. (1907). Catálogo de las plantas medicinales del Chaco Austral. *Revista Farmacéutica* (Buenos Aires) 47:220-221.

Ryder, N. (1985). Specific inhibition of fungal sterol biosynthesis by SF 86-327, a new allylamine antimycotic agent. *Antimicrob Agents Chemother* 27:252-256.

Ryder, N. (1987). Squalene epoxidase as the target of antifungal allylamines. *Pest Sci* 21:281-288.

Ryder, N. (1990). Inhibition of squalene epoxidase and sterol side-chain-methylation of allylamines. *Biochem Soc Trans* 18:45-46.

Ryder, N. (1992). Terbinafine: Mode of action and properties of the squalene epoxidase inhibition. *Br J Dermatol* 126, suppl 39:2-7.

Ryder, N., Dupont, M. (1985). Inhibition of squalene epoxidase by allylamine antimycotic compounds: A comparative study of the fungal and mammalian enzymes. *Biochem J* 230:765-770.

Ryder, N., Seidl, G., Troke, P. (1984). Effect of the amtimycotic drug naftifine on growth of and sterol biosynthesis in *Candida albicans. Antimicrob Agents Chemother* 25:483-487.

Ryder, N., Stütz, A., Nussbaumer, P. (1994). Inhibitors of squalene epoxidase. In *Design of enzyme inhibitors as drugs*, Volume 2, M. Sandler and J. Smith (Eds.). London: Oxford University Press, pp. 378-395.

Salie, F., Eagles, P., Leng, H. (1996). Preliminary antimicrobial screening of four South African Asteraceae species. *J Ethnopharm*, 52:27-33.

Sandven, P., Nilsen, K., Digranes, T., Tjade, T., Lassen, J. (1997). *Candida novergensis:* A fluconazole-resistant species. *Antimicrob Agents Chemother* 41:1375-1376.

Sangeorzan, J., Bradley, S., He, X., Zarins, L., Ridenour, G., Tiballi, R., Kauffman, C. (1994). Epidemiology of oral candidiasis in HIV-infected patients: Colonization, infection, treatment and emergence of fluconazole resistance. *Am J Med* 97:339-346.

Sanglard, D., Kuchler, K., Ischer, F., Pagani, J., Monod, M., Bille, J. (1995). Mechanisms of resistance to azole antifungal agents in *Candida albicans* isolates from AIDS patients involve specific multidrug transporters. *Antimicrob Agents Chemother* 39:2378-2386.

Santen, R., Vanden Bossche, J., Symoens, J., Brugmans, J., Decoster, R. (1983). Site of action of low-dose ketoconazole on androgen biosynthesis in men. *J Clin Endocrin Metab* 57:732-736.

Sawistowska-Schroder, E., Kerridge, D., Perry, H. (1984). Echinocandin inhibition of (1,3)-β-glucan synthase from *Candida albicans. FEBS Lett* 173:134-138.

Saxena, V., Sharma, R. (1999). Antimicrobial activity of the essential oil of *Toddalia asiatica. Fitoterapia* 70:64-66.

Schultes, E., Holmstedt, B. (1971). *De plantis toxicariis* e mundo novo tropicale commentations. VIII. Miscellaneous notes on Myristicaceous plants of South America. *Lloydia* 34:61-78.

Selitrennikoff, C. (1992). Screening for antifungal drugs. In *Biotechonology of filamentous fungi, technology and products*, Finkelstein, D., Ball, C. (Eds.). Boston: Butterworth Heinemann Press, pp. 189-217.

Selitrennikoff, C. (1995). *Antifungal drugs: (1,3)-β-glucan synthase inhibitors.* Heidelberg: Springer-Verlag, pp. 1-12.

Sentandreu, R., Mormeneo, S., Ruiz Herrera, J. (1994). Biogenesis of the fungal cell wall. In *The Mykota, growth, differentiation, and sexuality.* Wessels, Meinhardt (Eds.). Berlin: Springer-Verlag, pp.111-141.

Shaw, J., Mol, P., Bowers, B., Silverman, S., Valdivieso, H., Durán, A., Cabib, E. (1991). The function of chitin synthases 2 and 3 in the *Saccharomyces cerevisiae* cell cycle. *J Cell Biol* 114:111-123.

Shen, L. (1993). Molecular mechanism of DNA gyrase inhibition by quinolone antibacterial agents. In *Molecular biology of DNA topoisomerases and its application to chemotherapy,* Andoh, T., Ikeda, H., Oguro, M. (Eds.). Boca Raton: CRC Press, pp. 177-186.

Shen, L., Baranowsky, J., Pernet, G. (1989). Mechanism of inhibition of DNA gyrase by quinolones antibacterials: Specificity and cooperativity of drug binding to DNA. *Biochem* 28:3879-3885.

Smirnov, V., Bondarenko, A., Prikhodko, V. (1998). Antimicrobial activity of sesquiterpene phenol from *Bidens cernua. Fitoterapia* 69: 84-85.

Spengel, S. and Schaffner, W. (1993). Esculentoside S, a new saponin from the leaves of *Phytolacca acinosa. Nat. Prod. Letters* 2: 243-247.

Spengel, S.M., Luterbacher, S., and Schaffner, W. (1995). New aspects on the chemotaxonomy of *Phytolacca dodecandra* with regard to the isolation of phytolaccagenin, phytolaccagenic acid and their glycosides. *Planta Medica* 61: 385-386.

Strauss, A., Spengel, S., and Schaffner, W. (1995). Saponins from root cultures of *Phytolacca acinosa. Phytochem* 38: 861-865.

Sudoh, M., Yamazaki, T., Masubuchi, K., Taniguchi, M., Shimma, N., Yamada-Okabe, H., Yamada-Okabe, A. (2000). Identification of a novel inhibitor specific to the fungal chitin synthase; inhibition of chitin synthase I arrests the cell growth, but that of chitin syntase 1 and 2 is lethal in the pathogenic fungi *Candida albicans. J Biol Chem* 275: 32901-32905.

Suga, Y., Maruyama, Y., Kawanishi, S., and Shoji, J. (1978). Studies on the constituents of Phytolaccaceous plants. I. On the structures of phytolaccasaponin B, E, and G from the roots of *Phytolacca Americana. Chem Pharm Bull* 26: 520-525.

Sur-Altiner, D., Gurkan, E., Sarioglu, I., Tuzlagi, E. (1997). The antibacterial and antifungal effects of *Centaurea hermanii. Fitoterapia* 68:374-375.

Sur-Altiner, D., Gurkan, E., Sarioglu, I., Tuzlagi, S., Büyükbaba, Ö. (1997). The antibacterial and antifungal effects of *Daphne pontica. Fitoterapia* 68:547-548.

Taylor, A., Giles, K., Sarthy, A., McGonigal, T., Fostel, J. (1996). Identification of the gene encoding DNA topoisomerase I from *Candida albicans. FEMS Microb Lett* 138:113-121.

Thopp, J., Minija, J., Tajo, A. (1998). Antibacterial and antifungal activities of four varieties of *Ocimum basilicum. Fitoterapia* 69:191-192.

Timberlake, W., D'Enfert, C., Fonzi, W. (1994). Targets for inhibiting proliferation of virulence, 2nd NIAID Workshop in Medical Mycology, University of Arizona, Northern Arizona University, Flagstaff, AZ, June 8-11.

Tkacz, J. (1992). Glucan biosynthesis in fungi and its inhibition. In *Emerging targets in antibacterial and antifungal chemotherapy,* Sutcliffe, J. and Georgopapadakou, N. (Eds.). New York: Chapman and Hall, pp. 495-523.

Urbina, J., Cortés, J., Palma, A., López, S., Zacchino, S., Enriz, R., Ribas, J., Kouznetsov, V. 2000. Inhibitors of the fungal wall. Synthesis of 4-aryl-4-

N-arylamine-1-butenes and related compounds with inhibitory activities on β (1,3) glucan and chitin synthases. *Bioorg Med Chem* 8:691-698.

Vanden Berghe, D., Vlietink, A. (1991). Screening methods for antibacterial and antiviral agents from higher plants. In *Methods in plant biochemistry*, Volume 6, Hostettmann, K. (Ed.). London: Academic Press, pp. 47-70.

Varona, R., Perez, P., Durán, A. (1983). Effect of papulacandin B on β-glucan synthesis in *Schizosaccharomyces pombe*. *FEMS Microb Lett* 20:243-247.

Verma, D., Tripath, V., Rana, B., Taneja, V. (1998). Antifungal activity of triterpenoids from *Lantana indica* root. *Fitoterapia* 69:188-189.

Walsh, T. (1992). Invasive fungal infections: Problems and challenges for developing new antifungal agents. In *Emerging targets in antibacterial and antifungal chemotherapy*, Sutcliffe, J., and Georgopapadakou, N. (Eds.). New York: Chapman and Hall, pp. 349-373.

White, T., Marr, K., Bowden, R. (1998). Clinical, cellular and molecular factors that contribute to antifungal drug resistance. *Clin Microbiol Rev* 11:382-402.

Woo, W.S. and Kang, S.S. (1975). The occurrence and chemistry of *Phytolacca* triterpenoids. *J Pharm Soc Korea* 19: 189-208.

Woo, W.S. and Kang, S.S. (1976). Phytolaccoside B: Triterpene glucoside from *Phytolacca Americana*. *Phytochem* 15: 1315-1317.

Woo, W., Sick, W., and Kang, S.S. (1977). The structure of phytolaccoside G. *J Pharm Soc Korea* 212: 159-162.

Zacchino S., Badano, H. (1985). Stereoselective synthesis of 8.O.4'-neolignans: (±)-surinamensin and (±)-virolin. *J Nat Prod* 48:830-832.

Zacchino, S., Badano, H. (1988). Enantioselective synthesis and absolute configuration assignment of erythro-(3,4,5-trimethoxy-7-hydroxy-1'-allyl-3',5'- dimethoxy)-8.O.4'-neolignan, isolated from mace *(Myristica fragrans)*. *J Nat Prod* 51:1261-1265.

Zacchino, S., Badano, H. (1991). Enantioselective synthesis and absolute configuration assignment of erythro-(3,4-methylenedioxy-7-hydroxy-1'-allyl-3',5'- dimethoxy)-8.O.4'-neolignan and its acetate, isolated from nutmeg *(Myristica fragrans)*. *J Nat Prod* 54:155-160.

Zacchino, S., López, S., Pezzenati, G., Furlan, R., Santecchia, C., Muñoz, L., Giannini, F., Rodríguez, A., and Enriz, R. (1999). In vitro evaluation of antifungal properties of phenylpropanoids and related compounds acting against dermatophytes. *J Nat Prod* 62: 1353-1357.

Zacchino, S., Rodríguez, G., Pezzenati, G., Orellana, G., Enriz, R., Gonzalez Sierra, M. (1997). In vitro evaluation of antifungal properties of neolignans-8.O.4'. *J Nat Prod* 60:659-662.

Zacchino, S., Rodríguez, G., Santecchia, C., Pezzenati, G., Giannini, F., Enriz, R. (1998). In vitro studies on mode of action of antifungal 8.O.4'-neolignans occurring in certain species of *Virola* and related genera of Myristicaceae. *J Ethnopharm* 62:35-41.

Zacchino, S., Santecchia, C., Escalante, A., López, S., Gattuso, M., Gutiérrez Ravelo, A., Delle Monache, F., and Gonzalez Sierra, M. (in press). Isolation and in vitro antifungal evaluation of saponins from Phytolacca tetramera, submitted for publication October 2000.

Zacchino, S., Santecchia, C., López, S., Gattuso, S., Muñoz, J., Cruañes, A., Vivot, E., Cruañes, J., Salinas, A., Ruiz, R., Ruiz, S. (1998). In vitro antifungal evaluation and studies on mode of action of eight selected species from the Argentine flora. *Phytomed* 5:389-395.

Chapter 2

Natural Antimycotics from Croatian Plants

Stjepan Pepeljnjak
Ivan Kosalec
Zdenka Kalodera
Danica Kuštrak

INTRODUCTION

Floristic analyses show that there are about 500,000 plant species on our planet (Tivy, 1995). Out of these, about 12,000 plant species can be used to create biologically active products, which are used in treatment. About 500 of these plant species are used for this purpose in Croatia. About 160 to 170 native plants are used in different ailments.

In folk medicine, thousands of years ago medicinal herbs and plant products were used in treating a wide spectrum of infectious and other diseases. Today, a great number of different medicinal teas and other plant products are available on market (including cosmetics and pharmaceuticals), which contain biologically active substances (Nahrstedt, 2000). Scientific empirical investigations use experiences collected throughout history to find new and effective medicines. Why pursue plant products? Microorganisms develop resistance to antimicrobial drugs after continued and/or extensive use. Studies of products derived from plants, therefore, search for completely new, effective, nontoxic, and cheap antimicrobial drugs. With an increasing number of people having compromised immune systems, the number of fungal infections has also increased. Fungi, similar to bacteria and viruses, develop resistance toward existing antimicrobial drugs. The goal of many scientists is to find new antimycotic drugs. One of the richest sources of possibility is the plant world, and at present tens of thousands of plants have already been investigated.

The authors wish to thank Nikola Blaževic, PhD (Ireks-Aroma Ltd., Zagreb, Croatia), for providing analyses of essential oils and Martina Kranjec, MPharm, for design of Figure 2.7.

Our goal is to investigate the antifungal acitivity of the natural active substances of different plants from Croatia. Possibility of applications of different extracts, compounds, and essential oils will be determined. Also, the most efficient wide spectrum antifungal combination with the appropriate application will be pointed out. Apart from that, the defense of plants from fungal infections in natural habitats will also be explained.

BIOLOGICALLY ACTIVE COMPONENTS OF PLANT SPECIES

In many works of scientific research the intention is to get as thoroughly familiar as possible with the chemical composition of the natural "factory" of many hundreds of chemical compounds, to link them with the characteristics of the plant species, and to define their quantity in the specific areas in which these plants grow. Naturally, there are many differences among plant species, so the characteristics of all of these antifungal components contained in plants and their products will also be very different.

The main groups of compounds obtained from plant species that could have antimicrobial activity are as follows: essential oils (terpene compounds—hydrocarbons, alcohols, and phenols; ketones and aldehydes; acids, oxides, and peroxides; phenylpropane compounds); alkaloids, flavonoids, and similar compounds; saponins; polyphenols; and tannins. Of the groups mentioned, essential oils are particularly likely to show antimicrobial activity.

Essential Oils

Natural essential oils are volatile, scented plant products. They are mixtures of a large number of chemical compounds, the most common of which are terpenes (90 percent). Other components belong mainly to derivatives of phenylpropan, such as chain hydrocarbons and compounds with nitrogen and sulphur. They are derived from various plant extracts by distillation, extraction, and pressing. They can be defined as intensively aromatic, volatile, and nonviscous liquids (some essential oils are solid at room temperature and become liquid on heating). Essential oils, produced by distillation, are usually clear or colorless to slightly colored liquids. They are mostly characterized by a very strong and hot, burning taste, which becomes pleasant on dilution. They dissolve well in lipophilic solvents and slightly in water. Their specific weight is 0.84 to 1.18 and most of them are lighter than water.

Their boiling point is between 150 and 300°C. Essential oils are optically active (Wagner, 1993).

The ability to produce essential oils is not widespread in the plant world; it belongs only to seed plants. Investigation into the presence of essential oils in 295 families of higher plants showed that 87 (30 percent) of them produce essential oils. Plant families that contain high amounts of essential oils include the following: Apiaceae, Asteraceae, Brassicaceae, Lamiaceae, Lauraceae, Myrtaceae, Pinaceae, Piperaceae, Rutaceae, and Zingiberaceae (Gildemeister and Hoffmann, 1956; Wagner, 1993).

Essential oils appear as secondary products of metabolism. They are stored in prominent, morphologically differentiated containers, which are easily identifiable histologically. Their anatomical structure is characteristic of the whole genus or family. The oil containers can be situated in leaves, flowers, fruits, bark, roots, or underground stems. The plant stores the essential oil in special cells—the so-called *oil cells* (families Lauraceae, Piperaceae, Zingiberaceae)—which are found in intracellular gaps or channels (families Apiaceae, Rutaceae) or in containers between the cuticle and the cell membrane, or in glands, gland hairs, or gland shells (families Asteraceae, Lamiaceae) (von Denffer and Ziegler, 1991).

It is believed that the functions of essential oil in plants are to attract insects for pollination, to repel herbivores, to influence the reduction of transpiration, to protect the plant from phytopatogenic microorganisms and parasites, and to inhibit germination and growth of other plants as potential rivals for the same niche (Guenther, 1957; Schreier, 1984; Werker, 1993).

Components of Essential Oils

Essential oils are a mixture of a great number of compounds of chemically different structures. More than 500 components of essential oils are known. In some individual oils more than 60 components can be found. However, only a few components are present in sufficient quantities to determine the organoleptic, physical, and chemical characteristics and pharmacological effects of essential oils. The components belong to various groups of organic compounds. Depending on the structure of the hydrocarbons, we can distinguish terpenes, phenylpropanes, and other components (Guenther, 1957). They have been identified by different techniques; thin layer chromatography (TLC), gas chromatography (GC), gas chromatography–mass spectrometry (GC-MS).

Terpenes. The major components of essential oils are terpenes. Their structure is based on a molecule of isoprene (2-methyl-butadiene; C_5H_8). They can be classified into seven groups according to the number of carbon atoms: semiterpenes (C5), monoterpenes (C10), sesquiterpenes (C15), diterpenes (C20), triterpenes (C30), tetraterpenes (C40), and politerpenes. Semiterpene is a 3-isopentenil-pirophosphate ("active isoprene") out of which all terpene compounds, according to the so-called isoprene rule, are formed (Ružicka et al., 1953; Wagner, 1993).

Monoterpenes, sesquiterpenes, diterpenes, and triterpenes are present in essential oils. Monoterpenes (C10) are derivatives of 2- and 6-dimethyl-octane geraniol. About 150 different monoterpene components are known. They can be acyclic-, mono-, and bicyclic compounds. The biggest group among terpenes are sesquiterpenes (C15) as derivatives of 2-, 6-, and 10-trimethyl-N-dodecane farnesol. They can have acyclic, mono-, bi-, and tricyclic structures. Diterpenes (C20) and triterpenes (C30) are found in essential oils, which are obtained by extraction with organic solvents (Wagner, 1993).

Phenylpropanes. Derivatives of phenylpropane are natural products, with a C6-C3 hydrocarbon structure. They can be aldehydes, phenols, and phenylethers, which are derived from cinnamon acid (Wagner, 1993).

Other compounds. The components of essential oils can also include chain hydrogens and their derivatives with oxygen, and compounds with nitrogen and sulphur (sulphides, derivatives of anthranilic acid, and indole and acetylene derivatives) (Wagner, 1993; Vladimir-Kneževic, 1998).

ANTIMICROBIAL ACTIVITY OF ESSENTIAL OILS

Many biological effects of essential oils have been demonstrated by scientific investigations, including antibacterial, antifungal, antiviral, spasmolytic, antiphlogistic, diuretic, cytotoxic, and antihelmintic effects.

The first indications that essential oils have a toxic effect on the growth and reproduction of microorganisms originated from Robert Koch's 1881 test of the effect of terpentine oil on *Bacillus anthracis* (Kubeczka, 1982). Many investigations have been performed since then, which have mostly proven and explained the very old empiric observations regarding the antimicrobial effects of essential oils. In the large group of essential oils whose effect has been scientifically proven and which are widely in use in pharmacy, the most prominent are oil of thyme (Lens-Lisbonne et al., 1987), sage (Kuštrak and Pepeljnjak, 1989), lavender (Perucci et al., 1994), rosemary (Hetheleyi et al., 1989), eucalyptus (Hajji et al., 1993), cinnamon (Yousef and Tawil, 1980), clove (Kubeczka, 1982), and mint (Yousef and Tawil, 1980).

It has been proven that terpene components are mostly responsible for the antimicrobial effect of essential oils (Kellner and Kober, 1955; Goecheritz et al., 1974; Knoblock et al., 1989). Investigation of 32 terpene compounds showed that components with a methyl-isopropyl-cyclohexane ring, such as some alcohols and ketones (pulegone), have a strong antimicrobial effect which becomes stronger in the presence of an unsaturated cyclohexane ring in the structure of the molecule (terpinolene, terpineol, terpinenol). It has also been proven that cis-isomers (nerol) are less active than *trans*-isomers (geraniol). The same is said for α- compared to β-isomers (α- and β-pinene) (Hinou et al., 1989). It was found that the antimicrobial effect of terpene molecules is based on their ability to damage the biomembrane. In accordance with their lipophilic characteristics they are soluble in the phospholipid layer and they interact with membrane enzymes. This is how they inhibit processes that are necessary for the life and reproduction of microorganisms. The results of the investigation showed that phenol alcohols (thymol, carvacrol, eugenol) are the strongest inhibitors of enzyme processes. This is probably due to their lipophilic characteristics and the free hydroxy (–OH) group (for example: eugenol is better than methyl- or acetyleugenol). It was found that nonphenol alcohols (geraniol, linalool) have a weaker effect than phenol alcohols and their effect is reduced by esterification of the hydroxy group. The aldehydes show somewhat weaker effect than nonphenol alcohols, but their effect is stronger than that of the ketones. It was also proved that hydrocarbons, relative to other components of essential oils, are a weak inhibitor of enzyme processes (p-cymene, α- and β-pinene) (Knoblock et al., 1989).

METHODS OF ANTIMICROBIAL INVESTIGATION

Antimicrobial investigation of plant products (extracts, tincture of essential oils or their components) was carried out in several different phases. The aim was to prove the level of fungicidal activity of the plant sample. The plants (leaf, flower, stem, root, seeds, or fruits) were extracted in water or alcohol or with organic solvents.

Different approaches to the investigations derive from the defined scientific approach, from empiric observation, or from specific areas in which plants naturally grow or certain plant species are being cultivated. Internationally recognized microbiological methods can be used for fungicidal screening tests.

In Vitro Investigations

Cultures of Microorganisms

Antimycotic effect in vitro is tested on media of standard strains and clinical isolates of wild-type strains of different fungal species with data on their sensitivity to the existing antimycotics.

Monocellular fungi. Species of genera *Candida, Torulopsis, Geotrichum, Trichosporon* cause superficial skin and mucous mycosis, onychomycosis, diarrhea, and sepsis; *Malassezia* spp. are etiologically linked with seborrhoea of the scalp and skin and dandruff; and *Pityriasis versicolor* causes folliculitis and intertriginous infections.

Polycellular fungi. Fungi such as *Microsporum, Trichophyton,* and *Epidermophyton* cause mycosis of the superficial keratinous parts of skin are known as tinea, or dermatophytes and onychomycosis. Mold of the genus *Fusarium* and *Aspergillus* cause keratitis, onychomycosis, and systemic mycosis; *Curvularia* spp. (leaf-spot disease) and *Scopulariopsis* spp. (onychomycosis) cause phytopathogen molds and foodborne molds.

Methods Used to Prove Antimicrobial Activity

Methods used for in vitro investigations of antimicrobial activity of plant antimycotics comprise diffusion and turbidimetric methods. Bioautography is used to provide information on the antimicrobial activity of some of the components on thin layer chromatography (TLC).

Diffusion method for determining the zone of inhibition (ZI) (agar diffusion assay, agar overlay technique, paper disc agar diffusion technique, hole plate diffusion assay). The main principle of this method is measurement of diffusion within the nutrient medium of the tested substance from the point at which it was put into the inoculated medium with the specified fungal species. The application of the tested substance is done via metal cylinders (Brantner, 1997) making holes into the inoculated nutrient medium or by putting cylinders or paper discs (Brantner, 1997) onto the inoculated nutrient medium. These cylinders or discs were immersed earlier in the tested sample or a certain amount of it was sprayed on them. The nutrient medium is inoculated with the types of fungi that are being tested or with their spores, density 10^8 CFU (colony-forming units)/ml, given in 0.5 McFarland units, which corresponds to the cloudiness of the standard solution of barium sulphate (0.5 ml 1.175 percent w/v $BaCl_2 \times 2H_2O$ in 99.5 ml 1 percent w/v H_2SO_4). The density of the inoculum can be defined with a densitometer. Alternatively, the number of spores of dermatophytes and molds (10^6 CFU/ml) can be determined by volumetric unit using a microscopic scale to count them under the microscope or by counting the growing colonies after a certain amount of the suspension of spores/blastospores have been inoculated into a solid nutrient broth. The type of nutrient medium used in a diffusion test depends on the fungal species being tested; they are listed in Table 2.1.

Substances that have been put into the inoculated nutrient agar (thickness 4 ± 1mm) will then diffuse in all directions around the hole made by the metal cylinder (∅6 mm) or around the paper disc (∅6 mm) onto which a certain quantity of the substance to be tested was put. The concentration of

TABLE 2.1. Review of the nutrient mediums that are used in diffusion and dilution methods for the investigation of antifungal activity

Culture media name	Composition (per liter)*	pH
Sabouraud broth and agar	Pancreatic digest of casein (5 g), peptic digest of animal tissue (5 g), glucose or dextrose (20 g)	5.6 + 0.2
Mycophil broth	Peptic digest of animal tissue (5 g), pancreatic digest of casein (5 g), dextrose (40 g)	7.0 + 0.2 (4.7 + 0.2)
Czapek-Dox broth	Saccharose (30 g), sodium nitrate (3 g), magnesium sulphate (0.5 g), potassium chloride (0.5 g), ferro (II) sulphate (0.01 g), dipotassium hydrogen phosphate (1 g)	7.3 + 0.1
CASO (Casein-peptone soymeal-peptone) broth	Casaminoacid (2.5 g), dextrin (2.75 g), glycerol (2.35 g), pancreatic digest of gelatin (0.78 g)	4.6 + 0.2
RPMI 1640 broth and agar	highly defined synthetic media with L-glutamine, without sodium-bicarbonate	pH 7 (buffered with 0.165 M morpholine-propanesulfonic acid
Nutrient broth	Pancreatic digest of gelatin (5 g), beef extract (3 g), sodium chloride (65 g)	6.9 + 0.2
Brain heart infusion	Infusion from brain heart (6 g), peptic digest of animal tissue (6 g), sodium chloride (5 g), dextrose (3 g), pancreatic digest of gelatin (14.5 g), disodium phosphate (2.5 g)	6.9 + 0.2

*For broth add distilled water to 1,000 ml; for agar add 15 g agar and distilled water to 1,000 ml.

the tested substance will diminish as distance from the place where the test sample was originally put increase.

In order to accelerate the process of diffusion, the discs with the inoculated nutrient medium and the test sample were stored at 4°C for one to four hours. After acceleration of the diffusion process, the discs are incubated at 25 ± 2°C for 48 to 72 hours or more, depending on the fungi species tested. After that, the inhibition zone around the sample is measured. The data are given in millimeters and the diameter of the inhibition zone is measured (Photo 2.1).

Dilution method (microbroth dilution assay, serial dilution assay, agar dilution assay). Methods for determining the minimal inhibitory concentration (MIC) and minimal fungicidal concentration (MFC). This method is based on the dilution of the substance to be tested in the nutrient medium (solid or liquid). Different concentrations of the tested substance can be achieved by the method of serial dilutions in a liquid medium or by a solid broth, mostly with a twofold dilution. These methods offer the possibility of defining the minimal inhibitory concentration (MIC). The MIC is the small-

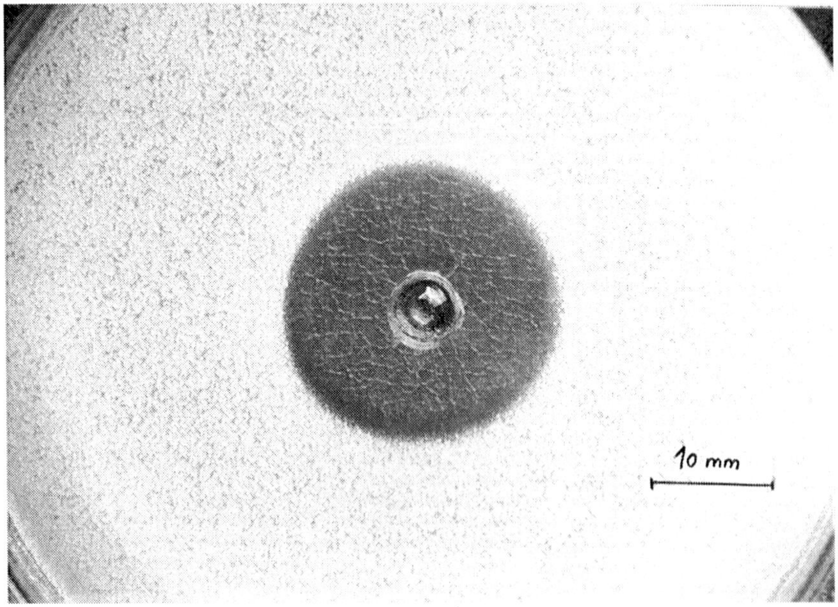

PHOTO 2.1. Results of the diffusion test: Inhibition zone (21.5 mm) of *Candida albicans* due to activity of essential oil of *Citrus aurantium* L ssp. *aurantium* (Copyright Department of Microbiology, University of Zagreb, Croatia.)

est concentration of the tested substance that inhibits the visible growth of the tested microorganism in a liquid medium. MFC is the minimal concentration of the tested substance that kills all microorganisms in the media. Detection of the breaking points of MIC and MFC is done visually, according to the cloudiness of each serial dilution (NCCLS, 1997); by turbidimetric analyses of each of the dilutions (Brantner, 1997); or by subcultivating each of the dilutions onto the solid nutrient broth, incubating them at 25-30°C, and then noting the degree of growth of microorganisms not visible to the naked eye which are apparent break points of certain dilutions (Pepeljnjak et al., 1996 (Photo 2.2). The data on MIC and MFC are given in μg/ml or vol%.

The test substance can be diluted and inoculated into the solid nutrient broth. This dilution method is used for a large number of test substances with a larger number of fungal species. It is used to determine the inhibitory activity of the test substance (Maggon et al., 1997; Hugo and Russell, 1997).

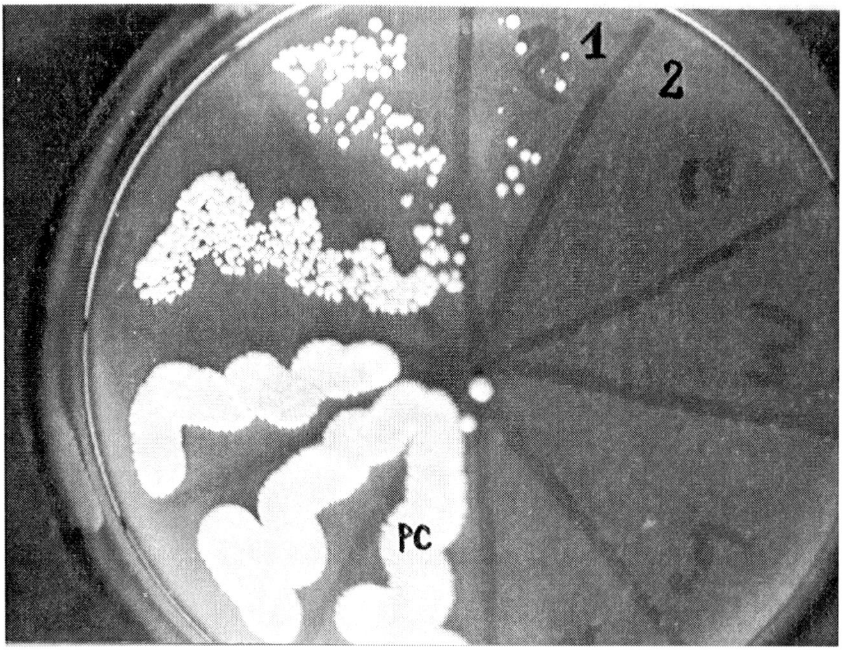

PHOTO 2.2. Borderline values of MIC and MFC achieved by subcultivation of serial dilutions on the Sabouraud agar (1 = MIC, 2 = MFC, PC = positive control) (Copyright Department of Microbiology, University of Zagreb, Croatia.)

Bioautographic method (bioautographic assay, bioassay, and mycoassay). A more recent method, which is not yet an official method in pharmacopeias, is the bioautographic method. It is convenient for the investigation of very small amounts of the test substance. The convenience of this method is that it combines TLC with bioassay in situ and enables the location of antimicrobial active components in a plant extract. The first step is to separate the plant extract components (depending on the solid or liquid phase used). The discs prepared in this way are dried and sprayed with the nutrient medium inoculated with a chosen microorganism. The inoculated discs are then stored in a humid atmosphere, and the zone of inhibition is read by spraying it with the appropriate reagent (water solution of 2,3,5-triphениltetrazolium chloride or alcohol solution of p-jodnitrotetrazolium violet or water solution of tetranitrotetrazolium blue or methylthiazolyltetrazoilum chloride) for easier viewing (Wang et al., 1989; Brantner, 1997; Hostettmann et al., 1998).

Diffusion methods show antimicrobial activity on each single test microorganism. On the basis of the preliminary results of the diffusion method, we can, for further investigation, examine those substances (essential oils, extracts) which have shown antifungal activity in these preliminary tests. However, we often find that we cannot compare the results achieved via the diffusion test (inhibition zone) with the results achieved via the dilution method (MIC).

Our modification of the diffusion method with microconcentrations of a specific extract was performed as an additional test for the verification of the antimicrobial activity of very small amounts of extracts or other substances (1-10 μl) also correspond with MIC. We directly put a specified amount of the test sample (plant extract, essential oil) into the inoculated nutrient broth (the same procedure as in the diffusion method), in which a hole was made with a warmed glass stick. After incubation, the degree of dilution that corresponded to MIC and MFC was ascertained by the naked eye or with a magnifying glass. In this way we can investigate very small concentrations of a certain substance separated or extracted from only one chromatographic location that contained only 0.5 to 1 μg of a specified active substance (Photo 2.3) (Pepeljnjak and Babic, 1991; Vladimir-Kneževic, 1998).

Dilution methods are more precise and give information on the minimal inhibitory concentration (MIC) and the minimal fungicidal concentration (MFC). If a particular test sample has shown inhibitory effects, we investigate MIC and MFC on a larger sample of microorganisms. Using a greater number of microorganisms enables us to produce a better statistical evaluation of the results and a clear assessment for possible further application.

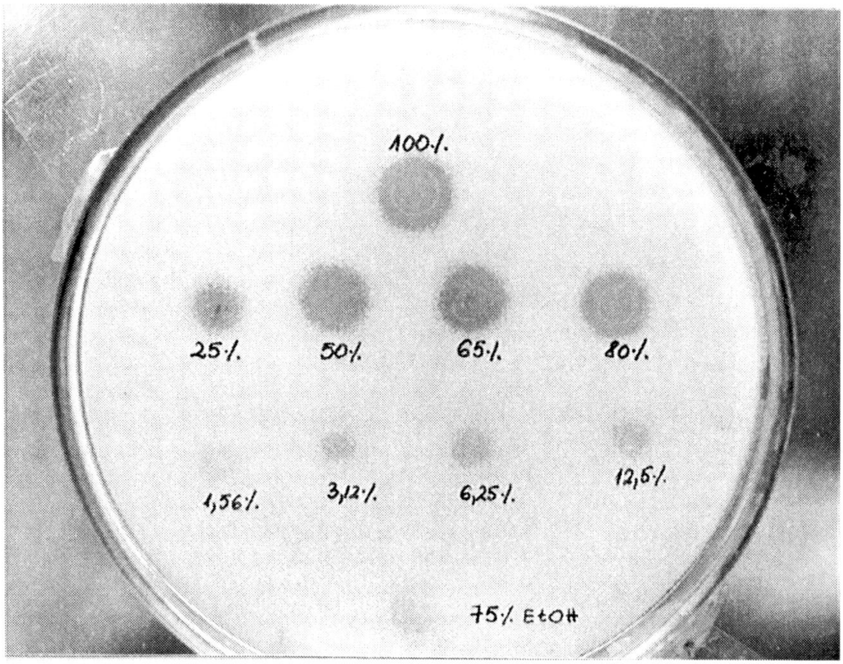

PHOTO 2.3. Antimycotic activity of *Micromeria thymifolia* (Scop.) Fritsch essential oil on *Candida kefyr* (Copyright Department of Microbiology, University of Zagreb, Croatia.)

The second step is to determine which components demonstrate antifungal activity. Analytical methods are used for isolation. Various physical and chemical methods are used for identification (GG/CM, HLPC, TLC), mostly directly with silica-gel discs in a bioautographic test. The reaction of the plant product or its single components to ultraviolet (UV) rays, temperature and time of storage, etc., as factors in product stability are also measured. Only after such investigations it is possible to undertake the preclinical and clinical investigations of single antimycotic, pure, or complex agents.

In Vivo Investigations of the Effectiveness of Plant Antimycotics

In vivo investigations for many antifungal substances have to be carefully interpreted because no adequate standardization of procedures exists in the applied investigations. The fungicidal activity of different substances in a biological experiment is influenced by the kind of laboratory animals (age, sex, weight), characteristics of the infectious fungal species, nature and place of the infection, immune status of the laboratory animal, and method of determining the end point of the experiment (cure). In vivo investigations on laboratory animals provide valuable data for later clinical investigations. However, the procedures are slow and expensive, and cannot be used for routine screening tests, except as corresponding scientific evaluation in certain investigations (Maggon et al., 1997). Rodents are the most-used experimental animals (mouse, rat, rabbit, guinea pig); domestic animals are used less often (horse, cow, sheep, and others). The rubbing technique is used in experimental provocation of mycosis: the agent is rubbed on/in/under the skin. Intraperitoneal, peroral, and intramuscular inoculation of fungi is also used. Experimental treatment of animals naturally infected with different kinds of mycosis is often undertaken.

ANTIFUNGAL ACTIVITY OF ESSENTIAL OILS AND PLANT EXTRACTS

The preservative effect of essential oils was already known in ancient Egypt 4,000 years B.C. This effect was also known to the ancient Greeks, Romans, and Indians. However, we do not know whether their knowledge extended to the application of essential oils to destroy fungi or treat fungal infections.

Investigations dealing exactly with this issue are currently being performed by many scientists, who are investigating hundreds of kinds of different essential oils. Most of the naturally active substances show fungistatic (FS) as well as fungicidal (FC) characteristics. Their very rich chemical composition of over 500 different components has not allowed the identification of exactly which one of the components present has the antifungal effect. In practice, numerous components have similar fungicidal characteristics, so the spectrum of their activity against different fungal species can be judged to be a synergistic effect.

Looking globally at the antimicrobial and antifungal activity of essential oils, we can presume that microbicidal activity depends on the dominant component of the essential oil, but its range of activity against a number of fungal species is enhanced by other components present in them. This is where the difference in antifungal activity of essential oils of different plant species is derived. Among the most active antifungal components of essential oils are thymol and carvacrol. These components have an excellent antifungal activity against *Malassezia furfur* and dermatophytes *Tricho-*

phyton rubrum, T. beigelli (Lacoste et al., 1996), *T. interdigitale, T. mentagrophytes,* and *Candida albicans* and *Scopulariopsis brevicaulis* (Pattnaik et al., 1997). Thymol belongs to the alcohol and phenol type of terpenes in essential oils. Because of its hydroxy group on the benzene ring, thymol can be compared with phenol as a standard disinfectant, and the effect of thymol is twenty times greater than phenol taken as a standard.

Similar activity was also proved on yeasts and Hyphomycetes by investigating the antifungal activity of citral and geraniol. Linalool, cineole, and menthol have a slightly weaker activity level (Pattnaik et al., 1997). Other components of essential oils, such as geraniol, citronellal, and citral, have a distinctly inhibiting activity against mycelial fungi with aerosol. Citral is the most effective agent against molds (Chaumont and Léger, 1992).

It is also known that phenol compounds eugenol, cavicol, and 4-alyl-2,6-dimethoxy-phenol inhibit the growth of yeast and yeastlike microorganisms, dermatophytes, and *Aspergillus fumigatus* (Chaumont and Léger, 1989). Eugenol, *O*-methoxy-cinnamaldehyde, and *trans*-cinnamaldehyde (from the bark of *Cinnamomum zeylanicum*) have a relatively low MIC against *Candida albicans* and *Torulopsis glabrata,* which are sensitive and resistant to azole antimycotics (Quale et al., 1996).

The old civilizations were familiar with the medicinal characteristics of plants, including the Greeks (Hippocrates, Theophrastus, Dioscurides) and Romans (Pliny, Galen). With the discovery of synthetic drugs in the nineteenth century, the empirical experience was rejected. The medical science of today has lost confidence in the healing properties of plants. However, our contemporary way of life has seen decreased resistance toward many sources of infectious diseases, including the causes of mycosis (Odds, 1996; Bossche et al., 1998). This is the result of the often-uncontrolled use of synthetic antimicrobial drugs, a higher incidence of people with lowered immunity, and better communications among different parts of the world. Motivated scientists are returning to the ancient search for valuable substances in the chemical composition of different plants.

In preliminary investigations, most authors are interested in the discovery of the antifungal activity of water, ethanol, and methanol extracts against different species of fungi, very often supported by the experience of the old empirical tradition.

The following substances have been investigated as potential antifungals: tannins (Bašic and Bosnic, 1989), saponins, polyphenol components (Baba Moussa et al., 1999), flavone glycosides (Serkedjieva, 1997; Amaro Luis et al., 1997), alkaloids (Navarro and Delgado, 1999), flavones and sesquiterpenes (Aljancic et al., 1999; Goren et al., 1996), and derivatives of benzofuran isolated from *Piper fulvescens,* which play the main role in antifungal activity against *Candida* and dermatophytes (Freixa et al., 1999).

Antifungal Activity of Essential Oils of Croatian Plants

Essential oils derived from Croatian plants show diversity in antifungal activity as well as in the spectrum of fungicidal activity against different fungal species, dermatophytes, and molds. Sensitivity to essential oils is different not only for different strains or genera of fungi but for strains within species (Table 2.2).

It is obvious that there is a wide range of the minimal inhibitory and minimal fungicidal concentrations, which start at 0.19 to 25 vol% (Pepeljnjak, 1996). Dermatophytes are the most sensitive pathogens to essential oils derived from Croatian native plants. The lowest fungistatic concentrations range from 0.019 to 3.125 vol%, while fungicidal concentrations are slightly higher and are different for each strain of dermatophyte. The most sensitive is *Epidermophyton floccosum* with MFC of 0.037 vol%. The species of *Trichophyton mentagrophytes* is slightly resistant with an MFC of 6.25 vol%.

Croatian Coastal and Island Plants

Lavender [*Lavandula officinalis* L. (family Lamiaceae)] contains 13 percent of essential oil; the dominant components are linalyl acetate and linalool. They, together with other components, have a low and relatively uneven activity against microorganisms similar to molds (MIC 3.12-25 vol%). Species of dermatophyte are slightly more sensitive (MIC 0.78-3.25 vol%). It is interesting that essential oils derived from *Lavandula hybrida* Reverch., which is cultivated on the island of Hvar (south Croatia), have a wide spectrum of activity against different mold species *(Aspergillus* spp., *Penicillium* spp., *Fusarium* spp., *Rhizopus* spp., *Thichoderma* spp.), although, the activity is also very uneven (MIC 1-100 vol%) (Figure 2.1). Its aerosol sterilizes the air in a closed space. The activity of essential oil aerosols can be linked with inhibited development of spores in filamentous fungi, probably through the inhibition of respiration (Inouye et al., 1998).

Rosemary [*Rosmarinus officinalis* L. (family Lamiaceae)] contains 1-2.5 percent of essential oil. Major components are 1,8-cineole (15-20 percent), borneole (10-20 percent), followed by bornyl acetate and camphor. This oil has a notable fungicidal aerosol against fungi and spores in the air. Nearly all the yeasts tested were very sensitive (MIC 0.09- 3.12 vol%) and so were dermatophyte spp. (MIC 1.56 vol%).

Essential oil of Dalmatian sage [*Salvia officinalis* L. (family Lamiaceae)] contains α-±β-thujone (to 68 percent), camphor (to 22 percent), 1,8-cineole

TABLE 2.2. Results of diffusion and dilution methods on some essential oils from plants growing in Croatia

	Plant / Microorganism Tested	Lamiaceae (Labiateae)												Asteraceae (Compositae)				Cupressa-ceae		Pinaceae		Rutaceae				Geraniaceae	
		Lavandula officinalis		*Salvia officinalis*		*Rosmarinus officinalis*		*Melissa officinalis*		*Mentha piperita*		*Micromeria thymifolia*		*Artemisia absinthium*		*Tanacetum parthenium*		*Juniperus communis*		*Picea abies*		*Citrus aurantium* subsp. *dulcis*		*Citrus limon*		*Pelargonium radula*	
		Z.I.	MIC	Z.I.	MIC	Z.I.	MIC	Z.I.	MIC	Z.I.	MIC	Z.I.	MIC	Z.I.	MIC	Z.I.	MIC	Z.I.	MIC	Z.I.	MIC	Z.I.	MIC	Z.I.	MIC	Z.I.	MIC
Yeasts	*Candida albicans*	-	5.00	-	0.05	15	0.09	29	0.19	20	0.78	-	3.1	13	1.56	0	-	9	1	16	0.05	21.5	1.56	12.3	1.56	-	0.6
	C. krusei	12	3.12	-	0.5	26	0.09	16	0.19	18	0.39	-	1.5	11	0.78	0	-	11	2	10	0.58	48.0	0.02	17.4	1.56	-	1.25
	C. tropicalis	18	12.5		0.25	23	0.09	25	2.0	30	0.39	-	2.0	10	3.12	7	6.25	10	3.12		0.15	44.5	0.05	10.9	1.56	-	0.6
	C. parapsilosis	19	-	-	-	25	3.12	30	-	22	0.78	-	3.1	13	0.19	0		0		14	0.29	16.0	0.09	12.8	0.19	-	1.25
	Torulopsis glabrata	13	6.25	-	1.5	16	0.09	15	0.19	18	0.09	-	6.2	16	0.39	9	2.5	9	3.12	15	-	30.9	0.05	9.5	6.25	-	2.5
	C. kefyr	23	25	-	6.25	50	0.8	31	0.78	19	0.09	-	2.0	22	0.39	9	6.25	10	0.78	13	0.78	16.9	0.09	10	1.56	-	0.6
	C. lusitaniae	15	25	-	6.25	38	0.39	19	0.19	29	0.09	-	-	0	-	-	-	0	-		-	38.7	0.39	11	0.7	-	-
	Hansenula anomala	-	25	-	6.25	-	-	-	-	-	-	-	-	16	0.78	7	6.25	8	8	-	-	32.7	0.05	11.9	1.56	-	1.25
	Geotrichum spp.	-	-	-	-	-	-	-	-	-	-	-	-	22	0.39	-	-	10	0.78	-	-	-	-	-		-	-
	Rhodotorula rubra	-	-	-	-		-	-	-	-		-	-	-	-	-	-	-	-	-	-	-	-	-		-	-
	Cryptococcus neoformans	-	-	-	-	-	-	-	-	-	-	-	-	14	3.12	6	6.25	8	10	16	0.29	40.9	0.02	15.0	1.25	-	-
	Saccharomyces cerevisiae	-		-	-	-	-	-	-	-	-	-	-	-	-	-	-	-	-	21	0.15	-	-	-		-	-
Dermatophytes	*Trichophyton rubrum*	-	-	-	-	-	-	-	-	-	-	-	0.2	26	0.39	-	-	14	0.39	21	0.15	65.0	0.19	20.0	0.39	-	-
	T. mentagrophytes	-	3.25	-	-	24	1.56	23	6.25	31	0.01	-	0.2	11	0.78	6	1.56	10	1.0	24	0.02	61.0	0.01	22.0	0.19	-	0.15
	Microsporum gypseum	-	0.78	-	-	-	1.56	42	8	32	0.02	-	0.9	43	0.09	5	3.12	11	2.0	26	0.07	54.5	0.02	21.6	0.19	-	1.25
	M. canis	-	-	-	-	-	-	-	-	-	-	-	0.4	13	3.12	-	-	0	-	25	0.01	44.0	0.01	23.0	0.05	-	-
	Epidermophyton floccosum	-	-	-	-	-	-	0.78	-	-	-	-	-	-	-	5	25	-	-	-	-	-	-	-	-	-	0.02

Z. I. inhibition zone (mm)
MIC minimal inhibitory concentration (vol%)
\- not perfomed

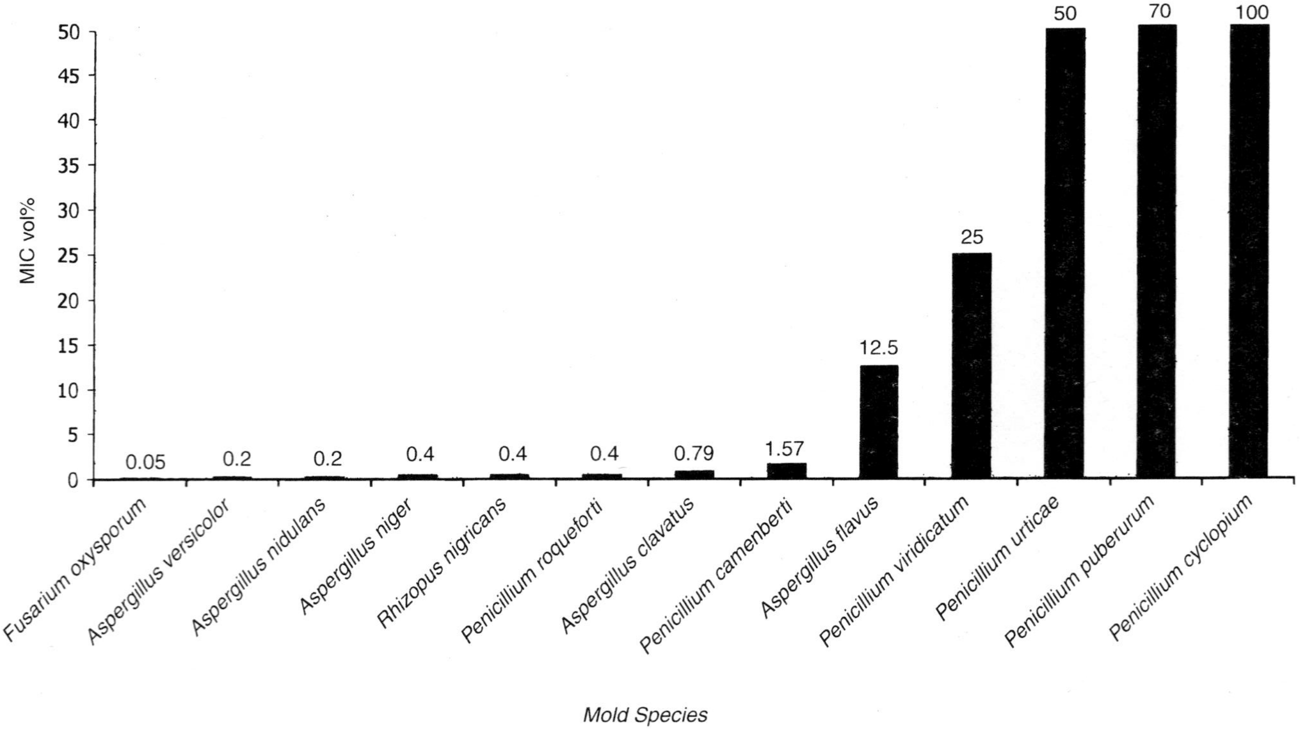

FIGURE 2.1. Fungitoxic effect of *Lavandula hybrida* Reverch. essential oil on some foodborne molds

(9 to 16 percent), and p-cymene (to 1.3 percent), as the most dominant components. The antimicrobial activity of Dalmatian sage oil depends on its chemical composition. The composition of the oil, on the other hand, is influenced by the local conditions of soil and climate. Variation in composition was observed in different localities along the Croatian coast, and it was previously found that since each of the chemical components has its own biological activity, the total antimicrobial activity of each given oil sample is different. Essential oil of *Salvia officinalis* showed strong activity against *Candida albicans* (MIC 0.05 vol%), dermatophytes (MIC 1.56 vol%), and *Aspergillus flavus* (MIC 1.56-3.125 vol%).

We have investigated the activity of sage oil as well as the effect of its components α-+β-thujone and 1,8-cineole against *Candida* spp., *Aspergillus flavus,* and dermatophyte strains. Furthermore, the influence of sage oil and 1,8-cineole on biosynthesis of aflatoxins was studied. The oil sample with more α-+β-thujone as compared with 1,8-cineole had a stronger inhibitory effect against *Candida* spp., *Aspergillus flavus,* and dermatophytes. 1,8-cineole also has some antifungal activity against *Aspergillus flavus* (MIC 6.25 vol%, MFC 12.5 vol%). The antimicrobial activity of sage oil samples from the coastal region of the north and south Adriatic is approximately the same as from the islands of Pag and Hvar, while the activity of Ciovo and Vis samples is extremely high (Kuštrak and Pepeljnjak, 1989).

Microcapsulated sage oil with gelatin acacia microcapsules (size from about 50 to 500 μm and 5 μm thick) contains all the same compounds as noncapsulated sage oil. The original composition of the essential oil was accurately preserved during the process of microcapsulation. The antibacterial activity of pure and microcapsulated oil was basically the same. In the case of microcapsules a definite time lag in achieving the full activity is more pronounced (72 h) than with the pure oil. On the other hand, the activity of microcapsulated sage oil against yeasts (*Candida* spp. and *Cryptococcus neoformans*) was lower after 72 hours (Jalšenjak et al., 1987).

The essential oil of the sweet orange [*Citrus aurantium* L., ssp. *dulcis,* (family Rutaceae)] contains about 90 percent of terpenes d-limonene, and more myrcene, decanol, geraniol, and neral than methylester of anthranilic acid. The activity of this essential oil is also very good against *Candida* spp. (MIC 0.048-1.56 vol%) and dermatophyte ssp. (MIC 0.0015-0.19 vol%) (Pepeljnjak et al., 1999).

The essential oil of lemon [*Citrus limon* L. (family Rutaceae)] contains as its major components citronellal, α-terpineol, geranyl acetate, linalyl acetate, limonene, and citral. Antifungal activity of the essential oil against *Candida* spp. is uneven (MIC 0.19-6.25 vol%) but very good against dermatophytes (MIC 0.048-0.39 vol%).

The antifungal activity of eucalyptus [*Eucalyptus globulus* Labill (family Myrtaceae)], which is not a native Croatian plant but grows on Adriatic islands, was also investigated. The essential oil contains about 70 percent of 1,8-cineole (eucalyptol), α-pinene, borneol, and myrten as major components. It also has considerable fungistatic and fungicidal activity against *Candida* spp. (MIC 0.16-1.0 vol%) and excellent activity against dermatophyte *Microsporum gypseuni* (MIC 0.195 vol%), but it is not effective against species of *Trichophyton* and *Epidermophyton.*

Plants from Inland Croatia

The essential oil of mint [*Mentha piperita* L. (family Lamiaceae)] contains about 40 percent of saturated terpene cyclic secondary alcohol-menthol, about 20 percent of menthone, and smaller percentages of menthofuran, cineole, and methyl acetate. It has a very good and wide spectrum of activity against *Candida* spp. and especially against dermatophytes (*Microsporum gypseum, Trichophyton mentagrophytes).* MIC for *Candida* spp. ranges from 0.098- 0.78 vol% and against dermatophytes *Microsporum gypseum* (MIC 0.024 vol%) and *Trichophyton mentagrophytes* (MIC 0.012 vol%).

Honey-plant [*Melissa officinalis* L. (family Lamiaceae)] contains a very low concentration of essential oil in leaves (0.02-0.2 vol%). More than 70 percent of compounds in this essential oil are terpenes (citral, citronellal, citronellol, geraniol, nerol, linalool, β-caryophyllene) that have a satisfactory fungicidal activity against *Candida* spp. (MIC 0.1-2.0 vol%) but slightly weaker activity against dermatophytes (MIC 6.2- 8.0 vol%) (Kosalec et al., 1998).

Wormwood [*Artemisia absinthium* L. (family Asteraceae)] contains 0.25-1.3 percent of essential oil. Its major components are 25 to 75 percent of thujone alcohol, and 3-10 percent of thujone, *cis*-epoxy-cynene, *trans*-sabinyl-acetate, and chrysanthenyl acetate. The essential oil has variable activity against *Candida* spp. (MIC 0.195-3.12 vol%) and dermatophytes (MIC 0.097-3.12 vol%).

Juniper tree [*Juniperus communis* L. (family Cupressaceae)] contains 0.5 to 2 percent of essential oil in its fruits. Major components are α- and β-pinene (16.5- 80 percent) and sabinene (0.2-50 percent) together with limonene, terpinene-4-ol and traces of borneole, α-terpineol, geraniol, and 1,8-cineole. Activity against *Candida* spp. is uneven (MIC 0.78-8 vol%). MIC for dermatophytes ranges from 0.39-2 vol%.

Spruce [*Picca abies* L. (family Karsten, Pinaceae)] contains the essential oil in its needles at 0.15-0.25 percent. The dominant components are bornyl-acetate (6-12 percent), followed by α-and β-pinene, 1-phellandrene, dipentene, cadinene, and terpineol. Its aerosol shows good activity on the growth of colonies of some *Candida* spp., but against others there is no activity (e.g., *C. lusitaniae*). For yeasts MIC ranges from 0.05-0.78 vol%, *Microsporum canis* is the most sensitive (MIC 0.012 vol%). It is slightly less sensitive but still effective against *Trichophyton rubrum* (MIC 0.146 vol%).

Feverfew [*Tanacetum parthenium* (L.) Schultz. (family Asteraceae)] contains 0.02-0.75 percent of essential oil with a high percentage of camphor (to 44 percent). Other less important components are *trans*-chrysanthenyl alcohol, α-thujone, α- and β-pinene, camphene, sabinene, α-phellandrene, p-cymene, limonene, τ-terpinene, borneol, bornyl acetate, and terpineol. The essential oil is active against species of *Candida* (MIC 6.25-12.5 vol%) and a dermatophyte, *Trichophyton mentagrophytes* (MIC 1.56 vol%). Other species have higher values of MIC (*Microsporum gypseum,* 3.12 vol%, and *Epidermophyton floccosum,* 25 vol%) (Kalodera et al., 1996).

Antifungal Activity of Croatian Plant Extracts

Table 2.3. shows the values of inhibition zones (ZI) and minimal inhibitory concentrations (MIC) for the plant extracts studied.

Savory [*Satureja montana* L., ssp. *montana* (family Lamiaceae)] is a wild aromatic plant, widespread on the Adriatic Coast and Croatian islands. Extracts of this plant contain lipophilic compounds (essential oil with carvacol, thymol, sterol, and triterpene) and hydrophilic compounds (flavonoids and tannins). A 75 percent extract has antifungal effect against *Candida* spp. (MIC 3.12-12.5 vol%). Dermatophytes are considerably more sensitive (MIC 4.0 vol%) (Pepeljnjak et al., 1999).

Pelargonium [*Pelargonium radula* (Cav.) L'Herit. (family Geraniaceae)] came from South Africa and is cultivated as a decorative plant in Croatia. Its ethanol extract contains flavonoids, saponins, and coumarins. The investigation of the flavonoid fractions enriched with isoquercetin and the flavonoid fraction enriched with rutin from the plant extract show different antifungal activity, so that among *Candida* spp. *C. tropicalis* is the only one sensitive to both tested fractions (MIC 6.1-6.25 vol%). Species of dermatophyte are sensitive to the flavonoid fraction enriched with isokvarcetin (MIC 6.25-15.0 vol%). The fraction with rutin is active against *Fusarium graminearum* (MIC 5.5 vol%) but has no activity against *Trichophyton mentagrophytes* while *Microsporum* and *Epidermophyton* are sensitive to it (MIC 6.25-9.0 vol%).

TABLE 2.3. Results of diffusion and dilution methods on some plant extracts from plants growing in Croatia

	Plant / Microorganism Tested	Essential Oil*				Alkaloids*				Iridoids*				Flavonoids*				Saponins*				Essential Oil*		Polysaccharides*	
		Micromeria thymifolia		*Satureja montana* subsp. *montana*		*Chelidonium majus*		*Berberis croatica*		*Centaurea rupestris*		*Vitex agnus-castus*		*Pelargonium radula*		flavonoid fraction with isoquercetin rutine		*Solidago gigantea*		*Solidago virgaurea*		*Tanacetum parthenium*		*Cetraria islandica*	
		Z.I.	MIC	Z.I.	MIC	Z.I.	MIC	Z.I.	MIC	Z.I.	MIC	Z.I.	MIC	Z.I.	MIC	Z.I.	MIC	Z.I.	MIC	Z.I.	MIC	Z.I.	MIC	Z.I.	MIC
Yeasts	*Candida albicans*	-	15.0	9	10	27	30	25	50	0	-	-	8	0	-	0	-	13	25	14	40	9	35	13	12.5
	C. krusei	-	20.0	14	10	12.5	28	13	50	0	-	-	15.6	0	-	0	-	9	27.5	8	50	15	12.5	12	9
	C. tropicalis	-	12.5	15	3.12	18.5	20	27	25	0	-	-	7	10	6.1	10	5	0	-	7	50	12.5	12.5	12	8.5
	C. parapsilosis	-	12.5	8	12.5	15	40	-	-	0	-		-	0	-	0		-	-			0	0	10.5	8.5
	Torulopsis glabrata	-	4.5	-	-	21.5	-	-	-	16	4.5	-	-	0	-	0	-	0	-	10	50	-	-	11	8
	C. kefyr	-	12.5	-	-	-	-	20	50	0	-	-	8.6	-	-	-	-	27	25	-	-	-	-	13	0
	C. lusitaniae	-	-	-	-	-	-	-	-	0	-		-	0	00	8	7	0	-	-	-	-	-	13	0
	Hansenula anomala	-	-	-	-	-	-	-	-	-	-	-	-		-	-	-	0	-	-	-	-	-	11.5	0
	Geotrichum spp.	-	-	-	-	-	-	-	-	-	-	-	-	-	-	-	-	10	-	0	-	-	-	-	0
	Rhodotorula rubra	-	-	-	-	-	-	-	-		-	-	-	-	-	-		-	-	-	-	-	-	-	0
	Cryptococcus neoformans	-	-	-	-	-	-	6	15	-	-	-	-	-	-	-	-	-	-	-	-	-	-	9	0
	Saccharomyces cerevisiae	-	-	-	-	-	-	-	-	-	-	-	-	-	-	-	-	-	-	-	-	-	-	0	-
Dermatophytes	*Trichophyton rubrum*	13	4.5	-	-	-	-		-	-	-	-	-	-	-	-	-	0	-	0	-	-	-	0	-
	T. mentagrophytes	19	4.5	10	4	27.2	5	-		-	-	-	3.5	12	15	0	100	17	29	30	14	28	4	0	-
	Microsporum gypseum	14	6.5	-	-	>30	8	-	-	18	1.4	-	5.3	14	6.25	11	6.25	15	17	15	11	-	-	0	-
	M. canis	15	4.5	-	-	>30	1.5	-	-	-	-	-	7.0	-	-	-	-	17	34	26	40	-	-	0	-
	Epidermophyton floccosum	-	-	-	-	>30	1.5	-	-	14	1.6	-	3.6	14	9	12	9	8.5	55	10	60	-	-	0	-

Z. I. inhibition zone (mm)
MIC minimal inhibitory concentration (vol%)
\- not perfomed
* the main compound

Feverfew [*Tanacetum parthenium* (L.) Schultz-Bip (family Asteraceae)] contains steroles, flavonoids, coumarins, and essential oil. Its ethanol extract has relatively weak activity against *Candida* spp. (MIC 12.5-35 vol%), while the species *Trichophyton mentagrophytes* and *Microsporum gypseum* are considerably more sensitive (MIC 4.5-6.5 vol%) (Kalodera et al., 1996).

Micromeria thymofolia (Scop.) Fritsch. (family Lamiaceae) is an endemic plant in Croatia, growing on the Croatian coastal mountains (Dinarids). The plant contains essential oil, flavonoids, saponins, and triterpene acid. Ninety percent ethanol extract shows a very good activity against dermatophyte spp. (MIC 4.5- 6.5 vol%) (Vladimir-Kneževic, 1998).

Christ's thorn [*Vitex agnus-castus* L. (family Verbenaceae)] is a widespread plant on the Adriatic Croatian coast and contains iridoides (agnusid, oucubin, eurostosid), flavonoids, plant hormones, and essential oil with α- and β-pinene and 1,8-cineole as major components. The antifungal activities of the plant material collected in different parts of Croatia were compared. Ninety percent ethanol extract shows activity against *Candida* spp. (MIC 6.5-37 vol%), and antifungal activity against dermatophyte spp. is slightly stronger, e.g., *Trichophyton mentagrophytes* (MIC 3-4.5 vol%), *Epidermophyton floccosum* (MIC 1.5-4 vol%), *Microsporum canis* (MIC 7-12 vol%), and *Microsporum gypseum* (MIC 4-6 vol%). MIC of mold *Penicillium viridicatum* is 7-19 vol% (Pepepljnjak et al., 1996).

Extracts of goldenrod [*Solidago virgaurea* L. and *S. gigantea* Ait. (family Asteraceae)] contain polysaccharides, flavonoids, glycosides, triterpene saponins, tannins, phenoglycosides, and essential oils. Seventy-five percent of ethanol plant extracts collected in the continental part of Croatia show antifungal activity in higher concentrations against *Candida* spp. (*S. virgaurea:* MIC 40-50 vol%; *S. gigantea:* MIC 25-27.5 vol%), and against dermatophyte: *Trichophyton mentagrophytes* (*S. virgaurea:* MIC 14-15 vol%; *S. gigantea:* MIC 28-30 vol%), *Microsporum canis* (*S. virgaurea:* MIC 39-40 vol%; *S. gigantea:* MIC 33-35 vol%), *M. gypseum* (MIC 13 vol%), and a high MIC against *Epidermophyton floccosum;* for both species MIC is 52-60 vol% (Pepeljnjak, Kuštrak, and Vukušic, 1998).

Celandine [*Chelidonium majus* L. (family Papaveraceae)] is a widespread plant in Croatia. It contains isoquinoline alkaloids, chelidonine, chelerythrine, and sanguinarine. Ethanol extract (1:1) has weak activity against *Candida* spp. (MIC 28-60 percent) and a slightly better activity against dermatophytes: *Microsporum canis* (MIC 1.5-2.5 vol%), *Epidermophyton floccosum* (MIC 1.5 vol%), and *Microsporum gypseum* (MIC 8-14 vol%). Antimicrobial activity depends on the vegetation period.

The flavonoid components from knapweed [*Centaurea rupestris* L. (family Asteraceae)] (7-*O*-β-d-glikopyranoside) exhibited outstanding antifungal activity against pathogenic fungi of dermatophyte spp. (*Trichophyton mentagrophytes, Microsporum gypseum,* and *Epidermophyton floccosum),* having a fungicidal effect with a concentration as low as 3 μg/ml. Flower extract in 45 percent ethanol showed greater pronounced antifungal effectiveness than the flavonoid component alone against dermatophyte spp., in which it produces a fungicidal effect in concentrations as low as 5-7 vol%.

Island moss [*Cetraria islandica* L. (family Acharius-Parameliaceae)] with major components polysaccharides (lichenin and isolichenin) and cetraric and usnic acid. Ethanol extract is effective against *Candida* spp. MIC 8.0-12.5 vol% and has no effect against dermatophytes (Pepeljnjak et al., 1999).

Barberry [*Berberis croatica* (Horvat) Kušan (family Berberidaceae)] contains berberine as the most important antimicrobial active component. Ethanol extract has a relatively weak antifungal activity against *Candida* spp. (MIC 12.5- 50 vol%) (Pepeljnjak and Petricic, 1992).

EXPERIMENTAL TREATMENT OF MYCOSIS WITH NATURAL ANTIMYCOTICS

Chronic fungal infections represent a serious therapeutic problem, particularly because such fungi have already developed resistance to many of the fungistatics currently used in their treatment. Therefore, biological experiments that can prove the therapeutical effect of essential oils or other natural antimycotics against clinically developed mycosis are especially interesting.

Tong et al. (1992) successfully treated 104 patients with tinea pedis with essential oil derived from tea tree [(*Melaleuca alternifolia* L. (family Myrtaceae)].

In the biological experiment as well as in the treatment of mycosis of guinea pig infected with *Microsporum audonii* and *Trichophyton mentagrophytes,* the treatment was successfully carried out with essential oil derived from *Chenopodium ambrosioides* L. (family Chenopodiaceae), *Artemisia nelagrica* L. (family Asteraceae), *Caesulia axillaris* Roxb. (family Asteraceae), *Cymbopogon citratus* L. (family Poaceae), and *Mentha arvensis* L. (family Lamiaceae) (Kishore et al., 1996). The fungicidal effect of this oil in vitro against *Aspergillus fumigatus* and *Cladosporium trichoides* was also demonstrated (Kishore et al., 1993).

Adam et al. (1998) successfully treated experimental dermatophytosis (*Trichophyton rubrum* and *T. beigelii*) in rats with thymol and carvacol. It is important that the result of the Ames test on mutations of essential oils *Lavandula officinalis* L., *Salvia fruticosa* L., and *Origanum vulgare* L. ssp. *hirium* (family Lamiaceae) was negative and that the same essential oils have a fungicidal effect against the causative agent of pityriasis *(Malassezia furfur)*.

Quale et al. (1996) treated stomatitis caused by *Candida* in AIDS patients with a drug based on cinnamon [*Cinnamomum zeylanicum* Blume (family Laureaceae)].

Pelargonium [*Pelargonium radula* (Cav.) L'Herit. (family Geraniaceae)], having 0.15-0.30 percent essential oil, was cultivated in the botanical garden of medicinal herbs 'Fran Kušan', at the Faculty of Pharmacy and Biochemistry, University of Zagreb. GC/MC analyses of the essential oil of this plant species *Pelargonium radula* (Cav.) L'Herit. proved that it contains geraniol, citronellol, linalool, citral, and menthol. Investigation of the antifungal activity of this essential oil showed that the best results were achieved against the strains *Microsporum gypseum* (MIC 1.25 vol%) and *Trichophyton mentagrophytes* (MIC 0.15 vol%). These results justified the production of the lotion, containing 3 percent essential oil, and the cream, containing 5 percent essential oil, which are used in the treatment of trichophytosis. The experiment was performed on seven cats spontaneously infected with a dermatophyte *(Thichophyton mentagrophytes)* in their street environment. The isolation of dermatophytes was done on Sabouraud dextrose agar with antibiotics (streptomycin and cycloheximide 0.5 g/L), incubation at 25 ± 2°C. The degree of the infection (changes on the skin and hair) was 20-70 percent. The lotion and cream was applied daily to 24 locations. The first positive results of the treatment with a 3 percent lotion and 5 percent cream respectively were observed after seven days. The skin stopped scaling; there was no redness; no hair fell out; and the infection did not spread. All experimental animals were completely cured after 30 days. New hair grew after 15 days. After that, the dermatophytes were not isolated anymore (Pepeljnjak el at., 1995) (Photos 2.4 and 2.5).

The investigations we have done on yeasts and dermatophyte species with extracts of parts of *Solidago virgaurea* L., *Chelidonium majus* L., and fruits of *Pimpinella anisum* L. encouraged us to proceed with the investigation in order to find an antimycotic for local use. The Department of Pharmacology, Medical Faculty, University of Zagreb, found out that the drug has no toxic effects. So, the clinical investigation has started on 30 patients between 15 and 70 years of age, both sexes, treated in the hospital. The therapeutic effect of the drug was confirmed on the patients treated. The clinical diagnosis of mycosis was *Dermatomycosis ingvinale, Erosia interdigitialis, Candidiomycetica,*

PHOTO 2.4. Cat before treatment (Copyright Department of Microbiology, University of Zagreb, Croatia.)

PHOTO 2.5. Cat after treatment (Copyright Department of Microbiology, University of Zagreb, Croatia.)

and *Pityriasis versicolor*. Thirty days after the treatment, drug was applied to the mycotic places twice a day, the symptoms (itching, erythema, scaling, rhagades, crust, erosion, rash specific for pityriasis) disappeared. This was a verification that the drug was appropriate for repeated use. It was also observed that the drug neither stained clothes nor irritated the skin. The added salicylic acid ensures its keratolitic effect, which enables good penetration of the antimycotic components into the skin.

Due to the yeast *Malassezia furfur*, the causative agent of *Pityriasis versicolor*, and the common seborrheic dermatitis of the scalp, a very efficacious hair shampoo based on a mixture of plant extracts was prepared. It has a good therapeutic and preventive effect, so it is recommended for the prevention of mycosis and seborrheic dermatitis of the scalp.

Both products are licensed for sale in Croatia under the name Izvanin and are produced by the pharmaceutical firm Belupo Ltd., Koprivnica.

CONCLUSIONS

Looking at the results of the antifungal investigations done by many authors, it becomes obvious that there is great variety in the results achieved. This variety derives from the difference in the applied diffusion and dilution methods. The type, composition, and pH of the nutrient broth varied significantly. Often there is a lack of a standardized inoculum density of the investigated fungi species, dermatophyte species, and yeast; the acceleration time of the diffusion process in a diffusion method at + 4°C is different; the time and incubation temperatures are different and so are the measurement of the break points.

It is known that a lower density of the inoculum as well as lower incubation temperature increases the inhibition zone (Galgiani, 1986; Galgiani, 1987; Brantner, 1997; Maggon et al., 1997). Diffusion methods are used for the primary screening test that tells us about the presence or absence of antimycotic activity in the investigated substance (Janssen et al., 1987).

The essential oil evaporates at the incubation temperature and very often saturates the air above the inoculated broth. Even this aerosol can inhibit the growth of some of the fungi species. In this case, it is necessary to decrease the amount of the essential oil until clear inhibition zones appear.

The problem during the investigation of essential oils with the dilution method is nonsolubility of essential oils in hydrophilic media, which makes it necessary to add polysorbate 20 or 80 (Tween) or dimethylsulfoxide (DMSO). During incubation, separation of the emulsion can often occur so it is recommended to shake the culture during incubation.

Regardless of these methodological difficulties in the investigation of antifungal effects of natural antimycotics and big differences in the biology of fungal species, all investigations done up to now show that plants contain a considerable wealth of sources of antifungal substances. Plant antimycotics are often linked in a compound with many components, which often react separately or synergistically against different fungal species. This is characteristic of many plant species, and their amount depends not only on the characteristics of the species but also on the surrounding environment in which the plant grows. Considering that the plants in their natural surrounding have a natural contact with a great number of fungi, dermatophytes, yeast, bacteria, viruses, and parasites, these substances will have a considerable protective role against infections as well as providing their physiological role (Figure 2.2).

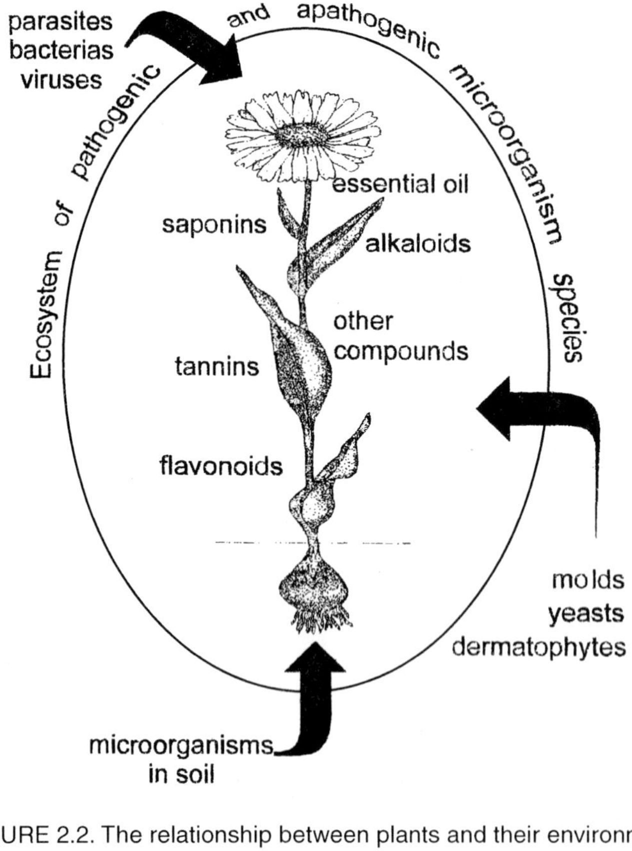

FIGURE 2.2. The relationship between plants and their environment

This relationship can be seen particularly in essential oil aerosols, which often decrease the number of spores of fungi in the surroundings of Mediterranean plants, which are rich in fungicidal essential oils (Pepeljnjak, Kuštrak, and Kalodera, 1998; Pepeljnjak and Šegvic, 2000).

All this tells us that numerous plant species represent very useful sources of interesting chemical substances which can be used in natural, semisynthetic, or even synthetic antimycotics.

Currently, the discovery and evaluation of antimicrobial as well as antifungal effects become an interesting line of direction in the development of a new science that deals with the increasing resistance of fungi to antimycotics and/or the discovery of useful natural substances which will enable the production of effective antimycotics in the treatment of mycosis.

Surely the same principal is valid for other microbicidal species. Nature, in its functional and protective role, contains a great many substances that can be useful to humankind in its fight against the numerous causative agents of infectious diseases.

In this context, Croatian aromatic plant species have an exceptionally rich potential. This is confirmed with our results of a relatively small number of plant extracts and essential oils.

REFERENCES

Adam, K., Sivropoulou, A., Kokkini, S., Lanaras, T., Arsenakis, M. (1998). Antifungal activities of *Origanum vulgare* subsp. *hirtum, Mentha spicata, Lavandula angustifolia* and *Salvia fruticosa* essential oils against human pathogenic fungi. *J of Agricult and Food Chem, 46*:1739-1745.

Aljancic, I., Vajs, V., Menkovic, N., Karadžic, I., Juranic, Milosavljevic, S., Macura, S. (1999). Flavones and sesquiterpene lactones from *Achillea atrata* subsp. *multifida:* Antimicrobial activity. *J of Natural Products* 62:909-911.

Amaro Luis, J.M., Adrian, M., Diaz, C. (1997). Isolation, identification and antimicrobial activity of ombuoside from *Stevia triflora. Annales Ph Fr* 55:262-268.

Baba Moussa, F., Akpagana, K., Bouchet, P. (1999). Antifungal activities of seven West African Combretaceae used in traditional medicine. *J of Ethnopharmacol* 66:335-338.

Bašic, F., Bosnic, T. (1989). The examination of antimicrobial activity of decocts of various plants and isolated tannins from *Corylus avellana. Farm Glas* 45: 243-250.

Bossche, H.V., Dromer, F., Improvisi, I., Lazono-Chiu, M., Rex, J.H., Sanglard, D. (1998). Antifungal drug resistance in pathogenic fungi. *Medical Mycology* 36 (Supplement I):119-128.

Brantner, A. (1997). Evaluation of pharmacopeia methods refering to the determination of antimicrobial activity and with special regard to natural substances. *Pharmazie* 52:34-40.

Chaumont, J.P., Léger, D. (1989). Propertietes antifongiques de quelques phenols et de composes chimiquement tres voisins. Relations structure-activite. *Plantes medicinales et phytotherapie* 2:124-128.

Chaumont, J.P., Léger, D. (1992). Campaign against allergenic molds in dwellings. Inhibitor properties of essential oils of *Geranium 'Bourbon', Citronellol, Geraniol* and *Citral. Ann Pharm Fr* 50: 156-166.

Freixa, B., Vila, R., Adzet, T., Ferro, E., Canigueral, S. (1999). Antifungal principles from *Piper fulvescens* CDC, 2000 Years of Natural Products Research, July 26-30, 1999, Amsterdam, The Netherlands.

Galgiani, J.N. (1986). Antifungal susceptibility testing. *The Antimicrobic Newsletter* 3:17-22.

Galgiani, J.N. (1987). Minireview—Antifungal susceptibility tests. *Antimicrobial Agents and Chemotherapy* 31: 1867-1870.

Gildemeister, E., Hoffmann, Fr. (1956). Die aetherische oelen. Berlin: Akademie-Verlag, pp. 233-245.

Goecheritz, O., Weuffen, W., Hoeppe, H. (1974). Terpene und terpenderivate vom carvon-und camphertyp—ihre antimicrobielle und verminoxen eigenschaften. *Pharmazie* 29:339-344.

Goren, N., Woerdenbag, H.J., Bozok-Johansson, C. (1996). Cytotoxic and antibacterial activities of sesquiterpen lactones isolated from *Tanacetum parthenium* subsp. *parthenium. Planta Medica* 62:419-422.

Guenther, E. (1957). *The essential oils*, Volume 1. Toronto-New York-London: D. Van Nostrad Company, pp. 77-83.

Hajji, F., Fkih-Tetouani, S., Tantaoui-Elaraki, A. (1993). Antimicrobial activity of 32 eucalyptus essential oils. *Fitoterapia* 64:71-77.

Hetheleyi, E., Koczka, I., Tetenyi, P. (1989). Phytochemical and antimicrobial analysis of essential oils. *Herba Hung* 28:99-115.

Hinou, J.B., Harvala, C.E., Hinou, B. (1989). Antibacterial activity screening of 32 common constituents of essential oils. *Pharmazie* 44:302-303.

Hostettmann, K., Potterat, O., Wolfender, J.L. (1998). The potential of higher plants as a source of new drugs. *Chimia* 52:10-17.

Hugo, W.B., Russell, A.D. (1997). *Pharmaceutical microbiology*. Oxford-London-Edinburgh-Melbourne: Blackwell Scientific Publications, p. 116.

Inouye, S., Watanabe, M., Nishiyama, Y., Takeo, K., Akao, M., Yamaguchi, H. (1998). Antisporulating and respiraton-inhibitory effects of essential oils on filamentous fungi. *Mycoses* 41:403-410.

Jalšenjak, V., Pepeljnjak, S., Kuštrak, D. (1987). Microcapsules of sage oil: Essential oils content and antimicrobial activity. *Pharmazie* 42:419-420.

Janssen, A.M., Scheffer, J.J.C., Baerheim Svendsen, A. (1987). Antimicrobial activities of essential oils (A 1976-1986 literature review on possible applications). *Pharm Weekbld (Sci)* 9:193-197.

Kalodera, Z., Pepeljnjak, S., Petrak, T. (1996). The antimicrobial activity of *Tanacetum parthenium* extract. *Pharmazie* 51:995-996.

Kellner, W., Kober, W. (1955). Moeglichkeit der verwendung aetherische ode zur raumdesinfection. *Arzneim-Forsch* 5:224-229.

Kishore, N., Chansouria, J.P.N., Dubey, N.K. (1996). Antidermatophytic action of the essential oil of *Chenopodium ambrosioides* and an ointment prepared from it. *Phytotherapy Research* 10:453-455.

Kishore, N., Mishra, A.K., Chansouria, J.P.N. (1993). Fungitoxicity of essential oils against dermatophytes. *Mycoses* 36:211-215.

Knoblock, K., Pauli, A., Ibler, B., Weigand, H., Weiss, N. (1989). Antibacterial and antifungal properties of essential oil components. *J Ess Oil Res* 1:119-128.

Kosalec, I., Pepeljnjak, S., Kalodera, Z. (1998). 1st Croatian Symposium on Aromatherapy, April 5-7, 1998, Opatija, Croatia.

Kubeczka, K.H. (1982). *Aetherische oele.* Stuttgart-New York: Georg Thieme Verlag, pp. 223-243.

Kuštrak, D. and Pepeljnjak, S. (1989). Antimicrobial activity of Dalmatian sage oil regions of the Yugoslav Adriatic Coast. *Acta Pharm Jugosl* 39:209-213.

Lacoste, E., Chaoumont, J.P., Mandin, D., Plumel, M.M., Matos, F.J.A. (1996). Les proprietes antiseptiques de l'huile essentielle de Lippia sidoides Cham. Application à la microflore cutanee. *Ann Pharmaceutiques Francaises* 5:228-230.

Lens-Lisbonne, C., Cremieux, A., Maillard, C., Balansrad (1987). Methods for the evaluation of the antibacterial acitivity of essential oils. *J Pharm Belg* 42: 297-302.

Maggon, K.K., Slee, A.M., Demos, C.H. (1997). Development of antifungal agents by the pharmaceutical industry. *Drugs of Today* 27:317-330.

Nahrstedt, A. (2000). Active constituents of traditional European medical plants—new developments. Natural products research in the new millenium, Int. Congr. and Ann. Meeting of the Soc. for Med. Plant Res. (GA), September 3-7, 2000, Zürich, Switzerland.

Navarro, V., Delgado, G. (1999). Two antimicrobial alkaloids from *Bocconia arborea. J of Ethnopharmacol 66:*223-226.

NCCLS (1997). Reference method for broth dilution antifungal susceptibility testing of yeasts; approved standard. NCCLS document M27-A, ISBN 1-56238-328-0, NCCLS. Wayne, USA.

Odds, F.C. (1996). Resistance of clinically important yeasts to antifungal agents. *In J Antimicrob Agents* 6:145-147.

Pattnaik, S., Subramanyam, V.R., Bapaji, M., Kole, C.R. (1997). Antibacterial and antifungal activity of aromatic constituents of essential oils. *Microbios* 89: 39-46.

Pepeljnjak, S. (1996). Ecosphere—medical plants and biologically active substances. In Seminar on Phytopharmacy, Faculty of Pharmacy and Biochemistry, University of Zagreb, Zagreb, Croatia.

Pepeljnjak, S., Antolic, A., Kuštrak, D. (1996). Antibacterial and antifungal activities of the *Vitex agnus-castus* L. extracts. *Acta Pharm* 56:201-206.

Pepeljnjak, S., Babic, A. (1991). Detection of trichotecenes mycotoxins T-2, HT-2, DON and DAS by thin-layer chromatograpy and biological methods. *Prehrambeno-Tehnološka Revija* 29:65-70.

Pepeljnjak, S., Kalodera, Z., Volenec, M. (1995). Antifungal activity of *Pelargonium radula* (Cav.) L'Herit essential oil. The 3rd International Symposium on

Cutaneous Fungal, Bacterial and Viral Infection and Therapy, September 14-17, 1995, San Francisco.

Pepeljnjak, S., Kuštrak, D., Kalodera, Z., Volenec, M. (1998). Antimicrobial investigation of essential oils. *1st Croatia Symposium on Aromatherapy,* Opatija, April 5-7, 1998, Croatia.

Pepeljnjak, S., Kuštrak, D., Vukušic, I. (1998). Investigation of the antimycotic activity *of Solidago viragaurea* and *Solidago gigantea* extracts. *Pharm Pharmacol Lett* 4:85-86.

Pepeljnjak, S., Petricic, J. (1992). The antimicrobic effect of berberin and preparation of *Tinctura berberidis. Pharmazie* 47:307-308.

Pepeljnjak, S., Šegvic, M. (2000). Occurrence of micromycetes in air and vegetation in continental and coastal region of Croatia. Symposium: Influence of pollutants of the environment on respiratory system, Zagreb, February 18, 2000, Croatia.

Pepeljnjak, S., Stanic, G., Potocki, P. (1999). Antimicrobial activity of the ethanolic extract of *Satureja montana* subsp. *montana. Acta Pharm 49:*65-69.

Pepeljnjak, S., Vrsalovic, M., Tuckar, D. (1999). Antiseptic activity of sweet orange essential oil and *Cetraria islandica* extract. 2000 Years of Natural Products, July 26-30, Amsterdam, The Netherlands.

Perucci, S., Mancianti, F., Cioni, P.L., Flamini, G., Morelli, I., Macchioni, G. (1994). In vitro antifungal activity of essential oils against some isolates of *Microsporum canis* and *Microsporum gypseum. Planta Medica* 60:184-187.

Quale, J.M., Landman, D., Zaman, M.Z., Burney, S., Sathe, S. (1996). In vitro activity of *Cinnamomum zeylanicum* against azole resistant and sensitive *Candida* species and a pilot study of cinnamon for oral candidiasis. *American Journal of Chinese Medicine* 24:103-109.

Ružicka, L., Eschenmoser, A., Heusser, H. (1953). The isoprene rule and the biogenesis of terpen compounds. *Experientia* 9:357-367.

Schreier, P. (1984). *Chromatographic studies of biosinthesis of plant volatiles.* Heidelberg-Basel-New York: Dr. Alfred Huethig Verlag, pp. 9-17.

Serkedjieva, J. (1997). Antiinfective activity of a plant preparation from *Geranium sanguieneum* L. *Pharmazie* 52:799-802.

Tivy, J. (1995). *Biogeography, a study of plants in the ecosphere.* Essex, England: Longman House, Burnt Mill, Harlow, p. 12.

Tong, M.M., Altman, P.M., Barnetson, R.S. (1992). Tea tree oil in the treatment of tinea pedis. *Australian J Derm* 33:145-149.

Vladimir-Kneževic S. (1998). *Micromeria thymifolia* (Scop.) Fritsch—composition of essential oil and biological effects. Dissertation, pp. 6-16. Faculty of pharmacy and biochemistry, University of Zagreb, Croatia.

von Denffer, D., Ziegler, H. (1991). *Lehrbuch der botanik fuer hochschulen.* Stuttgart: Gustav Fischer Verlag, p. 75 (Croatian translation, Third edition, Školska knjiga, Zagreb, Croatia).

Wagner, H. (1993). Pharmazeutische biologie 2. Drogen und ihre inhaltstoffe. Gustav Fischer Verlag, Stuttgart-New York, pp. 55-62.

Wang, Y., Hamburger, M., Gueho, J., Hostettmann, K. (1989). Antimicrobial flavonoids from *Psiadia trinervia* and their methylated and acetylated derivatives. *Phytochemistry* 28:2323-2327.

Werker, E. (1993). Function of essential oil-secreting glandular hairs in aromatic plants of the *Lamiaceae*—Review. *Flavour Frag J* 8:249-255.

Yousef, R.T., Tawil, G.G. (1980). Antimicrobial activity of volatile oils. *Pharmazie* 35:698-701.

Chapter 3

Antimycotic Activity of the Members of Meliaceae

María Cecilia Carpinella
Carlos Guillermo Ferrayoli
María Sara Palacios

INTRODUCTION

Fungi are considered to be beneficial organisms for several reasons, particularly for their role in the natural carbon cycle, and the parts they play in plant nutrition, in the industrial production of drinks and food, and in the production of chemical products, of which the most important are antibiotics.

Despite their many positive uses and properties, some fungi have a negative effect on the health of animals and plants. The occurrence of infections in humans caused by fungi has markedly increased in recent years, especially involving immunocompromised patients (Pitt, 1994; Kauffman and Hedderwick, 1997; Zacchino et al., 1999). The newest treatments, as well as recent diseases, such as AIDS (acquired immunodeficiency syndrome), have increased the incidence of hosts affected by this class of infections. *Candida* spp. accounted for 7.7 percent of the infections in clinics and hospitals of the United States between 1985 and 1988 (Shadomy and Pfaller, 1991). In addition, infections caused by *Aspergillus* spp., *Fusarium* spp., Zygomycetous, and other nonpathogenic fungi are being reported with increasing frequency.

It can be taken for granted that plant diseases caused by pathogenic agents already existed when *Homo sapiens* made their first appearance, and that bacteria, viruses, and fungi must have infected the vegetable kingdom throughout the millennia, but the definitive step toward recognizing the im-

---

The authors are grateful to the Agencia Córdoba Ciencia—Centro de Excelencia en Productos and Procesos de la Provincia de Córdoba (CEPROCOR) for financial support. M.C.C. gratefully acknowledges receipt of a fellowship from CONICET.

portance of pathogens was made when humans developed a more organized form of agriculture. The ancient Greeks already distinguished between anomalies and diseases in plants, and also described a series of different diseases, while the Arabs seem to have had good knowledge of the diseases of trees, describing, for example, what is now known as peach leaf curl (Fernandez Valiela, 1978).

The economic importance of plant diseases should be considered not only in terms of the damage they cause to the plants but also because of the additional expenses incurred through control measures, as well as through the limitations placed on the species and varieties of plants that can be grown only in particular farming areas.

In the United States, until race T of *Helminthosporium maydis* appeared, the genetic resistance of cultivated maize hybrids to fungi made it possible to obtain high yields. The susceptibility of maize to the pathogenicity of race T caused a crisis in maize cultivation in 1970-1971; the losses were higher than 11 million tons. The yellowing of sugar beet as a result of *Mycoplasma* is one of the most important diseases in the crops grown in southern and central Argentina (Fernandez Valiela, 1978). These examples show many effects produced by fungi in plants that cause serious problems in agricultural production. The search for new sources of control that are safe for humans and for the environment, and are also economically acceptable, is a daily task for many people interested in the welfare of their fellow human beings.

Control of Pathogenic Agents

The word *control,* both in phytopathology and in animal pathology, is synonymous with prevention, protection, struggle, or fight against the agents that produce diseases in plants and animals. It refers to measures designed to prevent the damage that these diseases may cause, in order to avoid their effects on human and animal health, and on the various species that humans cultivate for their own nourishment and well-being.

The continuous search for better and more economical methods of controlling pathogenic fungi is no longer directed toward synthetic compounds but rather focuses on new, harmless natural products that fit in with integrated pest management.

Even though many plants are affected by fungal diseases, this does not mean that they do not have the capacity to synthesize their own antifungal compounds in response to the attack. These, in turn, are able to be turned into natural, biodegradable fungicides in the near future. Many active substances belonging to the vegetable kingdom have been described (De Lucca et al., 1999; Phay et al., 1999; Wätcher et al., 1999). Among the popular remedies in northern Mexico and in the southwest United States, *Baccharis glutinosa* Pers, locally known as batamote, is of utmost importance. An infusion of this herb is applied to the feet in order to cure the well-known complaint of athlete's foot *(Tinea pedis)*. This plant, and especially its leaves and seeds, has a controlling effect on other dermatophytes, such as *Trichophyton mentagrophytes, T. rubrum, T. tonsurans,* and *Epidermophyton floccosum* (DiSalvo, 1974; Fernandez Valiela, 1978). Many essential oils, such as eucalyptus, lavender, lemongrass, rosemary, bergamot, cinnamon leaf, wormwood, turpentine, and others, have antibacterial and antifungal effects (Singh and Upadhyay, 1993). Antifungal effects of essential oils of palmarosa *(Cymbopogon martini),* vetiver *(Vetiveria zizaniodes),* and *Pimpinella anisum* have been detected on *Aspergillus niger, A. flavus, Fusarium oxysporum,* and *Penicillium* spp. Essential oils of *Pimpinella anisum* also has controlling effects on *Phytophthora parasitica* (Singh and Upadhyay, 1993).

Alkaloids isolated from *Chelidonium majus* are effective antifungals against strains of *Trichophyton, Microsporum canis, E. floccosum,* and *A. fumigatus* (Hejtmánková et al., 1984) and ether extracts of *Emblica officinalis* and *Trigonella foenum-graecum* demonstrate the controlling effect of *Candida albicans* (Khanna et al., 1971).

Antifungal Activity of Meliaceae

Meliaceae is a family of plants widely studied for their anti-insect effects. In connection with these studies, many researchers have concentrated their work on the controlling effects that specimens of this family have on different fungi. Many of the tests designed to measure these effects have been conducted in the laboratory or through the incorporation of vegetable material in the soil or plants. The active principle responsible has been identified in very few cases. On several occasions, the results showed the control of the microorganism, but it is not easy to determine the true cause of it, i.e., if it is due to the presence of antifungal compounds in the vegetable material used in the treatment, the control effect of other microorganisms on the pathogenic fungus, or a direct effect on the natural defenses of the plant.

AZADIRACHTA INDICA *A. JUSS*

The solids or extracts that come from various parts of the neem tree (*Azadirachta indica* A. Juss.), belonging to the Meliaceae family, are better known for their antifeedant, insecticide, and insect-growth regulator effects

(Mitchell et al., 1997). They show different controlling effects on other living organisms, including fungi, nematodes, viruses, and bacteria. Recent research has given a great boost to the study of the compounds derived from neem that could be used for the control of mycotic diseases (see Table 3.1).

Oil-cake solids (neem cake) are produced following the physical expression of the seed oil . This solid has been shown to be a good antifungal agent. In the cultivation of tomatoes, the application of neem cake suppressed the fungus *Rhizoctonia solani* (which causes the damping off and root and crown rots of many crops and fruits) as well as *Fusarium oxysporum* f. sp. *lycopersici* in 30 days posttreatment, while in a soilless medium system, it took two weeks of activity to reach the maximum reduction of the disease caused by *R. solani* on *Zinnia elegans* seedlings (Locke, 1995). The fungitoxic effect on tomato crops is attributed by Locke to antifungal substances released during neem cake breakdown, but the effect may be due to microbiological control. Neem cake also inhibited *R. solani, Fusarium* spp., and *Colletotrichum atramentarium* in field tests on crops of aubergines (eggplants), increasing the population of saprophytic fungi at the same time (Locke, 1995).

Helminthosporium nodulosum, Alternaria tenuis, Curvularia tuberculata, R. solani, and *F. oxysporum* f. sp. *lycopersici* were inhibited in vitro by aqueous extracts of neem oil cake, which showed a more fungistatic than fungicidal effect (Locke, 1995).

Rhizoctonia solani, which causes sheath blight in rice, has been controlled by preplanting soil amendments of neem cake (1 percent) and leaving them to decompose for a week. After decomposition, the rice seeds were sown, and on the fifteenth day after sowing the fungi were inoculated. After measuring the shoot and root growth, it was observed that seedling infection was reduced by 59 percent, while root and shoot length increased by 19 and 40 percent respectively in 15-day-old rice seedlings (Kannaiyan and Prasad, 1981). Similarly, the disease of cottonseeds and the putrefaction of their roots caused by the same fungus reduces crop populations by 30 percent. A test carried out in a pot culture experiment under laboratory conditions (Jeyarajan et al., 1987) designed to measure the effect of neem cake on *R. solani* was done by applying neem cake dosages equivalent to 2.5 and 5 tonnes per hectare (t/ha) in the soil. Pre- and postemergence mortality was measured and showed a reduction close to 35-45 percent for the first case and 55-60 percent for the second, after adding the neem. Populations of other fungi, bacteria, and actinomycetes increased in soils treated with *A. indica,* which probably means that antimycotic control takes place through the microorganisms mentioned previously, which are antagonistic to *R. solani*.

TABLE 3.1. Antimycotic effects of *Azadirachta indica* on different pathogenic fungi

Fungi	Crops or substrate	Neem material	Effectiveness	Concentration	Reference
Alternaria brassicicola	In vitro	Formulations of seed oils	No effect	31-1,000 ppm a.i.	Chianella and Rovesti, 1992
A. porri	In vitro	DCM phase from methanolic kernel extract	Inhibition	100 mg/ml	Steinhauer, 1994
		Isolated compound	Complete fungal inhibition	100 mg/ml	
A. tenuis	In vitro	Aqueous extract of oil cake	Inhibition		Locke, 1995
		Kernel extracts		ED_{50} = 0.1-1.0 mg/ml	Lehmann et al., 1993
Aphanomyces cochlioides	In vitro	Kernel extracts	Inhibition	ED_{50} = 0.1-1.0 mg/ml	Lehmann et al., 1993
Ascochyta lentis	In vitro	Commercial oils of seeds	Complete fungal inhibition	Neem Biagro 1%	Molera, 1994
			High inhibition	Neem-Azal-F 5%	
A. tritici	In vitro	DCM phase from methanolic kernel extract	Inhibition	100 mg/ml	Steinhauer, 1994
		Isolated compound	Inhibition	100 mg/ml	
Aspergillus flavus	In vitro	Leaf extract (nonvolatile comp.)	No effect 95%[b]	>10% of extract	Locke, 1995
	In vitro	Essential oils from leaves and seeds	Inhibition 100%[b]	500-1,000 ppm	Bankole, 1997
	Cotton bolls	Leaf extracts	No effect 98%[b]		Zeringue and Bhatnagar, 1990
A. parasiticus	In vitro	Fresh leaves	51%[a], 90%[b]	Volatiles from 20 g leaves	Zeringue and Bhatnagar, 1994
		2,3 butanediol and 1-heptanol	Partial fungal and aflatoxin inhibition	2.5-10 μM and 1.8-7 μM, respectively	
		4-pentenal and *trans*-2-heptenal	24%[a], 91%[b]	10 μM, and 7.6 μM, respectively	
		Ketones (C_2-C_{12})	Low inhibition		
		Leaf extract (nonvolatile comp.)	No effect 95%[b]	10% of extract	Locke, 1995
Aspergillus spp.	Soil/lime	Neem cake	Increased growth		Locke, 1995
	In vitro	Seed oil	Inhibition	0.2 ml oil/10 mm disc	Kher and Chaurasia, 1977

TABLE 3.1 *(continued)*

Fungi	Crops or substrate	Neem material	Effectiveness	Concentration	Reference
Botrytis cinerea	In vitro	Formulations of seed oils	Low effect	31-1,000 ppm a.i.	Chianella and Rovesti, 1992
	Apple	Seed oil	Low effect	2%	Moline and Locke, 1993
	In vitro	Commercial oils of seeds	Complete fungal inhibition	Neem-Azal-F 5%	Molera, 1994
Candida albicans	In vitro	Ethanol and acetone leaf extracts	Inhibition		Khan and Wassilew, 1987
		Aqueous leaf extract	No effect		
Cercospora beticola	Sugar beet	Aqueous seed extracts	No preventive effect		Chianella and Rovesti, 1992
Colletotrichum atramentarium	Soil/eggplant	Neem cake	Fungal inhibition/increased saprophytic fungi		Locke, 1995
C. lindemuthianum	In vitro	10-undecyn-1-ol	Complete fungal inhibition	MIC = 0.2 mg/cm^2	Govindachari, Suresh, and
		Leaf hexane extracts	Inhibition	MIC = 1 mg/cm^2	Masilamani, 1999
Curvularia lunata	In vitro	Seed oil	Inhibition	0.2 ml oil/10 mm disc	Kher and Chaurasia, 1977
C. tuberculata	In vitro	Aqueous extract of oil cake	Inhibition		Locke, 1995
Diclocarpon rosae	In vitro	Seed oil	Moderate fungal inhibition	1%	Locke and Carter, 1993
Dreschslera teres	In vitro	DCM phase from methanolic kernel extract	Inhibition	100 mg/ml	Steinhauer, 1994
		Isolated compound	Complete fungal inhibition	100 mg/ml	
Epidermophyton floccosum	In vitro	Petrol ether leaf extracts, MTB kernel extracts, and methanol kernel and leaf extracts	Inhibition		Khan and Wassilew, 1987
Erysiphe cichoracearum	Cucumber	Formulations of seed oils	Curative and preventive effect	250-350 ml a.i./100 liters	Chianella and Rovesti, 1992
E. graminis var. *hordei*	Barley	Aqueous seed extracts	Curative and preventive effect	500-1,500 ml a.i./100 liters	Chianella and Rovesti, 1992
		Formulations of seed oils			
E. graminis var. *tritici*	Wheat	Aqueaous seed extracts Formulations of seed oils	Curative and preventive effect	1,000 ml a.i./100 liters	Chianella and Rovesti, 1992
E. pisi	In vitro	Neem Azal	Inhibition	0.05 mg/ml	Singh et al., 2000

E. polygoni	Outdoor plots/herbaceous and woody nursery plants, hydrangea	Aqueaous emulsion of seed oil	Complete fungal inhibition	1%	Locke, 1992
Trichophyton rubrum, T. violaceum, T. mentagrophytes, T. concentricum, Epidermophyton floccosum, Microsporum canis, M. gypseum, Candida albicans, C. parapsilosis, Torulopsis glabrata, Trichosporon cutaneum, Fusarium spp., *Scopulariopsis brevicaulis,* and *Geotrichum candidum*	In vitro	Dried leaves, bark, kernels, flowers, drugs, and oils	No effect		Khan and Wassilew, 1987
Fusarium moniliforme	In vitro	Aqueous seed extracts	Inhibition	Neem-Azal-F 5%	Molera, 1994
	Commercial oils of seeds				
F. oxysporum	In vitro	Commercial oils of seeds	Inhibition	Neem-Azal-F 5%	Molera, 1994
		10-undecyn-1-ol	Complete fungal inhibition	MIC = 0.4 mg/cm^2	Govindachari, Suresh, Banumathi, et al., 1999
		Leaf hexane extract	Inhibition	MIC = 2 mg/cm^2	Govindachari, Suresh, Banumathi, et al.,1999
F. oxysporum f. sp. *chrysanthemi*	Soil	Aqueous seed oil emulsions	Increase population densities	1, 5, and 10%	Bowers and Locke, 2000
F. oxysporum f. sp. *ciceri*	In vitro	Seed oil, fruit pulp, leaf and bark extract	Complete fungal inhibition	5,000, 20,000, 25,000, and 30,000 ppm, respectively	Singh et al., 1980
F. oxysporum f. sp. *lycopersici*	Tomato crops	Neem cake	Inhibition/increased microflora population		Locke, 1995
	In vitro	Aqueous extract of oil cake	Inhibition		
F. solani	Soil/crossandra seedlings	Neem cake	Inhibition/increased microflora population	1% w/w	Jeyarajan et al., 1987
	Soil/chickpea seedlings		Inhibition	12 t/ha	

TABLE 3.1 *(continued)*

Fungi	Crops or substrate	Neem material	Effectiveness	Concentration	Reference
Fusarium spp.	Soil/eggplant	Neem cake	Inhibition/increased saprophytic fungi		Locke, 1995
	In vitro	Kernel extracts	Inhibition	ED_{50} = 1.0 mg/ml	Lehmann et al., 1993
	In vitro	Formulations of seed oils	No effect	31-1,000 ppm a.i.	Chianella and Rovesti, 1992
			Low effect	125-1,000 ppm a.i.	
F. oxysporum f. sp. *udum*, *vasinfectum*, or *lini*	Soil	Neem cake	Increased growth	0.5-5% w/w	Singh and Singh, 1990
Glomerella cingulata	Apple	Seed oil	Low effect	2%	Moline and Locke, 1993
Ganoderma lucidum	Soil/coconut palm	Neem cake	Inhibition/increased microflora population	5 kg/tree	Jeyarajan et al., 1987
Helminthosporium nodulosum	In vitro	Aqueous extract of oil cake	Inhibition		Locke, 1995
Macrophomina phaseolina	Field soil/soybean crop	Neem cake	Inhibition/increased microflora population	5 t/ha	Jeyarajan et al., 1987
		Leaves		10 t/ha	
Microsporum gypseum	In vitro	Ethanol and acetone leaf extracts	Inhibition		Khan and Wassilew, 1987
		Aqueous leaf extract	No effect		
Penicillium expansum	Apple	Seed oil	Very low effect	2%	Moline and Locke, 1993
Penicillium spp.	In vitro	Formulations of seed oils	No effect	31-1,000 ppm a.i.	Chianella and Rovesti, 1992
Phoma betae	In vitro	Kernel extracts	Inhibition	ED_{50} = 0.5-1.0 mg/ml	Lehmann et al., 1993
Phytophthora capsici	In vitro	Commercial oils of seeds	High inhibition	Neem-Azal-F 5%	Molera, 1994
	Soil/betelvine	Green leaves	Inhibition		Locke, 1995
	Soil/betelvine	Neem cake	Wilt incidence reduction	100 kg/ha	Jeyarajan et al., 1987
		Dried leaves		2.5 t/ha	
P. infestans	Tomato	Aqueous seed extracts	No preventive effect		Chianella and Rovesti, 1992
Plasmapara viticola	Grapevine leaves	Neem oil, Neem-Azal-S	Preventive effect	1:450	Achimu and Schlösser, 1992
		Seed extracts	Curative and preventive effect	1:1	

Puccinia antirrhini	Snapdragon	Aqueous emulsion of two hydrophobic solvent-extracted compounds from seeds	Inhibition	0.5% w/w	Locke and Walter, 1991
		Seed oil	Inhibition	1% v/v	Locke and Carter, 1993
	groundnut leaves	Extracts, polar extract, HPLC fractions, and nimonol and isomeldenin from uncrushed leaves	Inhibition	0.005-5.0 μg/cm^2	Suresh et al., 1997
P. arachidis	Detached groundnut leaves	Azadiradione	Inhibition	1 and 10 μg/cm^2	Govindachari et al., 2000
Pythium aphanidermatum	In vitro	Commercial oils of seeds	Complete fungal inhibition	Neem-Azal-F 0.5%, Neem Biagro 5%	Molera, 1994
	Soil	Green leaves	Inhibition	5% w/w	Singh and Pandey, 1966
	Natural soil	Alcohol-soluble component of oil cake	Stimulation		Locke, 1995
Pythium spp.	In vitro	Formulations of seed oils	No effect	31-1,000 ppm a.i.	Chianella and Rovesti, 1992
P. ultimum	In vitro	Kernel extracts	Inhibition	ED_{50} = 0.1-0.5 mg/ml	Lehmann et al., 1993
		ground seeds	60% increased in stand counts	2%	Locke and Carter, 1993
Rhizoctonia solani	Tomato, *Zinnia elegans*, soil/eggplant, cotton seedlings, and In vitro	Neem cake	Inhibition/increased microflora population		Locke, 1995; Jeyarajan et al., 1987
	Soil/rice seedlings	Neem cake	Inhibition	1%	Kannaiyan and Prasad, 1981
	Soil/cotton seeds	Neem cake	Inhibition/increased microflora population	2.5-5 t/ha	Jeyarajan et al, 1987
	In vitro	Kernel extracts	Inhibition	ED_{50} = 0.1-0.5 mg/ml	Lehmann et al, 1993
	In vitro	Seed oil, fruit pulp, leaf and bark extracts	Complete inhibition of fungi	500; 1,000; 10,000; 10,000 ppm	Singh et al, 1980
		Ground seeds	Increased stand counts	1 and 2% w/w	Locke and Carter, 1993
Rhizopus spp.	In vitro	Seed oil	Inhibition	0.2 ml oil/10 mm disc	Kher and Chaurasia, 1977
Sclerotinia sclerotiorum	In vitro	Seed oil, fruit pulp, leaf and bark extract	Complete fungal inhibition	2,000; 10,000; 15,000; and 20,000 ppm,. respectively	Singh et al., 1980

TABLE 3.1 *(continued)*

Fungi	Crops or substrate	Neem material	Effectiveness	Concentration	Reference
Sclerotium rolfsii	In vitro	Seed oil, fruit pulp, leaf and bark extract	Complete fungal inhibition	1,500, 12,000, 10,000, and 10,000 ppm, respectively	Singh et al., 1980
Septoria apiicola	Celery	Aqueous seed extracts	No preventive effect		Chianella and Rovesti, 1992
Sphaerotheca fuliginea	Cucumber	Aqueous extract of ground seeds	Curative and preventive effect	Undiluted	Häberle and Schlösser, 1993
	Vegetable	Aqueous seed extracts, formulations of seed oils		1,260-2,520 ml a.i./100 liters	Chianella and Rovesti, 1992
Uromyces appendiculatus	Pinto bean	Seed oil and seed wax	Inhibition	0.25% v/v	Locke and Stavely, 1992; Locke et al., 1993
U. dianthi	Carnation	Seed oil	Inhibition	1% v/v	Locke and Carter, 1993
Verticillium dahliae	In vitro	Commercial oils of seeds	Complete fungal inhibition	Neem-Azal-F 5%	Molera, 1994

[a]Fungi inhibition; [b]Inhibition in aflatoxin production; DCM: dichloromethane; MTB: methyl *tert*-butyl ether.

Another fungus that negatively affects crops is *Macrophomina phaseolina*, which causes 50 percent loss in average values of soybean crops. A comparative study was made to evaluate the effect of adding neem cake to the soil, to the leaves of the same tree, and to compost and sawdust to the order of 5 and 10 t/ha in a field experiment. These organic aggregates were compared with the fungicide quintozene (pentachloronitrobenzene) in a concentration of 5 kg/ha. The comparative results of this treatment on *M. phaseolina* are shown in Table 3.2.

After using neem cake, a lower incidence in root putrefaction was observed, with similar values to those obtained with quintozene ($p = 0.05$), and the yield of soybean per hectare was greater. The population of other fungi, bacteria, and actinomycetes was increased by the aggregate of the cake with a concomitant reduction in the spores of *M. phaseolina* (Jeyarajan et al., 1987), which suggests that these microorganisms have a controlling effect on the pathogenic fungus. These microorganisms increase in number three- to tenfold when the soil is treated with neem leaves, which results in about a threefold reduction in the units that form the spores of the pathogenic fungus; in the soil treated with neem cake, the microorganisms increase seven to twelve times in number, and the concomitant reduction in *M. phaseolina* is 4.7 times. These results show the greater effectiveness of control with neem cake. The methodology used in this case does not allow us to differentiate whether the control of the fungus is due to the existence of some antimycotic compound or through the increase of the antagonistic microbial population.

Cocoa palms infected with *Ganoderma lucidum* were treated with neem cake, with a dose equivalent to 5 kg/tree. For the comparison, a 1 percent

TABLE 3.2. The controlling effect of different amendments on *Macrophomina phaseolina* in soybean crop

Treatment	Root rot (%)	Yield (kg/ha)	*M. phaseolina* (cfu/g soil)
Neem leaves (10 t/ha)	15	1,488	33.7
Sawdust (10 t/ha)	17	2,438	36.9
Compost (10 t/ha)	14	2,348	28.7
Neem cake (5 t/ha)	10	2,750	22.4
PCNB (5 kg/ha)	11	2,488	14.6
Control	25	1,400	106.1

cfu = colony forming units

Bordeaux mixture was applied to the soil as positive control for the fungus in all the tests. The neem cake significantly reduced the disease and resulted in greater yields compared with the reference solution and with the control (Jeyarajan et al., 1987), as can be seen in detail in Table 3.3.

The effect of neem cake in this case was similar to the experiment on *M. phaseolina* in which the disease index was lower in those cases where the microbial population increased.

Chickpea plants of four to six weeks old are susceptible to *Fusarium solani,* which provokes the putrefaction of the roots. In a test designed to study the effect of neem, concentrations equivalent to 12 t/ha of neem cake were added to plastic containers with earth in which the chickpea seedlings were later placed. Fifteen-day-old seedlings were inoculated with the pathogen. At ten days posttreatment, the infection index was calculated, and the experiment showed a lower incidence of the disease in treated soils (42 percent) compared with the controls (90 percent) (Jeyarajan et al., 1987).

The controlling effect of neem on the fungus was also studied in crossandra seedlings. The effect of neem cake was compared with the aggregate of sawdust and compost, all at 1 percent w/w. With the use of the three amendments, a reduction was observed in the wilt incidence together with the increase of the microflora population in comparison to controls as had been seen in previous cases (Jeyarajan et al., 1987).

The attack of *Phytophthora capsici* on betelvine *(Piper betle)* was inhibited through the use of neem cake (100 kg/ha) and dried neem leaves (2.5 t/ha) bringing about a wilt incidence reduction of 63 percent and 58 percent compared to the control. At the same time, a 44 percent reduction was observed in parasitic nematode populations for both treatments (Jeyarajan et al., 1987).

However, the results in fungus control are not always satisfactory with neem products. It is suggested that an alcohol-soluble component of the oil cake is responsible for a stimulating effect on *Pythium aphanidermatum* populations in natural soil. Also, addition of neem cake to the soil to treat lime plants *(Citrus aurantifolia)* resulted in an increase of *Aspergillus* spp. in the rhizosphere soil compared to control over a period of six months

TABLE 3.3. Control of *Ganoderma lucidum* by different substrates in coconut palm

Treatment	Disease index	Nut yield/palm per year
Neem cake	12.3	39
Bordeaux mixture	69.8	30
Control	117.7	19

(Locke, 1995). When neem cake at 0.5-5 percent w/w was added to the soils previously infected with *F. oxysporum* f. sp. *udum*, *vasinfectum*, or *lini*, an increase in vegetative growth and sporulation of these fungi was observed in natural soil 15 days after amendment. The stimulatory effect was ascribed to the nutritive substances within the neem cake. *Fusarium oxysporum* f. sp. *udum* showed a better inhibition in autoclaved than in unautoclaved soil. These effects may be attributed to the suitability of the cake as a nutrient and to the presence or otherwise effect of antagonistic microflora within the amendment (Singh and Singh, 1990).

The neem fruit extracts and oils also show antifungal effects. Nonidentified active principles of neem kernel extracts showed good inhibition in the agar diffusion test on *Pythium ultimum*, *R. solani*, *Alternaria tenuis*, *Aphanomyces cochlioides*, *Phoma betae*, and *Fusarium* spp., which are responsible for the damping off, circular spots, root diseases, and mildew in the seedlings and rot. These were isolated from seedlings of sugar beet (*Beta vulgaris* L.). Their ED_{50} was between 0.1 and 1 mg/ml, depending on the fungus. Nonetheless, the ED_{50} for the fractions isolated against *R. solani* was equal to or greater than the ED_{50} for the extract that was the origin of these fractions ($ED_{50} = 0.1$ mg/ml) (Lehmann et al., 1993). This leads to the assumption of the presence of a more active compound in the extract not yet detected, or synergies between its different components. Some of the fractions tested (concentration = 1 mg/ml) showed growth inhibition of *A. cochlioides* similar to that reached by mancozeb [ethylenebis (dithiocarbamic acid) manganese zinc complex] (concentration = 0.1 mg/ml).

Liquid test material, such as extracts from an eluotropic solvent series, drugs (one recipe was based on fresh leaves mixed with curd and the other one on boiled neem leaves) and oils of neem, were assayed in order to find out their antimycotic effect. Solid test material, such as dried leaves, bark, kernels, and flowers, were also tested. The fungi tested were *Trichophyton rubrum*, *T. violaceum*, *T. mentagrophytes*, *T. concentricum*, *Epidermophyton floccosum*, *Microsporum canis*, *M. gypseum*, *Candida albicans*, *C. parapsilosis*, *Torulopsis glabrata*, *Trichosporon cutaneum*, *Fusarium* spp., *Scopulariopsis brevicaulis*, and *Geotrichum candidum* (Khan and Wassilew, 1987).

Dried plant material, drugs, and oils did not show any inhibitory effect on the growth of the test fungi; moreover, all these vegetable products became contaminated with molds. This finding agrees with the observation of Jacobson (1995) who reported contamination with mycotoxins, especially with aflatoxins, from neem oil, seed, or seed kernel extracts and neem seed cake. The ethanol and acetone leaf extracts showed an inhibitory effect on *M. gypseum* and *C. albicans*, while the aqueous extract developed no inhibitory effects. The effect of different neem extracts from an eluotropic solvent series had an inhibiting effect on dermatophytes, such as *E. floccosum,* in the agar diffusion test. The most effective was the petrol ether leaf extract, but methyl *tert*-butyl ether (MTB) kernel extract, methanol kernel extract, and methanol leaf extract also showed effectiveness. The antifungal effect could be attributed to low polarity substances as described by Lehmann et al. (1993) and Achimu and Schlösser (1992).

A polyherbal cream containing purified extract of neem seeds, quinine hydrochloride, and saponins extracted from the pericarp of *Sapindus mukerrossi* was used in clinical trials in order to control diseases caused by fungi. The total formulation completely inhibited the growth of *Candida albicans* on agar. Patients affected with *C. albicans* and *Chlamydia trachomatis* benefited clinically in trials. However, some of them remained culture positive. Due to the fact that quinine hydrochloride has a wide spectrum of action on microbes and viruses (Talwar et al., 1995), it cannot be clearly determined from the results found here whether neem has antifungal effects on *C. albicans* and *C. trachomatis.* Nevertheless, it was possible to confirm these effects when the polyherbal cream containing purified odorless and stain-free extract of neem (Praneem), saponins from *Sapindus mukorossi,* and *Mentha citrata* oil, and no quinine hydrochloride, inhibited the growth in culture of *C. albicans, C. krusei,* and *C. tropicalis* (Talwar et al., 1999).

The dichloromethane (DCM) phase from a methanolic neem kernel extract proved to have antifungal effects on *Dreschslera teres, Alternaria porri,* and *Ascochyta tritici* two or five days postinoculation. A later purification of the DCM phase revealed the presence of an unidentified compound that totally inhibits the growth of *D. teres* and *A. porri,* and to a lesser extent *A. tritici* (100 mg/ml). Even though the complete methanolic extract did not show a marked inhibition of the growth of the assayed fungi, the isolated compound did so, and both to the same concentration (Steinhauer, 1994). The percent of growth relative to control of the crude extract and the isolated compound can be compared (Table 3.4).

TABLE 3.4. Antimycotic effects of neem seed kernel extract and an isolated compound from *A. indica* at 100 mg/ml

	Growth (%)*	
Fungi	**Neem kernel extract**	**Compound**
Dreschslera teres	61.3	0.0
Alternaria porri	72.7	0.0
Ascochyta tritici	70.6	52.0

*Percent growth relative to control

Thionimone (a sulfur-containing component) obtained from neem seeds highly inhibited the growth of *F. oxysporum* f. sp. *lycopersici* (Zeringue and Bhatnagar, 1994).

Gedunin, a D-*seco* limonoid isolated from the oil of the dried and fresh fruits of neem (Lavie et al., 1971), was reported to have antifungal effects against *Polyporus* wood rot (van der Nat et al., 1991).

Foliar spray applications of 0.5 percent w/w of aqueous emulsions of two hydrophobic solvent-extracted fractions from neem seeds (oil and wax) showed to be effective against *Puccinia antirrhini* on greenhouse-grown snapdragons *(Antirrhinum majus)*. Although the wax and the oil compounds showed an equivalent level of effectiveness, the former appeared to have a longer residual activity (Locke and Walter, 1991). Solutions of 0.25 percent v/v of neem oil and neem wax showed to be effective against *Uromyces appendiculatus* (reducing pustule counts by at least 88 percent) when sprayed on to both upper and lower leaf surfaces of bean (*Phaseolus vulgaris* 'Pinto 111') (Locke and Stavely, 1992; Locke et al., 1993).

Ground neem seeds at 1 and 2 percent w/w increased stand counts over the *Rhizoctonia solani* check, and 2 percent of the same neem material resulted in ≤ 60 percent increase in stand counts for *Pythium ultimum* (Locke and Carter, 1993).

An aqueous extract of ground neem seeds proved to be effective in controlling *Sphaerotheca fuliginea,* one of the organisms responsible for causing powdery mildew on cucumber (Häberle and Schlösser, 1993). The undiluted extracts were applied to the upper leaf surfaces of three-week-old cucumber plants before and after inoculation of the fungi. Using the measure of the number of pustules per leaf, treatments with the extract were equally effective (percent efficacy 75.9-82.3 percent) in experiments pre- (one day before treatment) and postinoculation (2-6 days postinoculation [dpi]), showing a potent curative effect in the neem extract. The number of conidia per pustule and pustule size in mm^2 were considerably reduced as compared to control when the seed extract was applied 6 dpi, but the same effect was not observed when the extract was applied one day prior to the inoculation. However, pre- and postinoculation treatment with neem revealed a significant reduction in conidial production per leaf, and the 6 dpi application proved to be the most effective (92.7 percent). This study implies the presence of an antifungal substance in neem seed extracts. There was a significantly higher curative effect of the extract up to six days postinoculation as compared to an application one day prior to inoculation. This is due to the fact that the larger mycelial mass at 2-6 dpi absorbs more antifungal material than does germinating conidia. The observed curative effect of the neem seeds agrees with that found by Achimu and Schlösser (1992), who reported that, under laboratory conditions, commercial neem products such as Neem

Oil (0.08 percent azadirachtin), Neem-Azal-S (0.35 percent azadirachtin) and neem seed extracts reduced the area covered with *Plasmapara vitícola* on treated grapevine leaves. As a protective treatment, neem seed extracts and commercial neem products inhibited this organism by more than 99 percent when applied two days before the inoculation of the fungus. When neem seed extract was applied one to four days after inoculation, the efficacy was still high, with values of 94.6 and 84.2 percent respectively, thus indicating curative effects. It is thought that the antifungal properties of extracts are due to an inhibition of indirect germination of sporangia by preventing zoospore formation and/or release. The compounds responsible for this effect are quite stable and their efficacy persists (99.8 percent of efficacy) for 14 days. Another experiment was carried out in which sporangia were kept in contact with neem products for 12 hours, and later washed with water. It was found that the high degree of inhibition of sporangia continued (Achimu and Schlösser, 1992). This could be interpreted as the result of the presence of unidentified hydrophilic or hydrophobic fungicidal substances in the commercial products, or from water-insoluble fungicide or fungistatic compounds that remain in contact with the sporangia after the washing. This agrees with Lehmann et al. (1993) who reported activity of compounds against fungi in hydrophobic fractions of the extract of neem kernel. Locke and Walter (1991) reported antifungal effects of hydrophobic solvent-extracted compounds.

Interesting activity of some formulations of neem seed oils and aqueous seed extracts was observed on certain fungi. These showed both preventive and curative efficacy against *Sphaerotheca fuliginea* (in vegetables), *Erysiphe graminis* var. *tritici* (in wheat), and *Erysiphe graminis* var. *hordei* (in barley). The extract showed as much activity as sulphur when applied preventively or curatively. Both effects were also observed when formulations of the oil were applied only for the treatment of *Erysiphe cichoracearum* in cucumber. Although the oils show good activity, they also show phytotoxicity at high dosage rates on several crops, especially the cucurbits. When the aqueous extract was applied in crops inoculated with *Phytophthora infestans* (in tomato), *Cercospora beticola* (in sugar beet), and *Septoria apiicola* (in celery), no preventive effect was found. All these assays were made in crops. When the studies were carried out in vitro using the neem oil formulations no efficacy was found against *Penicillium* spp., *Alternaria brassicicola, Fusarium* spp., *Botrytis cinerea,* and *Pythium* spp. (Chianella and Rovesti, 1992).

Neem oil was evaluated in postharvest treatments to find out if it protects apples from the attack by *Botrytis cinerea, Glomerella cingulata,* and *Penicillium expansum* (Moline and Locke, 1993). The controlling effect of the oil was compared with $CaCl_2$. A 2 percent solution of neem oil was moderately antifungal against the first two fungi and showed little activity against the third, although the treatment reduced the decomposition of the fruit by 50 percent. The action of neem seed oil in reducing *Botrytis* decay may be explained by its nature of providing a protective barrier on the fruit surface (Moline and Locke, 1993).

The antifungal activity of neem seed oil was assayed in the experiment with paper discs together with other essential oils at a concentration of 0.2 ml oil/10 mm disc (Kher and Chaurasia, 1977). Fifteen fungi assayed belonged to *Aspergillus* spp., *Penicillium* spp., *Rhizopus* spp. and *Curvularia lunata*. Despite detecting zones of inhibition (2 to 6 mm) with neem oil at 48 hours, these were considerably less than the β-naphthol as standard reference.

Erysiphe polygoni commonly attack *Hydrangea macrophylla* under greenhouse conditions. Neem oil applied to individual rooted tip cuttings of *Hydrangea* in outdoor plots has shown efficacy against powdery mildew. When 1 percent aqueous emulsion of neem seed oil was applied on a 14-day schedule, complete protection against the fungi was observed (Locke, 1992; Locke et al., 1993). When 1 percent clarified neem seed oil was applied on *Phlox paniculata* in an outdoor situation, a reduction in the number of mildewed leaves and in the intensity of mildew development is observed (Locke, 1993). The use of this neem material in a foliar application at 1 percent v/v inhibited the growth of *Puccinia antirrhini* (on snapdragon), *Uromyces dianthi* (on carnation), and *Erysiphe polygoni* (on hydrangea), and was 81 to 84 percent more effective on *P. antirrhini* than soybean oil or a commercial fungicide (Locke and Carter, 1993).

In another laboratory assay (Molera, 1994) aqueous neem seed extracts and commercial oils of seeds (Neem Biagro [active compounds: azadirachtin (aza) 3,000 ppm, meliantriol and nimbidin-T] and neem azal-f [active compounds: azadirachtin 50,000 ppm]) were assayed under laboratory conditions against pathogenic fungi. The oils inhibited the growth of the fungi, including *Fusarium moniliforme, F. oxysporum, Verticillium dahliae, Phytophthora capsici, Pythium aphanidermatum, Ascochyta lentis,* and *Botritys cinerea*. The effects of the oils, according to Lehmann et al., are proportional to the concentration of aza, except for *A. lentis* and *P. aphanidermatum,* which are completely inhibited with the Neem Biagro oil at 5 percent. The latter suggests that the present findings corroborate statements by Lehmann et al. (1993).

An almost complete inhibition of the growth of soilborne pathogens, such as *F. oxysporum* f sp. *ciceri, R. solani, Sclerotium rolsfii,* and *Sclerotium sclerotiorum,* that incite wilt and rot in gram *(Cicer arietinum)* was observed in liquid medium when different aqueous neem extracts and seed oil were used (Singh et al., 1980). Complete inhibition of certain fungi was observed with extracts at different concentrations (Table 3.5).

Concentrations higher than 5,000 ppm of neem oil extracts exhibited inhibition of the germination of gram seeds (Singh et al., 1980). The black marks on roses caused by *Diclocarpon rosae* were moderately inhibited in vitro by 1 percent neem seed oil (Locke and Carter, 1993). However, the laboratory treatments and outdoor trials showed a delay in the development of the lesion and the subsequent abscission of the leaf.

Azadiradione, a tetranortriterpenoid from neem seed oil (0.21 percent), was observed to be effective in reducing rust pustule emergence of the groundnut caused by *Puccinia arachidis* in 76 and 77 percent (1 and 10 $\mu g/cm^2$ respectively) in a detached groundnut leaf bioassay 12 days after inoculation (Govindachari et al., 2000).

A combination of tither aerial sprays of *Pseudomonas fluorescens* and *P. aeruginosa* cell suspensions or neem azal proved to be more effective for the control of *Erysiphe pisi* of pea through inducing resistance in the host at all concentrations of neem azal (0.05 mg/ml being the optimum) than seed bacterization alone. Bacterization by both bacteria and the aerial spray of neem azal increased the dry weight of aerial parts, the number of nodes and pods, as well as the seed and pod weight of pea planes. Disease intensity was also reduced compared to control (Singh et al., 2000).

Fresh neem leaves and their extracts were also assayed as antifungal on various fungi. Incorporation of green neem leaves at 5 percent w/w results in a reduction of the population of pathogenic soil fungi as in the case of *Pythium aphanidermatum* (Singh and Pandey, 1966), and when it was applied near the base of betelvine, a reduction of *Phytophthora capsici* was achieved (Locke, 1995). Volatile compounds derived from fresh leaves produced an overall 90 percent reduction in aflatoxin production by *Aspergillus parasiticus* and a 51 percent reduction in the biomass of this fungus. These compounds were identified as a mixture of alcohols, aldehydes, ketones,

TABLE 3.5. Inhibition of fungal growth by different concentrations of neem extracts and oil

	Extract Concentration (ppm)			
Fungi	**Seed oil**	**Fruit pulp**	**Leaf**	**Bark**
F. oxysporum f. sp. *ciceri*	5,000	20,000	25,000	30,000
R. solani	500	1,000	10,000	10,000
S. rolsfii	1,500	12,000	10,000	10,000
S. sclerotiorum	2,000	10,000	15,000	20,000

and other compounds, such as terpenes, styrene, sulphur-containing compounds, and 2,5-dihydro-furan (Zeringue and Bhatnagar, 1994). Among the alcohols isolated, 2,3-butanediol and 1-heptanol, in concentrations of 2.5-10 μM and 1.8-7 μM respectively, had little effect on the fungus or the mycotoxin. C_2, C_5, C_6, C_7, and C_9 unsaturated aldehydes partially reduced radial growth of *A. parasiticus.* Ten μM of 4-pentenal and 7.6 μM of *trans*-2-heptenal reduced radial growth of the fungi by 26 percent and 23.2 percent respectively, but greatly reduced aflatoxin production (92.7 percent and 89.5 percent respectively), while the ketones (C_2 to C_{12}) showed slightly inhibitory effects either on the pathogen or on aflatoxin production.

Bhatnagar et al. (1990) studied efficacy of aqueous neem leaf extracts on the growth of two aflatoxigenic strains of *Aspergillus flavus* and *A. parasiticus* as well as on aflatoxin production. Nonvolatile constituents from the leaf extract (0 to 40 percent v/v) added to fungal growth medium prior to inoculation did not affect fungal growth (i.e., mycelial weight), but inhibited aflatoxin production in both *A. flavus* (100 percent) and *A. parasiticus* (95 percent), at concentrations of extract greater than 10 percent, suggesting the inhibition of aflatoxin synthesis in the earlier stage of the biosynthetic pathway. This was demonstrated not only in vitro but also under in vivo conditions in developing cotton bolls (Zeringue and Bhatnagar, 1990). When neem leaf extracts were injected into developing cotton bolls (30 days post-anthesis) 48 hours prior to inoculation of *A. flavus,* no inhibitory effects were shown on the growth of the fungus, but the seeds from the locules exhibited 98 percent inhibition in aflatoxin production compared to untreated bolls 13 days after spore inoculation. When the leaf extracts and the *A. flavus* spores were inoculated at the same time, the inhibition percentage was approximately 16 percent. It was demonstrated in a fungal growth medium in submerged culture that aflatoxin biosynthesis was irreversibly inhibited in the mycelia of both fungi, while the neem leaf extract did not kill the fungus. The effect on the inhibition of aflatoxin production in developing cotton bolls was as efficient as the commercial neem insecticide Margosan-O (0.3 percent azadirachtin and 3 to 5 percent neem oil).

Essential oils from leaves and seeds incorporated into a liquid medium decreased the growth of a toxigenic strain of *A. flavus*. In addition, the oils from seeds completely suppressed aflatoxin B-1 synthesis in inoculated maize grains at 500 and 1,000 ppm (Bankole, 1997).

10-undecyn-1-ol (90 percent purity) was isolated by Govindachari, Suresh, Banumathi, et al. (1999) from a leaf hexane extract. This compound was responsible for the total inhibition of the growth of *F. oxysporum* (MIC = 0.4 mg/cm^2) and *Colletotrichum lindemuthianum* (MIC = 0.2 mg/cm^2) on a glass plate surface.

The complete extract showed partial antifungal activity by inhibiting the spore germination of *F. oxysporum* and *C. lindemuthianum* only at 1-2 mg/cm^2, while mancozeb (ethylenebis [dithiocarbamic acid] manganese zinc complex) inhibited the fungi completely at 0.1 mg/cm^2.

Polar extracts, HPLC fractions, and two limonoids, nimonol and isomeldenin, isolated from uncrushed neem leaves, reduced the disease severity caused by *Puccinia arachidis* Sperg on groundnut leaves (*Arachis hypogaea* L.). With 5 μg/cm^2 of leaflet area of *n*-hexane extract of the neem leaves, an 80 percent reduction in the number of rust pustules per leaf was observed. A 50 percent reduction was noticed even at a concentration of 0.005 μg/cm^2. On the other hand, leaves treated with 0.005 μg and 0.05 μg/cm^2 *n*-hexane partitioned fraction from the previously mentioned extract did not show any disease reduction. The methanol partitioned fraction exhibited an increase in antifungal activity comparable to that of the hexane extract (0.005 μg and 0.05 μg/cm^2). Following isolation by HPLC, peaks 1-6 recorded an approximate 65-95 percent reduction in pustule number at the tested concentrations (0.005 μg and 0.05 μg/cm^2), while peaks 7 and 8 produced an 85 percent reduction at all doses. Nimonol and isomeldenin showed a disease reduction of 50-100 percent and 80-100 percent respectively at the same concentrations (Suresh et al., 1997) and were less active than peaks 7 and 8. It is also possible that the most active compound has not yet been isolated.

MELIA AZEDARACH *L.*

Melia azedarach, also known as chinaberry, cedro blanco, white cedar, persian lilac, pride of India in England, paraiso, mindi kechil in Malaysia, cinamomo in Brazil, among others, has been less studied than neem in relation to its bioactivities, and especially as an antimycotic (Bohnenstengel et al., 1999; Takeya et al., 1996; Valladares et al., 1997). The study of the antimycotic activity that the paraiso tree might have arose from the observation that solutions prepared with extracts from this tree and kept at room temperature showed no disease characterized by fungi. This led us to think that the extracts, and principally those from the fruit, must contain certain compounds able to inhibit the growth of these fungi, which are widely distributed in the environment in the form of spores.

Antifungal Activity of the Fruit Extract

The minimum inhibitory concentration (MIC) and the minimum fungicidal concentration (MFC) on different fungi and in different media of the ethanolic extracts of the *M. azedarach* fruits were measured. The extract was prepared by the Soxhlet extraction in which the ground fruit was defatted first with hexane to eliminate the lipids and then submitted to extraction with ethanol.

The fungi tested were *Aspergillus flavus, Fusarium moniliforme* Sheld (monosporic culture), *Microsporum canis, Candida albicans* isolated from human feces, and *C. albicans* (ATCC 10231). Two very rich media in terms of their nutrients were used to test *A. flavus* and *F. moniliforme:* carrot-potato-glucose agar and peanut agar. In the test carried out on *M. canis* and *C. albicans,* Sabouraud medium was used.

To test the fungicidal activity of the extract on *A. flavus* and *F. moniliforme,* a carrot-potato-glucose infusion was employed; for *M. canis* and *C. albicans,* a liquid peptone/glucose medium was used.

Minimum Inhibitory Concentration

The MIC, determined by an agar serial dilution test (Shadomy et al., 1991), was 300, 70, 50, and 100 mg/ml for *A. flavus, F. moniliforme, M. canis,* and *C. albicans,* respectively (Table 3.6) seven days from beginning the assay (Figure 3.1) after incubating in darkness at 28 ± 1°C and checking progress daily. In parallel, negative controls were carried out with miconazole nitrate which showed a MIC of 0.5, 3, 0.25, and 2 mg/ml for *A. flavus, F. moniliforme, M. canis,* and *C. albicans,* respectively, seven days

TABLE 3.6. MIC and MFC values of whole ripe fruit extract and miconazole nitrate in the moment of the growth of the control and at seven days from starting the assay

Fungi	MIC (mg/ml)				MFC (mg/ml)			
	Fruit Extract		Miconazole Nitrate		Fruit Extract		Miconazole Nitrate	
	24-72 h	168 h	24-72 h	168 h	24-72 h	168 h	24-72 h	168 h
Aspergillus flavus	200	300	0.25	0.50	250	500	0.50	0.50
Fusarium moniliforme	50	70	1.00	3.00	70	120	3.00	4.00
Microsporum canis	10	50	0.25	0.25	30	60	2.00	2.00
Candida albicans	50	100	0.25	2.00	120	120	0.25	2.00

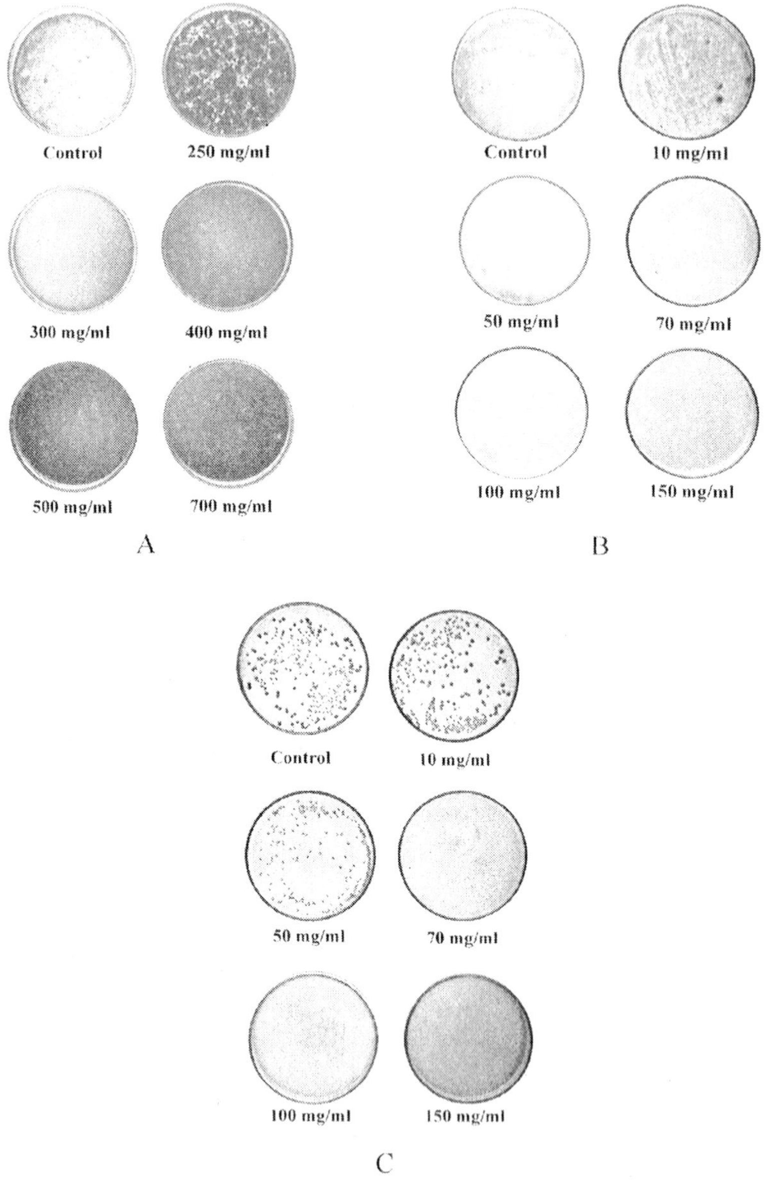

FIGURE 3.1. Minimum inhibitory concentration (MIC) of fruit extract from *M. azedarach* on *A. flavus* (carrot-potato-glucose agar) (A), *F. moniliforme* (carrot-potato-glucose agar) (B), and *C. albicans* (Sabouraud agar) (C)

from the beginning (Table 3.6). The MIC and the concentrations higher than this, which gave negative values of fungal growth, remained in incubation for four more weeks after the seven days, and at the end of this period they showed a total inhibition of growth (Carpinella et al., 1999).

Minimum Fungicidal Concentration (MFC)

The minimum fungicidal concentration of *M. azedarach* fruit extracts used in experiments on all the fungi were studied by a broth dilution test (Shadomy et al., 1991). Proper control tubes with miconazole nitrate were also maintained. All the tubes were observed and the microorganism counted with an optic microscope at 24 and 48 h. When fungal growth was observed in the control tube, 10 μl of each tested concentration was transferred to Sabouraud medium without extract and incubated for a period of seven days with checks every 24 hours.

The MFC values at seven days were 500, 120, 60, and 120 mg/ml for paraiso fruit extracts while for miconazole nitrate the MFC were 0.5, 4, 2, and 2 mg/ml for *A. flavus, F. moniliforme, M. canis,* and *C. albicans,* respectively (Table 3.6). The plates corresponding to these concentrations remained negative for a further period of four weeks. The values obtained in the moment in which the control becomes positive for both treatments may be seen in Table 3.6 (Carpinella, 1999).

Bioactive Compounds

After testing the antimycotic activity, a guided bioassay was made to isolate the active principles.

After processing the complete extract through column chromatographies, 32 fractions were obtained, and the antimycotic activity of each one was determined individually. For this measurement, bioautography was used in thin layer chromatography (TLC) (Homans and Fuchs, 1970) using *F. moniliforme* to reveal the activity. Its spores, dissolved in a medium of mineral salts containing glucose, were sprayed on the previously developed plates. At the end of the incubation period (seven days) the presence of a growth inhibition zone was observed in fractions 16 to 19 for the spot at Rf = 0.37. As from the fractions 20 to 24, the presence of a growth inhibition zone of the fungus was observed on a spot with Rf = 0.21.

In order to isolate these compounds, a series of purifications was then made in preparatory radial chromatography and from fractions 16 to 19 a white crystal was obtained (yield 0.02 g per 100 g of crushed seed kernels). Through spectroscopic techniques and by comparison with the bibliography (Pelter et al., 1982; Tsukamoto et al., 1984) it was determined that the isolated compound corresponded to (±) pinoresinol (compound 1) (Carpinella, 1999) (Figure 3.2).

(+) Pinoresinol was previously obtained from different plants belonging to such various families as, for example, the bark of *Fraxinus mandschurica* Rupr. var. *japonica* (Oleaceae) and of *Fraxinus japonica* (Tsukamoto et al., 1984), while Casabuono and Pomilio (1994) obtained the (–) pinoresinol form *Festuca argentina* [the (+) pinoresinol is the most widely distributed molecule in nature] (Casabuono and Pomilio, 1994). The compound was also found on one occasion from the extract of seeds of Brazilian *Melia azedarach* (Cabral et al., 1995), showing ecdysis inhibitory effects in larvae of *Rhodnius prolixus* (Cabral et al., 2000). However, as far as we know, its fungicidal activity has not been reported so far.

Fractions 20 to 24 were then isolated and a white crystalline compound was obtained, which corresponded to 0.0007 g% of the crushed seed kernels. According to nuclear magnetic resonance spectroscopy and mass spectroscopy, the compound corresponds to 3-metoxy 4-hydroxy cinnamaldehyde (compound 2) (Carpinella, 1999) (Figure 3.3).

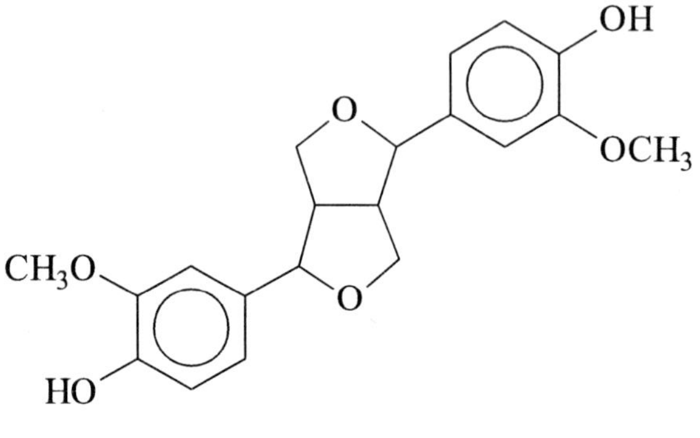

FIGURE 3.2. (±) Pinoresinol

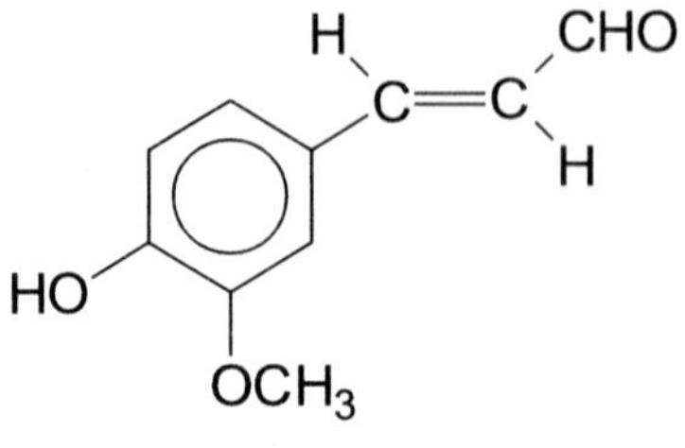

FIGURE 3.3. 3-metoxy 4-hydroxy cinnamaldehyde

Antifungal Effects of Pinoresinol

Various techniques were used to find out more about the antifungal activity of pinoresinol (Figure 3.2). The fungi on which these effects were tested were *Fusarium moniliforme* and *Candida albicans.*

Fusarium moniliforme

Different masses of the compound were placed in the center of a TLC on which 12 hours later they were sprayed with the mineral salts containing glucose medium, to which the spores of the mold had been inoculated. Upon daily checks, the plates showed the following profile after seven days (Table 3.7).

A broth dilution test was also made to determine the MIC and MFC of pinoresinol.

The pinoresinol solubilized in dimethylsulfoxide (DMSO) was added to the mineral salts with glucose medium until the desired concentrations were achieved. A control was also maintained.

The fungus inocula was added to each tube and they were then incubated in a dark chamber at 28°C. They were checked with optic microscope daily up to seven days, observing MIC at 4.0 mg/ml concentration (Table 3.8).

To measure the fungicidal activity, a concentration of pinoresinol at 4.0 mg/ml was tested in a liquid medium of mineral salts, and after 168 hours of incubation with the fungus, 10 μl of it was subcultured on a plate with solid

TABLE 3.7. Determination of the presence and size of the growth inhibition zone of *Fusarium moniliforme* in plates of TLC sown with pinoresinol

Mass (mg)	Growth Inhibition Zone	Growth (cm)
0.400	+++	1.1
0.200	+++	1.1
0.100	++	1.1
0.050	+	1.1
0.025	–	–
Control	–	–

– = Inhibition zone absent
+++ = Complete inhibition of fungal growth
++ = Moderate inhibition of fungal growth
+ = Low inhibition of fungal growth

TABLE 3.8. Minimum inhibitory concentration (MIC) of pinoresinol on *Fusarium moniliforme* in a medium of mineral salts with glucose

	Time (hours)						
Concentration (mg/ml)	**24**	**48**	**72**	**96**	**120**	**144**	**168**
4.0	NG	NG	NG	NG	NG	NG	NG
1.5	NG	NG	NG	+	+	++	++
Control	NG	+	+	+	++	+++	+++

\+ = Growth
NG = No growth

V_8 medium free of pinoresinol. Only this concentration was measured since it was the only one that exhibited no growth for seven days in the study previously mentioned. The subcultures of the 4.0 mg/ml concentration showed no growth at seven days while those of the control showed some growth within 48 hours. This shows that pinoresinol is an effective fungicide for *F. moniliforme* at a concentration of 4.0 mg/ml, and this strength is the same level as that observed by Carpinella (1999) for miconazole nitrate (Tables 3.6 and 3.9).

The study was also carried out with the so-called filter paper assay discs technique. Paper discs with the necessary microliters of pinoresinol in acetone were placed in a solid V_8 medium previously inoculated with a spore suspension of fungus. The presence of the growth inhibition zone was observed daily for seven days. A retraction in fungal growth was recorded when it was confronted with concentrations greater than 0.050 mg/cm^2 of pinoresinol up to 144 hours, with zones of inhibition of 1-2.5 cm^2 (Table 3.10).

Candida albicans

A broth dilution test was applied in the same way as for *F. moniliforme*, adding both the pinoresinol as well as the viable cells of *C. albicans* to the culture medium.

After incubation it was noted that pinoresinol delays the growth of *C. albicans* for a period of 48 hours, at a concentration of 4.35 mg/ml (Table 3.11). After this period, the yeast begins to grow in this concentration (Carpinella, 1999).

TABLE 3.9. Minimum fungicidal concentration (MFC) of pinoresinol on *Fusarium moniliforme* in the V_8 medium free of substance

	Time (hours)						
Concentration (mg/ml)	**24**	**48**	**72**	**96**	**120**	**144**	**168**
4.0	NG	NG	NG	NG	NG	NG	NG
Control	NG	+	+	+	+	+	+

\+ = Growth
NG = No Growth

TABLE 3.10. Determination of the presence and measurement of the diameter of the growth inhibition zone of *Fusarium moniliforme* in the Test with diffusion discs containing pinoresinol

	Time (hours)						
Concentration (mg/cm²)	**24**	**48**	**72**	**96**	**120**	**144**	**168**
	Growth (cm)						
0.400	NG	2.0	1.0	1.0	1.0	–	–
0.200	NG	2.0	1.0	1.0	1.0	–	–
0.100	NG	2.5	1.1	1.0	1.0	–	–
0.050	NG	2.0	1.1	1.0	1.0	–	–
0.025	NG	–	–	–	–	–	–
Control	NG	–	–	–	–	–	–

NG = No growth
– = Inhibition zone absent

TABLE 3.11. Minimum inhibitory concentration (MIC) of Pinoresinol on *Candida albicans* in a peptone meat broth medium with glucose

	Time (hours)		
Concentration(mg/ml)	**24**	**48**	**72**
4.35	NG	NG	+
1.50	+	+	+
Control	+	+	+

\+ = Growth
NG = No growth

It can be observed that pinoresinol is not a strong fungistatic on *C. albicans* and does not work as a fungicide on it, since after subculturing 10 μl of the concentrate, equivalent to 4.35 mg/ml in the moment of growth of the control on a plate of Sabouraud medium free of compound, growth occurred after 24 hours.

It may be inferred from these studies that compound 1 has proven fungistatic and fungicidal effects on *F. moniliforme* in a medium of mineral salts with glucose. Bearing in mind that the extract of the fruit of *M. azedarach* shows fungicidal effects on *C. albicans,* we may suppose that another fungistatic and/or fungicidal compound must also exist for that yeast.

Antifungal Effects of 3-Methoxy 4-Hydroxy Cinnamaldehyde

To evaluate the activity of compound 2 individually, a filter paper disc assay was carried out on *F. moniliforme.* From a solution of this compound solubilized in ethyl ether, the microliters necessary to obtain concentrations of 0.400, 0.200, 0.100, 0.050, and 0.025 mg/cm^2 were added to each paper disc, 1 cm in diameter, and after the total evaporation of the solvent, they were then placed in the center of an agar V_8 plate.

In this preliminary test a fungistatic effect can be noted for the concentration of 0.200 mg/cm^2 or greater up to 144 hours, which indicates a controlling but not a fungicidal effect, as seen in Table 3.12 (Carpinella, 1999). It must be remembered that the culture medium is V_8 medium, which is very

TABLE 3.12. Determination of the presence and size of the diameter of the growth inhibition zone of *Fusarium moniliforme* in the test with diffusion discs with 3-methoxy 4-hydroxy cinnamaldehyde

Concentration (mg/cm^2)	Time (hours)						
	24	48	72	96	120	144	168
	Growth (cm)						
0.400	NG	NG	2.2	1.6	1.3	1.0	–
0.200	NG	NG	1.6	1.4	1.1	1.0	–
0.100	NG	NG	–	–	–	–	–
0.050	NG	NG	–	–	–	–	–
0.250	NG	NG	–	–	–	–	–
Control	NG	NG	–	–	–	–	–

NG = No growth
– = Inhibition zone absent

rich in ingredients, and which, as would happen with pinoresinol, could have an antagonistic effect, just as occurs with the Sabouraud (also very rich in components) with miconazole (Shadomy et al., 1991).

On the other hand, as the concentrations of compound 2 assayed were not very high, we think that the inhibited effect over the organism could be reached at higher doses.

The substances isolated (1 and 2) up to the present as antifungal components of extracts of fruit of *M. azedarach* do not totally account for the high activity found for the extracts, and this would suggest either the existence of synergies between them, or the presence in the extract of one or more natural antifungal compounds that can act individually or in synergy. To the best of our knowledge, there is no other report of antifungal activity of *Melia azedarach* extracts; neither is any isolated compound responsible for this effect.

Up to this point, part of the antimycotic effect that *Azadirachta indica* and *Melia azedarach* exercise on different pathogenic fungi has been described. These trees belonging to the Meliaceae family are not the only kinds that enjoy these characteristics. *Swietenia mahogani, Khaya senegalensis, Toona ciliata,* and *Aglaia odorata,* which belong to the same family, also have these fungitoxic properties.

OTHER MELIACEAE MEMBERS WITH FUNGITOXIC PROPERTIES

Swietenia mahogani *and* Khaya senegalensis

Some compounds such as B, D-seco limonoids, for example, mexicanolide, 3 β-acetoxy-mexicanolide, 3 β-hydroxy mexicanolide, 2-α,3-β-dihydroxy mexicanolide from *Swietenia mahogani,* and 6-acetyl swietenine and 6-acetyl-3-tigloyl swietenolide from *Khaya senegalensis* have been isolated and are known to be effective in reducing the number of pustules of *Puccinia arachidis* on detached groundnut leaves (Govindachari, Suresh, Banumathi, et al., 1999).

Toona ciliata

A tretanortriterpenoid from powdered wood of *Toona ciliata,* cedrelone (0.4 percent w/w), reduced the rust pustule counts of *Puccinia arachidis* in a detached groundnut leaf bioassay after 12 days postinoculation. The inhibition percentages were 98 and 93 percent at 1 and 10 $\mu g/cm^2$ respectively.

These figures were higher than those exhibited for azadiradione, a compound isolated from *A. indica* and very similar to those obtained by mancozeb.

Replacement or modifications of the functional groups of ring A or B in cedrelone showed a reduction in their effectiveness against the mentioned fungi. Dihydrocedrelone, dihydrocedrelone diacetate, isocedrelonic acid and methyl isocedrenolate showed a rust pustule reduction of the fungi at 1 and 10 $\mu g/cm^2$ of 43 and 67 percent, 78 and 87 percent, 37 and 91 percent, and 8 and 2 percent, respectively (Govindachari et al., 2000).

Aglaia odorata

A 10 percent extract solution of *Aglaia odorata* was evaluated for its antifungal activity against *Botrytis cinerea* by a microtiter plate technique. After 24 hours of incubation of the plant extract and spore suspension of the fungi in each well, the density of fungal growth as absorbency was measured, as was spore germination, which was evaluated with an inverted light microscope after 12 and 48 hours. Extracts from *A. odorata* showed intermediate antifungal activity, which means an optical density reading under 40.00 and less than 10 percent *B. cinerea* spore germination after 48 hours (Wilson et al., 1997).

Chisocheton paniculatus

Meliacin type compounds such as 1,2-dihydro-6α-acetoxyazadirone have been isolated from the fruits of *Chisocheton paniculatus*. These compounds showed strong inhibitory properties against *Curvularia verruciformis, Dreschlera oryzae,* and *Alternaria solani* (Bordoloi et al., 1993).

Last, the results presented in this chapter open new doors in the field of control of fungi pathogenic to plants and mammals, which is calling for new substances innocuous to health and to the environment, and for which resistance has not been generated.

REFERENCES

Achimu P. and Schlösser E. (1992). Effect of neem seed extracts (*Azadirachta indica* A. Juss) against downy mildew *(Plasmopara viticola)* of grapevine. In *Practice Oriented Results on Use and Production of Neem-Ingredients. Proceedings of First Workshop* (Ed., Kleeberg H.). Giessen: Trifolio M GmbH, Druck and Graphic, pp. 99-107.

Bankole S.A. (1997). Effect of essential oils from two Nigerian medicinal plants (*Azadirachta indica* and *Morinda lucida*) on growth and aflatoxin B-1 production in maize grain by a toxigenic *Aspergillus flavus*. *Lett. Appl. Microbiology* 24: 190-192.

Bhatnagar D., Zeringue H.J., and McCormick S.P. (1990). Neem leaf extracts inhibit aflatoxin biosynthesis in *Aspergillus flavus* and *A. parasiticus*. In *Neem's Potential in Pest Management Programs. Proceedings of USDA Neem Workshop* (Eds., Locke J.C. and Lawson R.H.). USDA/ARS publication 86. Washington, DC: U.S. Department of Agriculture, pp. 118-127.

Bohnenstengel F.I., Wray V., Witte L., Srivastava R.P., and Proksch P. (1999). Insecticidal meliacarpins (C-*seco* limonoids) from *Melia azedarach*. *Phytochemistry* 50: 977-982.

Bordoloi M., Saikia B., Mathur R.K., and Goswami B.N. (1993). A meliacin from *Chisocheton paniculatus*. *Phytochemistry* 34: 583-584.

Bowers J.H. and Locke J.C. (2000). Effect of botanical extracts on population density of *Fusarium oxysporum* in soil and control of *Fusarium* wilt in the greenhouse. *Plant Disease* 84: 300-305.

Cabral M.M.O., Azambuja P., Gottlieb O.R., and Garcia E.S. (2000). Effects of some lignans and neolignans on the development and excretion of *Rhodnius prolixus*. *Fitoterapia* 71: 1-9.

Cabral M.M.O., Garcia E.S., and Kelecom A. (1995). Lignanes from the Brazilian *Melia azedarach*, and their activity in *Rhodnius prolixus* (Hemiptera, Reduviidae). *Mem. Inst. Oswaldo Cruz* 90: 759-763.

Carpinella M.C. (1999). Aislamiento e identificación de principios pesticidas en *Melia azedarach* y estudio de su toxicidad. Doctoral thesis, Universidad Nacional de Córdoba, Argentina, pp. 133-189.

Carpinella M.C., Herrero G.W., Alonso R.A., and Palacios S.M. (1999). Antifungal activity of *Melia azedarach* fruit extracts. *Fitoterapia* 70: 296-298.

Casabuono A.C. and Pomilio A.B. (1994). Lignans and a stilbene from *Festuca argentina*. *Phytochemistry* 35: 479-483.

Chianella M. and Rovesti L. (1992). First experiences with neem in Italy and its potential uses in plant protection. In *Practice Oriented Results on Use and Production of Neem-Ingredients. Proceedings of First Workshop* (Ed., Kleeberg H.). Giessen: Trifolio M GmbH, Druck and Graphic, pp. 57-64.

De Lucca A.J., Jacks T.J., and Broekaert W.J. (1999). Fungicidal and binding properties of three plant peptides. *Mycopathologia* 144: 87-91.

DiSalvo A.F. (1974). Antifungal properties of a plant extract. I. Source and spectrum of antimicrobial activity. *Mycopathologia et Mycologia applicata* 54: 215-219.

Fernandez Valiela M.V. (1978). *Introducción a la Fitopatología,* Volume III., pp. 3-153, Colección Científica del INTA, Buenos Aires.

Govindachari T.R., Suresh G., Banumathi B., Masilamani S., Gopalakrishnan G., and Krishna Kumari G.N. (1999). Antifungal activity of some B,D-*seco* limonoids from two Meliaceous plants. *J. Chem. Ecol.* 25: 923-933.

Govindachari T.R., Suresh G., Gopalakrishnan G., Masilamani S., and Banumathi B. (2000). Antifungal activity of some tetranortriterpenoids. *Fitoterapia* 71: 317-320.

Govindachari, T.R., Suresh G., and Masilamani S. (1999). Antifungal activity of *Azadirachta indica* leaf hexane extract. *Fitoterapia* 70: 417-420.

Häberle R. and Schlösser E. (1993). Protective and curative effects of neem seed extracts against *Sphaerotheca fuliginea* on cucumber. In *Practice Oriented Results on Use and Production of Neem-Ingredients and Pheromones. Proceedings of Second Workshop* (Ed., Kleeberg H.). Giessen: Trifolio M GmbH, Druck and Graphic, pp. 107-112.

Hejtmánková N., Walterová D., Preininger V., and Simánek V. (1984). Antifungal activity of quaternary Benzo[*c*] phenanthridinealkaloids from *Chelidonium majus*. *Fitoterapia* 55: 291-294.

Homans A.L. and Fuchs A. (1970). Direct bioautography on thin-layer chromatograms as a method for detecting fungitoxic substances. *J. Chromatography* 51: 325-327.

Jacobson M. (1995). Toxicity of neem to vertebrates and side effects on beneficial and other ecologically important non-target organisms. In *The Neem Tree. Source of Unique Natural Products for Integrated Pest Management, Medicine, Industry and Other Purposes* (Ed., Schmutterer H.). Weinheim: VCH Verlagsgesellschaft mbH, pp. 484-495.

Jeyarajan R., Doraiswamy S., Bhaskaran R., and Jayaraj S. (1987). Effect of neem and other plant products in the management of plant diseases. In *Natural Pesticides from the Neem Tree and Other Tropical Plants. Proceedings of the Third International Neem Conference* (Eds., Schmutterer H. and Ascher K.R.S.). Eschborn: Deutsche Gesellschaft für Technische Zusammenarbeit (GTZ) GmbH, pp. 635-644.

Kannaiyan S. and Prasad N.N. (1981). Effect of organic amendments on seedling infection of rice caused by *Rhizoctonia solani*. *Plant and Soil* 62: 131-133.

Kauffman C.A. and Hedderwick S. (1997). Opportunistic fungal infections: Filamentous fungi and cryptococcosis. *Geriatrics* 52: 40-42, 47-49.

Khan M. and Wassilew S.W. (1987). The effect of raw material from neem tree, neem oil, and neem extracts on fungi pathogenic to humans. In *Natural Pesticides from the Neem Tree and Other Tropical Plants. Proceedings of the Third International Neem Conference* (Eds., Schmutterer H. and Ascher K.R.S.). Eschborn: Deutsche Gesellschaft für Technische Zusammenarbeit (GTZ) GmbH, pp. 645-650.

Khanna P., Mohan S., and Nag. T.N. (1971). Antimicrobials from plant tissue culture. *Lloydia* 168-169.

Kher A. and Chaurasia S.C. (1977). Antifungal activity of essential oils of three medicinal plants. *Indian Drugs* 15: 41-42.

Lavie D., Levy E.C., and Jain M.K. (1971). Limonoids of biogenetic interest from *Melia azadirachta* L. *Tetrahedron* 27: 3927-3939.

Lehmann W., Ibenthal W.-D., and Heitefuss R. (1993). Fungitoxic activity of neem kernel extracts against damping off of sugarbeets. In *Practice Oriented Results on Use and Production of Neem-Ingredients and Pheromones. Proceedings of*

Second Workshop (Ed., Kleeberg H.). Giessen: Trifolio M GmbH, Druck and Graphic, pp. 113-120.

Locke J.C. (1992). Comparison of neem oil, sunspray 6E plus horticultural oil and benlate for control of powdery mildew on greenhouse grown *Hydrangea*. *Phytopathology* 82: 1121 (Abstr.).

Locke J.C. (1993). Field evaluation of clarified neem seed oil, sunspray 6E plus horticultural oil and Funginex for control of powdery mildew on perennial garden phlox. *Phytopathology* 83: 1337 (Abstr.).

Locke J.C. (1995). Fungi. In *The Neem Tree. Source of Unique Natural Products for Integrated Pest Management, Medicine, Industry and Other Purposes* (Ed., Schmutterer H.). Weinheim: VCH Verlagsgesellschaft mbH, pp. 118-127.

Locke J.C. and Carter M.R. (1993). Evaluation of natural products for innovative disease management. *Phytopathology* 83: 695 (Abstr.).

Locke J.C., Larew H.G., and Walter J.F. (1993). Efficacy of clarified neem seed oil against foliar fungal pathogens and greenhouse whiteflies. In *ACS Conference Proceedings Series. Pest Management: Biologically Based Technologies* (Eds., Lumsden R.D. and Vaughn J.L.). American Chemical Society, pp. 287-289.

Locke J.C. and Stavely J.R. (1992). Efficacy of neem seed extraction products and two petroleum oils against bean rust. *Fungicide and Nematicide Tests* 47: 76.

Locke J.C and Walter J.F. (1991). Use of two natural compounds extracted from neem seed to control rust on greenhouse grown snapdragon. *Phytopathology* 81: 1177 (Abstr.).

Mitchell M.J., Smith S.L., Johnson S. and Morgan E.D. (1997). Effects of the neem tree compounds azadirachtin, salannin, nimbin and 6-desacetylnimbin on ecdysone 20-monooxygenase activity. *Archives of Insect Biochemistry and Physiology* 35: 199-209.

Molera A. (1994). Efectos fungicidas de diferentes extractos de Nim. In *Practice Oriented Results on Use and Production of Neem-Ingredients and Pheromones. Proceedings of Third Workshop* (Ed., Kleeberg H.). Giessen: Trifolio M GmbH, Druck and Graphic, pp. 93-97.

Moline H.E. and Locke J.C. (1993). Comparing neem seed oil with calcium chloride and fungicides for controlling postharvest apple decay. *Hortic. Science* 28: 719-720.

Pelter A., Ward R.S., Watson D.J., Collins P., and Kay I.T. (1982). Synthesis of 2,6-diaryl-4,8-dihydroxy-3,7-dioxabicyclo[3.3.0]octanes. *JCS Perkin* 1: 175-181.

Phay N., Higashiyama T., Tsuji M., Matsuura H., Fukushi Y., Yokota A., and Tomita F. (1999). An antifungal compound from roots of Welsh onion. *Phytochemistry* 52: 271-274.

Pitt J.I. (1994). The current role of *Aspergillus* and *Penicillium* in human and animal health. *J. Med. Vet. Mycol.* 32: 17-32.

Shadomy S., Espinel-Ingroff A., and Cartwright R.Y. (1991). Estudios de laboratorio con agentes antifúngicos: Pruebas de susceptilidad y bioensayos. In *Manual de Microbiología Clínica* (Eds., Lennette E.H., Balows A., Hausler W.J., and H.J. Shadomy). Buenos Aires: Editorial Médica Panamericana S.A., pp. 1229-1238.

Shadomy S. and Pfaller M. (1991). Laboratory studies with antifungal agents: Susceptibility tests and quantitation in body fluids. In *Manual of Clinical Microbiology*, Fifth Edition (Eds., Balows A., Hausler W.J., Herrmann K.L., Isenberg H.D., and H.J. Shadomy), pp. 1173-1183.

Singh R.S. and Pandey K.R. (1966). Effect of green and mature plant residues and compost on population of *Pythium aphanidermatum*. *Indian Phytopath.* 19: 367-371.

Singh U.P., Prithiviraj B., Singh K.P., and Sarma B.K. (2000). Control of powdery mildew *(Erysiphe pisi)* on pea *(Pisum sativum)* by combined application of plant growth-promoting rhizobacteria and Neemazal. *Journal of Plant Diseases and Protection* 107: 59-66.

Singh R.S. and Singh N. (1990). Effect of oil-cake amendment of soil on populations of some wilt causing species of *Fusarium*. *Phytopathology Z.* 69: 160-167.

Singh, U.P., Singh, H.B., and Singh, R.B. (1980). The fungicidal effect of neem *(Azadirachta indica)* extracts on some soil-borne pathogens of gram *(Cicer arietinum)*. *Mycologia* 72: 1077-1093.

Singh G. and Upadhyay R.K. (1993) Essential oils: Potent of natural pesticides. *J. Scientific and Industrial Res.* 52: 676-683.

Steinhauer B. (1994). Antifungal compounds from *Azadirachta indica*. In *Practice Oriented Results on Use and Production of Neem-Ingredients and Pheromones. Proceedings of Third Workshop* (Ed., Kleeberg H.). Giessen: Trifolio M GmbH, Druck and Graphic, pp. 93-97.

Suresh G., Narasimhan N.S., Masilamani S., Partho P.D., and Gopalakrishnan (1997). Antifungal fractions compounds from uncrushed green leaves of *Azadirachta indica*. *Phytoparasitica* 25: 33-39.

Takeya K., Qiao Z-S., Hirobe C., and Itokawa H. (1996). Cytotoxic azadirachtin-type limonoids from *Melia azedarach*. *Phytochemistry* 42: 709-712.

Talwar G.P, Garg S., Dhar V., Chabra R., Ganju A., and Upadhyay S.N. (1995). Praneem polyherbal cream and pessaries with dual properties of contraception and alleviation of genital infections. *Current Science* 68: 437-440.

Talwar G.P, Raghuvanshi P., Dhar R., Dawar H., and Kumar N. (1999). Two medical applications of neem. In *World Neem Conference*, S-2-4, Vancouver, Canada (Abstr.).

Tsukamoto H., Hisada S., and Nishibe S. (1984). Lignans from bark of *Fraxinus mandshurica* var. *japonica* and *F. japonica*. *Chem. Pharm. Bull.* 32: 4482-4489.

Valladares G., Defagó M.T., Palacios S.Y, and Carpinella M.C. (1997). Laboratory evaluation of *Melia azedarach* (Meliaceae) extracts against the elm leaf beetle (Coleoptera: Chrysomelidae). *J. Econ. Entomol.* 90: 747-750.

van der Nat J.M., van der Sluis W.G., de Silva K.T.D., and Labadie R.P. (1991). Ethnopharmacognostical survey of *Azadirachta indica* A. Juss (Meliaceae). *J. Ethnopharmacol.* 35: 1-24.

Wätcher G.A., Hoffmann J.J., Furbacher T., Blake M.E., and Timmermann B.N. (1999). Antibacterial and antifungal flavanones from *Eysenhardtia texana*. *Phytochemistry* 52: 1469-1471.

Wilson C.L., Solar J.M., El Ghaouth A., and Wisniewski M.E. (1997). Rapid evaluation of plant extracts and essential oils for antifungal activity against *Botrytis cinerea*. *Plant Disease* 81: 204-210.

Zacchino S.A, Lopez S.N., Pezzenati G.D., Furlán R.L., Santecchia C.B., Muñoz L., Giannini F.A., Rodriguez A.M., and Enriz R.D. (1999). In vitro evaluation of antifungal properties of phenylpropanoids and related compounds acting against dermatophytes. *J. Nat. Prod.* 62: 1353-1357.

Zeringue H.J. and Bhatnagar D. (1990). Inhibition of aflatoxin production in *Aspergillus flavus* infected cotton bolls after treatment with neem *(Azadirachta indica)* leaf extracts. *J. Amer. Oil Chem. Soc.* 67: 215-216.

Zeringue H.J. and Bhatnagar D. (1994). Effects of neem leaf volatiles on submerged cultures of aflatoxigenic *Aspergillus parasiticus*. *Appl. Environ. Microbiol.* 60: 3543-3547.

Chapter 4

Plant-Derived Extracts and Preparations As Antimycotics

Tadeusz Wolski
Kazimierz Glowniak

INTRODUCTION

Human and animal mycoses are among the diseases caused by conditionally pathogenic microorganisms. The development of systemic and local mycoses is favored by a weakened immune system. Therefore, in the treatment of mycoses it is very important that the drug has no immunosuppressive effects on the body in addition to its effects on the pathogens. By changing the body's resistance, immunomodulators either enhance or weaken immunity, expressed as humoral and cellular response. Immunomodulatory effects are of special importance when the drugs are used to treat diseases associated with immunosuppression. As a general rule, a patient's immune state is the primary risk factor responsible for the incidence of mycosis—over and above mycotic virulence and the infectious dose. Epidemiological data show that the natural defense mechanism, both in humans and in animals, effectively protects against mycotic infections (Smith et al., 1986; Kowszyk-Gindifer and Sobiczewski, 1986; Glinski and Wolski, 1997). However, therapeutic effects largely depend on the activity of the drug used to kill or inhibit the growth of the pathogens in the infected body, as well as patient's immunity and the drug's side effects (Glinski and Wolski, 1997).

The search for new antifungal agents, or modifications of those already known, is not easy, and producing an effective antimycotic drug is even more difficult than finding an antibacterial preparation. The number of antimycotic drugs used in practice is small, and even those often have serious adverse side effects (Walsh and Pizzo, 1988). Synthetic chemotherapeuticals, in addition to antibiotics, are used to treat systemic and surface mycoses. They include, among others, amphotericin B, nystatin, allylamines, amorolphin, polyoxins, nikkomicin, and β-glucosan synthesis inhibitors such as echinocandins, benamomycins, and 5-fluorocytosine (Hunter et al., 1995). Recently, certain antimycotic drugs have been used together with immunostimulators. The preparations potentiating or controlling immune reactivity, specifically or nonspecifically, enhance the humoral and cellular responses involved in fighting mycotic invasion. The safest currently available antimycotic substances, in terms of their range of activity, include imidazole derivatives (Foil, 1994; Glinski and Wolski, 1997).

Apart from zoopathogenic fungal strains, many entomo- and phytopathogenic strains pose a hypothetical threat to insects and plants. Entomopathogenic fungi are commonly found in the biosphere of various climatic zones. Many of them develop outside the host body, in the soil, or on plants, but they retain the ability to invade and infect insects. Currently, about 1,000 fungal species pathogenic to insects have been identified. The vast majority of fungi causing mycoses in insects belong to a group of opportunistic microorganisms that develop and act pathogenically in the weakened host's body (Glinski et al., 1998). The incidence of bee mycoses has shown a rapid increase in recent years, a phenomenon not easily explained. The incidence of mycoses correlates with the presence of immunosuppressive factors in the insect habitat, which include such factors as environmental pollution with pesticides and heavy metals. Climatic conditions, overexploitation, abused hygienic standards, *Varroa jacobsoni* invasion, immunosuppressive factors—all of these predispose to mycoses. *Ascosphaera apis,* the etiological factor of cystoid mycosis, and the *Aspergillus*-type fungi accounting for aspergillosis reveal their activity in bees as soon as favorable endogenous and exogenous or environmental factors make their appearance (Glinski et al., 1998).

Recently, great expectations have been aroused by natural substances, especially biologically active compounds produced by higher plants, due to their multidirectional biological activity and few side effects (Georgijewskij et al., 1990; Glinski and Wolski, 1997). In addition to having antimycotic properties, many substances of natural origin stimulate natural and potential immunity. A separate problem is the appearance of drug-resistant strains; however, this phenomenon is less common in the case of medicines derived from plants than from chemotherapeuticals and antimycotic antibiotics.

Extracts from fruits, seeds, flowers, bulbs, and leaves of higher tropical plants exhibit antimycotic activity. Most often they are volatile oils containing, among other substances, phenols and polyphenols (Pandey et al., 1982; Bourrel et al., 1993; Glinski and Wolski, 1997). The exception among them is a strongly fungistatic volatile oil isolated from fresh *Ageratum haustonianum* leaves which lacks phenol derivatives. Sulphides and sulphoxides in garlic extracts (Maleszka et al., 1991; Wolski et al., 1996a) are also likely to show antifungal activity, as do alkaloids, e.g., berberin, in the extract of *Berberis aristata* leaves, or chaksin and isochaksin in *Cassia albus* and

Cassia tora (Mukherjee et al., 1996). Essential oils from the tea tree *(Melaleuca alternifolia)* (Konopacka-Brud, 1995), and *Melissa officinalis* L. (Kedzia et al., 1990) exhibit interesting bacteriostatic and mycostatic properties as well as the ability to stimulate the immune system.

One of the important groups of biologically active substances is the coumarins (Kaminski et al., 1978; Murray et al., 1982). These substances are used as anticoagulants, coronary relaxants, spasmolytics, blood vessel sealants, and cholagogues (Murray et al., 1982; Cisowski, 1983, 1986). Because of their photosensitizing properties, numerous furanocoumarins, such as xanthotoxin, imperatorin, bergapten, and isopimpinellin, are used in the treatment of vitiligo and psoriasis. Those substances also act as antimycotics, especially on human and animal dermatophytes and *Ascophaera apis* (Ericsson and Sherris, 1971; Glinski et al., 1988; Kedzia et al., 1996; Wolski et al., 1996; Wolski, Glinski, Holderna-Kedzia, et al., 1997; Glinski and Wolski, 1997).

Interest has recently been increasing in coumarins from lower organisms showing a range of valuable pharmacological activities. They are synthesized by Actinomycytes, fungi, and numerous species of Spermatophyta, mainly from families Umbellifereae, Asteraceae, and Rutaceae (Kaminski et al., 1978; Murray et al., 1982; Glowniak, 1988; Pieta et al., 1995).

Furanocoumarins are obtained from plants by extraction with organic solvents (Glowniak, 1988, Glowniak et al., 1991; Wolski et al., 1996). Recently, supercritical fluids have been used to extract substances from plants (Stahl et al., 1986; Wenclawik, 1992; Witkiewicz, 1995). Owing to its physicochemical properties and low toxicity, as well as cost, liquid CO_2 is considered the best supercritical fluid (Gawdzik et al., 1994; Peplonski, 1994; Gawdzik et al., 1995; Gawdzik et al., 1996). Another advantage of CO_2 is its easy removal from extracts, which eliminates the chances of contamination—often a problem in the case of extraction by organic solvents.

The richest source of furanocoumarins is the Umbellifereae, especially the genera *Pastinaca, Ammi, Angelica, Heracleum,* and *Peucedonaum,* as well as some others. Furanocoumarins are obtained from their fruits and roots, and rarely from bark or leaves (Kaminski et al., 1978; Wawrzkiewicz et al., 1990; Glowniak et al., 1991; Wolski et al., 1993).

The highest antimycotic activity of furanocoumarins is against dermatophytes (Cisowski, 1986; Wolski et al., 1996a). The most active compounds include psoralen, bergapten, imperatorin, isopimpinellin, and angelicin. They strongly inhibit the growth and development of *Trichophyton* and *Microsporum* at concentrations of 7.8-125 μg/ml (Honda, 1984; Georgijewskij et al., 1990). Their structures are presented in Table 4.1.

As already mentioned, higher plants are a valuable source of many biologically active substances. Garlic (*Allium sativum* L.) is one of them. Since

TABLE 4.1. Basic furanocoumarin structures

No.	Name	Structure Type	Substituents
1	Psoralen	I	$R_5 = R_8 = H$
2	Bergapten	I	$R_5 = OCH_3$ $R_8 = OCH_3$
3	Xanthotoxin	I	$R_5 = H$ $R_8 = OCH_3$
4	Imperatorin	I	$R_5 = H$ $R_8 = O{-}CH_2{-}CH = C\langle^{CH_3}_{CH_3}$
5	Isoimperatorin	I	$R_5 = O{-}CH_2{-}CH = C\langle^{CH_3}_{CH_3}$ $R_8 = H$
6	Xanthotoxol	I	$R_5 = R_8$ $R_8 = O\text{-}H$
7	Isopimpinellin	I	$R_5 = R_8 = OCH_3$
8	Angelicin	II	$R_5 = R_6 = H$
9	Isobergapten	II	$R_5 = OCH_3$ $R_6 = H$
10	Sfondyna	II	$R_5 = H$ $R_6 = OCH_3$
11	Pimpinellin	II	$R_5 = R_6 = OCH_3$

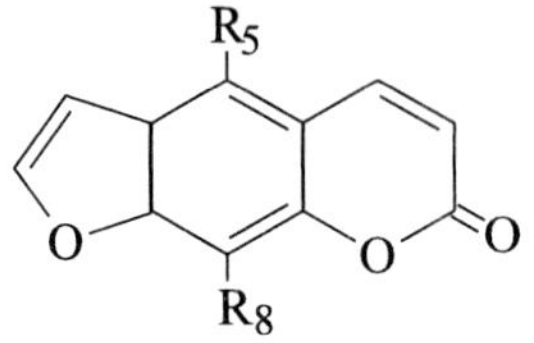

Type I—linear

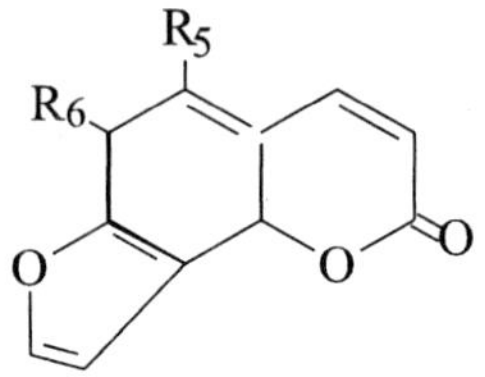

Type II—angular

ancient times garlic has been used worldwide as a popular remedy for colds, hypertension, and other conditions. In addition to antimycotic and antibacterial activity, garlic also exerts a synergistic influence, enhancing the potential of administered antibiotics. Kabelik (1970) found that substances isolated from garlic act much more strongly on certain fungi than upon bacteria. This observation was confirmed by studies of antimycotic activity of garlic extracts and preparations used in the therapy of dermatomycosis pedis (Kadziela-Wypyska et al., 1995a) and their effects upon other dermatophytes and phytopathogens (Venugopal and Venugopal, 1995; Singh et al., 1995).

Modern studies on the composition of garlic (*Allium sativum* L.) and its therapeutic properties have fully confirmed its bacteriocidal, bacteriostatic, antiviral, and antimycotic properties (Black et al., 1993; Calvey and Koach,

1994; Lutomski, 1987; 1995; Machowicz-Stefaniak et al., 1995). Apart from that, garlic acts synergistically with common antibiotics.

At Lublin, Poland, systematic research has been carried out to obtain furanocoumarin extracts from plants and garlic preparations, and to evaluate their antibacterial and antimycotic activity, their photo- and allergenic activity, and their immunostimulating influence and therapeutic effectiveness in the treatment of human and animal mycoses (Glinski et al., 1988; Glowniak, 1988; Wawrzkiewicz et al., 1990; Kuczynska et al., 1992; Gawdzik et al., 1995; Kedzia et al., 1996; Wolski et al., 1996).

The aim of the present study was a survey of methods to isolate furanocoumarins from plants, as well as from garlic extracts and garlic preparations, and to determine their antimycotic potential. The activity of antimycotic substances used to treat mycoses was examined in vitro, and their effectiveness was evaluated in clinical studies.

MATERIALS AND METHODS

The material used in the examination was *Angelica archangelica* (*Archangelica officinalis* L. Hoffm.) fruits, *Pastinaca* fruits (grown at the Central Seed Station, Bydgoszcz, Poland) and *Heracleum sosnowskyi* Manden (grown at the Institute of Soil Science and Plant Cultivation, Pulawy, Poland). The material for examination was dried and ground to a semipowdered state (FP IV, 1970; FP V, 1990). The water content was 6.5 percent, 6.0 percent, and 6.8 percent respectively.

Furanocoumarin complexes were isolated by continuous extraction in a closed system, the Soxhlet apparatus, with petroleum ether, paraffin ether, dichloromethane, and their mixtures used as extracting solvents. The crude extracts obtained after solvent evaporation were purified by fractional crystallization (Glowniak et al., 1991). The furanocoumarins thus obtained were filtered, washed with heptane, dried, weighed, and finally concentrated in a rotary evaporator (Type 350, Unipan, Poland) (see Table 4.2). The residue was placed in a cooler for two days to cool down to –5°C. The crystallized furanocoumarins were filtered, washed with heptane, and then dissolved in the mixture of solvents containing n-heptan-dichloromethane-diisopropyl ether (5:47:3, v/v). The solution was applied to a silica gel chromatographic column, 200 × 500 mm (Kiesegel, Merck, 40-60 mm grain diameter). The column was conditioned by mobile phase of the composition as above, and eluted with a dichloromethane-diisopropyl ether gradient (47 percent, 57 percent, and 77 percent). The elution process was then completed by pouring dichloromethan-diisopropyl ether (97:3, v/v) through the column. The mobile phase flow rate was 6 ml/min. The four furanocoumarin fractions obtained in this way from *Archangelica officinalis* fruits were designated as AO_1, AO_2, AO_3 and AO_4.

For direct isolation of pure furanocoumarin fractions from *Archangelica officinalis* fruits, supercritical fluid extraction (SFE) has recently been used. Supercritical furanocoumarin extraction from *A. officinalis* fruits was performed on a micropreparative scale, with CO_2 fluid in the supercritical state as extractant according to the method by Gawdzik et al. (1994, 1995, 1996). With that method it was possible to extract furanocoumarins from about 0.5 g of source material. SFE by the micropreparative method was performed at pressures of 160 and 260 bar, at 40, 48, and 60°C, at a flow rate of 0.8-1.4 ml/min, and with a CO_2 fluid density of 0.3-0.8 g/l (Table 4.3).

Apart from that, SFE in the preparative scale was performed with the use of a thermostatic extractor on a water bath, vol. 350 ml. For examination we used a sample of 147 g of medium-ground dried fruits of *A. officinalis*, a flow rate of 75-85 g CO_2/h, temperature 40°C, pressure 100 and 200 bar, modified by ethyl alcohol used at the dose of about 50 g at a flow rate of about 0.5 g/h, which gave ARC1, ARC2, and ARC3 fractions.

Both the qualitative and quantitative extract compositions were determined by high performance liquid chromatography (HPLC) (Table 4.3a). Individual fractions and furanocoumarin complexes were analyzed by micro-HPLC on a chromatograph, type 100 SLK (Selko Inv. Warsaw) with a UV 254 nm detector and computerized system of data selection and analysis (CHROM 1). Analysis parameters: Altech column C18, 250 × 1mm, dp = 5 μm, mobile phase: methanol + water (65 + 35, v/v), mobile phase flow rate 50 μl/min. Furanocoumarins were identified by comparing the retention times of standard peaks and model substances. Concentrations were determined on the basis of the dependence of the peak area upon the concentration of the model substance (Kedzia et al., 1996; Wolski et al., 1996; Wolski, Glinski, Buczek, et al., 1997; Wolski, Glinski, Holderna-Kedzia, et al., 1997). Quantitative and qualitative compositions of the furanocoumarin complexes that occur in garden archangelica and *Heracleum sosnowskyi* Manden and of four fractions AO_1-AO_4 obtained by a preparative method from *A. officinalis* fruits (Tables 4.3a, 4.4) were determined by the HPLC method.

Garlic extracts (GE) were obtained by repeated maceration and extraction of garlic bulbs with plant oil, or by pressing juice out of fresh garlic bulbs (*Allium sativum* and *A. ursinum*). Garlic preparations (GP) from *A. sativum* were obtained with the use of *Oleum thymi*. GP used in the examinations contained 5 percent of *Oleum thymi*. GE components were identified by the HPLC method (Europaisches Artzneibuch Nachtrag, 1998).

MICROBIOLOGICAL EXAMINATIONS

The examinations used furanocoumarin extracts from *A. officinalis* of known in vitro activity upon dermatophytes isolated from humans and animals, such as *Trichophyton mentagrophytes, T. verruscosum,* and *Microsporum canis,* as well as *Candida albicans, Pityrosporum pachydermatis, Aspergillus niger,* and *Ascophaera apis,* the etiological factor of cystoid mycosis in helminth.

The values of minimum inhibitory concentrations (MIC) and minimum fungicidal concentration (MFC) of furanocoumarin extracts for *Ascopharea apis* strains were determined according to Fritz (1976) and for dermatophytes according to Ryley (1990) and Wawrzkiewicz et al. (1990).

The subsequent experiments used eight dermatophyte strains from clinical material, including two strains of *Trichophyton mentagrophytes,* one strain of *T. mentagrophytes* var. *granulosum,* one *T. mentagrophytes* var. *interdigitalis,* one *T. mentagrophytes* var. *asteroides,* one *T. tonsurans,* and two strains of *Microsporum gypseum* (Kedzia et al., 1996).

The experiments were carried out on solid (agar) Sabouraud's medium (Difco). Furanocoumarin complexes and fractions were dissolved in DMSO, 10 mg/ml. Further dilution was on Petri plates, 10 cm in diameter, on which 10 ml of agar medium containing 5, 10, 25, 50, 75, 100, 125, and 250 mg furanocoumarin in 1 ml were placed. Dermatophyte cultures were maintained on Sabouraud's agar medium and were suspended in physiological NaCl solution. One milliliter of suspension contained about 5×10^6 of live mycelium and spores. Later, dermatophyte suspensions were cultured on the surface of agar-medium-containing particular furanocoumarin concentrations by the linear method. MICs of the examined substances were determined after four to seven days of plate incubation at 37°C.

The influence of different surrounding conditions (temperature and storage time for antimycotic activity of the juice obtained from garden garlic bulbs [*A. sativum*] was determined by examination of its activity depending on the juice storage time at different temperatures: –20, 4, 20, and 37°C, at elevated temperatures (56, 70, and 100°C) and at different pH values (5-9). Apart from that, the examination evaluated the influence of additives, such as cysteine or dithiothreitol, upon the antimicrobial activity of the resulting juice. For the sake of comparison, a similar examination was conducted with the juice from *A. ursinum*. The examination determined maximal inhibitory dilution (MID) by the Mueller Hinton technique on agar medium.

CLINICAL STUDIES

Clinical studies evaluating the antimycotic activity of furanocoumarin and garlic preparations were carried out in the Dermatological Clinic of the Medical University, Lublin, Poland. They included both in vitro tests and clinical applications (Kuczynska et al., 1992). In vitro tests used strains isolated from patients' pathologically changed skin and nails. The activity of tested preparations was evaluated by the serial dilution method: 1:10, 1:100, 1:200, 1:400, 1:600, and 1:1,000, with the culture in Sabouraud's fluid medium composed of peptone 10.0 g/1,000ml, glucose 40.0 g/1,000 ml, actidione 500 mg/1,000 ml, and detreomycin 50 mg/1,000 ml. The concentrations of antimycotic preparations were 0.5-500 mm/ml. For the sake of comparison, the sensitivity of the tested strains to griseofulvin of a similar dilution was determined. The control group included analogous media without fungistatic additives. The cultures were incubated at 37° C for three to seven days, i.e., the time necessary to obtain visible growth in control tests. MICs of the preparations were determined. The lowest concentration of active substance at which no increase of fungus growth was observed was assumed to be the MIC value (Table 4.7).

The clinical applications concerned dermatomycoses and onychomycoses. In the treatment of dermatomycoses 1 mg/ml furanocoumarin preparation was applied to nine patients and 1.5 mg/ml preparation in 15 patients. To treat onychomycoses, a 2 mg/ml preparation in 50 percent DMSO solution was applied: fingernail mycoses were treated in seven patients and toenail mycoses were treated in eight patients. The examination covered 39 patients, aged 17-68 years, 21 men and 18 women. Pathological skin foci and the free edge of the affected nail plate were brushed twice daily. Periodically, once a week, the softened, cornified mass was mechanically removed from under the detached nail plate and from the surface of the exposed nail bed.

Examinations in vitro used oil garlic extracts (GE) and garlic preparations (GP) containing 0.5 percent addition of *Oleum thymi*. The activity of the tested preparations was evaluated by serially increased dilutions: 1:10, 1:100, 1:1,000 and 1:10,000. The control group included analogous media with no fungistatic additives. MICs of the examined preparations were determined. The lowest GE or GP concentration at which no macroscopic fungus growth was observed was assumed to be the MIC.

The study group consisted of 10 patients, 6 women and 4 men, 22-64 years old (mean age 49.8 years). All patients had had their clinical diagnosis confirmed by direct and cultured mycological examinations. The pathogeny were cultured on Sabouraud's medium with chloramphenicol and actidione. Clinically the patients presented with *Tinea pedium* var. *interdigitalis*. The following pathogens were recovered: *Trichophyton rubrum* in three patients; T. mentagrophytes var. *interdigitalis* in two patients; *T. mentagrophytes* var. *granulosum* in three patients; *Epidermophyton floccosum* in two patients. The affected areas were rubbed with GE or GP twice daily.

RESULTS

Owing to the significant activity in vitro, absence of toxic effects at therapeutic concentrations, and the need for active and safe antimycotic drugs, *Archangelica officinalis* L. may be viewed as a potential source of active substances that can be used to produce a local antimycotic. Previously industrial cultivation had been used to grow *A. officinalis* only for its roots, a source of volatile oils. It now seems purposeful, however, to grow that plant for its fruits, which contain large amounts of furanocoumarins (4-5 percent) that could serve as a source of coumarin antimycotic preparations (Kedzia et al., 1996; Wolski et al., 1996a).

The applied extraction method, which is simple and provides reproducible results, makes the production of biologically active furanocoumarin complexes possible. Table 4.1 presents their structures.

The results of comparative studies on the kinesis of the furanocoumarin mixtures extracted from *A. officinalis* fruits by organic solvents in the Soxhlet apparatus are presented in Table 4.2. The results of extraction by CO_2 fluid in the supercritical state (SFE) on the micropreparative scale are assembled in Tables 4.3a and 4.3b.

The extraction of crude coumarin complexes is more effective when performed in the Soxhlet apparatus with the use of mixed solvents, such as paraffin ether containing 10 percent and 30 percent dichloromethane, and extracting petrol with the addition of 20 percent and 50 percent dichloromethan and a three-component extractant composed of extracting petrol + dichloromethane and diisopropyl ether (7+2+1, v/v) (Table 4.2). The percentage effectiveness of extraction with mixed solvents ranged from 3.262 to 4.061 percent. The most effective extractant was paraffin ether containing 30 percent dichloromethan; the extraction product amounted to 4.061 percent of the original material used. The kinetic studies showed that an optimal and sufficient extraction time of crude coumarin complexes was three hours. During that time, 93.4 to 96.1 percent furanocoumarin complexes were obtained from *A. officinalis* fruits.

By means of fraction of extraction with CO_2 fluid in the supercritical state by the SFE method on the micropreparative scale, the content of xanthotoxin, bergapten, and imperatorin in individual fractions of the furano-

TABLE 4.2. Weight of furanocoumarin complexes obtained from the *Archangelica officinalis* Hoffm. fruits in the Soxhlet apparatus using different solvents

Extraction solvents	3 h extraction (mass, g)		Exhausting extraction (mass, g) (%)	
	mass (g)	%	mass (g)	%
EN + 10% CH_2Cl_2	1.9574	3.262	0.0828	0.138
EN + 20% CH_2Cl_2	2.4368	4.061	0.0912	0.152
BE + 30% CH_2Cl_2	2.1952	3.659	0.0544	0.090
BE + 50% CH_2Cl_2	2.3562	3.925	0.0780	0.130
BE + CH_2Cl_2 + iPr_2O 7+2+1	2.3810	3.858	0.0882	0.145
Heksan	0.3800	0.640	0.4000	0.670
BE	1.0600	1.430	0.6600	1.110
EN	1.6600	2.770	0.3200	0.540
CH_2Cl_2	1.2600	2.100	0.7500	1.230
$CHCl_3$	0.9900	1.650	0.2800	0.470
CCl_4	0.3900	0.640	0.5400	0.840

EN = ether; BE = extr. petrol; iPr_2O = isopropylene ether

coumarin complex was determined for different extraction parameters, e.g., pressure, temperature, CO_2 flow rate, and density. The results of extraction by the SFE method for three typical furanocoumarins were compared with those obtained through exhaustive extraction with chloroform (Table 4.3a). Imperatorin appeared to be the dominant furanocoumarin in all tested fractions obtained by SFE independently of supercritical extraction conditions, as well as in the furanocoumarin complex obtained by continuous extraction with chloroform. Optimal micropreparative supercritical extraction parameters: pressure 200 bar, 60°C, CO_2 flow rate 0.8 ml/min, CO_2 density 0.8 g/ml. Most of the furanocoumarins appeared in the first ten fractions obtained, and imperatorin was the dominant component (Table 4.3a).

SFE on the preparative scale was performed over a period of 25 h on 147 g of material *(A. officinalis)* at 40°C, pressure 200 bar. Three furanocoumarin complex fractions were obtained: ARC1, ARC2, and ARC3. Analysis of the individual extract masses and percentage content of the furanocoumarin complex revealed 0.867 g of ARC1, of which furanocoumarins constituted 64.7 percent; 6.276 g of ARC2, furanocoumarin content 47.5 percent; and 8.437 g of ARC3, furanocoumarin content 20.7 percent. Those values imply that the optimal fraction with the biggest

TABLE 4.3a. Content of furanocoumarins in individual fractions of extract obtained from *A. officinalis* fruits by SFE macropreparative method and by extraction with CO_2

Substance	Furanocoumarin in individual fractions (mg/g source material)				Extraction parameters
	0-10	10-20	20-30	Σ	
Xanthotoxin A	A 1.39 B 1.05 C 7.15	1.66 1.23 8.83	1.89 1.96 7.15	4.49 3.27 26.37	p = 160 atm v = 1.4 cm^3/min t = 40°C d_{CO2} = 0.3 g/dm^3
	A 2.03 B 2.30 C 3.21	0.87 1.43 8.83	0.62 0.12 0.49	2.95 3.85 22.01	p = 160 atm v = 1.0 cm^3/min t = 40°C d_{CO2} = 0.3 g/dm^3
Bergapten B	A 3.48 B 7.90 C 22.40	0.47 1.04 6.00	0.05 0.03 0.17	3.97 3.85 28.49	p = 160 atm v = 1.0 cm^3/min t = 48°C d_{CO2} = 0.75 g/dm^3
	A 2.87 B 3.08 C19.92	0.75 1.24 8.54	0.05 0.05 2.91	3.67 4.37 31.75	p = 260 atm v = 1.4 cm^3/min t = 60°C d_{CO2} = 0.8 g/dm^3
Imperatorin C	A 3.95 B 4.34 C 29.45	0.00 0.00 3.01	0.00 0.00 0.00	3.67 4.37 31.75	p = 260 atm v = 0.8 cm^3/min t = 60°C d_{CO2} = 0.8 g/dm^3
	A 4.34 B 2.88 C 24.38			4.34 2.88 24.38	Exhausting extraction with chloroform

TABLE 4.3b. Carbon dioxide supercritical fluid extraction on preparative scale of *A. officinalis* fruits (ARC); raw fruit weight: 147 g, CO_2 flow about 80 g/h

Extract Symbol	Extraction time (h)	Extraction parameters	CO_2 use (g)	Extract mass (g)	Furanocoumarin content (%w)
ARC 1	5	40°C/100 atm	451	0.867	64.7
ARC 2*	10	40°C/200 atm	727	6.276	47.5
	10	40°C/200 atm	792	8.437	20.7
ARC 3	Σ = 25	+EtOH 10%	Σ = 1970		

*Composition of furanocoumarin complex present in the optimal extract ARC2: xanthotoxin (5.6 percent); pimpinellin (1.1 percent); bergapten (3.7 percent); imperatorin (81.7 percent); isopimpinellin (6.9 percent); isoimperatorin (1.0 percent).

furanocoumarin content was ARC2. The addition of ethanol as a modifier increased the extraction effectiveness to 8.473 g; however, the furanocoumarin content of the crude extract decreased to 20.7 percent (Table 4.3b).

The use of HPLC to analyze and evaluate the quantitative and qualitative composition of the crude furanocoumarin extract obtained by both extraction methods from *A. officinalis* (continuous extraction by organic solvents and SFE) made it possible to compare separately the activities of xanthotoxin, bergapten, pimpinellin, isopimpinellin, and isoimperatorin from furanocoumarin complexes.

SFE is a much more beneficial method as regards the purity of furanocoumarin complexes obtained by fractionated extraction compared with chloroform extraction. Chloroform extracts contain inactive compounds both before and after the desired furanocoumarin complex. Thus, the SFE method augurs well for obtaining pure furanocoumarin complexes with a low content of inactive substances and no solvent contaminants.

The furanocoumarin complex of *A. officinalis* fruit was divided by means of column chromatography into four fractions (AO_1-AO_4). An analysis of the percentage content of typical furanocoumarins in *A. officinalis* and *H. sosnowskyi* complexes and in *A. officinalis* fractions is shown in Table 4.4. The furanocoumarin fractions and complexes examined contained psoralen and its derivatives, i.e., bergapten, xanthotoxin, isopimpinellin, and imperatorin, as well as derivatives of angelicin, i.e., isobergapten and pimpinellin. The furanocoumarin complex from *A. officinalis* contained mainly imper-

TABLE 4.4. Furanocoumarin content (percent) in complexes and fractions obtained from *Archangelica officinalis* Hoffm. and *Heracleum sosnowskyi* Manden fruits

Compound	Furanocoumarin complexes		Furanocoumarin fractions from *A. officinalis*			
	A. officinalis	*H. sosnowskyi*	AO_1	AO_2	AO_3	AO_4
Psoralen	trace		37.3			
Bergapten	7.3	63.2	22.5		86.3	79.2
Xanthotoxin	8.2	12.4	2.2		4.1	12.7
Isopimpinellin	0.5	4.1	6.7			
Imperatorin	73.8	14.8		96.6	0.7	
Isobergapten	trace			3.4		
Pimpinellin	1.2					
Unidentified	9.0	16.7	32.3		8.9	8.1

atorin (73.8 percent) and the *H. sosnowskyi* complex contained bergapten (63.2 percent). In fraction AO_2 from *A. officinalis* imperatorin (96.6 percent) was the predominant component, and in fractions AO_3 and AO_4 it was bergapten (86.3 percent and 79.2 percent respectively). In fraction AO_1 there were almost equal quantities of psoralen, bergapten, and unidentified furanocoumarins (Table 4.4).

The studies of fungicidal activity showed that the fungistatic effect of plant extracts depended on the kind of plant material used containing furanocoumarins *(A. officinalis, H. sosnowskyi)* or compounds of sulphur *(A. ursinum, A. sativum),* on the extraction and division methods, and on the susceptibility of the mycotic strains used in reference studies.

The values of MIC and MFC for fungi belonging to the *Trichophyton, Microsporum, Candida, Pityrosporum, Aspergillus,* and *Ascosphaera apis* species are presented in Table 4.5. The fungistatic effect of *A. officinalis* extracts clearly depends on the species of fungus, and dermatophytes appear to be very sensitive. The MIC value for *Microsporum canis, Trichophyton verrucosum,* and *T. mentagrophytes* ranges from 6.25 to 12.5 mg/ml, and for *Ascosphaera apis* it is 2.0 μg/ml (Wawrzkiewicz et al., 1990; Wolski et al., 1993, 1996; Kedzia et al., 1996; Glinski and Wolski, 1997). The studies of antimycotic activity of extracts of furanocoumarin complexes from *A. officinalis* and *H. sosnowskyi* and fractions AO_1-AO_4, whose results are presented in Table 4.6, indicated that fraction AO_2 shows the greatest activity against yeasts, molds, and dermatophytes.

TABLE 4.5. The values of MIC (μg/ml) and MFC (μg/ml) of the extracts from *Archangelica officinalis* Hoffm., amphotericin B, and clotrimazole

Fungi	*A. officinalis*		Amphotericin B		Clotrimazole	
	MIC	MFC	MIC	MFC	MIC	MFC
T. mentagrophytes 1	12.5	100.0	125.0	125.0	3.9	31.0
T. mentagrophytes V85/51	12.5	100.0	3.9	7.8	1.95	1.95
T. mentagrophytes var. *granulosum*	12.5	100.0	15.6	15.6	0.9	3.9
T. verrucosum	6.25	25.0	7.8	7.8	0.9	0.9
Microsporum canis	6.25	50.0	7.8	15.6	0.9	0.9
Candida albicans	250.0	1000.0	3.9	3.9	0.9	7.8
Pityrosporum pachydermatis	250.0	1000.0	3.9	3.9	1.95	3.9
Aspergillus niger	25.0	100.0	3.9	3.9	1.95	31.0
Ascophaera apis	2.0	5.0	238.0	350.0	–	–

TABLE 4.6. Activity of furanocoumarin complexes and fractions obtained from *Archangelica officinalis* Hoffm. and *Heracleum sosnowskyi* Manden fruits on dermatophytes

Dermatophytes	Furanocoumarin complexes		Furanocoumarin fractions from *A. officinalis*				Imperatorin
	A. officinalis	*H. sosnowskyi*	AO_1	AO_2	AO_3	AO_4	
T. mentagrophytes M4	10*	75	50	<5	<5	<5	25
T. mentagrophytes J3	50	50	100	<5	25	25	25
T. mentagrophytes var. *granulosum* 3M	50	100	100	10	25	25	25
T. mentagrophytes var. *interdigitalis* M15	50	75	50	<5	<5	<5	10
T. mentagrophytes var. *asteroides* J9	50	50	100	<5	50	25	10
T. tonsurans 5M	50	75	100	<5	10	<5	25
M. gypseum K1	10	50	<5	<5	10	25	10
M. gypseum K2	10	100	100	10	10	10	10

*MIC (μg/ml)

The results collected in Table 4.6 show that furanocoumarin complexes obtained from *A. officinalis* and *H. sosnowskyi* were active on dermatophytes in concentrations ranging between 10 and 100 μg/ml. A similar activity was demonstrated by fraction AO_1 from *A. officinalis,* whereas furanocoumarin fractions AO_2, AO_3, and AO_4 showed considerable antimycotic activity. They inhibited the development of the dermatophytes studied in concentrations 5-50 μg/ml, and fraction AO_2 was the most active, inhibiting the growth of dermatophytes in concentrations of 5-10 μg/ml. This fraction contained 96.6 percent of imperatorin and 3.4 percent of isobergapten (Table 4.4). By comparison, pure imperatorin showed average antimycotic activity: it checked the growth of dermatophytes in concentrations from 10 to 25 μg/ml (Table 4.6).

The results of the in vitro studies aimed at determining the MIC values of furanocoumarin preparations and griseofulvin for 11 strains of fungi are presented in Table 4.7a. The activities of the furanocoumarin preparation and griseofulvin were very similar in most strains studied.

On the strength of the studies in vitro, attempts were made to treat some varieties of mycoses with a furanocoumarin preparation from *A. officinalis*

TABLE 4.7a. The value of the minimum inhibitory concentration (MIC) of furanocoumarin extracts and griseofulvin

Dermatophytes	Number of strains	Concentration of the preparation in µg/1 ml of the liquid medium	
		Furanocoumarin extract	Griseofulvin
T. mentagrophytes var. *granulosum*	1	100	10
T. mentagrophytes var. *granulosum*	2	100	100
T. mentagrophytes var. *interdigitalis*	2	10	10
T. mentagrophytes var. *interdigitalis*	1	100	10
T. rubrum	2	100	100
Epidermophyton floccosum	3	10	10

applied externally. The results of the treatment of dermato- and onychomycoses with furanocoumarin extracts are presented in Table 4.7b.

Among the 24 patients with dermatomycoses, there were nine cases of *Tinea interdigitalis* (athlete's foot), six cases of ringworm of the feet and hands, one case of ringworm of the hands, four cases of *Tinea glabrosa,* one case of trichophytosis profunda, and three cases of *Tinea cruris.* The treatment period for patients with nail mycoses was 2-18 months (mean: 8 months). The treatment period for patients with dermatomycoses ranged from 10 to 28 days (mean 19 days). In the initial studies nine patients with dermatomycoses were treated with the coumarin preparation at a concentration of 1 mg/ml; the next 15 patients were similarly treated at 1.5 mg/ml, and the preparation with 2 mg/ml was used on nails.

The activity of furanocoumarins was similar to griseofulvin, with the exceptions of *Trichophyton mentagrophytes* var. *granulosum* and *Trichophyton mentagrophytes* var. *interdigitale* strains. An external solution with the concentration of 1.5 mg/ml proved effective in 80 percent of cases. A combination of *A. officinalis* extracts and dimethylsulphoxide (DMSO) enhances treatment, owing to easier penetration of the furanocoumarin complexes into tissues and the mycelium. It is especially useful in the treatment of nail mycosis, where a solution of a furanocoumarin complex in a 50 percent solution of DMSO with a concentration of 2mg/ml was used. A preparation on a furanocoumarin base exhibits low toxicity and is easy to prepare.

TABLE 4.7b. The results of the treatment of dermatophytoses and onychomycoses with furanocoumarin extracts

Dermatophytosis	Concentration (mg/ml)	Recovery		Without improvement	
		number	percent	number	percent
Cutaneous mycoses	1.0	3	33	6	67
	1.5	12	80	3	20
Nail mycoses of hand	2.0	3	43	4	57
Nail mycoses of foot	2.0	0	0	8*	100

*Negative laboratory analyses in two patients though clinical improvement was not observed

There has been a remarkable worldwide increase in the incidence of *Ascosphaera apis* affecting honeybees in the past few years (Glinski and Jarosz, 1988; Glinski et al., 1998). The observed increase is connected with the invasion of an external mite called *Varroa jacobsoni* Oud, which plays an important role as a vector of *A. apis* (Glinski and Jarosz, 1988; Glinski, 1988). Studies were undertaken to assess the fungicidal activity on *A. apis* and toxicity of furanocoumarin extract from the fruit of *A. officinalis*. The extract turned out to have only a low toxicity for the honeybee brood and the worker bees. The value of LD_{50} (Table 4.8), which exceeds over fiftyfold the MIC value, is only slightly higher than the minimum inhibitory concentration (Glinski et al., 1988). Because of such extensive therapeutic activity, the medicine is safe and overdosage will not bring about adverse side effects. The results of toxicity studies were confirmed by a safety test in which worker bees were sprayed with or fed a 0.5 percent saccharose solution with the addition of *A. officinalis* extract at a concentration of 1:100-1:1,000. Neither spraying nor feeding adversely affected the queen's brood, the numbers of the brood, or its rearing. Moreover, the tested solutions of saccharose containing the extract did not repel the bees. The results obtained gave a basis for a clinical evaluation of the preparation, which was performed outside the laboratory during the treatment of bees infected with *A. apis*.

The extract from the fruit of *A. officinalis* produces a fungistatic and fungicidal effect on *A. apis* in vitro, due to the presence of furanocoumarins in suitable proportions. The values of MIC and MFC are 1.5-2.0 μg/ml and 4.5-5.0 μg/ml, respectively (Table 4.9). The activity of the extract is greater than those of nystatin, amphotericin B, and griseofulvin, as well as of the choline salt of N-glucosylpolyfungin and the furanocoumarin extracts from

TABLE 4.8. Toxicity of the extract from fruits of *Archangelica officinalis* Hoffm. and antifungal antibiotics for honeybee brood and worker bees (µg/insect)

Drug	LD_{50}	
	Brood	**Worker bees**
Extract from *A. officinalis*	145.5	125.0
Nystatin	128.0	93.0
Amphotericin B	190.0	118.0
Choline salt of *N*-glucosylpolyfungin	320.0	210.0

TABLE 4.9. Effects of the extracts from *A. officinalis, P. sativa, L. intermedia,* and two antimycotics on *Ascophaera apis* (brood and worker bees)

Drug	MIC (µg/ml)	MFC (µg/ml)	LD_{50} (µg/insect)	
			brood	**workers**
A. officinalis	1.5 - 2.0	4.5 - 5.0	145.0	125.0
P. sativa	35.0 - 38.0	41.0 - 49.0	310.0	280.0
L. intermedia	25.0 - 40.0	400.0	105.0	80.0
Nystatin	328.0 - 340.0	350.0	128.0	93.0
Choline salt of *N*-glucosylpolifungin	5.5 - 15.0	10.0 - 20.0	320.0	210.0

the fruit of *Pastinaca sativa* and *Libanothis intermedia* used by way of comparison (Glinski et al., 1988; Glinski and Wolski, 1996). As a result of a small difference between the values of MIC and MFC, the extract in the applied concentrations is well tolerated by both bees and brood. It produces a disinfecting effect, destroying the *A. apis* contaminating the inside of the hive and the combs.

One year after the treatment there were relapses of the disease in 3.5 percent of the treated families. These were most probably caused by mycelium and spores of *A. apis* persisting in the combs. What is more, the fungus can survive outside the bee and brood in pollen reserves and combs for several years. It is to be expected that eradication of the source of infection in the hive, i.e., the mummified brood, the combs, and the honey and pollen reserves infected with *Ascosphaera apis,* will increase the effectiveness of treatment and limit possible relapses to a minimum.

With a constantly growing incidence of a large variety of mycoses, it seems advisable to look for new antimycotic drugs of vegetable origin. One of them is common garlic (*Allium sativum* L.), a plant containing a large number of compounds with antibacterial and antimycotic properties. The experiments carried out by Tokin (1953), now of historical interest only, proved that fresh pulp of *A. sativum* inhibits the growth of numerous dermatophytes and yeastlike fungi. Since that time the antimycotic properties of *A. sativum* have been repeatedly stressed in the literature (Tansley and Appelton, 1975; Maleszka et al., 1991).

Studies in vitro of fresh juice obtained from two varieties of garlic, i.e., common garlic *(Allium sativum)* and *A. ursinum*, collected in natural locations showed a comparable efficiency in inhibiting the growth of mycelium of some strains examined. The inhibition of growth occurred with identical high dilutions MID (maximum inhibitory dilution) of 1:1,200 of juice added to the nutrient, which successfully inhibited the growth of the strains of *Candida albicans* NCTC 10716, *Geotrichum candidum, Aspergillus fumigatus, Penicillium* spp., *Trichophyton mentagrophytes,* and *T. rubrum*. With the same dilution MID = 1:8,000 of both examined juices growth inhibition was observed in *Candida albicans, Scopulariopsis* spp., and *Trichophyton galinnae* (Table 4.10).

The studies showed a considerably higher activity of the fresh juice of *A. ursinum* compared with *A. sativum* on the strains of *Cryptococcus neoformans* and *Microsporum gypseum*. In the case of *Epidermophyton floccosum* strains, the activity of *A. ursinum* was twice that of common garlic.

TABLE 4.10. A comparison of maximum inhibitory dilution (MID) of extracts from *A. sativum* and *A. ursinum* for different species of fungi

Fungi	**MID (Maximum Inhibitory Dilution)**	
	A. ursinum	***A. sativum***
Candida albicans	1: 800	1: 800
C. albicans NCTC 10716	1: 200	1: 200
Cryptococcus neoformans	1:1,200	1: 800
Geotrichum neoformans	1:1,200	1:1,200
Aspergillus fumigatus	1:1,200	1:1,200
Penicillium spp.	1:1,200	1:1,200
Scopulariopsis spp.	1: 800	1: 800
Trichophyton mentagrophytes	1:1,200	1:1,200
T. gallinae	1: 800	1: 800
T. rubrum	1:1,200	1:1,200
Epidermophyton floccosum	1:1,200	1: 600
Microsporum gypseum	1: 600	1: 400

Examination of the influence of storage conditions (temperature) upon the activity of the juice revealed that it retained its activity when stored at –20°C for eight weeks. Juice stored at 20°C lost 60 percent of its activity, and at 37°C it was devoid of activity after eight weeks.

As a consequence of the studies in vitro indicating that inhibition of growth of a fungus was observed by macroscopy at a dilution of 1:100, clinical tests were performed. In the clinical evaluation of patients the following symptoms were noted: itching, exfoliation, and maceration of epidermis. A three-stage scale of clinical symptoms was established: mild, average, and intensified. A clinical and mycological evaluation was performed before treatment was introduced, again in the second and third week of treatment, and then two weeks after the end of treatment. We excluded patients who used topical antimycotics in the final two weeks of treatment (Kadziela-Wypyska et al., 1995a,b).

Among the intense clinical symptoms occurring before commencement of treatment were the following: pruritus in all patients, exfoliation of epidermis in eight, and maceration in four patients. Those symptoms subsided gradually over the course of treatment. After two weeks of treatment, only mild itching was observed in four patients, exfoliation in two, and maceration in one person.

In the third week of treatment mild itching persisted in this last patient, but maceration and exfoliation had subsided completely. Two weeks after the end of therapy there were neither relapses nor clinical symptoms in any patient, as confirmed by mycological examinations, both direct and cultures. In two cases we noted symptoms of secondary allergization, yet they subsided after the administration of general antiallergics (clemastin).

Positive results of direct examinations persisted in only two cases and the results of cultures were positive in one case. After three weeks of treatment the results of both direct examinations and cultures were negative for all patients.

DISCUSSION

Mycotic infection is a difficult problem of the skin that constitutes a considerable proportion of dermatological diseases and frequently presents a difficult therapeutic problem. Since these infections not only require treatment but also exclude the patients from professional activity for long periods of time, it is essential both to treat these conditions early and to prevent them (Walsh and Pizzo, 1988). The treatment of mycoses is complicated by considerable local or general toxicity of antimycotic drugs that sometimes occurs, weak penetration of the preparation into the tissues (Dismukes, 1991), and a limited spectrum of antifungal activity. The existence of drug-resistant dermatophyte strains makes it necessary to search for new effective medicinal products.

There is a growing interest in the possibilities of combining antifungal preparations on the basis of polytherapy, replacing monotherapy to obtain new, synthetic, or natural preparations that would be both effective against fungi and safe to use.

The results obtained are fairly promising in view of the possible applications of furanocoumarin complexes and fractions in dermatological practice. This applies in particular to the furanocoumarin fraction AO_2 obtained from the fruit of *A. officinalis*. Its fungicidal activity may be compared with that of the antibiotics used to fight dermatophyte infections, such as naftifin, clotrimazole, etc. (Nolting and Seebacher, 1993; Macura et al., 1995) (MIC for all dermatophyte strains 100 μg/ml), and surpasses the antibiotic activity of griseofulvin (MIC for 16 percent of dermatophyte strains 100 μg/ml) and printaricin (MIC for 10 percent of dermatophyte strains 100 mg/ml) (Macura et al., 1995).

Fraction AO_2 contains 96.6 percent of imperatorin and 3.4 percent of isobergapten. Imperatorin alone did not display such strong fungicidal properties; in our studies it inhibited the growth of dermatophytes at a rate of 10-25 μg/ml, whereas fraction AO_2 was active against dermatophytes at concentrations of 5 to 10 μg/ml. These observations prompt the conclusion either that the potency is due to isobergapten or that the two furanocoumarins act on dermatophytes synergistically. The problem requires further research.

As with the furanocoumarin-containing *A. officinalis* extract previously examined (Wawrzkiewicz et al., 1990), other furanocoumarin fractions and complexes described in this chapter also deserve attention. The inhibition of the growth of dermatophytes at concentrations $\geq$ 100 μg/ml means that their antimycotic activity is comparable to, and in some cases even higher than, that of antibiotics used in the treatment of dermatophytoses (Macura et al., 1995).

It is worth stressing again that attempts to treat some varieties of mycoses with an extract of *A. officinalis* containing furanocoumarins should be made (imperatorin 60.0 percent, xanthotoxin 9.7 percent, and bergapten 2.9 percent). Out of 15 cases of dermatomycosis caused by dermatophytes, recovery was observed in 12 cases (80 percent), and out of seven cases of onychomycosis of the hands recovery was observed in three cases (43 percent). Moreover, the extract did not show any toxicity and was well tolerated by the patients (Kuczynska et al., 1992). It is more remarkable that many synthetic drugs used in the therapy of mycoses (amorolfin, griseofulvin, cyclopirox, tolnaftate) produce adverse side effects, such as irritation and itching of the skin (Nowicki, 1995).

According to research by Glinski et al. (1988), coumarin fractions obtained from the fruit of *A. officinalis* are also highly effective against *Ascosphaera apis,* are practically nontoxic for the brood and bees, and, what is more, do not repel the bees. These results suggest a possible application of furanocoumarin extracts in the treatment of this disease *(A. apis),* and they form a basis for designing a method of obtaining a fungicidal preparation for bees (Wolski et al., 1992). Compounds displaying therapeutic effectiveness on a particular species in the therapeutic dose combined with immunostimulating properties have vital importance in fighting mycoses. This postulate is fulfilled to a large extent for bees with imidazole derivatives and furanocoumarins which, acting as cellular and humoral immunostimulators, enhance the defense powers of a bee family in fighting mycoses.

The different activities of *A. officinalis* extracts containing furanocoumarin fractions may be the result of variations in the content of individual furanocoumarins (bergapten, xanthotoxin, imperatorin, and others) in the fractions and extracts studied. Hence, there is a need for choosing optimal conditions for the extraction of plant material to obtain the most active fractions or extracts.

Common garlic *(Allium sativum),* which has been considered the most popular plant in the world since antiquity, is a potential source of numerous biologically active compounds with antimycotic properties. Apart from being antifungal and antibacterial, garlic also acts synergistically in combination with commonly used antibiotics (Maleszka et al., 1991; Kadziela-Wypyska et al., 1995a,b). The substances contained in garlic act many times more powerfully on certain species of fungi than on bacteria (Tansley and Appelton, 1975; Maleszka and Maleszka, 1988; Maleszka et al., 1991). As demonstrated in clinical tests, the oil extract of *A. sativum* is highly effective in the treatment of dermatomycosis pedis (Kadziela-Wypyska et al., 1995a,b). Thus, it seems essential to continue the research on a larger number of clinical cases, as well as to introduce garlic preparations with the addition of other substances, such as furanocoumarins or essential oils, to the treatment of dermato- and onychomycoses.

Evaluating the results of the conducted studies, we should like to stress the effectiveness of the *Allium sativum* preparation in the treatment of athlete's foot. It seems necessary to continue research on a larger number of cases, as well as to apply garlic preparations containing a small amount of other substances, such as furanocoumarins. A beneficial activity of *A. officinalis* furanocoumarins has already been reported by Kuczynska et al. (1992) and others (Glinski et al., 1988; Wawrzkiewicz et al., 1990).

The lipophilic properties of furanocoumarin structures result in a low solubility in polar solvents and aqueous solutions, whereas pharmacological effects are related to a compound's solubility in body fluids. To take advantage of furanocoumarins' affinity for lipophilic solvents, vegetable fats and oils were used as vehicles for emulsions and ointments as external forms of antimycotic drugs. Thus, methods of obtaining new antimycotic preparations were found (Wolski et al., 1996a,b; Wolski and Kawka, 1997a,b).

CONCLUSIONS

The use of mixed solvents in the extraction of furanocoumarin complexes from the fruit of *Archangelica officinalis* Hoffm. gave the best results. Over 96 percent extraction efficiency of raw extracts was attained by means of solvent extractions with petroleum ether + dichloromethane and purified benzin + dichloromethane over three hours. Petroleum ether containing 30 percent of dichloromethane proved the most beneficial extractant.

Qualitative evaluation by means of high performance liquid chromatography (HPLC) of the composition of furanocoumarin complexes obtained by extraction with organic solvents in the Soxhlet apparatus and by supercritical fluid extraction with liquid carbon dioxide (SFE) resulted in a high purity of furanocoumarin complexes obtained by the SFE method.

The furanocoumarin complexes obtained by extraction with organic solvents show a fungicidal activity on human and animal dermatophytes and on *Ascosphaera apis*, a pathogen of the honey bee. The results of clinical tests confirmed the usefulness and effectiveness of preparations containing *A. officinalis* extracts in the treatment of human dermato- and onychomycoses and *A. apis* in honeybee.

Owing to its high activity against *A. apis*, low toxicity, and lack of repellent effect or contamination of the bees' products, the extract of the fruit of *A. officinalis* in the suggested doses applied as described previously may be recommended for the treatment of *A. apis*.

Clinical studies indicate a possible use of furanocoumarin complexes in the treatment of dermato- and onychomycoses. Development of a furanocoumarin preparation combined with oil extract of garlic would appear worthwhile.

The use of organic solvents mixed in the closed system of the Soxhlet apparatus for the extraction of the fruit of *A. officinalis* Hoffm. shortens the extraction time to three hours. The use of liquid carbon dioxide under supercritical conditions makes isolation of furanocoumarin complexes from the fruit of *A. officinalis* possible. On the micropreparative scale, the extracts obtained in this way are characterized by a higher purity, compared with extracts obtained with the use of organic solvents. A possibility of extracting furanocoumarin complexes from *A. officinalis* by means of the SFE method on a preparative scale was confirmed and optimal extraction parameters were selected. However, the content of furanocoumarin complexes observed (about 50 percent) indicates a need for further studies to improve this result.

The results obtained form a basis for further studies on a more extensive microbiological evaluation and on the technology of obtaining furanocoumarin antimycotic preparations. The fruit of *A. officinalis* and *H. sosnowskyi* may be considered the raw materials for obtaining furanocoumarin complexes and fractions. The activities of the latter fractions depend on the qualitative and quantitative composition. Clinical evaluation will allow a decision to be made about extending and completing the selection of antifungal drugs with furanocoumarin preparations.

The garlic preparation shows well-marked antimycotic activity. A three-week period of application seems sufficient to produce subsidence of symptoms, and is comparable to the period of application of other antifungal preparations. The applied preparations are stable and easy to use, and they produce no side effects.

REFERENCES

Black E., Naganathan S., Putman D., Zhao S.H. (1993). *Allium* chemistry: HPLC analysis of thiosulfinates from onion, garlic, wild garlic (ramsoms), leek, scallion, shallot, elephant (great-headed) garlic, chioe and Chinese chioe. Uniquely high allyl to methyl ratios in some garlic samples. *J Agric Food Chem* 40(12): 2418-2430.

Bourrel C., Vilarens G., Perineau F. (1993). Chemical analysis of bacteriostatic and fungistatic properties of the essential oil of elecampane (*Inula helenium* L.). *J Essent Oil Res* 5: 411-413.

Calvey E.M., Roach J.H.G. (1994). Supercritical fluid chromatography of garlic *(Allium sativum)* extracts with mass spectrometric identification of allicin. *J Chromatogr Sci* 32(3): 93-96.

Cisowski W. (1983). Biologiczne wlasciwosci kumaryn. Dzialanie na rosliny oraz wlasciwosci farmakologiczne i przeciwbakteryjne. *Herba Pol.* 29: 301-318.

Cisowski W. (1986). Badania fitochemiczne niektórych roslin leczniczych z rodziny Umbelliferae w zakresie pochodnych benzo-γ-pirenu. *Rozprawa habilitacyjna. AM Wroclaw,* 114 pp.

Dismukes W.E. (1991). The present status of antifungal treatment, with particular reference to systemic infections: A review. *Current Med Red Opinion Suppl* 12: 49-52.

Ericsson H.M., Sherris J.C. (1971). Antibiotic sensitivity testing. *Acta Pathol Microbiol Scand Suppl* 11, 217-219.

Europaisches Artzneibuch Nachtrag (1998). Deutsche Apotheker Verlag Studgrad, 470-471.

Farmakopea Polska IV (FP IV.) (1970). *PZWL*, Warszwa, t. II, 872 pp.

Farmakopea Polska V (FP V) (1990). *PZWL*, Warszwa, t. I, 172 pp.

Foil C.S. (1994). Fungal diseases. *Clinics in Dermatology* 12: 529-530.

Fritz R. (1976). Action qelques fungicides sur la croissnce mycelienne trois aspects d'entomophthoerales. *Entomophaga* 21: 239-241.

Gawdzik J., Kawka S., Mardarowicz M., Suprynowicz Z., Wolski T. (1994). Supercritical fluid extraction of biological active compounds from plants. I. The isolation and characterization of furanocoumarins from the fruits of *Archangelica off.* Hoffm. 16th International Symposium on Capillary Chromatography. Riva del Garde (Italy) September 27-30, pp. 1740-1749.

Gawdzik J., Kawka S., Mardarowicz M., Suprynowicz Z., Wolski T. (1995). Supercritical fluid extraction of furanocoumarins from the fruit of *Archangelica officinalis* Hoffm. *J High Resol Chromatogr* 18: 781-783.

Gawdzik J., Kawka S., Mardarowicz M., Suprynowicz Z., Wolski T. (1996). Carbon dioxide fractionated supercritical fluid extraction of furanocoumarins from the fruit of *Archangelica officinalis* Hoffm. *Herba Pol* 42(1): 26-35.

Georgijewskij W.P., Komissarenko N.F., Dmitruk S.F. (1990). Biologiczeski aktiwnyje wieszczestwa lekarstwiennych rastienij. *Nauka*, Nowosybirsk, pp. 113-115.

Glinski Z. (1988). The effect of *Varroa jacobsoni* on the incidence and course of chalkbrood disease in *Apis mellifera*. *Annales UMCS*, sec DD, 43: 23-32.

Glinski Z., Jarosz J. (1988). *Varroa jacobsoni* invasion and the level of cell-free immunity in upright larvae of the worker honey bee, *Apis mellifera* L. *Folia Vet* (Kosice) 32: 39-42.

Glinski Z., Rzedzicki J., Wolski T., Kauko L. (1998). Immunomodulatory effects of antifungal drugs and prospects in therapy of chalkbrood of the honey bee. *Annales UMCS* Sec. DD, 53(20): 201-216.

Glinski Z., Wolski T. (1996). Complexes of plant furanocoumarin useful in therapy of chalkbrood disease of the honey bee. *Umbelliferae Improv Newsletter* 6: 28-32.

Glinski Z., Wolski T. (1997). Plant extracts containing furanocoumarins in therapy of mycoses in man and in animals. *Annales UMCS* Sec. DD 52(16): 143-154.

Glinski Z., Wolski T., Chmielewski M. (1988). The "in vitro" studies on antifungal action or *Archangelia officinalis* Hoffm. seed extract against *Ascosphera apis*. *Med Wet* 44(9): 552-556.

Glowniak K. (1988). Badania i izolacja zwiazków kumarynowych z krajowych surowców roślinnych. *Rozprawa habilitacyjna*, AM Lublin, 230 pp.

Glowniak K., Wolski T., Dragan T. (1991). Sposób wyodrebniania kumaryn z surowców roślinnych zwlaszcza z owoców arcydziegla i olesnika. *Pat Pol* 152163, 1-5, *Med Univ Lublin*, Poland.

Honda G. (1984). Shoyakugaku Zasshi 38: 221. In *Med Aromat Plant Abstr* (1995) 5: 2714.

Hunter P.A., Darbyu G.K., Ryssell L.J. (Eds.) (1995). *New developments in non-azole antifungals for human disease.* Cambridge: Cambridge University Press.

Kabelik J. (1970). Antimikrobielle Eigenschaften des Knoblauch. *Die Pharmazie* 25: 266-267.

Kadziela-Wypyska G., Wolski T., Lecewicz-Torun B., Kawka S. (1995a). Antimycotic effectivity of the *Allium sativum* derivatives. *Materialy Ogólnopolskiej Konferencji Naukowej, Nauka w praktyce ogrodniczej* AR Lublin, 733-736.

Kadziela-Wypyska G., Wolski T., Lecewicz-Torun B., Kawka S. (1995b). Zastosowanie preparatów czosnkowych w leczeniu niektórych grzybic. *Materialy XVI Naukowego Zjazdu PTFarm.* 20-21.10.1995, Warszawa, 84.

Kaminski B., Glowniak K., Majewska A. (1978). Poszukiwanie zwiazków kumarynowych w owocach i nasionach. I. Owoce z rodziny baldaszkowatych *(Umbelliferae). Farm Pol* 34: 25-26.

Kedzia B., Siegied-Kujawa E., Holderna E., Krzyzaniak M. (1990). Chemical contents and antimicroorganism activity of sage essential oil *(Ol. Salviae). Herba Pol* 36(4): 155-164.

Kedzia B., Wolski T., Kawka S., Holderna-Kedzia E. (1996). Activity of furanocoumarin complexes and fractions from *Archangelica officinalis* Hoffm. and *Heracleum sosnowskyi* Manden. fruits on dermatophytes. *Herba Pol* 42(1): 47-54.

Konopacka-Brud I. (1995). Olejek drzewa herbacianego. *Aromaterapia* 1: 5-7.

Kowszyk-Gindifer Z., Sobiczewski W., (Eds.) (1986). Grzybice i sposoby jej zwalczania. *PZWL, Warszawa* 257-268.

Kuczynska L., Lecewicz-Torun B., Wolski T., Kosinska J., Kawka S. (1992). Experimental external treatment of some forms of mycosis with furano-kumarin extracts. *Postepy Dermatologii* 9: 149-156.

Lutomski J. (1987). Components and biological properties of some *Allium* species. *Institute of Medicinal Plants, Poznan,* 58 pp.

Lutomski J. (1995). Szanse ziololecznictwa. Od tradycyjnej medycyny do oficjalnej terapii. *Farm Pol* 51: 953-957.

Machowicz-Stefaniak Z., Pieta D., Wolski T. (1995). In vitro effect of garlic extract on growth of some phytopathogens. *Pestycydy* 3: 23-28.

Macura A.B., Pawlik B., Filip E. (1995). A comparison of glass slide microculture dilution method and disk-diffusion test applied in assessment of the susceptibility of dermatophytes to antifungal drugs. *Mikol Lek* 2: 5-10.

Maleszka R., Lutomski J., Swiatlowska-Górna B., Rzepecka B. (1991). Study on extending of the activity spectrum of a garlic preparation against candidasis. *Herba Pol* 37(2): 85-88.

Maleszka R., Maleszka A. (1988). New possibilities of prevention of occupational *Candida albicans* infection. *Przegl Dermatol* 75(6): 425-427.

Mukherjee P.K., Kahahi Saha, Saha B.P., Pal M., Das J. (1996). Antifungal activities of the leaf extract of *Cassia tora* Linn. *(fam. Leguminosae). Phytotherapy Research* 10(6): 521-522.

Murray R.D.H., Mendez J., Brown S.A. (1982). The natural coumarins. *Occurrence chemistry and biochemistry.* R.D.H. Murray (Ed.). New York-Brisbane-Toronto-Singapore: John Willey and Sons, 702 pp.

Nolting S., Seebacher C. (1993). Ciclproxolamin. Wegweister topischer Mycose-Therapie. *Universitstverlag GmbH*. Jena, 176 pp.

Nowicki R. (1995). Curent antifunglas in dermatology. *Mikol Lek* 2: 111-119.

Pandey D.K., Chandara H., Tripathi N.N. (1982). Volatile fungitoxic activity of some higher plants with special reference to that of *Callistemon lanceolatus* D.C. *Phytopathologische Zeischrift* 105(2): 175-178.

Peplonski R., Kwiatkowski J., Jarosz M., Lisiecki Z. (1994). Extraction of some components for flavouring, colouring and preservation of food using supercritical carbon dioxide. Proceedings of the Third International Symposium. *Supercritical Fluids*. Strassburg October 17-18, 2: 435-437.

Pieta D., Machowicz-Stefaniak Z., Wolski T., Kawka S. (1995). Furanocoumarine preparation and attempt to use it as a mean of controlling phytophatogens. *Pestycydy* 3: 29-36.

Ryley J.F. (1990). Screening and evaluation in vitro. In *Chemotherapy of fungal diseases* (Ed. J.F. Ryley). Berlin: Springer-Verlag, pp. 129-147.

Singh U.P., Prithiviraj B., Wagner K.G., Plank-Schumacher K. (1995). Effect of ajoene, a constituent of garlic *(Allium sativum)*, on powdery mildew *(Erysiphe pisi)* of pea *(Pisum sativum)*. *Zeitschrift fur Pflanzenkrankheiten und Pflanzenschutz* 102(4): 339-406.

Smith K.J., Warnock D.W., Kennedy C.T.C. (1986). Azole resistance in *Candida albicans J Med Vet Mycol* 24(2): 133-144.

Stahl E., Quirin K.W., Gerard D. (1986). *Dense gases for extraction and refining*. Berlin, Heidelberg, NewYork, London, Paris, Tokyo: Springer Verlag.

Tansley M., Appelton J. (1975). Inhibition of fungal growth by garlic extracts. *Mycologia* 67: 409-413.

Tokin B. 1953. *Fitoncydy. PWRiL*. Warsaw, 115 pp.

Venugopal P.V., Venugopal T.V. (1995). Antidermatophytic activity of garlic *(Allium sativum)* in vitro. *International Journal of Dermatology* 34(4): 278-279.

Walsh T.J., Pizzo A. (1988). Treatment of systemic fungal infections: Recent progress and current problems. *Eur J Clin Microbiol Inf Dis* 7: 461-464.

Wawrzkiewicz K., Wolski T., Ziólkowska G., Kawka S. (1990). Antifungal properties of the extracts from the fruits of *Archangelica officinalis* Hoffm. *Med Wet* 46(8): 289-292.

Wenclawik B. (1992). *Analysis with supercritical fluids: Extraction and chromatography*. Berlin, Heidelberg, New York, London, Paris, Tokyo, Hong Kong, Barcelona, Budapest: Springer-Verlag.

Witkiewicz Z. (1995). Podstawy chromatografii. *WNT*. Warsaw, 253 pp.

Wolski T., Glinski Z., Buczek K., Wolska A. (1996). Preparation and characteristic of furanocoumarin plant extracts of antifungal activity. *Herba Pol* 42(3): 168-173.

Wolski T., Glinski Z., Buczek K., Wolska A. (1997). Control of fungal diseases with plant extracts. *Herba Pol* 43(1): 47-52.

Wolski T., Glinski Z., Chmielewski M. (1993). Sposób otrzymywania preparatów zapobiegajacych powstawaniu grzybic oraz niszczacych *Varroa jacobsoni* Qud. *Pat Pol*, 157485 oraz 157486, *Med Univ Lublin*, Poland, 1-3; 1-3.

Wolski T., Glinski Z., Glowniak K., Chmielewski M. (1991). Sposób oznaczania aktywnosci grzybobójczej substancji otrzymywanych z surowców farmakognostycznych, a zwlaszcza z owoców arcydziegla, pasternaku i barszczu. *Pat Pol* 151877, *Med Univ Lublin,* Poland, 1-3.

Wolski T., Glinski Z., Glowniak K., Chmielewski M. (1992). Sposób otrzymywania preparatu grzybobójczego dla pszczól. *Pat Pol* 155126, *Med Univ Lublin,* Poland, 1-3.

Wolski T., Glinski Z., Holderna-Kedzia E., Rzedzicki J. (1997). Extraction methods of furanocoumarin production from plant materials. *Annales UMCS* Sec. DD, 52(17): 155-165.

Wolski T., Kawka S. (1997a). Sposób otrzymywania nowego preparatu grzybobójczego i nowy preparat grzybobójczy, *Pat Pol* 171060, 1-3, *Med Univ Lublin,* Poland.

Wolski T., Kawka S. (1997b). Sposób otrzymywania nowego preparatu grzybobójczego i nowy preparat grzybobójczy, *Pat Pol* 171521, 1-4, *Med Univ Lublin,* Poland.

Wolski T., Kawka S., Wolski J. (1996a). Sposób otrzymywania nowego preparatu przeciwgrzybiczego i nowy preparat przeciwgrzybiczy, *Pat Pol* 169921, 1-4, *Med Univ Lublin,* Poland.

Wolski T., Kawka S., Wolski J. (1996b). Sposób otrzymywania nowego preparatu przeciwgrzybiczego i nowy preparat przeciwgrzybiczy, *Pat Pol* 169922, 1-4, *Med Univ Lublin,* Poland.

Chapter 5

Dermatophytes As a Tool in the Discovery of New Natural Substances to Be Used As Antimycotics: Use of *Microsporum cookei* As a Vehicle for Reflections on the Protoanemonin Action Mechanism

Donatella Mares

INTRODUCTION

The use of medicinal plants to relieve illness can be traced back over five millenia to written documents in China, India, and the Near East, although the art is undoubtedly as old as humankind. Even today, plants provide the sole source of drugs for most of the world's population, and even in industrialized countries substances derived from higher plants constitute approximately 25 percent of all prescribed medicines. Bioassaying plant constituents for use as drugs requires the cooperation of various research groups—i.e., biochemists, biologists, and pharmacologists—but, above all, it requires facilities for cell culture. Therefore, a test that is easy to culture enables easy and reproducible experiments.

In an attempt to elucidate the mechanism behind the action of plant extracts and/or plant-derived secondary metabolites, dermatophytes—a group of fungi that has been relatively neglected by researchers—as test fungi were chosen. These fungi are etiological agents of dermatophytosis and are classified in three anamorphic genera (asexual or imperfect status), identified as *Microsporum, Epidermophyton,* and *Trichophyton,* belonging to the anamorphic class Hyphomycetes, group Deuteromycota *(Fungi imperfecti).* A description of the genera is given following the Emmons (1934)

The author is grateful to the staff of the Electron Microscopic Center of Ferrara University for the use of instruments.

classification based on conidial morphology and updated subsequent to the discovery of new species (Table 5.1).

Some dermatophytes—particularly the zoophilic species *Microsporum* and *Trichophyton*—are also able to reproduce sexually, producing ascocarps and ascospores. These fungi are classified, respectively, in the genera *Nannizzia* and *Arthroderma*, family Arthrodermataceae, order Onygenales, class Ascomycetes, phylum Eumycota.

Dermatophytes have long been divided into anthropophilic, zoophilic, and geophilic on the basis of their primary habitat. The anthropophilic dermatophytes are primarily associated with humans and rarely infect ani-

TABLE 5.1. Anamorph genera and species of dermatophytes

***Epidermophyton* Sabouraud**	***Microsporum* Gruby**	***Trichophyton* Malmsten**
E. floccosum (Harz) Langeron et Milochevitch	*M. audouinii* Gruby	*T. concentricum* Blanchard
	M. canis Bodin	*T. equinum* (Matruchot & Dassonville) Gedoelst
	M. cookei Ajello	*T. gourvilii* Catanei
	M. equinum (Delacroix et Bodin) Gueguén	*T. kanei* Summerbell
	M. ferrugineum Ota	*T. megninii* Blanchard
	M. fulvum Uriburu	*T. mentagrophytes* (Robin) Blanchard
	M. gallinae (Megnin) Grigorakis	*T. raubitschekii* Kane, Salkin, Weitzman, Smitka
	M. gypseum (Bodin) Guiart & Grigorakis	*T. rubrum* (Castellani) Sabouraud
	M. nanum Fuentes	*T. schoenleinii* (Lebert) Langeron & Milochevitch
	M. persicolor (Sabouraud) Guiart & Grigorakis	*T. simii* (Pinoi) Stockdale, Mackenzie, & Austwick
	M. praecox Rivalier, ex Padhye, Ajello, & McGinnis	*T. soudanense* Joyeux
	M. racemosum Borelli	*T. tonsurans* Malmsten
	M. vanbreudeghemii Georg, Ajello, Friedman, & Brinkman	*T. verrucosum* Bodin
		T. violaceum Bodin
		T. yaoundei Cochet & Doby Dubois

mals. The zoophilic dermatophytes normally affect animals and only occasionally humans; they are most often associated with keratinous material (i.e., hair, nails, horns, and feathers) after it has been shed by a living animal. The geophilic species are considered the progenitors of those pathogens that affect the skin thanks to their ability to decompose keratin. The geophilic species are often isolated from the ground and keratinic remains lying on the ground, ecologically overlapping with the zoophilic dermatophytes. They differ from the zoophilic and anthropophilic dermatophytes (Table 5.2).

Traditionally, the infections caused by dermatophytes are defined according to the anatomic position involved by adding the Latin term for that specific area after the word *Tinea,* for example, *Tinea capitis, Tinea cruris, Tinea corporis, Tinea pedis,* etc. In the present study we used dermatophyte *Microsporum cookei*—imperfect form of *Nannizzia cajetani*—as the test organism. This fungus, which is occasionally a pathogen for humans and pets, is used routinely in our laboratory prior to screening a wider range of a major pathogen, such as *Trichophyton rubrum, T. mentagrophytes,* and *Epidermophyton floccosum.* For some time we have been studying the cytological

TABLE 5.2. Ecological classification of dermatophytes

Anthropophilic species	**Zoophilic species**	**Geophilic species**
E. floccosum	*M. canis*	*E. stockdaleae*
M. audouinii	*M. equinum*	*M. amazonicum*
M. ferrugineum	*M. gallinae*	*M. boullardi*
T. concentricum	*M. persicolor*	*M. cookei*
T. gourvilii	*T. equinum*	*M. gypseum*
T. kanei	*T. mentagrophytes*	*M. nanum*
T. megninii	*T. sarkisorii*	*M. praecox*
T. mentagrophytes	*T. simii*	*M. racemosum*
T. raubitschekii	*T. verrucosum*	*M. rapariare*
T. rubrum		*M. vanbreuseghemii*
T. schoenleinii		*T. ajelloi*
T. soudanense		*T. flavescens*
T. tonsurans		*T. gloriae, T. longifusom*
T. violaceum		*T. phaseoliforme*
T. yaoundei		*T. terrestre*
		T. vanbreuseghemii

events that occur in dermatophytic fungi as result of treatment with natural substances. A comparison of control hyphae, untreated hyphae, and treated hyphae, has enabled us to advance precise hypotheses on the action mechanism of various substances.

In this work protoanemonin (PrA), an unsaturated lactone present in many species of *Ranunculaceae*—plants commonly used in popular household medicine of the past, is studied in depth. This vesicant, blistering compound occurs in high concentrations in several genera of the tribe Anemonae: *Helleborus, Anemone, Clematis,* and *Ranunculus.* It is generated from a glucoside called ranunculin, and may vary, even considerably, within a single species (Bonora et al., 1985, 1987). Protoanemonin is an antibiotic of proven effect versus a number of prokaryotic and eukaryotic organisms (Thimann and Bonner, 1949; Caltrider, 1967; Misra and Dixit, 1980; Martin et al., 1990; Didry et al., 1993). Our group has, in particular, assayed the biological activity of PrA on algae (Mares et al., 1997) and on fungi of varying organizations: filamentous (Mares, 1987, 1989; Mares and Fasulo, 1990) and yeast (Mares et al., 1992).

The general spectrum of activity suggests that the site of action is common among biological systems. These studies have shown that PrA mainly acts on cellular structures rich in thiolic groups, as is the case of numerous other unsaturated cytotoxic lactones (Hall et al., 1980). In our previous study on the effect of PrA in *Microsporum cookei* Ajello, we observed a series of ultrastructural alterations that could be attributed to an interaction with the sulphydryl groups in cytoplasmic microtubules (Mares, 1989).

The present work considers the youngest parts of the mycelium, i.e., the terminal edges, where fungal hyphae growth and wall deposition takes place. The variations of the main wall components are studied using precise cytochemical methods.

THE TECHNIQUES

Phytochemicals

Aqueous fractions of *Ranunculus bulbosus,* collected by steam distillation of fresh leaves (purity = 98 percent), were used as stock solutions and added to the culture medium after Millipore 0.22 μm filtration. The amount was assessed and the active principle characterized by ultraviolet spectrophotometry using a Perkin-Elmer 554UV spectrophotometer (Bonora et al., 1985).

Culture Techniques

Microsporum cookei Ajello, strain CBS n° 202.66, was employed as the test organism. This was obtained from Centraalbureau voor Schimmelcultures, Baarn, The Netherlands, and was maintained in vials containing the culture medium plus 10 percent glycerin kept at –190°C in liquid nitrogen. For the experiments, cultures of *Microsporum* were obtained by transplanting mycelium disks from a single stationary phase culture. These cultures were then incubated at 26 ± 1°C on free medium (Sabouraud dextrose agar (SDA) (Difco) until they reached the midlog phase. They were subsequently transferred to Petri plates containing subinhibitory concentration of the drug (19 μg/ml) (Mares, 1989). The growth rate was measured daily as colony diameter in millimeters up to seven days.

Optical and Electron Microscopy

For optical microscopy (OM), transmission electron microscopy (TEM), and scanning electron microscopy (SEM), the youngest *M. cookei* hyphae were chosen from the margin of the untreated colony and from colonies treated for 24 h with 19 μg/ml of PrA.

For optical microscopy three types of stains were used: (1) Fluoresceina Isothiocyanate (FITC) alone (purchased from Merck); (2) FITC in conjunction with Concanavalin A lectin, specific for alpha mannans and alpha glucans (Goldstein and Hayes, 1978) (purchased from Sigma); (3) Uvitex, diaethanol (Ciba-Geigy A1207), specific for chitin (Marichal et al., 1984) (kindly provided by Professor F. J. Schwinn, Research and Development Department, Agricultural Division, CIBA-GEIGY Limited, Basle, Switzerland). The last two techniques were used to highlight the main parietal components using specific chitin, α-mannan, and α-glucan binding substances. The samples were incubated in PBS (phosphate buffer saline) with FITC, or with FITC-ConA, or with Uvitex, at a concentration of 100 μg/ml. The controls of the staining were performed by adding the specific haptens—i.e., α-methyl-D-mannoside (0.1M) for ConA and *N*-acetyl-D-glucosamine (0.1M) for Uvitex—to the solutions. After washing in PBS, the samples were stained for two hours in the dark with FITC, or with ConA-FITC, or alternatively with Uvitex. After washing briefly in PBS and in water, the samples were observed under a Zeiss Axiophot microscope with suitable filters. For FITC the following filter combination was used: $\lambda_{ex.}$: 485 – . λ_{em}: 520 nm; for Uvitex the combination was: $\lambda_{ex.}$: 365 – . λ_{em}: 460 . The pellicle used was a HP5,27 DIN.

The samples to undergo TEM were routinely fixed with 6 percent glutaraldehyde (GA) in 0.1 M phosphate buffer (pH 7.1) for 3 h at 4°C, washed with the same buffer, and postfixed overnight with 1 percent osmium tetroxide (OsO_4). Only in a few samples was 2 percent (W/v) tannic acid added to the GA mixture to detect proteinaceous material (Horobin, 1982). The fungi were then dehydrated in a graded series of ethanol solutions and embedded in Epon-Araldite resin. Sections were cut with an LKB Ultratome III, stained with uranyl acetate and lead citrate, and observed with a Hitachi H 800 electron microscope at 100 kV.

For SEM, the hyphae were fixed with 6 percent GA in a phosphate buffer, briefly postfixed in 1 percent OsO_4, rapidly dehydrated in acetone, critical-point dried, and gold coated in an S 150 Sputter coater. Observations were made with a Cambridge Stereoscan 360 scanning electron microscope at an accelerating voltage of 20 kV. Both microscopes are owned by the Electron Microscopic Center of Ferrara University.

Cytochemical reactions

The Thiéry (1967) method was applied for the visualization of polysaccharides. This reaction (PATAg) was processed on gold grids exposed to periodic acid (PA), thiocarbohydrazide (TCH) and silver proteinate (Ag) in order to detect periodate-sensitive materials. To detect proteinaceous substances, a pronase (0.5 percent) digestion was performed on grids following the method suggested by Monneron and Bernard method (1966) or, in some samples, adding tannic acid 2 percent to the fixative mixture (Horobin, 1982).

MICROSPORUM COOKEI (= NANNIZZIA CAJETANI) *AS A PRIMARY TEST DERMATOPHYTE*

Microsporum cookei is used for the screening of natural antifungal substances. Hence, the morphology is given as follows:

When cultured in vitro, the mycelium of this fungus is made of long, linear hyphae showing particularly ordered outer layers (Photo 5.1). Magnification of one of these hyphae shows the typical organization: initially a more or less hemispherical apical zone followed by a straight hyphal tube (Photo 5.2). In this fungus, saprophytic macroconidia can also be seen, which appear to individually free themselves of the hyphae (Photo 5.3). In the oldest portion of the mycelium, many multicellular resistance spores, with a thickened, characteristically echinulate wall, are present: the macro-

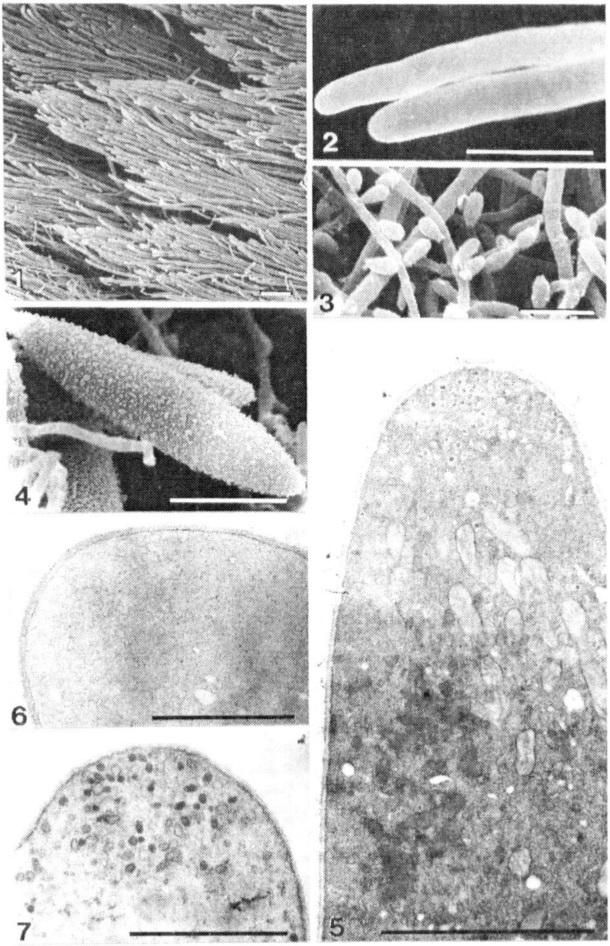

PHOTOS 5.1.-5.7. (1) The youngest portion of control *Microsporum cookei* mycelium; scale bar = 10 µm. (2) Detail of the previous photo at greater magnification showing the apical dome and the straight subapical hyphal tube; scale bar = 5 µm. (3) Control *M. cookei:* individual single-cell saprophytic conidia emerging from conidiophore hyphae; scale bar = 10 µm. (4) Control *M. cookei:* in the innermost, oldest portion of the mycelium there are macroconidia with the characteristic echinulate surface; scale bar = 10 µm. (5) Apex of untreated *Microsporum* showing the Spitzenkörper in the apical dome and all other organules in the subapical zone; scale bar = 10 µm. (6) Microvesicles (20-50 nm in diameter): the carbohydrate content is evidenced using the PATAg method; scale bar = 0.5 µm. (7) Two types of apical vesicles (70-120 nm in diameter) with electron-opaque and electron-translucent matrix; scale bar = 1 µm.

conidia (Photo 5.4). However, the most interesting part is the small zone located at the extreme apex, the true and proper hyphal elongation zone. As in most fungi, in *M. cookei* the apical cytoplasmic organization is characterized by the occurrence of a large number of apical vesicles and microvesicles, together representing the ultrastructural equivalent to the Spitzenkörper. These vesicles are located just beneath the apical membrane, in the apical-most region at the tip cell, measuring only about 1.5-2 μm. In the subapical region, extending back to the first septum, all typical cytoplasmic organules—i.e., nuclei, mitochondria, Golgi equivalent, and vacuoles—are present and generally oriented along the longitudinal axis of the hypha. Microtubules are axially arranged and mitochondria are aligned with them (Photo 5.5), Whereas all microvesicles (between 20 and 50 nm in diameter) have similar electron opaque content (Photo 5.6), another group of apical vesicles (70-120 nm in diameter) vary both in size and contrasting contents. Of the latter, after conventional chemical fixation (GA-OsO_4) and heavy metal (Au-Pb) staining, two types of vesicles can be distinguished on the basis of their internal granular matrix: (1) with an electron-opaque matrix, and (2) with an electron-translucent matrix (Photo 5.7).

When 2 percent tannic acid was added to fixative mixture to detect proteinaceous material, many of the apical vesicles showed an evident central electron-opaque core (Photo 5.8). Moreover, the same vesicles were deeply digested when an enzymatic protease (pronase 0.5 percent) was applied to the sections, thus indicating the proteinaceous nature of their contents (Photo 5.9). On the other hand, the apical vesicles are highly reactive to PATAg, which also indicates the presence of polysaccharides with vicglycol groups accessible to a periodic oxidation (Photo 5.10).

Both types of vesicles are found free in the apical cytoplasm, migrating toward the apex, then fusing with the plasma membrane (Photo 5.10). This area is, in fact, the point where apical exocytosis of the vesicles takes place, permitting normal elongation of the hyphae and proper deposition of the wall components. The apical wall is thinner and different from the wall along the trunk. When the hyphae are observed under an optical microscope after FITC staining, they show an intense, steady fluorescence, both cytoplasmic and parietal (Photo 5.11). Since it is assumed that the main constituents of the *M. cookei* wall are approximately the same as those found in other Ascomycetes—i.e., chitin, β-glucans, and mannoproteins, where the base monomers are mannose and glucose—we sought to visualize them using fluorochromes which, alone or in conjunction with lectin, specifically bind with the various wall components.

Under fluorescence microscopy, cells stained with Fluoresceina Isothiocyanate in conjunction with Concanavalin A (FITC-ConA), a lectin specific for mannose-glucose residues, showed a particularly concentrated fluores-

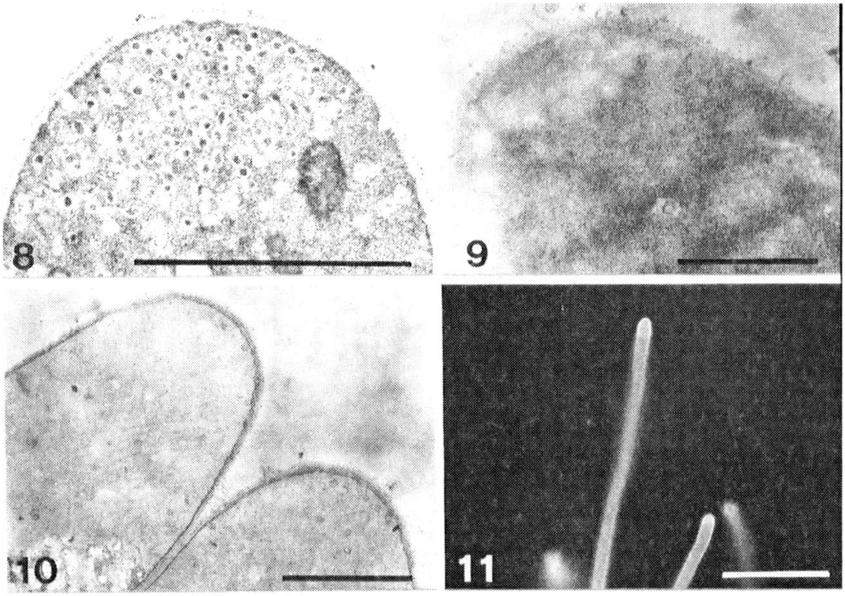

PHOTOS 5.8.-5.11. (8) The vesicles show a central electron-opaque core after fixation with GA+ tannic acid/OsO_4; scale bar = 1 μm. (9) After application of the protease pronase to sections, digestion of the vesicles indicates the protein nature of their contents; scale bar = 1 μm. (10) Both vesicles, free in the cytoplasm and fusing with the plasma membrane, are reactive to the Thièry reaction, indicating the presence of polysaccharides with vic-glycol groups (arrows); scale bar = 1 μm. (11) Hyphae of *M. cookei* controls after staining with FITC: cytoplasm and wall appear intensely fluorescent; scale bar = 10 μm.

cence at the apex (Photo 5.12); when α-methylmannose (a lectin-specific sugar) was added, the fluorescence disappeared from the outer portion of the wall (Photo 5.13). Using Uvitex (a fluorescent stilbene brightener known to bind to chitin), we found widespread, bright fluorescence, particularly in the apical tips and the septa, thus demonstrating the presence of this wall component along the entire hyphal tube (Photo 5.14).

This normal aspect of the youngest parts of *M. cookei* was markedly altered when the fungus was treated with PrA at the subinhibitory concentration of 19 μg/ml for 24 hours. The treated fungus showed a marked slowdown in growth: in the control cultures the growth rate was 5.5 μm/min, whereas in the PrA-treated specimens the growth rate was reduced to 2.7 μm/min. Moreover, there are strong morphological alterations, many of which have been described previously (Mares, 1989). Here we will describe, in particular, those alterations encountered in the apical zone. The

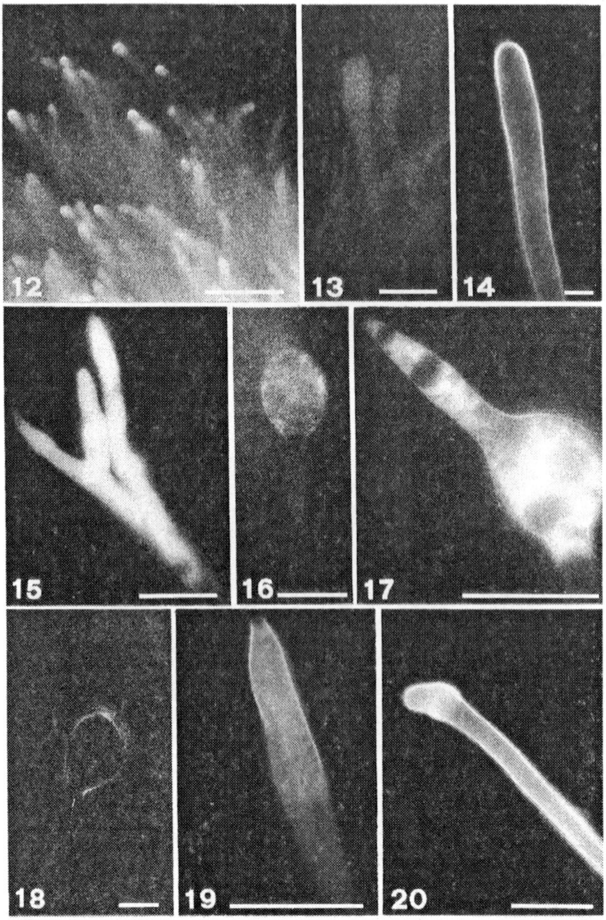

PHOTOS 5.12.-5.20. (12) After staining with FITC-ConA, a mannose- and glucose-specific lectin, the fluorescence is only parietal and principally concentrated on the apices in untreated hyphae; scale bar = 10 μm. (13) The same as Photo 5.12 plus the addition of α-methyl mannose, a sugar specific for the lectin: the fluorescence disappears; scale bar = 5 μm. (14) After staining with Uvitex, specific for the chitin, the same sample as in Photo 5.12 shows brightly fluorescent apical and lateral walls; scale bar = 1 μm. (15-17) Various aspects of *M. cookei* apices after PrA treatment: forked (15), globular (16), and swollen in the subapical region (17); FITC staining; scale bar = 5 μm. (18) PrA-treated hypha after staining with FITC-ConA: the fluorescence due to wall mannoprotein is extremely weak; scale bar = 5 μm. (19-20) Same samples as in Photo 5.18, stained with Uvitex: the fluorescence is absent from the apex (19) or concentrated subapically (20); scale bar = 5 μm.

apices were normally tapered in the control specimens, but in the PrA-treated specimens they became forked (Photo 5.15) or globular as balloons (Photo 5.16), often becoming thickened in the subapical region (Photo 5.17). Following FITC-ConA staining, the mannoprotein-induced fluorescence in the treated specimens was quite weak (and subapical) (Photo 5.18), and the chitin, revealed using Uvitex, was either absent from the apical position (Photo 5.19) or showed subapical accumulation (Photo 5.20). TEM showed the apexes with broad vacuolization, which is highly unusual in this area (Photo 5.21).

Posttreatment (PrA) Changes

After PrA treatment, one aspect that was always observed in *Microsporum* was the lateral extrusion of abundant material wrapped around the hyphae (Photo 5.22). At times it proved possible to photograph vesicles running laterally from the center of the hypha, extruding their content at the sides rather than at the tips (Photo 5.23); at times this material even cemented one hypha to another (Photo 5.24). As regards the chemical nature of this material, in the samples fixed with tannic acid it showed a characteristic, intensely stained blebbing (Photo 5.24); the same was poorly evidenced using the PATAg method, with only its outlines being clearly marked (Photo 5.25). The protein digestion with pronase provides clearer information about the chemical nature of this material since it is partially digested (Photo 5.26). The cytochemical characteristics of the apical vesicles matrix (which are sensitive to protease, tannic acid reactive, and PATAg positive) suggest that they are most likely composed of proteins and polysaccharides. As regards chitin, in the same PrA-treated samples stained with diaethanol, fluorescence at the tip was absent (Photo 5.19) and there was a strong subapical fluorescence at the wall (Photo 5.20). This fact seems to demonstrate the absence of the polymer in the apical position and its abnormal subapical accumulation in the PrA-treated specimens.

MECHANISM OF ACTION

The increase in the number of mycoses seen in the past 20 years because of the increased number of people whose immune systems are compromised by HIV/AIDS, aging, organ transplantation, or cancer therapy and the lack of safe, effective drugs has led to the search for new antifungal drugs with low host toxicity. Previous works have shown that PrA has a fairly good activity versus dermatophytes (Mares, 1987) since the MIC

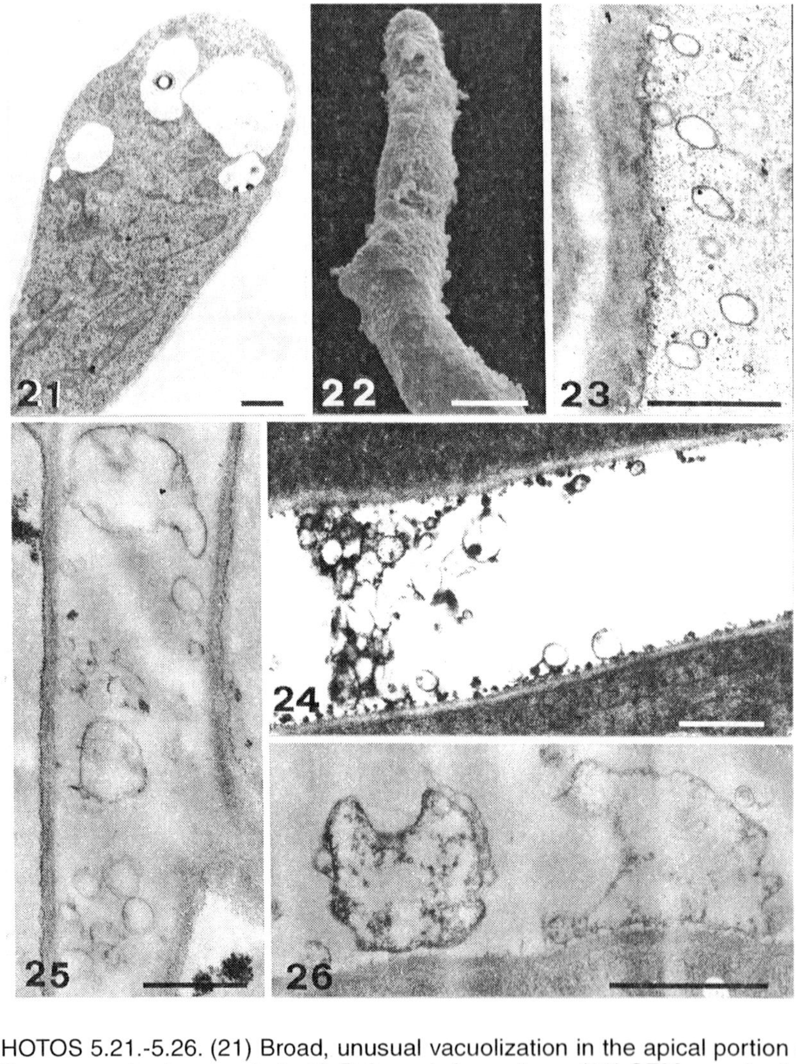

PHOTOS 5.21.-5.26. (21) Broad, unusual vacuolization in the apical portion of *Microsporum* hyphae treated with PrA; scale bar = 1 μm. (22) SEM examination of treated fungi: a hypha with contorted shape and covered with significant amounts of extruded material; scale bar = 2 μm. (23) TEM detail of the vesicles pouring their content out along the lateral wall in PrA-treated fungi; scale bar = 1 μm. (24) Laterally extruded material, cementing one hypha with the next, shown in samples fixed with tannic acid; scale bar = 1 μm. (25) The same as in Photo 5.24 after PATAg method staining: only the outlines are clearly marked; scale bar = 1 μm. (26) The same as in Photo 5.25 after digestion with pronase: the materials are partially digested; scale bar = 1 μm.

(minimum inhibitory concentration) ranges between 47.5 μg/ml for *Epidermophyton floccosum* and 142.4 μg/ml for *Trichophyton mentagrophytes* while the MLC (minimum lethal concentration) is between 95 and 200 μg/ml. Nevertheless, despite its effective antifungal activity, its relative toxicity and irritant properties do not support its use in clinical medicine. This compound has, however, proved useful in understanding the normal growth mechanism of the dermatophyte *Microsporum cookei,* observing the alterations encountered following treatment with a dose equivalent to one-third the MIC (19 μg/ml). The abnormalities are encountered in the youngest parts, the apical portion of the mycelium, which accounts for the strong growth inhibition induced in the treated fungi as compared to the controls.

As with many compounds of plant origin—i.e., antitumoral sesquiterpene lactones, cardiac glucosides, coumarinic derivatives—the basis for protoanemonin cytotoxicity lies in the presence of a lactone ring in the molecule.

In a previous work (Mares, 1987), it was supposed that the action mechanism of substances containing an

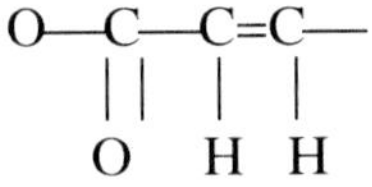 group (such as coumarin and protoanemonin)

is based on the ability to interact with –SH groups, through a Michael-type addition (Hall et al., 1977), thus determining its biological action (Thimman and Bonner, 1949; Caltrider, 1967). In particular, it has been hypothesized that PrA interacts with the –SH groups of microtubules, both in fungi (Mares, 1989; Mares et al., 1992) and in algae (Mares et al., 1997), thus affecting normal cell development.

Because the microtubules in the fungi are difficult to view, the treatment-induced change in gross hyphae morphology can offer indirect evidence of their involvement. In fact, it is known that growth and development are normally regulated by signals received through a cytoskeletal organization. In eukaryotic cells there is a characteristic geometry recognized by the position of the organules and by their outer surface features. All the components of the cytoskeleton—microtubules (MTs), microfilaments (MFs), and intermediate filaments (IFs)—reflect this geometry, but it is determined by the MTs. In fact, the MTs are aligned following the greater cell axis, and their presence is essential for maintaining the asymmetry of the hyphal cell shape. Moreover, the MTs also affect how other cytoskeletal elements are distributed. Therefore, other cytoskeletal components can also be affected by treatment. In this respect we observed that the characteristic cytoplasmic streaming—i.e., the process seen in leaves of the aquatic weed *Elodea canadensis,* and supported by the presence of MFs—stopped about one hour after the addition of protoanemonin (personal observation). This could mean that PrA has a certain direct or indirect action on these cytoskeletal elements as well. On the other hand, there is a great deal of evidence for the presence of an F-actin rich apical cytoskeleton in fungi (Heath, 1990a). It may be that a deregulation of the F-actin hood of the growing fungus by PrA causes the alterations observed here in *M. cookei.*

The cytoskeleton plays a central role in defining the shape and governing how growth takes place: it guides the progression of the Spitzenkörper and orders the pattern of exocytosis and the manner of hyphal extension. In fact, the hyphae of *Microsporum cookei,* as those of all fungi, extend at the apex: they represent a polarized secretory system. Vesicles originate in the deepest part of the cytoplasm from Golgi-like bodies. They carry precursors and enzymes for wall biosynthesis and are brought forward by cytoskeleton and exocytosed at the apex. It is in this zone that the following activities take place: fusion of the vesicles with the plasmalemma, extrusion of the various wall constituents or their precursors, and incorporation of the vesicle membranes themselves, thus causing an increase in plasma membrane (Wessels 1986, 1994, 1999; Wessels and Meinhardt, 1994; Heath 1990b; Bartnicki-Garcia et al. 1989, 1995). In the opinion of Bartnicki-Garcia (1995), the vesicles supplied by the Spitzenkörper are discharged in all directions, but most frequently they point in the direction of vectorial translocation. The shape of the growing tip is defined by the field of exocytosis sites (Heath and van Rensburg, 1996).

TEM fixation technique with tannic acid and pronase digestion indicates that a protein component is present in the apical vesicles of *M. cookei*. At the same time, the PATAg reaction confirms that there are also sugars with free vic-glycol bonds (mannose and glucose). The OM data for the control fungi show, moreover, that these manno-glucoproteins, seen with FITC-ConA, are particularly abundant in the extreme tip of the apices, i.e., at the same point where the chitin concentration peaks, as shown by Uvitex. It is true that autoradiographic studies have revealed that, in growing apices, the chitin is in a nonfibrillar conformational state, whereas in the nongrowing apices and in the older parts there is a network of chitin microfibrills (Vermeulen and Wessell, 1984). Therefore, hyphal growth is the result of an orderly apical exocytosis requiring fully functional organules and a precise subcellular organization. This is ensured by the cytoskeletal components and, in particular, by the microtubules that not only enable vesicle transport but permit perfect cell polarity as well. This means that the loss of polarity can interfere with efficient transport. In a structure such as the fungal hypha, efficient transport of the vesicle containing the parietal precursors would be impossible unless the internal organization is kept strictly parallel to the transport direction. Following treatment with PrA, the cytoskeletal elements would lose their capacity for orderly distribution of vesicles containing the parietal precursors throughout the apical dome. This would lead to an abnormal swelling of the apex, similar to what was described by Gupta and Heath (1997) in *Saprolegna ferax* treated with an actin inhibitor, latruncolin B.

After PrA treatment, we often observed a swelling that was not at the extreme tip; rather, it was a subapical swelling, at the base of the extension zone. Indeed, according to the Wessels steady-state theory (1988), this is the weakest point at which the wall has the minimal rigidity to withstand normal turgor pressure. This is coherent with the previously mentioned hypothesis, since, as we move downward from the apical to subapical wall, we find a gradual rebuilding of the wall through a progressive process of crossed chitin-β-glucan bonds. In fact, once released, the main constituents of the wall integrate until they become strictly intermingled, so much so that they come to form a complicated scaffolding made up of fibrillar zones (chitin and β-glucans) and amorphous zones (glycoprotein complexes) (Wessels et al., 1990). By affecting the normal cytoskeletal organization, PrA alters the direction in which the vesicles run, replacing them with random movements. This condition is proved both by the decreased hyphal extension rate and by changes in the appearance of the wall.

The analyses at the optical fluorescence microscope, using fluorochrome diaethanol (Uvitex) to visualize chitin (Marichal et al., 1984), are interesting enough to warrant the consideration that chitin synthetase has a different localization. It is well known that, in filamentous fungi, chitin is synthesized only at the tip of the cell (Gooday and Schofield, 1995) and not over the entire surface; moreover, it is known that the hyphal extension is maintained by a precise sythetase-active chitin gradient, which is quickly deactivated during the transition from the apex to the lateral walls. The subapical chitin deposition, observed in *M. cookei* hyphae treated with PrA and evidenced under the OM with Uvitex, may be explained by postulating that latent chitin-synthetase, occurring uniformly along the hyphae rather than only at the apex as happens in the controls, is activated following a loss of control of polarity.

The present work has shown that there are two types of apical vesicles in the apical tips of untreated *M. cookei:* (1) small vesicles, which respond to the Thièry reaction and most likely carry glycoprotein material, and (2) larger vesicles, with more or less dense contents, perhaps matching parietal precursors and/or their zymogens. In untreated fungi, the apical cytoplasm shows both types of vesicle freely migrating toward the apex to fuse with the plasma membrane. Proper apical exocytosis leads to the formation of a wall where electron-translucent and electron-opaque layers alternate. The opaque layers are most likely made up of glucans and mannans with various bonds (β-1,4; β-1,6) and the presence of free vic-glycol groups makes them highly reactive to PATAg (Thièry, 1967). On the other hand, the electron-translucent layers are made up of chitin, which is preferentially located in the innermost layers of the wall. The outermost layer of the wall is made up of mannoproteins and is thus revealed by OM with FITC-ConA and TEM reactive to tannic acid. This layer is most likely related to recognition of the host to parasite (trichophytin).

The various polysaccharides, therefore, perform specific functions in the dermatophyte walls: while chitin and β-glucans make up the fibrillar, mechanical portion, the amorphous polysaccharides, often in association with proteins, act as cement, making up the carbohydrate portion for the extracellular enzymes and parietal antigens. By altering the cytoskeletal component needed for normal apical exocytosis, PrA causes mannoprotein to be released along the lateral subapical walls. Since the walls in this area are already rigid, normal integration of these components into the wall proves impossible; instead their release to the outside medium is promoted. Moreover, it may be that in *M. cookei* the mannoproteins of the outermost layer are bound together by disulfide linkages between their protein moieties, as hypothesized by Valentin et al. (1984) for *Saccharomyces cerevisiae*. Thus we can see how, by breaking down these bonds, PrA treatment induces the release into the medium of those substances normally integrated into the outermost parietal layer. That these substances are glycoproteins is shown by their reactivity with PATAg and by the fact that they are digested by protease pronase. Due to the aspecific lateral exocytosis, endocytosis reabsorption of the membranes of the vesicles, containing the extruded material, is also missing. This leaves "empty shells" between one hypha and the next. The presence of membrane-limited cytoplasmic materials sticking to, or in the vicinity of, intact cell walls brings about the characteristic cell-to-cell adhesion induced by treatment. A similar blebbing material was observed even after treatment of other fungi with known antimicrotubular substances, such as griseofulvin (Gull and Trinci, 1974) or colchicin and coumarin (Welker, 1982).

If the delicate balance that leads to elongation of the hyphal tube is disturbed, morphological aberrations can occur. Such changes as curling, abnormal spherical growth, frequent branching, and increased wall thickness have been observed when different antibiotics were applied to growing fungi (Farkás, 1979). Whatever the case, the result is that the apical elongation loses its normal support and the parietal precursors and/or enzymes are randomly distributed, thus stimulating isotropic growth of the hyphal cells.

On the basis of the results observed here, it is reasonable to suggest that PrA interferes with the MTs, as has been assumed (Mares, 1989). Impaired functioning of the cytoskeleton can result in altered metabolism, in smaller growth (due to a change in the transport of the substances at the apex), and in the observed alterations of the hyphal shape, particularly at the tips.

Any sort of interference exerted by such a plant lactone on cytoskeletal function would not only be considered a potential fungicide but, above all, a tool with which to study the structure and function of MTs in fungi.

CONCLUSIONS

From the obtained results and from studying dermatophytes for years, I believe that these kinds of fungi represent an interesting system to process in-depth studies of the basic structural features of fungal cells and, moreover, to better understand the real target and the mechanism of action of antimycotic molecules through the observation of structural modifications induced.

Among the 250,000 species of fungi that are widespread in nature, about 200 are recognized as pathogens responsible for diseases in humans and animals. Regarding this problem, some opportunistic fungi, such as the yeasts *Candida* and *Cryptococcus,* or the filamentous fungi, *Aspergillus, Mucor, Fusarium,* and some others, are known to be responsible for deep and systemic mycoses. Other dimorphic fungi cause important mycoses, such as histoplasmosis by *Histoplasma capsulatum* or coccidioidomycosis by *Coccidioides immitis.* Finally, other mycota are the agents of superficial mycoses; among these pathogenic agents are the yeastlike fungi, invading mucous membranes, skin, and nails, and all the filamentous fungi, called dermatophytes, affecting the keratinous tissues, skin, hair, and nails.

I made the choice to research—through the means reported in this chapter—the dermatophyte *Microsporum cookei* because of its low pathogenic potential by direct contact with humans, being a geophilic fungus and, above all, for its easily growing and subculturing in vitro without nutritional requirements. All these biological properties allow researchers to easily and routinely process reproducible experiments on possible antimycotic activity both of synthetic and natural compounds, and to compare growth and morphology of the untreated fungus with that of the treated one. Because of all these reasons, *M. cookei* is the microorganism of choice in our laboratories, employed as primary test fungus prior to performing deep investigations by screening a wider range of other and potentially more pathogenic fungi.

FUTURE DIRECTIONS

Among the antifungal compounds employed in fungal disease treatments, azoles are those of main interest—even if employing them can cause severe hepatotoxicity. Although the polyenes nystatin and amphotericin are used as antifungals, they can provoke important toxicological effects; and griseofulvin is also employed, even though it is known to induce serious headache, memory lapses, or impairment of judgment. Because of all these reasons, researchers today stretch to turn attention toward molecules of new synthesis that can link a good disease treatment with few minor or without harmful secondary effects.

The plant kingdom may be an important source of safer or more effective substitutes for synthetically produced pharmaceuticals. In fact, natural products represent a rich source of active substances that can be employed against diseases caused by fungi. Currently, plant-derived chemicals play an important role in the discovery and development of modern drugs. In spite of this, plant products are still only poorly explored as a source of biologically active substances. These chemicals may serve as natural compound resources with antifungal effects, or as new leads for synthesis of antifungal compounds.

Our purpose is to extend the screening of the biological effects of chemicals derived from some indigenous and foreign plants employing a dermatophytic fungus—easy to culture and subculture in vitro—as a model for our research. This project aims to characterize new crude plant extracts or single plant extract compounds as possible pharmaceuticals against mycological diseases, and to utilize them instead of the synthetic fungicides usually employed in current Western medicine.

REFERENCES

Barnitcki-Garcia, S., Hergert, F., and Gierz, G. (1989). Computer simulation of fungal morphogenesis and the mathematical basis for hyphal (tip) growth. *Protoplasma* 153: 46-57.

Barnitcki-Garcia, S., Hergert, F., and Gierz, G. (1995). Determinants of fungal cell wall morphology: The vesicle supply center. *Can. J. Bot.* 73 (suppl. 1): S372-S378.

Bonora, A., Dall'Olio, G., and Bruni, A. (1985). Separation and quantitation of protoanemonin in Ranunculaceae by normal—and reversed-phase HPLC. *Planta Med.* 5: 364-367.

Bonora, A., Dall'Olio, G., Donini, A., and Bruni, A. (1987). An HPLC screening of some Italian Ranunculaceae for the lactone protoanemonin. *Phytochem.* 26: 2277-2279.

Caltrider, P.G. (1967). Protoanemonin. In *Antibiotics, Mechanism of Action,* Volume 1 (Eds. Corcoran, J.W. and Hahn, F.). Berlin, Heidelberg, New York: Springer-Verlag, pp. 671-673.

Didry, N., Dubreuil, L., and Pinkas, M. (1993). Microbiological properties of protoanemonin isolated from *Ranunculus bulbosus. Phytother. Res.*7: 21-24.

Emmons, C.W. (1934). Dermatophytes: Natural grouping based on the form of the spores and accessory organs. *Arch. Dermatol. Syphilol.* 30: 337-362.

Farkás, V. (1979). Biosynthesis of cell walls in fungi. *Microbiol. Rev.* 43: 117-144.

Goldstein, J.J. and Hayes, C.E. (1978). The lectins: Carbohydrate-binding properties of plants and animals. *Adv. Carbohydrate Chem. Biochem.* 35: 127.

Gooday, G.W. and Schofield, D.A. (1995). Regulation of chitin synthesis during growth of fungal hyphae. The possible participation of membrane stress. *Can. J. Bot.* 73: S114-S121.

Gull, K. and Trinci, A.P.J. (1974). Ultrastructural effects of griseofulvin on the myxomycete *Physarum policephalum.* Inhibition of mitosis and production of microtubule crystals. *Protoplasma* 81: 37-48.

Gupta, G.D. and Heath, I.B. (1997). Actin disruption by latrunculin B causes turgor-related changes in tip growth of *Saprolegnia ferax* hyphae. *Fungal Genet. Biol.* 21: 64-75.

Hall, I.H., Lee, K.H., Mar, E.C., Starnes, C.O., and Waddell, T.G. (1977). Antitumoral agents. 21. A proposed mechanism for inhibition of cancer growth by tenulin and helenalin and related cyclopentenones. *J. Med. Chim.* 20: 333-336.

Hall, I.H., Lee, K.H., Starnes, C.O., Muraoka, O., Sumida, Y., and Waddell, T.G. (1980). Antihyperlipidemic activity of sesquiterpene lactones and related compounds. *J. Pharm. Sci.* 69: 694-696.

Heath, I.B. (1990a). The role of actin in hyphal tip growth. *Int. Rev. Cytol.* 123: 95-127.

Heath, I.B. (1990b). *Tip growth in plants and fungal cells.* San Diego: Academic Press, pp. 351.

Heath, I.B. and van Rensburg, E.J.J. (1996). Critical evaluation of the VSC model for tip growth. *Mycoscience* 37: 1-10.

Horobin, R.W. (1982). *Histochemistry: An Explanatory Outline of Histochemistry and Biophysical Staining* (Ed. Fischer, G.). Stuttgart-New York: Butterworth, pp. 179-180.

Mares, D. (1987). Antimicrobial activity of protoanemonin, a lactone from Ranunculaceous plants. *Mycopathol.* 98: 133-140.

Mares, D. (1989). Electron microscopy of *Microsporum cookei* after "in vitro" treatment with protoanemonin: A combined SEM and TEM study. *Mycopathol.* 108: 37-46.

Mares, D., Bonora, A., Sacchetti, G., Rubini, M., and Romagnoli C. (1997). Protoanemonin-induced cytotoxic effects in *Euglena gracilis. Cell Biol. Internat.* 21: 397-404.

Mares, D. and Fasulo, M.P. (1990). Ultrastructural alterations in *Epidermophyton floccosum* and *Trichophyton mentagrophytes* exposed in vitro to protoanemonin. *Cytobios* 61: 89-95.

Mares, D., Romagnoli, C., Fasulo, M.P., and Bruni, A. (1992). Morphological observations on possible calcium-protoanemonin interactions in the yeast *Rhodotorula glutinis*. *Biochem. Physiol. Pflanzen.* 188: 273-278.

Marichal, P., Gorrens, J., and Vanden Bossche, H. (1984). The action of itraconazole and ketonazole on growth and sterol synthesis in *Aspergillus fumigatus* and *Aspergillus niger*. *Sabouraudia* 22: 13-21.

Martin, M.L., San Román, L., and Domínguez, A. (1990). In vitro activity of protoanemonin, an antifungal agent. *Planta Med.* 56: 66-69.

Misra, S.B. and Dixit, S.N. (1980). Antifungal principle of *Ranunculus sceleratus*. *Econ. Bot.* 34: 362-367.

Monneron, A. and Bernard, W. (1966). Action de certaines enzymes sur de tissues inclus in epon. *J. Microsc.* (Paris) 5: 697-714.

Thièry, Y. (1967). Mise en évidence des polysaccharides sur coupes fines en microscopie électronique. *J. Microsc.* (Paris) 6: 987-1019.

Thimann, K.V. and Bonner, W.D. (1949). Inhibition of plant growth by protoanemonin and coumarin and its prevention by BAL. *Proc. Natl. Acad. Sci.* (US) 35: 272-276.

Valentin, E., Herrero, E., Pastor, F.I.J., and Sentendreu, R. (1984). Solubilization and analysis of mannoprotein molecules from the cell wall of *Saccharomyces cerevisiae*. *J. Gen. Microbiol.* 130: 1419-1428.

Vermeulen, C. A. and Wessels, J.G.H. (1984). Ultrastructural differences between wall apices of growing and non-growing hyphae of *Schizophyllum commune*. *Protoplasma* 120: 123-131.

Welker, D.L. (1982). Coumarin and antimicrotubular agents as probes of microtubule function in *Dictyostelium discoideum*. In *Microtubules in Microorganisms* (Eds. Cappuccinelli, P. and Morris, N.R.). New York/Basel: M. Dekker Inc., pp. 99-108.

Wessels, J.G.H. (1986). Cell wall synthesis in apical hyphal growth. *Int. Rev. Cytol.* 104: 37-79.

Wessels, J.G.H. (1988). A steady-state model for apical wall growth in fungi. *Acta Bot. Neerl.* 37: 3-16.

Wessels, J.G.H. (1994). Developmental regulation of fungal cell wall formation. *Annu. Rev. Phytopathol.* 32: 416-437.

Wessels, J.G.H. (1999). Fungi in their own right. *Fungal Genet. Biol.* 27: 134-145.

Wessels, J.G.H. and Meinhardt, F. (1994). *The Mycota.* Volume I: *Growth, Differentiation and Sexuality.* Berlin: Springer.

Wessels, J.G.H., Mol, P.C., Sietma, J.H., and Vermeulen, C.A. (1990). Wall structure, wall growth, and fungal cell morphogenesis. In *Biochemistry of Cell Walls and Membranes in Fungi* (Eds. Kuhn, P.J., Trinci, A.P.J., Jung, M.J., Goosey, M.W., and Copping, L.G.) Berlin, Heidelberg, New York, London, Paris, Tokyo, Hong Kong: Springer-Verlag, pp. 81-95.

Chapter 6

Plant-Derived Antimycotics: Potential of Asteraceous Plants

Mahendra Rai
Deepak Acharya
Prasad Wadegaonkar

INTRODUCTION

The treatment of human mycosis was a great challenge before the advent of clinicians and dermatologists. On one hand, opportunistic fungal infections are increasing at an alarming rate; on the other hand, allergic reactions of the skin are increasing day by day. The latter is due to greater sensitivity in the present generation to antimycotic agents. Potential human pathogenic fungi in general and opportunistic fungal infections in particular are usually treated with drugs belonging to the imidazole family. However, these potent antimycotics are not responding to the new spectrum of opportunistic fungal infections that are common in immunocompromised hosts. Therefore, to combat these new fungal infections, steps should be taken to make the benefits of successful pharmaceutical research available to all, and especially to those who are in the greatest need. It is time to search for new antifungal agents of herbal origin that are relatively affordable, safer, and widely available to the public. Moreover, sometimes imidazole derivatives are not effective so alternative drugs are required (Lowey et al., 1985). A perusal of the literature indicates that many investigators have reported fungistatic and bacteriostatic properties of extracts of higher plants (Bhakuni et al., 1969, 1971; Dhar et al., 1974; Ray and Majumdar, 1975, 1976; Tansey and Appleton, 1975; Jain and Agarwal, 1976; Dhawan et al., 1977; Misra and Dixit, 1979; Jain, Charia, et al., 1980; Jain, Jain, et al., 1980; Wahab et al., 1981; Barde and Singh, 1983; Ikram and Inam-ul-haq, 1984; Singh and Deshmukh, 1984; Rai, 1987; Rai and Upadhyay, 1988, 1989; Mares, 1989; Mares and Fasulo, 1990; Yadav and Saini, 1990; Lima et al., 1992, 1993; Perrucci et al., 1994; Rahalison et al., 1994; Villarreal et al., 1994; Grosvenor et al., 1995; Rai, 1995; Adams et al., 1996; Jain, 1996; Narayanarao et al., 1996; Gopallakrishnan et al., 1997; Rai et al., 1997, 1999; Rai and Acharya, 1999; Singh, 2000). Many antimycotic agents were introduced for the treatment of mycosis during the recent past. Most of these active agents are useful for topical application only because of their toxic nature. These broad-spectrum antimycotics are effective against fungi that infect human beings. In the treatment of mycosis, such active agents are preferred because a broad spectrum therapy is required (Gellin et al., 1972; Emmett and Marrs, 1973; Tronnier and Kosen, 1985).

The antifungal activity of nonvolatile constituents of higher plants was earlier reviewed by some researchers (Dekker, 1969; Thapliyal and Nene, 1969; Calpouzos, 1969; Fawcett and Spencer, 1970; Dixit and Tripathi, 1982; Mahadevan, 1982). Many plants produce essential oils as secondary metabolites, but their exact role in the life processes of plants is unknown. A review of the literature reveals that a large number of essential oils was reported to possess fungitoxic activity (Maruzzella et al., 1960; Barnes, 1963; Korta and Starzyk, 1963; Maruzzella, 1963; Hiller, 1964; Birch, 1966; Korbely and Florian, 1971; Garg, 1974; Zutschi et al., 1975; Overeem, 1976; Gautam et al., 1980; Jain, Charia, et al., 1980; Jain, Jain, et al., 1980; Ikram and Inam-ul-haq, 1980, 1984; Pandey et al., 1983b; Deshmukh et al., 1986; Singh et al., 1986; Kishore and Dwivedi, 1991; Jain and Agarwal, 1992; Perrucci et al., 1994; Mwosu and Okafor, 1995; Goren et al., 1996; Gopallakrishnan et al., 1997; Rai et al., 1999). Most members of family Asteraceae are known to contain essential oils that usually have antifungal/cytotoxic sesquiterpene lactons.

International Status

A survey of the literature indicates that many investigators abroad have studied herbal antifungal agents in the recent past (Maruzzella and Logeuri, 1959; Maruzzella and Balter, 1959; Maruzzella et al., 1959; Tokin, 1960; Fawcett and Spencer, 1970; Birner and Nicolis, 1973; Burden and Bailey, 1975; Daphne et al., 1982; Rahalison et al., 1993; McCutchen et al., 1994; Villarreal et al., 1994; Mwosu and Okafor, 1995; Demchenko et al., 1995; Alkofahi et al., 1996; Goren et al., 1996; Khan and Evans, 1996; Navarro et al., 1996; Achola et al., 1997; Al Magboul et al., 1997). Maruzzella and colleagues from 1956 to 1963 did a major work in this direction and tested about 119 essential oils, out of which 59 were reported as very effective antimicrobial agents. Tokin (1960) studied the antibiotic substance produced by higher plants in detail. He proposed the name *phytoncide* for biologically active substances produced by higher plants, and he studied the phytoncides of onion, garlic, and other plants that contained the strongest antibiotic properties. Aizeman (1978) studied antibiotic properties of about 1500 varieties of higher plants selected from the former USSR. Ikram and Inam- ul-haq (1984) screened about 100 medicinal plants of Pakistan for their antimicrobial activity, but only a few exhibited remarkable properties. Mares (1987) found lactones to be antidermatophytic. Extracts of rhuburb were reported to be effective against *Trichophyton, Microsporum,* and *Epidermophyton* (Itsuo, 1985). McCutchen et al. (1994) screened more than 100 methanolic plant extracts for antifungal activity against nine fungal species. Eighty-one were found to have some antifungal activity and 30 extracts showed activity against four or more of the fungi assayed. They reported *Artemisia ludoviciana* and *A. tridenta* to be active against nine fungi.

National Status

Although several investigators have contributed to the subject of the antimycotic activity of medicinal plants, the work is very fragmentary and meager (Jain and Agarwal, 1976; Rao, 1976; Thind and Dahiya, 1976; Sharma et al., 1978; Misra and Dixit, 1979; Mishra et al., 1979; Tripathi and Dixit, 1981; Alankara et al., 1981; Asthana et al., 1982; Chandra et al., 1982; Singh and Deshmukh, 1984; Chauhan and Saxena, 1985; Deshmukh et al., 1986; Rai, 1987, 1988, 1993; Tripathi et al., 1988; Radhika Iyer, 1992; Rai and Vasanth, 1995; Pattnaik et al., 1996; Rai et al., 1997; Rai and Acharya, 1999, 2000; Singh, 2000). Screenings of Indian plants for a wide range of activities (antimalarial, antiprotozoic, antiviral, antihelminthic, anticancer, antifungal, etc.) have been carried out by various investigators. A perusal of the literature indicates that many investigators have reported fungistatic and bacteriostatic properties of extracts of higher plants (Dhar et al., 1968; Bhakuni et al., 1969, 1971; Dhar et al., 1973, 1974; Ray and Majumdar, 1975, 1976; Jain and Agarwal, 1976; Dhawan et al., 1977; Misra and Dixit, 1979; Dhawan et al., 1980; Jain, Jain, et al., 1980; Wahab et al., 1981; Barde and Singh, 1983; Aswal et al., 1984; Singh and Deshmukh, 1984; Abraham et al., 1986; Rai, 1987; Rai and Upadhyay, 1988; Yadav and Saini, 1990; Rai, 1995; Jain, 1996; Narayanarao et al., 1996; Gopallakrishnan et al., 1997; Rai et al., 1997, 1999; Rai and Acharya, 1999). However, Bhakuni et al. (1971) reported that antifungal activity could be observed in only three extracts out of 300 plants, which indicates that more thorough investigation is required in the search for antifungal activity.

The investigators of the Central Institute of Medicinal and Aromatic Plants (CIMAP), Lucknow, India, have screened about 3,231 materials from 3,051 plants for their biological activity (Dhar et al., 1968; Bhakuni et al., 1969, 1971; Dhar et al., 1973, 1974; Dhawan et al., 1977, 1980). Only ten plants exhibited activity against pathogens of superficial mycosis. Out of these plants only *Artemisia dranunculus* (family Asteraceae) showed fungitoxic activity against pathogenic fungi (Table 6.1).

Noteworthy Contributions in the Field

A review of literature reveals that a significant contribution has been made regarding the antimycotic potential of family Asteraceae (Rao, 1976; Mathela and Sinha, 1978; Geda and Bokadia, 1979; Chandra and Dikshit, 1981; Wahab et al., 1981; Chandra et al., 1982; Daphne et al., 1982; Devi and Nandkumar, 1983; Singh et al., 1986; Rai and Upadhyay, 1990; Yadav and Saini, 1990; Kishore and Dwivedi, 1991; Rai, 1993; Villarreal et al., 1994; Alkofahi et al., 1996; Khan and Evans, 1996; Miguel et al., 1996; Romanelli et al., 1996; Al Magboul et al., 1997; Rai et al., 1997; Singh, 1999; Rai and Acharya, 1999). Rao (1976) studied the antimicrobial effect of the essential oil of *Ageratum conyzoides*. Mathela and Sinha (1978) tested some indigenous essential oils for their antibacterial and antifungal activity. *Aster beduncularis* and *A. thomsonii* were found to be antifungal. Geda and Bokadia (1979) reported antifungal activity of *Blumea membranacea*. Pandey et al. (1982) recorded fungitoxic and phytotoxic properties of the essential oil of *Caesulia axillaris*. Chauhan and Saxena (1985) studied the antifungal activity of the leaves of *Inula cuspidata*. Antimycotic potential of *Parthenium hysterophorus* against human pathogenic fungi was investigated by Rai and Upadhyay (1990) and later again by Rai (1993, 1994, 1995). Rai and Vasanth (1995) evaluated the sensitivity of three keratinophilic fungi to some vicolides isolated from *Pentanema indica*. Goren et al. (1996) reported cytotoxic and antibacterial activities of sesquiterpene lactones isolated from *Tanacetum praeteritum*. Romanelli et al. (1996) observed pharmacological activities of methanolic extracts of *Centaurea deusta* and *Crepis lacera*. A notable contribution has been made by Alkofahi et al. (1996), who reported antimicrobial activity of 52 medicinal plants of Jordan. *Achillea aleppica* extract showed highly antifungal activity. Miguel et al. (1996) isolated chemical constituents from *Lychnophora salicifolia* which later exhibited antifungal activity by the disc diffusion methods. Zheng et al. (1996) isolated two new flavones from *Artemisia giraldii* and observed antifungal activity against *Aspergillus flavus* and *Trichoderma viride*. *Artemisia mexicana* posessed strong in vitro antifungal activity against *Candida albicans,* found by Navarro et al. (1996). Al Magboul et al. (1997) isolated vernolepin and vernodalin from *Vernonia amygdalina* and studied fungicidal activity against *Aspergillus niger* and

TABLE 6.1. Plants screened, part used, biological tests carried out, and antimycotic activity observed by CIMAP

Plant	Part used	Biological test carried out	Antimycotic activity observed
Acanthospermum hispidum DC.	PL	B 1-5	–
Achillea millefolium L.	PL	B 1-5	–
Adenostemma reticulatum DC. Syn. *Adenostemma viscosum* J. R. and G. Forst. var. *reticulatum* DC.	PL	B 1-3, 6, 7	–
Ageratum conyzoides L.	PL	B 1-5	–
Ainsliaea aptera DC.	PL	B 1-5	–
A. pteropoda DC.	PL	B 1-2	–
Amberboa ramosa Roxb.	PL	B 1-5	–
Anaphalis cinnamomea C. B. Clarke	PL	B 1-5	–
A. contorta Hook.	PL	B 1-5	–
A. cuneifolia DC.	PL	B 1-5	–
A. marcescens (Wight) C. B. Clarke	PL	B 1-3, 6, 7	–
A. neelgerryana DC.	PL	B 1-3, 6, 7	–
Arctium lappa L.	PL	B 1, 3, 6	–
Artemisia dranunculus L.	PL	B 1-3, 6, 7	B 2, 3, 6, 7 +
A. laciniata Willd.	PL	B 1-3, 5, 7	–
A. maritima L.	PL	B 1, 3, 6	–
A. roxburghiana Besser.	PL	B 1-5	–
A. tournefortiana Reichb.	PL	B 1-3, 4, 5	–
A. vestita Wall.	PL	B 1-5	–
A. vulgaris L.	PL	B 1-5	–
Aster albescens DC.	PL	B 1-5	–
Bidens biternata (Lour.) Merr. and Sherff.	PL	B 1-5	–
B. lacera DC.	PL	B 1-5	–
B. malcolmii (Cl.) H. K. F.	PL	B 1-3, 6, 7	–
B. obliqua (L.) Druce. var. *arenaria* (DC.) Mahesh (*B. amplectens* DC. var. *arenaria* DC.)	PL	B 1-3, 6, 7	–
B. tripartita L.	PL	B 1-5	–

TABLE 6.1 *(continued)*

Plant	Part used	Biological test carried out	Antimycotic activity observed
B. triplinervia H. B. K. var. *macrantha* (Wedd.) Sherrif	PL	B 1-3, 6, 7	–
Caesulia axillaris Roxb.	PL	B 1-5	–
Calendula officinalis L.	PL	B 1-5	–
Callistephus chinensis Linn. (Nees)	PL	B 1, 3, 6	–
Carduus nutans Hook.	PX	B 1, 3, 6	–
Carpesium abrotanoides L.	PL	B 1-5	–
Carthamus oxyacantha Bieb.	PL	B 1-5	–
C. tinctorius L.	SD	B 1-5	–
Centipeda minima (L.) A. Br. and Archers.	PL	B 1-5	–
Centratherum anthelminticum Willd. Kuntze.	SD	B 1-5	–
Chrysanthemum leucanthemum L.	PL	B 1-5	–
Cichorium intybus L.	PL	B 1-5	–
Circium arusus (L.) Scop.	PL	B 1-5	–
C. arvense (L.) Scop.	PL	B 1-5	–
Cnicus wallichii DC.	PL	B 1-5	–
C. wallichii DC. var. *wightii* Hook.	PX	B 1-5	–
Conyza ambigua DC.	PL	B 1-5	–
C. stricta Willd.	PL	B 1-5	–
C. viscidula Wall.	PL	B 1-5	–
Coreopsis grandiflora Hogg.	PL	B 1-5	–
Cotula hemispherica Roxb. (Wall. ex Benth. and Hook.)	PL	B 1-5	–
Cousinia thomosonii Clarke.	PL	B 1-3, 6, 7	–
Crepis sancta Linn. Babc. ssp. *bifida* (Vis.) Thell ex. Syn. *Pterethica felconeri* H. K. F.	PL	B 1-3, 6, 7	–
Cyathocline purpurea (Doh.) Kuntze.	PL	B 1-5	–
Cynara scolymus L.	PL	B 1-5	–

Plant	Part used	Biological test carried out	Antimycotic activity observed
Dahlia pinnata Cav.	PX	B 1, 3, 5	–
Echinops. cornigerus DC.	PL	B 1-5	–
E. echinatus DC.	RT	B 1-3, 5, 7	–
E. rehinatus Roxb.	PL	B 1-5	–
Eclipta alba (L.)	PL	B 1-5	–
Elephantopus scaber L.	PL	B 1-5	–
Enhydra fluctuans Laur.	PL	B 1-5	–
Erechtites valerianaefoila (Wolf) DC.	PL	B 1-5	–
Erigeron annuus (L.) Pers.	PL	B 1-3, 6, 7	–
E. asteroides Roxb.	PL	B 1-5	–
E. rarvinskianus DC. Syn. *Erigeron mucronatus* DC.	PL	B 1-3, 5, 7	–
Eupatorium adoratum L.	PX	B 1-5	–
E. chinense L. Syn. *Eupatorium reevessi* DC.	PX	B 1-3, 6, 7	–
E. glandulosum DC.	PL	B 1-5	–
E. wallichii DC.	PX	B 1-3, 6, 7	–
Gaillardia pulchella Foug.	PL	B 1-3, 6, 7	–
Galinsoga ciliata Raf.	PL	B 1-5	–
G. parviflora Cav.	PL	B 1-5	–
Gerbera lanuginosa Benth.	PL	B 1-5	–
Gnaphalium luteo-album L.	PL	B 1-5	–
G. pensylvanicum Willd. (*G. purpureum* Hook.)	PL	B 1-3, 6, 7	–
G. pulvinatum Delile.	PL	B 1-5	–
Grangea maderaspatana (L.) Poir.	PL	B 1-5	–
Gynura cusimbua (D. Don) Moore. Syn. *G. angulosa* DC.	PL	B 1-5	–
Helichrysum bracteatum Andr.	PL	B 1-5	–
H. buddleioides DC.	PX	B 1-3, 5, 7	–
Hypochoeris glabra L.	PX	B 1-3, 6, 7	–
Inula cappa DC.	PL	B 1-5	–
I. cuspidata DC. Clarke.	PL	B 1-5	–
Jurinea macrocephala (Royle) C. G. Clarke	PL	B 1-5	–
Lactuca longifolia DC.	PL	B 1-5	–

TABLE 6.1 *(continued)*

Plant	Part used	Biological test carried out	Antimycotic activity observed
L. sativa L.	PL	B 1-5	–
L. trunoniana Wall. Clarke.	PL	B 1-5	–
Laggera pterodonta DC.	PL	B 1-5	–
Launaea resedifolia L. O. Kuntze. Syn. *Launaea chondrilloides* DC. H. K. F.	PL	B 1, 3, 6	–
Leucomeris spectabilis D. Don.	PL	B 1-5	–
Ligularia sibirica (L. F.) Cass. var. *kitam* Syn. *Senecio ligularia* H. K. F.	PL	B 1, 3, 5	–
Mikania cordata Burm.	PX	B 1-5	–
Parthenium hysterophorus L.	PL	B 1-5	–
Picris hieracioides L. Sp. *Kaimaensis kitam* (*P. hieracioides* L.)	PL	B 1, 3, 6	–
Pluchea indica Less.	PX	B 1-5	–
P. lanceolata DC. Cl.	PX	B 1, 3, 6	–
Pulicaria angustifolia DC.	PL	B 1, 3, 6	–
P. foliolosa DC.	PL	B 1-5	–
Saussurea albescens (DC.) Sch.	PL	B 1-3, 5, 7	–
S. heteromalia D. Don.	PL	B 1-5	–
S. jacea Clark.	PL	B 1-3, 6, 7	–
S. obvallata Wall.	PL	B 1-5	–
Senecio chrysanthemoides DC.	PL	B 1-5	–
S. connatus Balakr. Syn. *Senecio arnicoides* (Royle) Cl. *Cremanthodium arnicoides* Wall. R. Good.	PL	B 1-5	–
S. grahamii H. K. F.	PL	B 1, 3, 5	–
S. kunawarensis Nair. Syn. *Senecio arnicoides* Wall. var. *firigida* H. K. F.	PL	B 1-3, 5, 7	–
S. kunthianus Wall. ex. DC.	PL	B 1-3, 5, 7	–
S. rufinervis DC.	PL	B 1-5	–
S. tenuifolius Burm. f.	PL	B 1-5	–
Solidago virga-aurea L.	PL	B 1-5	–

Plant	Part used	Biological test carried out	Antimycotic activity observed
S. virgaurea L. var. *pubescence* (Wall.) Clarke.	PL	B 1, 3, 6	–
Sonchus asper Vill.	PL	B 1-5	–
S. brachyotus DC.	PL	B 1-5	–
Siegesbeckia orientalis L.	PL	B 1-5	–
Silybum marianum L.	PL	B 1-3, 6, 7	–
Spilathes peniculata Wall. ex. DC.	PL	B 1-5	–
Tagetes erecta L.	PL	B 1-5	–
Taraxacum officinale Weber.	PL	B 1-5	–
Tithonia diversifolia (Grey.)	PL	B 1-3, 6, 7	–
T. rotundifolia Black.	PL	B 1, 3, 6	–
T. tegetiflora Desf.	PX	B 1-5	–
Tricholepis elongata DC.	PL	B 1-5	–
Tridax procumbens L.	PL	B 1-5	–
Vernonia altenuata DC.	PL	B 1, 3, 6	–
V. arborea Ham.	PX	B 1-5	–
V. cinerea Less.	PL	B 1-5	–
V. elegans Gandn.	PX	B 1-3, 6, 7	–
V. malaberica Hook. f.	PX	B 1-5	–
V. pectiniformis DC.	PX	B 1-3, 6	–
V. volkameriaefolia DC.	PX	B 1-5	–
Vicoa indica (Willd.) DC.	PX	B 1-5	–
V. vestita (Wall ex. DC.)	PL	B 1-3, 6 ,7	–
Vittadinia australis A. Rich.	PL	B 1-5	–
Wedelia calendulacea Less.	PL	B 1-5	–
W. urticaefolia DC.	PL	B 1, 3, 6	–
Xanthium strumarium L.	RT	B 1-5	–
Zinnia elegans Jacq.	PL	B 1, 3, 6	–

Note: Plant names arranged alphabetically. Plant part used: bulb (BU), flower or inflorescence (FL), fruit (FR), leaf (LF), entire plant (PL), plant excluding root (PX), root bark (RB), rhizome (RH), root (RT), stem bark (SB), seeds (SD), stem (ST), wood of root (WR), wood of stem (WS). Biological test carried out: B = antifungal. The following two stains of yeastlike fungi, two dermatophytes, and three plant pathogenic fungi were employed: *Candida albicans* (B1), *Cryptococcus neoformans* (B2), *Trichophyton mentagrophytes* (B3), *Microsporum canis* (B4), *Aspergillus niger* (B5), *Aspergillus fumigatus* (B6), and *Sporotrichum schenckii* (B7).

Candida albicans. Vasanth (1996) studied the chemistry and biology of *Pentanema indica* and observed very good antifungal activity of the plant against *Candida albicans* and *C. krusei.* Hamsaveni et al. (1992) also studied the antimicrobial efficacy of *Pentanema indica.* Achola et al. (1997)

observed pharmacological activities of *Gutenbergia cordifolia*. Rai and Acharya (1999) evaluated in vitro activity of 31 plant extracts of family Asteraceae against *Fusarium oxysporum* and *Trichophyton mentagrophytes*. The maximum antimycotic activity against both the fungi was exhibited by the flower extract of *Tagetes erecta*.

ANTIFUNGAL POTENTIAL OF SOME IMPORTANT ASTERACEOUS PLANTS

Achillea *Species*

Antifungal activity in the leaf oil of *A. fragrantissima* was reported against *C. albicans*, a causal organism of common *Tinea pedis* (Barel et al., 1991). Similarly, Kedzia et al. (1990) found strong activity of leaf oil extracted from *A. millefolii* against *Candida albicans*. Abbasoglu and Kusmenoglu (1994) recorded slight antifungal activity in the aerial parts of the four species of *Achillea*.

Ageratum *Species*

A. conyzoides

Rao (1976) and Sharma et al. (1978) reported antimycotic activity in leaf oil of *A. conyzoides* against various plant pathogens, namely, *Alternaria helianthi, Colletotrichum capsici, Fusarium moniliforme, F. solani, Helminthosporium oryzae, H. turcicum, Penicillium chrysogenum, P. javanicum, Pyricularia setariae, P. oryzae, Pythium vexans, Rhizoctonia bataticola,* and *R. solani*. The fungistatic nature of the oil at 0.2 percent (2.0×10^3 µl/l) concentration against various fungi was reported by Chandra and Dikshit (1981). Mycelial growth of dermatophytes, namely, *Epidermophyton floccosum, Microsporum canis,* and *Trichophyton mentagrophytes* was completely checked when dipped in a 4,000-ppm (4.0×10^3 µl/l) dose of the oil (Singh et al., 1986).

A. houstonianum

Pandey et al. (1983a) found antifungal activity in essential oil of leaf at 100 ppm (0.1×10^3 µl/l) concentration, possessing fungistatic nature against *Blastomyces dermatidis, Epidermophyton floccosum, Histoplama capsulatum, Trichophyton mentagrophytes, T. simii, T. terrestre, T. tonsurans, T. verrucosum,* and *T. vialaceum*. However, it killed *Microsporum gypseum* within one second at a 300-ppm (0.3×10^3 µl/l) dose by contact. The chemical of the oil that is responsible for antifungal activity was found to be thermostable and fungistatic against *Fusarium lateritium* at 0.05 percent (0.5×10^3 µl/l) concentration.

Anaphalis contorta

Saxena et al. (1984) reported that the leaf oil of the plant showed antifungal activity against *Aspergillus niger, Microsporum gypseum,* and *Penicillium notatum.*

Arnica latifolia

A significant antimycotic activity of oil was reported against *Candida albicans, Cephalosporium sacchari, Ceratocystis paradoxa, Curvularia lunata, Epidermophyton floccosum, Fusarium moniliforme, Helminthosporium sacchari, Physalosporo tocumenensis, Trichophyton rubrum,* and *Sclerotium rolfsii* (Singh, 1976).

Artemisia *Species*

A. absinthium

Kaul et al. (1976) noted the toxicity of oil at 1:1,000 (1.0×10^3 µl/l) dilution against *Candida* species and *Aspergillus niger.*

A. afra

The essential oil present in leaf possesses a high degree of fungitoxicity (Graven et al., 1992).

A. capillaris

The antimycotic activity in essential oil of *A. capillaris* is due to capillin (Imai, 1956). Ikenaka et al. (1956) reported that the oil showed toxicity to *Alternaria kikuchiana, Aspergillus awamori, A. niger, A. oryzae, Gibberella fujikuroi, G. bataticola, G. sanbinetti, Colletotrichum miyabeanus, Penicillium chrysogenum, Pyricularia oryzae,* and *Rhizopus javanicus.*

A. cina

The oil of *A. cina* was found to be active against *Candida albicans* and *Microsporum* species (Vichkanova et al., 1972).

A. giraldi

Zheng et al. (1996) reported toxicity in oil of *Artemesia giraldi* against *Aspergillus flavus* and *Trichoderma viride*. This must be due to the flavones present in the oil.

A. martima

The antimycotic activity of *A. martima* against *Microsporum gypseum, Trichophyton equimum,* and *T. rubrum* was reported by Dikshit and Hussain (1984).

A. mexicana

Artemesia mexicana possessed strong in vitro antifungal activity against *Candida albicans* (Navarro et al., 1996).

A. pallens *and* A. vulgaris

Fungitoxicity in essential oil of *A. pallens* and *A. vulgaris* was reported against *Fusarium moniliforme, Helminthosporum longisporum, Trichoderma viride,* and *Colletotrichum* species (Laxmi and Rao, 1991).

A. parviflora

Mehrotra et al. (1993) found fungitoxicity in essential oil against *Candida albicans* and *Sporotrichum* species.

A. vestita

The antifungal activity of the oil at 1:1,000 (1.0×10^3 µl/l) dilution was reported against *Nannizzia fulva, N. gypsia,* and *N. incurvata* (Kaul et al., 1976; Gautam et al., 1980).

A. judaica *and* A. herba-alba

Charchari et al. (1996) reported remarkable antifungal activity in essential oil of *A. herba-alba* and *A. judaica* against *Candida albicans, C. stellatoidea, C. tropicalis, Microsporum canis, M. gypseum, Trichophyton interdigitale,* and *Aspergillus terreus.*

Blumea membranacea

The antimycotic activity of the oil was reported against *Alternaria helianthi, Fusarium moniliforme, Helminthosporum oryzae, H. turcicum, Pyricularia setariae,* and *Rhizoctonia solani* (Zutschi and Mehta, 1977; Sharma et al., 1978). Later, its toxicity was reported against *Aspergillus luchuensis, A. sydowi,* and *Cladosporium cladosporioides* (Geda and Bokadia, 1979).

Caesulia axillaris

Zutschi et al. (1975) found antifungal activity in essential oil of *Caesulia axillaris*. Pandey et al. (1982) also reported antimycotic potential of the plants against *Helminthosporium oryzae.* The oil showed a broad fungitoxic spectrum as well as superiority over eight synthetic fungicides.

The antifungal factor of the oil was thermostable, durable up to 180 days of storage, and could bear increased inoculum density (Pandey et al., 1982).

Cirsium dipsacolepsis

Cyperenal isolated from the plant *Cirsium dipsacolepsis* possessed strong antibiotic activity agaist bacteria and fungi. The antifungal activity was measured against *Rhizoctonia solani* (Achenbach and Benirschke, 1994).

Eupatorium *Species*

Eupatorium ayapana

Antimycotic activity in oil was recorded against species of *Aspergillus, Curvularia,* and *Penicillium* (Chourasia and Kher, 1978). Further, antifungal activity of the oil was also noted against *Alternaria* spp., *Aspergillus* spp., *Cladosporium herbarum, Cunninghamella echinulata, Fusarium* spp., *Helminthosporium sacchari, Microsporum* spp., *Mucor mucedo, Penicillium digitatum, Rhizopus* spp., and *Trichophyton* spp. (Sharma and Singh, 1979).

E. cappilifolium

The leaves yielded 1 percent oil, which was fungistatic to various fungi at a 1,000-ppm (1.0×10^3 µl/l) dose (Chandra et al., 1982). They further reported that the fungitoxicity of the oil was enhanced at pH 7 and pH 9. Rao et al. (1992) observed antifungal activity in the leaf oil against *Colletotrichum falcatum, Curvularia pallescens,* and *Periconia atro-purpurea.*

E. triplinerve

Antifungal activity in oil extracted from the leaves of the plant was recorded against *Aspergillus, Curvularia, Fusarium, Paecilomyces, Trichurus,* and *Helminthosporium* (Garg, 1974; Yadav and Saini, 1990).

Glossocordia bosvallia

Antimycotic activity of the essential oil against plant pathogenic as well as human pathogenic fungi, namely, *Aspergillus niger, Botryodiplodia theobromae, Botryothichum keratinophilum, Chrysosporium tropicum, Microsporum gypseum, Malbranchea pulchella, Phytophthora parasitica* var. *piperina,* and *Rhizopus nodosus,* was reported by Pathak and Dixit (1984).

Helianthella quiquenervis

Castaneda et al. (1996) evaluated the antifungal efficacy of this plant against three fungi: *Candida albicans, Aspergillus niger,* and *Trichophyton mentagrophytes.*

Helichrysum italicum

Chirkina and Osipova (1974) isolated eugenol, furfurol, geraniol, neurol, and α-pinene in oil extracted from flowers and found to be antifungal against *Candida albicans.*

Inula ***Species***

Inula cuspidata

The oil is highly antifungal against *Aspergillus* species (Chauhan and Saxena, 1985).

I. helenium

Bourrel et al. (1993) reported antimycotic activity in essential oil against seven test fungi.

I. racemosa

The toxicity in oil of *I. racemosa* was reported against *Alternaria helianthi, Colletotrichum capsici, Fusarium moniliforme, F. solani, Helminthosporium turcicum, H. oryzae, Pyricularia setariae, Pythium vexans, Rhizoctonia bataticola,* and *R. solani* (Misra and Dixit, 1978; Mishra et al., 1979).

Jasonia *Species*

Hammerschmidt et al. (1993) found antifungal activity in essential oil extracted from *J. candicans* and *J. montana* against *Trichophyton mentagrophytes, Cryptococcus neoformans,* and *Candida albicans.*

Parthenium *Species*

P. hysterophorus

Sharma and Singh (1979) reported a remarkable antimycotic activity in oil. The oil completely inhibited the growth of *Alternaria* spp., *Aspergillus* spp., *Cladosporium herbarum, Cunninghamella echinulata, Fusarium* spp., *Helminthosporium sachhari, Microsporum* spp., *Mucur mucedo, Penicillium digitatum, Rhizopus* spp., *Trichophyton* spp., and *Trichothecium roseum.* Antimycotic potential of different extracts of *Parthenium hysterophorous* against human pathogenic fungi was also investigated by Rai and Upadhyay (1990) and later by Rai (1993, 1994, 1995).

P. tomentosa *and* P. argentatum

Fungistatic sesquiterpenoids isolated from the plants demonstrated significant antifungal activity against *A. niger* (Maatooq and Hoffmann, 1996).

Pectis elongata

Prudent et al. (1995) found the leaf oil toxic against six fungi.

Pentanema indica

Rai and Vasanth (1995) reported the sensitivity of vicolides isolated from *Pentanema indica* to *Microsporum gypseum, Chrysosporium tropicum,* and *Trichophyton mentagrophytes* (Table 6.2).

TABLE 6.2. Sensitivity of vicolides isolated from *Pentanema indica* to *Microsporum gypseum*, *Chrysosporium tropicum*, and *Trichophyton mentagrophytes*

Test Fungi	Drug			
	MG (μg)	CT(μg)	TT(μg)	Average
VA	62.50	15.62	62.50	46.87
VB	125	31.25	31.25	62.50
VC	62.50	15.62	62.50	46.87
VD	125	31.25	31.25	62.50
(Control)	7.81	15.62	15.62	13.01
Average	76.562	21.872	40.624	

Source: Adapted from Rai and Vasanth (1995).
MG = *Microsporum gypseum*
CT = *Chrysosporium tropicum*
TT = *Trichophyton terrestre*

Sphaeranthus indicus

The oil showed antimycotic activity against *Cephalosporium sacchari*, *Ceratocystis paradoxa*, *Curvularia lunata*, *Fusarium moniliforme*, *Helminthosporium sacchari*, *Physalospora tucumanensis*, and *Sclerotium rolfsii* (Rao and Joseph, 1971).

Tagetes erecta

Kishore and Dwivedi (1991) reported that the essential oil of leaf completely inhibited the mycelial growth of *Pythium aphanidermatatum* at 200-ppm (2.0×10^3 μl/l) dilution. Rai and Acharya (1999, 2000) recorded maximum antimycotic activity of the aqueous extract and essential oil of the plant against *Fusarium oxysporum* and *Trichophyton mentagrophytes* (Tables 6.3 and 6.4).

SESQUITERPENES AS ANTIMYCOTICS

Dermatophytes and other fungal pathogens have been found to be sensitive to sesquiterpene lactones, which are present as active agent in Asteraceous plants (Al Magboul et al., 1977; Lima et al., 1993; Mares et al., 1993; Maatooq and Hoffmann, 1996).

TABLE 6.3. In vitro evaluation of different extracts of plants of Asteraceae against *F. oxysporum* and *T. mentagrophytes*

		Dry weight in mg	
Name of plant	**Plant part used**	**FO**	**TM**
Acanthospermum hispidum	WP	358.6 (14.65%)	266.4 (23.83%)
Ageratum conyzoides	WP	265.2 (36.88%)	255.2 (27.89%)
Blainvillea acmella	WP	357.5 (14.9%)	290.5 (16.9%)
Blumea balsamifera	WP	232.2 (44.73%)	236 (32.52%)
B. mollis	WP	233.5 (44.42%)	239.3 (31.57%)
Caesulia axillaris	WP	332.4 (20.88%)	252.2 (27.89%)
Carthamus tinctorius	WP	335.1 (20.25%)	271.5 (22.37%)
Cichorium intybus	WP	312.8 (25.5%)	295.3 (15.58%)
Cyathocline purpurea	WP	220.4 (47.55%)	245 (29.95%)
Dahlia pinnata	WP	378.5 (9.91%)	267 (23.65%)
Echinops echinatus	WP	287.3 (31.62%)	285 (18.51%)
Eclipta alba	WP	275.5 (34.42%)	248.4 (28.97%)
E. prostrata	WP	258.3 (38.52%)	280 (19.94%)
	L	357.5 (14.91%)	275.5 (21.22%)
Elephantopus scaber	WP	327 (22.17%)	398.5 (−13.93%)
Emilia sonchifolia	WP	395.4 (5.8%)	323.5 (7.5%)
Erigeron asteroides	WP	337.7 (19.62%)	247 (29.37%)
Eupatorium triplinerve	WP	231.4 (44.92%)	225.5 (35.52%)
Galinsoga perviflora	WP	265.2 (36.88%)	285.5 (18.37%)

TABLE 6.3 *(continued)*

Name of plant	Plant part used	Dry weight in mg	
		FO	TM
Gnaphalium albo-lutium	WP	315.2 (24.98%)	235 (32.8%)
G. purpureum	WP	310.5 (26.09%)	255.5 (26.94%)
Launaea acaulis	WP	332.5 (20.86%)	432.5 (−23.65%)
Parthenium hysterophorus	WP	220.8 (47.44%)	230.7 (34.03%)
Spilanthes acmella	WP	409.5 (2.53%)	266.4 (23.8%)
	F	348 17.17%)	252 (27.94%)
Synedrella nodiflora	WP	285.4 (32.07%)	280.2 (19.88%)
Tagetes erecta	WP	232.2 (44.73%)	220 (37.09%)
	L	218.2 (48.06%)	23 (31.66%)
	F	197.9 (52.9%)	201 (42.6%)
T. patula	WP	215.2 (48.78%)	211 (37.67%)
	L	229.3 (45.42%)	237.8 (32%)
	F	215.2 (48.78%)	220 (37.09%)
Tithonia diversifolia	WP	245.9 (41.47%)	285.4 (18.4%)
Tridax procumbens	WP	360 (14.31%)	315.3 (9.84%)
Vernonia cineria	WP	368.4 (12.31%)	285 (18.5%)
V. divergens	WP	387.2 (7.84%)	270 (22.8%)
Xanthium strumarium	WP	311.5 (25.86%)	239.5 (31.52%)
Control		420.16	349.75

Source: Adapted from Rai and Acharya (1999).
Note: WP = whole plant, L = leaf, F = flower. Percentage inhibition is given in parentheses, FO = *F. oxysporum*, TM = *T. mentagrophytes.*

TABLE 6.4. In vitro evaluation of essential oils of some plants of Asteraceae against *Fusarium oxysporum* and *Trichophyton mentagrophytes*

		Inhibition zone in mm	
Plant species	**Plant part used**	***Fusarium oxysporum***	***Trichophyton mentagrophytes***
Ageratum conyzoides	WP	10.66 (±0.57)	10.00 (±0.5)
Blumea mollis	WP	21.00 (±0.86)	25.33 (±0.5)
B. balsamifera	WP	21.33 (±0.74)	27.83 (±0.20)
Caesulia axillaris	WP	19.83 (±1.56)	18.86 (±0.13)
Chrysanthemum indicum	WP	11.00 (±0.68)	7.83 (±0.11)
Cyathocline purpurea	WP	25.33 (±0.58)	27.83 (±0.15)
Dahlia pinnata	WP	8.66 (±0.28)	6.33 (±0.30)
	F	15.16 (±1.02)	16.83 (±0.78)
Eupatorium triplinerve	WP	25.00 (±0.5)	25.33 (±0.35)
Spilanthes acmella	WP	16.83 (±0.78)	18.33 (±0.35)
Tagetes erecta	WP	25.00 (±0.95)	27.90 (±0.34)
	F	26.66 (±0.38)	28.33 (±0.07)
Tagetes patula	WP	24.33 (±0.29)	27.50 (±0.5)
	F	26.06 (±0.79)	28.00 (±0.5)
Control (miconazole)		28.45 (±0.25)	30.05 (±0.13)

Source: Adapted from Rai and Acharya (2000).
Note: WP = whole plant, F = flower. The numbers in parentheses are standard deviation.

Magboul et al. (1977) reported antimycotic activity of vernolepin and vernodalin isolated from *Vernonia amygdalina* Del. They found that *Aspergillus niger* and *Candida albicans* were sensitive to both the pigments. Villarreal et al. (1994) stated that only one sesquiterpene lactone, namely, taraxasterol, showed antimycotic activity against *Candida albicans.* Maatooq and Hoffmann (1996) reported fungistatic activity of partheniol and guayulone (a new dinorsesquiterpenoid diketone), two pigments isolated from *Parthenium argentatum* x *P. tomentosa* (guayule hybrid). Artemisinin, a sesquiterpene isolated from *Artemisia annua* (sweet wormwood) was found to be strongly antifungal. Two new flavones, 4',6,7-trihydroxy-3',5'-dimethoxy-flavone and 5',5-dihydroxy-3',4',8-trimethoxyflavone, were isolated from *Artemisia giraldii,* and their structures were identified by spectroscopic methods. These two new flavones also showed antimycotic activity against *Aspergillus flavus* and *Trichoderma viride* (Zheng et al., 1996). Sesquiterpene lactones isolated from various plants have shown their potential against the infections caused by various fungi in general, and dermatophytes in particular as is obvious from Table 6.5.

MECHANISM OF ACTION

The actual mechanism of action of unsaturated sesquiterpene lactones is not yet clearly known. However, some reports indirectly suggest its action as an auxin inhibitor. Cavallito and Haskell (1945) suggested that the action of lactones is due to their specific reactivity with the sulphydryl (–SH) group. Later, their study was supported by many other researchers (Thimann and Bonner, 1949; Hall et al., 1980). It is assumed that the inhibitory action of unsaturated lactones should be prevented by BAL (2,3-dimercaptaol-propanol) or cysteine, a compound known to protect the –SH group from inactivating the substances.

TABLE 6.5. Antimycotic potential of some sesquiterpene lactones

Plant	Sesquiterpene lactone	Fungal pathogen
Artemisia giraldii	Flavones	*Aspergillus flavus* *Trichoderma viride*
Chamomilla reticulata (L.) Rauschert	Herniarin	*Microsporum cookei*
Inula racemosa Hook. f.	Alantolactone, isoalantolactone	*Fusarium solani*, *Rhizoctonia* spp., *Paecilomyces lilacinus*, *Trichophyton mentagrophyes*, *Microsporum canis*
Parthenium hysterophorus L.	Parthenin	*Microsporum gypseum*, *T. mentagrophytes*, *T. rubrum*, *Epidermophyton floccosum*, *Aspergillus niger*, *A. sulphureus*, *Rhizopus oryzae*
Vicoa indica DC.	Vicolides (A,B,C,D)	*T. terrestris*, *Chrysosporium tropicum*, *M. gypseum*
Vernonia amygdalina Del.	Vernolepin and vernodalin	*Aspergillus niger*, *Candida albicans*

Mares (1987) also worked on mode of action of protoanemonin, a sesquiterpene lactone of the family Ranunculaceae. She reported that *Rhodotorula* was the most sensitive yeast and *Epidermophyton floccosum* was the most sensitive dermatophyte. The variation in sensitivity may be due to varying pemeability of the mycelial and spore walls of different fungi tested. However, it is known that unsaturated lactones act as inhibitory substances against several microorganisms due to the ability of the molecule to penetrate the microbial cell.

According to Hall et al. (1980), the mechanism of action of anticancerous sesquiterpene lactones such as protoanemonin is based on the capacity of substances to react with –SH groups by a Michael type of addition. Mares (1987) stated that the moiety of the molecules containing protoanemonin, coumarin, and its derivatives may react with sulphydryl groups.

Many key regulators, e.g., DNA polymerase, phosphofructokinase, and microtubular proteins of the mitotic apparatus, contain exposed –SH groups, which could be susceptible to interaction with this type of substance (coumarin). Mares (1987) assumed that inhibitory activity of protoanemonin on growth is due to inactivation of sulphydryl-containing enzymes necessary for cellular replications. Evidence for this hypothesis is the lack of inhibition of yeast's growth when L-cysteine is added to the culture medium. In fact, L-cysteine could remain bound to the antibiotic, thus preventing the binding and the inhibition of several metabolic enzymes.

Protoanemonin induced ultrastructural modifications in *Microsporum cookei* that may be attributed to an interaction of molecule with thiolic groups. Further, it is assumed that the alterations in shape and polarity of the hyphae must be due to the effect of protoanemonin on the –SH group.

CONCLUSIONS AND FUTURE PERSPECTIVES

Emergence of dreaded diseases such as AIDS and cancer are responsible for an increase in number of secondary infections generally caused by opportunistic fungi due to their immunocompromising capacity. The azoles and other antifungal drugs often fail to respond well to these infections. Therefore, there has been greater need to search for alternative antifungal agents from microbes or plants. Asteraceae, being the largest family of the plant kingdom and also owing to presence of essential oils in them, may prove to be the best natural alternative for discovering antifungal drugs. Traditional utilization of some of these plants against skin infections provides evidence that they contain antifungal properties. Most members of the family are great reservoirs of precious volatile oils containing sesquiterpenes as active chemicals. The antimycotic nature of most volatile oils has been established by researchers all over the world.

The existing costly therapy for fungal infections does not bode well for millions of individuals, particularly those in the developing world. Plant extracts or essential oils are easily available secondary metabolites and are within the reach of underprivileged and poor people.

There are some basic advantages of utilization of Asteraceous plant-derived products/essential oils for immunocompromised patients suffering from cancer or severe burn. These oils will provide soothing fragrance, tone infected skin, and also ameliorate the areas of the body colonized by secondary pathogens in general and mycotic infections in particular. These natural products are extremely useful for cancer patients not only for fighting against secondary mycotic infections but also for alleviating the severity of the disease, as most antifungal sesquiterpenes are anticancerous, too.

Further, it is necessary to expose the secret of proper mechanism of action of sesquiterpene on fungal pathogens as well as on hosts. Tremendous therapeutic and commercial potential exists in antimycotic agents (essential oils) of Asteraceous plants, but the present need is to tap these valuable natural resources. Revitalization of the natural curative power of plants such as *Tagetes erecta, T. patula, Eupatorium triplinerve,* and *Tridax procumbens* will generate awareness among the public of the utility of these plants. Finally, there is a pressing need to search for more plants containing volatile substances that can be useful in combating mycotic infections. In new antifungal drug targeting strategies, Asteraceous essential oils should be given prime importance because the majority of the promising antifungal essential oils may generate new drug candidates. In twenty-first century, discovery of plant-based antifungal drugs might be a biotechnologically driven process—with increasing importance attached to the discovery of new drugs from Asteraceous plants.

REFERENCES

Abbasoglu, U. and Kusmenoglu, S. (1994). Antibacterial and antifungal studies on *Achillia* L. species. *J. of Faculty of Pharmacy of Gazi University* 11: 177-181.

Abraham, Z., Bhakuni, D.S., Garg, H.S., Goel, A.K., Mehrotra, B.N., and Patnaik, C.K. (1986). Screening of Indian plants for biological activity. Part (XII). *Ind. J. Exp. Bio.* (24): 46-68.

Achenbach, H. and Benirschke, G. (1994). P38 chemical and biological activity of cyperenal and structurally related sesquiterpenes. *Euro. J. Pharm. Sci.* 2: 117-194.

Achola, K.J., Mwangi, J.W., Senie, K.A., Munenge, R.W., and Mwaura, A.M. (1997). Pharmacological activities of *Gutenbergia cordifolia*. *International Journal of Pharmacognosy* 35(1): 60-62.

Adams, S., Kunz, B., and Weidenborner, M. (1996). Mycelial deformations of *Cladosporium herbarum* due to the application of eugenol or carvacrol. *J. Essent. Oil Res.* 8: 535-540.

Aizeman B.E. (1978). *Microbiol zh (Kiev.)* 40: 233.

Alankara, G.S., Rao, J.G., and Prasad, Y.R. (1981). Investigation on the essential oil from *Artemisia pallens* Wall and *A. vulgaris* L. *Indian Perfum.* 25: 12.

Alkofahi, A., Batshoun, R., Owais, W., and Najib, N. (1996). Biological activity of some Jordanien medicinal plants extracts. *Fitoterapia* 67(5): 435-442.

Al Magboul, A.Z.I., Bashir, A.K., Khalid, S.A., and Farouk, A. (1997). Antimicrobial activity of vernolepin and vernodalin. *Fitoterapia* 68(1): 83-84.

Asthana, A., Chandra, H., Dikshit, A., and Dixit, S.N. (1982). Volatile fungitoxicants from leaves of some higher plants against *Helminthosporium oxyzae*. *Pflakarnakh Pflschutz.* 89: 475-479.

Aswal, B.S., Bhakuni, D.S., Goel, A.K., Kar, K., Mehrotra, R.N., and Mukherjee, K.C. (1984). Screening of Indian plant for biological activity (Part X). *Indian Journal of Experimental Biology* 22: 312-332.

Barde, A.K. and Singh, S.M. (1983). Activity of plant extracts against scytalidium anamorph of *Hendersonula toruloidea* causing nail diseases in man. *Indian Drugs* 20(9): 362-364.

Barel, S., Segal, R., and Yashphe, J. (1991). The antimicrobial activity of the essential oil from *Achillea fragrantissma*. *Journal of Ethnopharmacology* 33(1-2): 187-191.

Barnes, G.L. (1963). In vitro toxicity of various fixed and essential oils of pecan scab fungus *Fusicladium effusum*. *Plant. Dis. Reptr.* 47: 114-117.

Bhakuni, D.S., Dhar, M.L., Dhar, M.M., Dhawan, B.N., Gupta, B., and Srimal, R.C. (1971). Screening of Indian plants for biological activity, Part III. *Indian J. Experimental Biol.* 9: 91-102.

Bhakuni, D.S., Dhar, M.L., Dhawan, B.N., and Mehrotra, B.N. (1969). Screening of Indian plants for biological activity, part III. *Indian J. Experimental Biol.* 7: 250-262.

Birch, A.J. (1966). Some natural antifungal agents. *Chem. Indust.* 1173-1176.

Birner, J. and Nicolis, J.M. (1973). *Antimicrob. Agents. Chemothr.* 3(1): 105-109.

Bourrel, C., Vilarem, G., and Perineau, F. (1993). Chemical analysis, bacteriostatic and fungistatic properties of the essential oil of elecmpane (*Inula helenium*). *J. of Essential Oil Research* 5: 411-417.

Burden, R.S. and Bailey, J.A. (1975). Capsidiol: An antifungal compound produced in *Nicotiana tabacum* and *N. clovolandii* following infection with tobacco necrosis virus. *Phytochemistry* 14(5-6): 1989-1990.

Calpouzos, L. (1969). Oils. In *Fungicides,* Vol II. Torgeson, D.C. (Ed.). London: Academic Press, Inc. Ltd., pp. 367-393.

Castaneda, P., Gomez, L., and Mata, R. (1996). Phytogrowth-inhibitory and antifungal constituents of *Helianthella quinquenervis*. *J. Nat. Prod.* 59: 323-326.

Cavallito, C.J. and Haskell, T.H. (1945). The mechanism of action of antibiotics. The reaction of unsaturated lactones with cystein and related compounds. *J. Ana. Chem. Soc.* 67: 1991-1994.

Chandra, H. and Dikshit, A. (1981). Volatile fungitoxicants from from the leaves of *Ageratum conyzoides* against *Colletotrichum capsici* and *Penicillium italicum. J. Indian Bot. Soc.* 60: 23.

Chandra, H., Pandey, D.K., Tripathi, N.N., and Dixit, S.N. (1982). Toxicity of essential oil of *Eupatorium capillifolium* against *Helminthosporium oryzae. Indian Perfum.* 26: 231-236.

Charchari, S., Dahoun, A., Bachi, F., and Benslimani, A. (1996). In vitro antimicrobial activity of essential oils *Artemisia herba-alba* and *Artemisia judaica* from Algeria. *Rivista Italiana EPPOS* 18: 3-6.

Chauhan, P. and Saxena, V.K. (1985). Antifungal activity of essential oil of the leaves of *Inula cuspidata. Ind. J. Pharm. Sci.* 47: 160-161.

Chirkina, N.N. and Osipova, E.A. (1974). Antimicrobial properties of essential oils from everlasting flower *(Helichrysum italicum).* cultivated in the cremea. *Biol. Nauki.* 17: 86-89.

Chourasia, S.C. and Kher, A. (1978). Inhibitory effect of essential oils of three medicinal plants against various fungi. *Indian Drugs Pharm.* 14: 3-6.

Daphne, M., Felipe Balza, G.H., Towers, N., and Levy, J.G. (1982). Antitumour effects of the sesquiterpene lactone parthenin. *Planta Medica* 45: 23-27.

Dekker, J. (1969). Antibiotics. In *Fungicides* Vol II. Torgeson, D.C. (Ed.). New York and London: Academic Press, pp. 580-625.

Demchenko, N.P., Bugayenko L.A., Serkova A.A., and Domidov, L.V. (1995). Investigation of essential oils, spice-aromatic and medicinal plants in IEOMP. *Proceedings of 13th International Congress of Flavour, Fragrances and Essential oils,* Vol. 1. October 15-19. Istanbul, Turkey: K.H.C. Baser, p. 65.

Deshmukh S.K., Jain, P.C., and Agrawal, S.C. (1986). A note on mycotoxicity of some essential oils. *Fitoterapia* 67: 295-296.

Devi, Y.P. and Nandkumar, N.V. (1983). Herbicidal properties of *Parthenium hysterophorus* root extracts. *Geobios* 10: 134.

Dhar, M.L., Dhar, M.M., Dhawan, B.N., Mehrotra, B.N., and Ray, C. (1968). Screening of Indian plants for biological activity: Part I. *Indian J. Experimental Biol.* 6: 232.

Dhar, M.L., Dhar, M.M., Dhawan, B.N., Mehrotra, B.N., Srimal, R.C., and Tandon, J.S. (1973). Screening of Indian plants for biological activity: Part IV. *Indian J. Experimental Biol.* 11: 43-54.

Dhar, M.L., Dhawan, B.N., Prasad, C.R., Rastogi, R.P., Singh, K.K., and Tandon, J.S. (1974). Screening of Indian plants for biological activity: Part V. *Indian J. Experimental Biol.* 12: 512-523.

Dhawan, B.N., Dubey, M.P., Mehrotra, B.N., Rastogi, R.P., Singh, K.K., and Tandon, J.S. (1980). Screening of Indian plants for biological activity: Part IX. *Indian J. Experimental Biol.* 18: 594-606.

Dhawan, B.N., Patnaik, G.K., Rastogi, R.P., Singh, K.K., and Tandon, J.S. (1977). Screening of Indian plants for biological activity: Part V and VI. *Indian J. Experimental Biol.* 15: 208-219.

Dikshit, A. and Hussain, A. (1984). Antifungal action of some essential oils against animal pathogens. *Fitoterapia* 55: 171-176.

Dixit, S.N and Tripathi, S.C. (1982). Antifungal antibiotics from higher plants. In *Recent advances in the biology of micro-organisms.* Bilgrami, K.S. and Vyas, K.M. (Eds.). Dehra Dun, India: Bisen Singh Mahendra Pal Singh, pp. 519-525.

Emmett, E.A. and Marrs, J.M. (1973). *Dermatol.* 108: 98-99.

Fawcett, C.H. and Spencer, D.M. (1970). Plant chemotherapy with natural products. *Ann. Rev. Phytopathol.* 8: 403-418.

Garg, S.C. (1974). Antifungal activity of some essential oils. *Indian J. Pharm.* 36: 46-47.

Gautam, M.P., Jain, P.C., and Singh, K.V. (1980). Activity of some essential oils against dermatophytes. *Indian Drugs* 17: 267-270.

Geda, A. and Bokadia, M.M. (1979). Antifungal activity of the essential oil of *Blumea membranacea. Indian Drugs Pharm.* 14: 21-22.

Gellin, G.A., Maibach, H.I., and Wachs, G.N. (1972). Contact allergy to tolnaphtate. *Dermatol.* 106: 715-716.

Gopallakrishnan, G., Banumathi, B., and Suresh, G. (1997). Evaluation of the antifungal activity of natural xanthones from *Garcinia mangostana* and their synthetic derivatives. *Journal of Natural Products* 60(5): 519-524.

Goren, N., Woerdenbag, H.J., and Bozok-Johansson, C. (1996). Cytotoxic and antibacterial activities of sesquiterpene lactones isolated from *Tanacetum praeteritum* subsp. *praeteritum. Planta Medica* 62: 419-422.

Graven, E.H., Deans, S.G., Svoboda, K.P., Mavi, S., and Gundidza, M.G. (1992). Antimicrobial and antitoxidative properties of the volatile (essential) oil of *Artemisia afra. Flavour and Fragrance Journal* 7: 121-123.

Grosvenor, P.W., Supriona, A., and Gray, O.O. (1995). Medicinal plants from Riau province, Sumatra, Indonesia, Part 2: Antibacterial and antifungal activity. *Journal of Ethnopharmacology* 45(2): 97-111.

Hall, I.H., Lee, K.H., Starnes, C.O., Muraoka, O., Sumida, Y., and Waddell, T.G. (1980). Antitumoral agents. 21. A proposed mechanism for inhibition of cancer growth by tenulin and helenalin and related cyclopentenones. *J. Pharm Sc* 69: 694.

Hammerschmidt, F.J., Clark, A.M., Soliman, F.M., El- Kashoury, E.A., El- Kawy, M.M., and El- Fishawy, A.M. (1993). Chemical composition and antimicrobial activity of essential oils of *Jasonia candicans* and *J. montana. Planta Medica* 59: 68-70.

Hamsaveni Gopal, Vasanth, S., and Kundu, A.B. (1992). In vitro antimicrobial efficacy of *Vicoa indica. Fitoterapia* 63: 546-550.

Hiller, K. (1964). Antimicrobial substances in flowering plants—A review. *Pharmazie* 19: 167-188.

Ikenaka, T., Ikeda, N., Taneka, K., and Sugawara, S. (1956). Studies on the essential oil of *Artemisia capillaris* Thunb. I. Antifungal activity of essential oil. *Bayer Synp.* 5: 325-343.

Ikram, M. and Inam-ul-haq (1980). Screening of medicinal plants for antimicrobial activity, *Fitoterapia* part I, 51(5): 231-235.

Ikram, M. and Inam-ul-haq (1984). Screening of medicinal plants for antimicrobial activity, *Fitoterapia* part III, 55(1): 62-64.

Imai, K. (1956). Studies on the essential oil of *Artemisia capillaris* Thunb. III. Antifungal activity of essential oil. Structure and antifungal principal capillin. *J. Pharm. Soc. Japan* 76: 405-408.

Itsuo, Nishioka (1985). Discovery of a new biological activities and the active components in rhubarb. In *Herbal Medicine: Kampo, Past and present.* Taro Takemi, Mitsuto Hasegawa, Akira Kumagai, and Yasuo Otsucak (Eds.). Tokyo: Tsumura Juntendo Inc., p. 103.

Jain, P.C. (1996). Antimycotic components from higher plants: Assessment of present status and future strategies. In *Herbal Medicines, biodiversity and conservation strategies.* R.C. Rajak and M.K. Rai (Eds.). Dehradun, India: International Book Distributors, p. 292.

Jain P.C. and Agarwal, S.C. (1976). Activity of plant extract against some keratinophilic species of *Nannizzia. Indian Drugs* 13(12): 25-26.

Jain, P.C. and Agarwal, S.C. (1992). Sporostatic effect of some oils against fungi causing otomycosis. *Indian J. Medical Sci.* 46: 1-6.

Jain, P.C., Charia, A.K., Sharma, S. K., and Bokadia, M. M. (1980). Study of some essential oils for their antifungal activities. *Indian Drugs* 16: 122-123.

Jain P.C., Jain, C.K., and Jain, K. (1980). A note on the activity of odoriferous compounds against dermatophytes. *Indian Drugs* 17(12): 397-398.

Kaul, V.K., Nigam, S.S., and Dhar, K.L. (1976). Antimycotic activities of essential oils of *Artemisia absinthium* Linn., *A. vestita* Wall., and *A. vulgaris* Linn. *Indian. J. Pharm.* 38: 21.

Kedzia, B., Krzyzaniak, M., and Holderna, E. (1990). Effect of Yarrow essential oil *(A. millefolii)* and its components on pathogenic microorganisms. *Herba Polonica* 36(3): 117-125.

Khan, M.T. and Evans, F.J. (1996). Clinical evaluation of *Tagetes erecta* in the treatment of parakeratosis. *Phytotherapy Research* 10(1): 186-188.

Kishore, N. and Dwivedi, R.S. (1991). Fungitoxicity of the essential oil of *Tagetes erecta* L. against *Pythium aphanidermatum* Fitz., the damping-off pathogen. *Flavour and Fragrance Journal V* 6(4): 291-294.

Korbely, I. and Florian, E. (1971). Effects of essential oils on *Candida albicans. Gyogyszereszet* 15: 462-471.

Korta, J. and Starzyk, J. (1963) Investigation on antibiotic properties of essential oils of certain species of Umbelliferae. *Acta Botinica Cravoviensia* 6: 149-160.

Laxmi, U.S. and Rao, J.T. (1991). Antimicrobial properties of the essential oil of *Artemisia pallens* and *Artemisia vulgaris. Perfumerie und Kosmetik* 72(8): 510-511.

Lima, E.O., Gompertz, O.F., Giesbrecht, A.M. and Paulo, M.Q. (1993). In vitro antifungal activity of essential oils obtained from officinal plants against dermatophytes. *Mycoses* 36: 333-336.

Lima, E.O., Gompertz, O.F., Paulo, M.Q., and Giesbrecht, A.M. (1992). In vitro antifungal activity of essential oils against clinical isolates of dermatophytes. *Rev. Microbiol. Sao Paulo* 23(4): 235-238.

Lowey, M., Parent, D., and Ledoux-Corbusier, M. (1985). A new combination of tolnaphtata and methylparticine (SPA-5-345, Triol). *Mycosen* 28(9): 452-456.

Maatooq, G.T. and Hoffmann, J. J. (1996). Fungistatic sesquiterpenoids from *Parthenium. Phytochem* 43(1): 67-69.

Mahadevan, A. (1982). *Biochemical aspects of plant disease resistance.* Part I. *Performed inhibitory substances—Prohibitins.* New Delhi: Today and Tomorrow's Printers and Publishers.

Mares, D. (1987). Antimicrobiol activity of protoanemonin, a lactose from Ranunculaceous plant. *Mycopathologia* 98(3): 133-140.

Mares, D. (1989). Electron microscopy of *Microsporum cookei,* in vitro treatment with protoanemonin: A combined SEM and TEM study. *Mycopathologia* 108: 37-46.

Mares, D. and Fasulo, Maria P. (1990) Ultrastructural alterations in *Epidermophyton floccosum* and *Trichophyton mentagrophytes* exposed in vitro to protoanemonin. *Cytobios* 61: 89-95.

Mares, D., Romagnoli, C., and Bruni, A. (1993). Antidermatophytic activity of *Hernarin* in preparations of *Chamomilla reticulata* (L.) Rauschert. *Plantes Medicinale et Phytoterapie* 26(2): 91-100.

Maruzzella, J.C. (1963). The effect of perfume oils on the growth of phytopathogenic fungi. *Plant Dis. Rep.* 47: 756-757.

Maruzzella, J.C. and Balter, F. (1959). The action of essential oils on phytopathogenic fungi. *Plant Disease Res.* 43: 1143.

Maruzzella, J.C., Balter, F., and Katz, A. (1959). Further studies on the action of perfume oil vapors microorganisms. *Perfum Ess. Oil Res.* 50: 955.

Maruzzella, J.C. and Logeuri, F. (1959). The in vitro antifungal activity of essential oils. *J. Am. Pharm. Ass.* 47: 250.

Maruzzella, J.C., Serandis D., Serandis, J.B., and Grahm, G. (1960). Action of odoroferous organic chemicals and essential oils on wood destroying fungi. *Plant Diseases Res.* 44: 789.

Mathela, C.S. and Sinha, G.K. (1978). Antibacterial and antifungal study of some indiginous essential oils. *J. Res. Indian Med. Yoga Homoeopathy* 13: 122-124.

McCutchen, A.R., Ellis, S.M., Hancock, R.E.W., and Towers, G.H.N. (1994). Antifungal screening of medicinal plants of British Columbian native peoples. *Journal of Ethnopharmacology* 44(3):157-169.

Mehrotra, S., Rawat, A.K.S., and Shome, U. (1993). Antimicrobial activity of the essential oils of some Indian *Artemisia* species. *Fitoterapia* 64: 65-68.

Miguel, O.G., Lima, E.O., Morais, V.M.F., Gome, T.A., Monache, F.D., Cruz, A.B., Cruz, R.C.B., and Filho, V.C. (1996). Antimicrobial activity of constituents isolated from *Lychnophora salicifolia* (Asteraceae). *Phyto. Res.* 10(8): 694-696.

Mishra, S.H., Chaturvedi, S.C., and Dixit, V.K. (1979). Antimicrobial activity of essential oil of *Inula racemosa. Indian Drugs* 16: 141-143.

Misra, S.B. and Dixit, S.N. (1978). Fungitoxic activity of volatile fraction of three plants. *Indian Perfumer* 22: 61-62.

Misra, S.S. and Dixit, S.N. (1979). Antifungal activity of leaf-extracts of some higher plants. *Acta Bot. Indica* 7: 147.

Mwosu, M.O. and Okafor, J.L. (1995). Preliminary studies of the antifungal activities of some medicinal plants againts *Basidiobolus* and some other pathogenic fungi. *Mycoses* 38(5-6): 191-195.

Narayanarao, A., Padmasri, K., and Ananadkumar, R.C. (1996). Antifungal activity of extracts from *Azadirachta indica* Juss. against *Drechslera oryzae* Breda de Hann. *Geo-Bios New Reports* 15: 126-128.

Navarro, V., Villarreal, M.L., Rajas, G., and Lozaya, X. (1996). Antimicrobial evaluation of some plants used in Mexican traditional medicine for the treatment of infection as diseases. *Journal of Ethnopharmacology* 53(3): 143-147.

Overeem, J.C. (1976). In *Biochemistry of host parasite relationship*. Friend, J. and Threlfall, D.R. (Eds.). London: Academic Press, pp. 159-206.

Pandey, D.K., Chandra, H., Tripathi, N.N., and Dixit, S.N. (1983a). Mycotoxicity in leaves of some higher plants with special reference to that of *Ageratum haustonianum* Mill. *Mykosen* 26: 565-573.

Pandey, D.K., Chandra, H., Tripathi, N.N., and Dixit, S.N. (1983b). Toxicity of the essential oil of *Ageratum haustonianum* against *Fusarium lateritium* f. sp. *cajani. Beitr. Biol. Pflanz.* 58: 115-122.

Pandey, D.K., Tripathi, N.N., Tripathi, R.D., and Dixit, S.N. (1982). Fungitoxic and phytotoxic properties of the essential oil of *Caesulia axillaris* Roxb. (Compositae). *Angew. Botanic.* 56: 259-267.

Pathak, A. and Dixit, V.K. (1984). Antifungal and antibacterial studies on essential oil of *Glossocardia bosvallia. Indian Perfumer* 28: 91-94.

Pattnaik, S., Subramanyam, V.R., and Role, C. (1996). Antibacterial and antifungal activity of ten essential oils in vitro. *Microbios* 86(349): 237-245.

Perrucci, S., Mancianti, F., Cioni, P.L., Flamini, G., Morelli, I., and Macchioni, G. (1994). In vitro antibacterial activity of essential oils against some isolates of *Microsporum canis* and *Microsporum gypseum. Planta Med.* 60: 184-187.

Prudent, D., Perineau, F., Bessiere, J.M., and Michel, G. (1995). Analysis of essential oil *Pectis elongata* Kunth from Martinique. Evaluation of its bacteriostatic and fungistatic properties. *Journal of Essential Oil Research* 7: 63-68.

Radhika Iyer, S. (1992). Ethnobotany of certain medicinal plants used by tribals of India against skin infections. *Ancient Sci. of Life* 11(3/4): 143-152.

Rahalison, L., Hamburger, M., Hostettman, K., Monod, M., Frenk, E., Gupta, M.P., Santana, A.I., Correa, A.M., and Gonzalez, A.C. (1993). Screening for antifungal activity of Panamanian plants. *International Journal of Pharmacognosy* 31(1) : 68-76.

Rahalison, L., Hamburger, M., Monod, M., Frenk, E., and Hostettmann, K. (1994). Antifungal tests in phytochemical investigations: Comparison of bioautotrophic methods using phytopathogenic and human pathogenic fungi. *Planta Medica* 60(1): 41-44.

Rai, M.K. (1987). A first report of *Curvularia verruculosa,* an opportunistic human pathogen from scalp and in vitro sensitivity of this fungus to different plant extracts. *Indian Drugs* 24(11): 518-520.

Rai, M.K. (1988). In vitro sensitivity of *Microsporum nanum* to some plants extract. *Indian Drugs* 25: 521-523.

Rai, M.K. (1993). Laboratory evalution of fungitoxic activity of crude extract of *Parthenium hysterophorus. J. Environ Biol* 14(1): 41-44.

Rai, M.K. (1994). In vitro evaluation of aqueous extract of *Parthenium hysterophorus* against *Rhizophus oryzae*: A causal organism of otomycosis in a college student. *Indian Medicine* 44 (2): 4-5.

Rai, M.K. (1995). Comparative antimycotic activity of different parts of *Parthenium hysterophorus L. World Weeds* (2): 53-57.

Rai, M.K. and Acharya, D. (1999). Screening of some Asteraceous plants for antimycotic activity. *Compositae Newsletter* 34: 37-43.

Rai, M.K. and Acharya, D. (2000). Search for fungitoxic potential in essential oils of Asteraceous plants. *Compositae Newsletter* 35: 18-23.

Rai, M.K., Qureshi, S., and Agrawal, S.C. (1997). In vitro evaluation of inhibitory nature of extracts of 18 plant species of Chhindwara against 3-keratinophilic fungi. *Hindustan Antibiotics Bulletin* 39(1-4): 56-60.

Rai, M.K., Qureshi, S., and Pandey, A.K. (1999). In vitro susceptibility of opportunistic *Fusarium* spp. to essential oils. *Mycoses* 42(1,2): 97-101.

Rai, M.K. and Upadhyay, S.K. (1988). Screening of medicinal plants of Chhindwara district against *T. mentagrophytes,* causal organism of Tinea-pedis. *Hindustan Antibiotics Bulletin* 30(1-2): 33-36.

Rai, M.K. and Upadhyay, S.K. (1989). In vitro sensitivity of some plant extracts against *T. mentagrophytes. Indian Medicine* (1-4), 1-3.

Rai M.K. and Upadhyay, S.K. (1990). In vitro efficacy of different extracts of *Parthenium hysterophorous* Linn. against human pathogenic fungi using different techniques. *Indian J. Pathol. and Microbiol.* 33(2): 179-181.

Rai M.K. and Vasanth, S. (1995). Laboratory evaluation of sensitivity of three keratinophilic fungi to some vicolides. *Hindustan Antibiotics Bull.* 37(1-4): 48-50.

Rao, B.G.V.N. and Joseph, P.L. (1971). Activity of some essential oils towards phytopathogenic fungi. *Riechest Aromas. Koer.* 21: 405-410.

Rao, G.P., Singh, M., and Singh, H.N. (1992). Fungitoxic evolution of essential oils extracted from higher plants against some sugarcane pathogens. *Tropical Science* 32: 377-382.

Rao, J.R. (1976). The in vitro antimicrobial effect of the essential oil of *Ageratum conyzoides. Aromas. Koer.* 26: 50.

Ray P.G. and Majumdar, S.K. (1975). New antifungal substance from *Alpinia officinarum. Indian J. Exp. Biol.* 13(4): 489.

Ray P.G. and Majumdar, S.K. (1976). Antifungal flavonoid from *Alpinia officinarum. Indian J. Exp. Biol.* 14(6): 712-714.

Romanelli, L., Moronne L.A., Amino M.C., Mattioli, F., Menichini, F., and Valeri, P. (1996). Pharmacological activities of methanolic extract of *Centaurea deusta* and *Crepis lacera* in isolated rabbit jejanum. *Phyto. Res.* 10(Suppl.): 36-38.

Saxena, V.K., Sahai, A., and Sanadya, G. (1984) Studies on antimicrobial efficacy of essential oil of the leaves of *Anaphalis contorta. Indian Perfumer* 28: 177-178.

Sharma, G.P., Jain, N.K., and Garg, B.D. (1978). Antifungal activity of some essential oils. *Indian Drugs* 16: 21-23.

Sharma, S.K. and Singh, V.P. (1979). The antifungal activity of some essential oils. *Indian Drugs Pharm.* 14: 3-6.

Singh, G. (1999). Studies on the biocidal activities of certain essential oils. *JMAPS* 21: 1119-1130.

Singh, H.B. (2000). Fungitoxicity in volatile plant products: A review. In *Integrated management of plant resources.* M.K. Rai, Ajit Varma, and R.C. Rajak (Eds.). Jodhpur, India: Scientific Publisher, pp. 191-244.

Singh, K.V. and Deshmukh, S.K. (1984). Volatile constituents from members of Liliaceae and spore germination of *Microsporum gypseum* complexes. *Fitoterapia* 55(5): 297-299.

Singh, L. (1976). Effect of some oils on in vitro growth of three phytopathogenic fungi. *J. Indian. Bot. Soc.* 55: 186-189.

Singh, S.P., Shukla, H. S., Singh, R.S., and Tripathi, S.C. (1986). Antifungal properties of essential oil of *Ageratum conyzoides* L. *Nat. Acad. Sci Letters* 9: 97-99.

Tansey, M.R. and Appleton, J.A. (1975). Inhibition of fungal growth by garlic extract. *Mycologia* 67: 409-413.

Thapliyal, P.N. and Nene, Y.L. (1969). Inhibition of plant pathogens by higher plant substances. *J. Sci. Indian Res.* 26: 289-299.

Thimann, K. V. and Bonner, W.D. (1949). Inhibition of plant growth by protoanemonin and coumarin and its prevention by BAL. *Proc. Nati. Acad. Sci.* 35(45): 272-276.

Thind, T.S. and Dahiya, M.S. (1976). Inhibitory effect of essential oils of some medicinal plants against soil inhibiting dermatophytes. *Indian Drugs* 14: 17-18.

Tokin, B.P. (1960). Phytoncides, the destroyers of microbes. *Moscow Izdat.* Sov. Russia, 198.

Tripathi, R.D. and Dixit, S.N. (1981). *Fungitoxicity of lawsone, isolated from leaves of* Lawsonia inermis *and some related napthaquinones of angiosperm origin* (abstract). Paper presented at Third International Symposium on Plant Pathology, IARI, New-Delhi, December 14-18.

Tripathi, R.N., Pandey, D.K., Tripathi, N.N., and Dixit, S.N. (1988). Antifungal activity of some higher plants. *Indian Phytopathol.* 35: 346.

Tronnier, H., and Kosen, Entzundliche Dermatomy (1985). Thetrapievergleich zwisenen naftifin und linem costeroidhaltgen Imidazoli derivat (III Internationales Nurnberger Expertengesprach, Nurnberg. *Mykosen* 28: 98.

Vasanth, S. (1996). The chemistry and biology of *Pentanema indicum,* a folklore antifertility drug. In *Compositae: Biology and utilization. Proceedings of the International Compositae Conference,* Kew, 1994. D.J.N. Hind (Ed.). Volume 2. Royal Botanic Garden: Kew, pp. 423-429.

Vichkanova, S.A., Adqina, V.V., and Izosimova, S.B. (1972). Essential oils as a new source of antifungal preparations. *Fitontsidy Mater-Sovesbch. Naukova Dumka, Kiev,* 123-126.

Villarreal, M.L., Alvarez, Laurea, Alonso, D., Navarro, V., Garcia, P., and Delgado, G. (1994). Cytotoxic and antimicrobial screening of selected terpenoids from Asteraceae species. *J. of Ethnopharmacology* 42: 25-29.

Wahab, S., Tandon, R.N., Jacob, Z., Sagar, P., and Srivastava, O.P. (1981). In vitro activity of a phytochemical, alantolactone from *Inula racemosa* Hook. f. against some pathogenic and opportunistic fungi. *J. Indian Bot. Soc.* 60: 278-281.

Yadav, R.N. and Saini, V.K. (1990). In vitro antimicrobial efficacy of the essential oil of *Eupatorium triplinerve* leaves. *Indian Perfumer* 34: 61-63.

Zheng, W.F., Tan, R.X., Yang L., and Liv, Z.L. (1996). Two flavones from *Artemesia giraldii* and their their antimicrobial activity. *Planta Medica* 62(2): 160-162.

Zutschi, S.K., Joshi, S.K., and Bokadia, M.M. (1975). Antimicrobial activity of some Indian essential oils. *Indian J. Pharm.* 37: 129-130.

Zutschi, S.K. and Mehta, S.C. (1977). Screening of some essential oils for antifungal properties. *Indian Drugs and Pharm.* 12: 13-14.

Chapter 7

Recent Advances in the Search for Antimycotic Activity in South American Plants

Elena Mongelli

INTRODUCTION

Human and animal fungal infections pose serious medical and veterinary issues, whereas fungal infections of plants represent significant losses of agricultural products. There is general consensus among researchers, clinicians, and pharmaceutical and agrochemical companies that new, potent, effective, and safe antifungal drugs are needed (Selitrennikoff, 1992). Historically, many bioactive substances have been natural products. Therefore, it is logical that the search for new prototype antifungal products should also include natural products. The major advantage of this approach over chemical synthesis or modification of existing agents is the likelihood of similar toxicities, cross-resistance, or even mechanism of action. Over the past several decades the search for antifungals has been limited primarily to microorganisms, particularly actinomycetes and some fungi.

In designing a search for novel prototype antifungals, it seems reasonable to assume that if new agents are to be found that have different structures and different or supplemental activities from those in current use or development, sources other than the more traditional microorganisms must also be investigated. In particular, higher plants are a logical choice, chiefly because of their seemingly infinite variety of novel molecules, which are often referred to as "secondary metabolites" (Clark and Hufford, 1992). Antifungal agents are widely distributed among higher plants (Cáceres et al., 1991), but only a few have been evaluated for their activity against human, animal, and plant pathogenic fungi.

The present work was supported by the CONICET (grant PIP 775/98). The author wishes to thank Mrs. Mary Dickhuth for her assistance.

The AIDS crisis has served to focus renewed interest on the discovery of agents to treat opportunistic infections in immunocompromised hosts (Eckerman and Graham, 2000). These infections rarely occur in patients with normally functioning immune systems, but are a leading cause of morbidity and mortality in AIDS patients. Opportunistic infections also occur commonly in patients with any type of immunosuppression, i.e., cancer patients and organ transplant recipients, both of whom receive drugs to prevent life-threatening opportunistic infections. Therapy for these kinds of infections is complicated by several factors, including the relative ineffectiveness of available agents, the relative severe toxicities of such agents, the development of resistance to existing agents, and the underlying immune disorder of the patient. All of these factors contribute to the urgent need for the discovery of new prototype antibiotics.

Among the most common opportunistic infections are the disseminated mycotic infections cryptococcosis, candidiasis, and histoplasmosis. In the past, some of the most intensive efforts to develop new antifungal agents have been centered on the synthesis of analogs of the existing synthetic azole antifungal agents. However, the fundamental mechanisms remain the same. Thus, all the azoles are fungistatic in their action, requiring long-term use. Therefore, it seems unlikely that efforts to modify the structures of existing agents will provide any new products that do not share similar toxicity or resistance problems.

The role of prototype agents in the future of drug discovery in general, and antibiotics in particular, is crucial. Novel antibiotics with novel chemical structures can serve two important functions: (1) as "lead" compounds for structure-activity relationship studies and subsequent development of improved agents and (2) as probes for molecular targets (Clark and Hufford, 1992). Plants possess a variety of antimicrobial defenses, up to 10 percent of their dry mass, ranging from small molecules (for example, phytoalexines) to proteins (for example, chitinases, glucanases). However, only a small fraction of known plant species have been examined for the presence of antimicrobial products (Balandrin et al., 1992).

South America is an extremely interesting region for investigating new bioactive molecules. Some of the reasons are the existence of geographic areas with highly diverse flora, the unique ecological and physiological conditions under which many of the plants grow, and the presence of human societies with a strong tradition in the use of plant resources as medicinal agents.

The aim of the present chapter is to describe the advances in the search for antimycotic compounds and antimycotic activity in plants used for different purposes in South America.

ISOLATION OF ANTIFUNGAL COMPOUNDS FROM PLANTS

The process of searching for new antibiotic prototypes generally begins with sample acquisition. The selection of plants for evaluation can be based on traditional medicinal uses for diseases states that can be interpreted as infectious diseases. For example, the utilization of a particular plant for the treatment of such ailments as fever, diarrhea, or skin abrasions can be due to the presence of antimicrobial constituents in this plant. In addition, either random collection of samples or selection of plants based on chemotaxonomy, on an ecological basis, or from a combination of literature reports on plant extract, pharmacology, and ethnomedical claims associated with the same medicinal use may also lead to the identification of active plants (Farnsworth, 1994).

Once plants to be studied have been selected, plant extracts are subjected to evaluation in primary in vitro assay in order to assess inhibitory activity against fungi. Some of the fungi more commonly used are *Candida albicans, Cryptococcus neoformans, Aspergillus niger, A. fumigatus, A. flavus,* and *Saccharomyces cerevisiae.* One assay procedure commonly used during primary screening of activity is the agar well diffusion assay (Clark and Hufford, 1992), which is a qualitative method. The agar or broth dilution method (Zacchino et al., 1999), which allows quantification of the antifungal activity by means of calculating the minimum inhibitory concentration (MIC, the lowest amount of drug required to prevent growth), has also became a standard procedure. The purpose of the primary assay is to identify plant extracts that possess activity, after which a second stage of screening can be used. During secondary screening, one or more assays of higher specificity than the primary biossay is used. The target can be either whole cells or enzymes. In a whole-cell target screen, the effect of samples on an intact test microorganism is determined. One example is the whole cell *Neurospora crassa* hyphal growth inhibition method. This assay is an agar diffusion method useful to detect agents whose mode of action is associated with cell wall polymer synthesis or assembly. In enzyme-targeted screens, samples are tested for their ability to inhibit an enzyme activity in vitro. Because all human pathogens and most plant pathogens contain chitin and β-linked glucans in their cell walls, inhibition of chitin synthase and glucan synthase, respectively, are the most commonly used enzyme-targeted screens. Other mechanistic targets may be cell wall polymer cross-linking, mannoprotein synthesis, DNA topoisomerases, membrane sterols, microtubules, actine filaments, and energy generation.

To isolate the active constituents, active crude extracts are submitted to bioassay-directed fractionation. Once the active compounds are isolated, structural modifications for structure-activity relationship studies and total synthesis may be carried out. Finally, the promising candidates are assayed using a more detailed evaluation, also called tertiary evaluation, that can include mechanistic assays and the study of efficacy and toxicity using in vivo assays in appropriate animal models.

SEARCHING FOR ANTIMYCOTIC ACTIVITY IN SOUTH AMERICAN PLANTS

The search for antifungal activity in South American plants includes both studies that deal with the activity of crude extracts and reports of the isolation of the active compounds. The former are generally carried out in order to validate the ethnomedical uses of medicinal plants employed for disease states bearing some correlation with an infection. There are a large number of plants used in South America as antifungals, mainly related to fungal skin infections. A list of antifungal South American plants is summarized in Table 7.1 (Schultes and Raffauf, 1990; Lahitte et al., 1998).

It has already been stated that South America is a promising region for the study of biological activity in plants, mainly due to the presence of human societies with a strong tradition in the uses of plant resources as medicinal agents and the existence of highly diverse and little-studied flora. Of the 250,000 species of higher plants known to exist on Earth, a great proportion grows in South America. On the other hand, it is known that about 80 percent of the inhabitants of the world live in the less developed countries, and the World Health Organization estimates that a high percentage of these people rely almost exclusively on traditional medicine for their primary health care needs (Phillipson, 1999). Because medicinal plants are the backbone of traditional medicine, there is a need to study them on a regular basis. Since, this is the case for South America, the study of bioactivity of medicinal plants is extremely important in order to validate the ethonobotanical uses claimed. Furthermore, active plant extracts could be worthy of the search for new drugs.

Lorenti et al. (1981) tested the activity of Argentine plants *Bauhinia candidans* Benth, *Nierembergia hippomanica* Miers., *Gomphrena martiana* Moquin, and *Wedelia buphthalmiflora* Lorentz against strains of *Aspergillus niger* and *Saccharomyces cerevisiae*. The results indicated that the ethanolic extract of *Gomphrena martiana* was active against both microorganisms, probably because of its high content of flavonoid aglycones. On the other hand, the petrol extract of *Bauhinia candidans* showed activity against *Aspergillus niger*.

In a screening study of 46 South American plants employed in folk medicine, using *Aspergillus niger*, *Mucor mucedo*, and *Candida albicans* as test

TABLE 7.1. South American plants used as antifungals

Plants	Uses
ACANTHACEAE	
Justicia cabrerae Leonard	The roots are steamed for skin afflictions of the groin, probably of fungal origin.
Justicia schultessi Leonard	Same purposes as *J. cabrerae*
AMARANTACEAE	
Alternanthera pungens Kunth	Made into an infusion, the whole plant is used to treat skin diseases.
ANNONACEAE	
Fusaea decurrens R.E. Fries	The fruits are dried and ground and the powder is used to dust bleeding areas between the toes.
APIACEAE	
Hydrocotyle bonariensis Lam.	The juice of the plant is used against dermatosis.
APOCYNACEAE	
Aspidosperma schultesii Woodson	The latex is used between the toes on sores of fungal origin.
Mandevilla vanheurkii (Muell.-Arg.) Markgraf	The latex is applied to fungal infections of the skin.
Spongiosperma macrophyllum (Muell.-Arg.) Zarucchi	The latex is used externally to treat fungal infections of the scalp.
COMPOSITAE	
Baccharis articulata (Lam.) Pers	The decoction of the stem is used for treating wounds and skin diseases.
Baccharis trimera (Less.) DC.	Same purposes as *B. articulata*
CURCUBITACEAE	
Gurania bignonacea (P. & E.) C. Jeffrey	The leaves are rubbed on areas of skin affected by fungal infections.
EUPHORBIACEAE	
Chamaesyce hirta (L.) Millspaugh	The latex is used to treat fungal infections between the toes.
Euphorbia serpens Kunth	The infusion of the whole plant is used against dermatosis.
Micrandra minor Bentham	The latex is applied for fungal skin infections.
Sebastiana brasiliensis Spreng.	The latex is used for dermatosis.
GUTTIFERAE	
Caraipa laxiflora Bentham	The sap is used to treat fungal infections of the skin.

TABLE 7.1 *(continued)*

Plants	Uses
Vismia angusta Miquel	The latex is used for cutaneous fungal infections.
LEGUMINOSAE	
Enterolobium spp.	Made into an infusion, the bark is also applied to the skin as a fungicide.
Erythrina crista-galli L.	The fresh bark is crushed and applied for treating dermatosis.
MALPIGHIACEAE	
Tetrapterys silvatica Cuatrecasas	The leaves are burned and mixed with oil or grease to apply to fungal infections of the skin.
MENISPERMACEAE	
Curarea tecunarum Barneby & Krukoff	The stem is used to treat fungal and other skin infections.
MORACEAE	
Ficus gemina Ruiz ex Miquel	The latex is spread on the skin to relieve itching probably caused by fungal infection.
Ficus niceforoi Dugand	The latex is used to treat culebrilla, a fungal inflammation, especially of the feet.
Helicostylis scabra (Macbr.) C. C. Berg	The latex is used as antifungal agent when repeatedly painted and dried on the affected parts of the skin.
MYRISTICACEAE	
Dialyanthera parvifolia Markgraf	The bark is crushed and rubbed on the skin for treating infections caused by mites and fungi.
Iryanthera juruensis Warburg	The soft inner bark is scraped off and rubbed on areas of fungal infection of the skin.
Iryanthera paraensis Huber	The inner bark is topically applied to eliminate mites and treat fungal skin infections.
Otoba parvifolia (Mkf.) A. Gentry	The bark and red resin are rubbed on affected areas of skin.
Virola peruviana (A. DC.) Warburg	The inner bark and resin-like liquid is used to treat fungal infections of the skin.
ONAGRACEAE	
Ludwigia peploides (Kunth) P.H.Raven	The infusion of the plant is used to treat dermatosis.
POLIGONACEAE	
Poligonum punctatum Elliot.	The juice of the plant is used against dermatosis.

Plants	Uses
RANUNCULACEAE	
Clematis bonariensis Juss ex DC.	Made into an infusion, the leaves are used against dermatosis.
Clematis montevidensis Spreng	Made into an infusion, the leaves are used against skin afflictions.
RUBIACEAE	
Borreria laevis (Lam.) Grisebach	The plant is valued for treating fungal infections of the skin and for "white skin blotches."
Calycophyllum acreanum Ducke	The shredded bark is placed in water and applied to fungal infections called *wangga.*
Calycophyllum spruceanum (Benth.) Hooker Fil. ex K. Schumann	The powdered bark is put on to the skin for fungal infections.
Geophila cordifolia Miquel	The plant is employed for fungal infections.
Geophila repens (L.) I. M. Johnston	The plant is employed for fungal infections.
Warszewiczia coccinea (Vahl) Klotzch	The root is powdered and applied with oil to skin problems of apparent fungal origin.
SALICACEAE	
Salix humboldtiana Willd.	The ashes of the bark are macerated and applied on the skin for treating dermatosis.
SAPOTACEAE	
Chrysophyllum cainito L.	The latex is a cure for what appears to be a fungal infection of the groin.
STYRACACEAE	
Styrax tessmannii Perkins	The crushed leaves are applied between the toes to treat fungal infections.
SOLANACEAE	
Cestrum parqui L'Herit	The decoction of the leaves is used against dermatosis.
Physallis viscosa L.	Same purpose as *C. parqui*
Salpichroa origanifolia (Lam.) Tell.	The decoction of the root is employed for skin afflictions.
ULMACEAE	
Celtis tala Gillies ex Planch.	The fresh leaves are applied on the skin to treat skin diseases.

organisms, the extract of *Prosopis ruscifolia* Gris. showed activity against *Mucor mucedo* and *Candida albicans,* while the extract of *Usnea campestris* Santesson was active against all the microorganisms tested (Gutkind et al., 1981).

Another screening of antifungal activity of 56 higher Colombian plants showed that only nine plants demonstrated significant activity against the fungi tested (Sanabria and Mantilla, 1986).

Further studies using *Gomphrena martiana* Moquin showed that the ethanolic extract was active against *Candida albicans* and the ethanolic extract of *Gomphrena boliviana* Moquin showed activity against this yeast and *Saccharomyces cerevisiae* (Pomilio et al., 1992).

On the other hand, it is interesting to pointing out that *Tabebuia* spp. (family Bignoniaceae) bark products, in addition to their folk medicinal use against cancer, are being used against candidiasis-type fungal infections (Gentry, 1992).

Another study, conducted with 122 plants used in Argentina for the therapy of diverse infectious states, showed that *Punica granatum* L., *Citrus sinensis* Osbeck, and *Allium sativum* L. were active against *Aspergillus niger* (Anesini and Pérez, 1993). On the other hand, Penna et al. (1994), in an antifungal screening study, determined that different extracts of *Heimia salicifolia* Lk. Otto demonstrated significant activity against Candida *albicans.*

A series of Brazilian medicinal plant extracts were screened for in vitro antimicrobial activity. Among the plants tested, *Bryophyllum pinnatum* Aschers. & Schweinf and *Kalanchoe brasiliensis* Cambess displayed good fungicidal activity (Cunha et al., 1995).

The crude extract of the Peruvian plant *Mutisia acuminata* var. *acuminata* Ruiz & Pavon is a good example for the potential treatment of fungi crop pests, since in vitro assays showed that was quite potent against *Botrytis cinerea,* indicating its possible use as a natural pesticide to prevent storage rot (Catalano et al., 1998).

Several compounds isolated from South American plants have also been studied for their antimycotic activity. Some examples are listed next.

Epiphorellic acid 1 isolated from the lichen *Cornicularia epiphorella* (Nyl.) Dr. (Figure 7.1), was analyzed against a series of 27 strains of pathogenic and opportunistic fungi, showing different levels of susceptibility (Ugarte et al., 1987). This compound was active against all the fungal species tested—except *Chaetomium globosum,* which was resistant at dose levels of 100 μg/ml. On the other hand, *Cryptococcus neoformans* and *Aspergillus fumigatus* were susceptible at all the concentrations assayed (up to 20 μg/ml).

Xanthoxyline (2-hydroxy-4,6-dimethoxyacetophenone), an active constituent isolated from the leaves and stems of *Sebastiana schottiana* Muell. Arg. and *Phyllanthus sellowianus* Muell. Arg., was found to possess antifungal activity against *Candida albicans, Microsporum canis, Trichophyton rubrum, Aspergillus parasiticus, A. flavus,* and *Penicillium* spp. The results

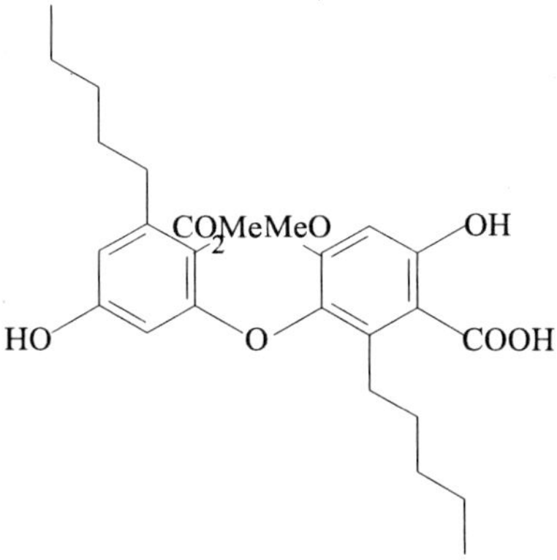

FIGURE 7.1. Ephiphorellic acid 1 from *Cornicularia epiphorella* (Me: methilo)

obtained could therefore support the popular use of these plants in South American folk medicine (Lima et al., 1995).

The known polyine phenylhepta-1,3,5-tryine (Figure 7.2), obtained from the methanolic extract of the whole plant of the subtropical American weed *Bidens pilosa* L., was found to be active against the dermatophytes *Trychophyton mentagrophytes* and *Microsporum gypseum,* while the new compound β-D-glucopyranosyloxy-3-hydroxy-6(E)-tetradecen-8,10,12-tryine (Figure 7.2) exhibited moderate activity against *T. mentagrophytes* (Alvarez et al., 1996).

Piper angustifolium Lam. is a plant native from Peru that is used as a decoction for external disinfection of wounds and sores. The essential oil exhibited fungistatic activity against *Candida albicans, Trychophyton mentagrophytes, Aspergillus flavus,* and *Aspergillus fumigatus.* The gas chromatography-mass spectrometry (GC-MS) analysis showed that camphor and camphene were the main constituents. Nevertheless, this compound exhibited only a moderate fungistatic activity against *C. albicans*. The higher fungistatic activity obtained with the oil, compared with those of the two major components, can be explained either by the synergistic effect of the different components in the oil and/or by the presence of other constituents that may be active even in small concentrations (Tirillini et al., 1996).

Bioactivity-guided fractionation of a dichloromethanic extract from leaves of the Brazilian plant *Piper hispidum* H.B.K. yielded a new pyrrolidine amide, N-[7-(3',4'-methylenedioxyphenyl)-2(Z), 4(Z)-heptadienoyl] pyrrolidine (Figure 7.3), which showed antifungal activity against *Cladosporium sphaerospermum* (Alécio et al., 1998).

Chemical investigations on species of *Virola* and related genera of Myristicaceae from the Amazonian region led to the hypothesis that the al-

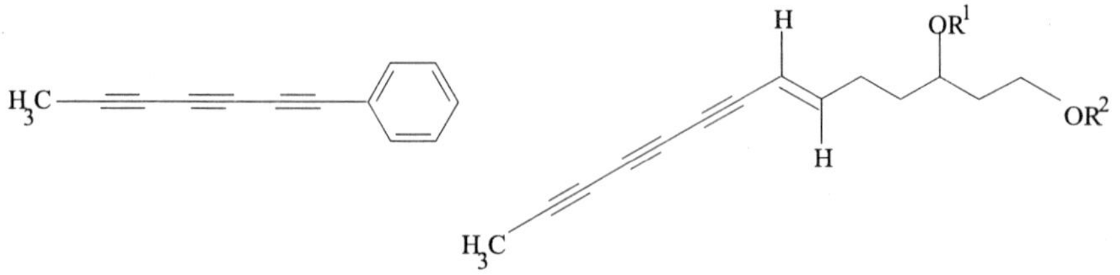

FIGURE 7.2. Phenylhepta-1,3,5-tryine and compound β-D-glucopyranosyloxy-3-hydroxy-6(E)-tetradecen-8,10,12-tryine from *Bidens pilosa* L.

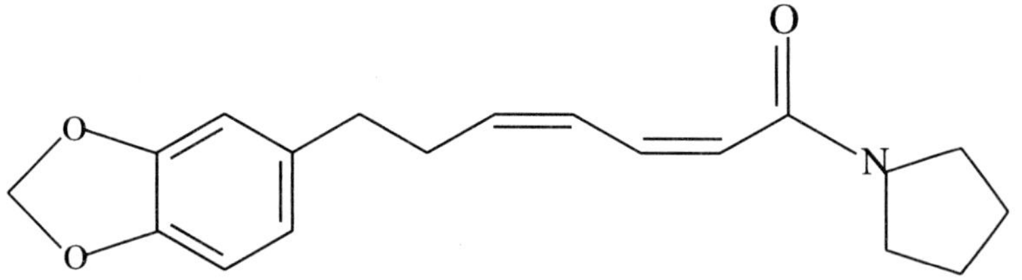

FIGURE 7.3. Chemical structure of N-[7-(3',4'-methylenedioxyphenyl)-2(Z), 4(Z)-heptadienoyl] pyrrolidine from *Piper hispidum* H.B.K.

leged usefulness of plasters made from their leaves or bark resin in the treatment of skin fungal infections may be due to the fungistatic or fungitoxic activity of neolignans. Among the wide variety of known neolignans, the 8.O.4'-type represents a small group whose members were isolated exclusively from plants of Myristicaceae. Different ketones and alcohols of *threo* and *erythro* relative configuration have been isolated from *Virola surinamensis* (Roland) Warb., *V. carinata* (Benth) Warb., *V. pavonis* (A. DC.), *V. elongata* (Benth) Warb., and *Myristica fragans* Houtt. An evaluation carried out with 18 racemic 8.O.4' neolignans with six different substitution patterns in rings A and B (Compounds 1-18, Figure 7.4), in their ketone and in their *erythro* and *threo* forms, against *Microsporum canis, Microsporum gypseum, Trichophyton mentagrophytes, T. rubrum,* and *Epidermophyton floccosum* using the agar dilution showed that alcohols (1-12) but not ketones (13-18) possess significant antifungal activity. This activity is dependent upon relative stereochemistry (*erythro* up to three times more active than *threo* alcohols) and upon the substitution patterns of rings A and B (Zacchino et al., 1997).

To know the mode of action of the active compounds, the activity in whole cell *Neurospora crassa* hyphal growth inhibition was determined. The results obtained strongly suggested that these compounds could act by

	R1	R2	R3	Conf.
1	OMe	OMe	H	erythro
2	OMe	OMe	H	threo
3	OMe	OMe	OMe	erythro
4	OMe	OMe	OMe	threo
5	-O-CH2-O-		H	erythro
6	-O-CH2-O-		H	threo

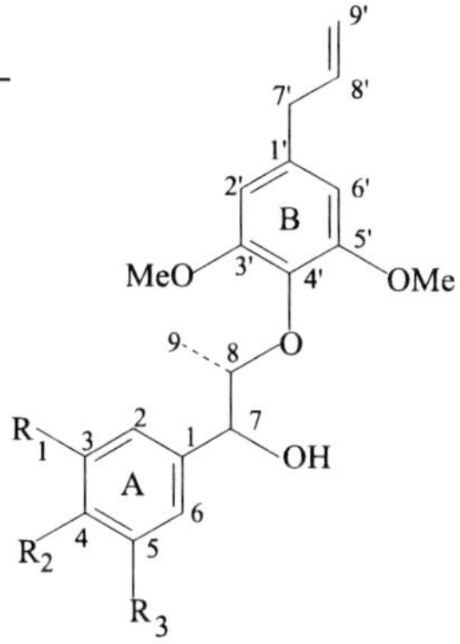

	R1	R2	R3	R4	Conf.
7	OMe	OMe	H	allyl	erythro
8	OMe	OMe	H	allyl	threo
9	OMe	OMe	H	transpropenyl	erythro
10	OMe	OMe	H	transpropenyl	threo
11	OMe	OMe	OMe	transpropenyl	erythro
12	OMe	OMe	OMe	transpropenyl	threo

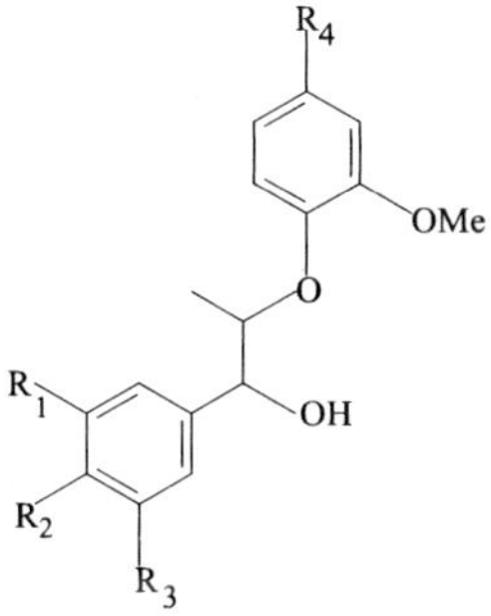

	R1	R2	R3	R4	R5
13	OMe	OMe	H	OMe	allyl
14	OMe	OMe	OMe	OMe	allyl
15	-O-CH2-O-		H	OMe	allyl
16	OMe	OMe	H	H	allyl
17	OMe	OMe	H	H	transpropenyl
18	OMe	OMe	OMe	H	transpropenyl

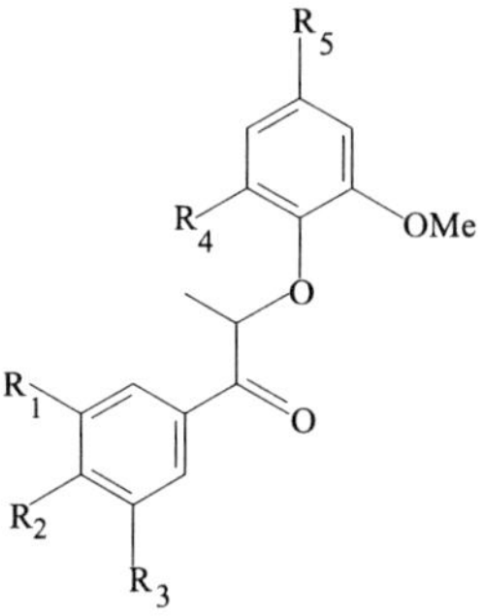

FIGURE 7.4. Chemical structures of the 8.O.4' neolignans

inhibiting cell wall polymer synthesis or assembly. Active compounds were tested for their inhibitory activities against (1,3)-β-glucan synthase. Although the compounds were found to be inhibitors of the enzyme, comparison of these results with those from agar dilution assays allowed the inference that they do not act via the inhibition of glucan synthase as a primary mode of action (Zacchino et al., 1998).

Zacchino et al. (1999) also studied the activity of a series of phenylpropanoids against the dermatophytes mentioned previously in order to establish a structure-activity relationship α-Halopropiophenones exhibited a broad spectrum of activities, with MIC values between 0.5 and > 50 μg/ml. Keto, alcohol, and α-haloketopropyl derivates of naphtalene and phenanthrene were also very active, indicating that in these series a halogen atom is not necessary for activity. Phenantryl derivates were the most active when compared to naphthyl derivatives. According to the results obtained, the authors concluded that the antifungal activities reported for 8.O.4' neolignans could not be attributed to phenylpropanoid moieties.

CONCLUSIONS AND FUTURE DIRECTIONS

The quest for antifungal drugs is critical for several reasons. Immune suppression causes susceptibility to fungal infections. The number of immune-suppressed individuals continues to rise as the planet faces an aging population, an increase in HIV-infected patients, and medical advances. Most drugs used to treat mycological infections have low availability or are too toxic for prolonged use. Also, many new fungal strains are emerging with drug resistance, as fungal pathogens are exposed to extended pharmaceutical treatment. On the other hand, validation of the ethnomedical use of plant extracts used as antifungals is also important for South America, since this region has a high percentage of inhabitants that rely almost exclusively on traditional medicine for their primary health care needs.

The search for new active natural products such as antimycotics has, in many aspects, only just begun. The poorly known flora of the drylands of "Patagonia," located in the far south of Argentina and Chile; the xerofitic vegetation of the "Caatinga" region, which covers almost 1 million km^2 in northeastern Brazil and is extensively used in local popular medicinal practice, particularly among the African-derived populations that inhabit the region; the flora of the "Pantanal," a swamp ecosystem in west central Brazil, which is highly preserved and has scarce registered ethnobotanical data; and the Chocoan rainforest in northwest Colombia, no doubt with pharmacological secrets waiting to be unveiled.

Approximately two-thirds of the planet's biodiversity can be found in the southern hemisphere (Prance, 1977). However, the industrialized countries are mainly in the northern hemisphere. The access to genetic resources in the South has become a necessary step in the drug development process that takes place in industrialized countries. The world's genetic resources, especially plants, have always been regarded as a common heritage of humankind, i.e., for all to use without restrictions. However, this situation changed after the Convention on Biological Diversity that occurred in Rio de Janeiro in 1992. At the convention, developing countries successfully argued for national sovereignty over their genetic resources, and for a right to negotiate for benefits of biological diversity. With this development, a new era came about in which the establishment of partnerships between institutions in the countries supplying genetic resources and organizations from the recipient countries has brought together their genetic and technological resources in collaborative ventures. Furthermore, benefits of biological resources should not only be shared at a national level but also with indigenous and local communities since in many cases it is realized that these groups have played an important role over the centuries in the selection and propagation of current genetic resources.

In the future, the increase of this collaborative research will be one of the most important challenges for humankind in the search for new and more effective bioactive compounds from natural products.

REFERENCES

Alécio, A., Da Silva Bolzani, V., Marx Young, M.C., Kato, M., and Furlan, M. (1998). Antifungal amide from leaves of *Piper histidum*. *J. Nat. Prod.* 61: 637-639.

Alvarez, L., Marquina, S., Villarreal, M.L., Alonso, A., Aranda, E., and Delgado, G. (1996). Bioactive polyacetylenes from *Bidens pilosa*. *Planta Med.* 62: 355-357.

Anesini, C. and Pérez, C. (1993). Screening of plants used in Argentine folk medicine for antimicrobial activity. *J. Ethnopharmacol.* 39: 119-128.

Balandrin, M.F., Kinghorn, A.D., and Farnsworth, N.R. (1992) Plant-derived products in drug discovery and development. In *Human Medicinal Agents from Plants* (Eds., Kinghorn, A. and Balandrín, M.). American Chemical Symposium series 534. New York: American Chemical Society, pp. 2-12.

Cáceres, A., Jauregui, E., Herrera, D., and Logemann, H. (1991). Plants used in Guatemala for the treatment of dermatomucosal infections. 1: Screening of 38 plants extracts for anticandidal activity. *J. Ethnopharmacol.* 33: 277-283.

Catalano, S., Cioni, P.L., Paniz, L., and Morelli, I. (1998). Antimicrobial activity of extracts of *Mutisia acuminata* var. *acuminata*. *J. Ethnopharmacol.* 59: 207-209.

Clark, A.M. and Hufford, C.D. (1992). Discovery and development of novel prototype antibiotics for opportunistic infections related to acquired immunodeficiency syndrome. In *Human Medicinal Agents from Plants* (Eds., Kinghorn, A. and Balandrín, M.). American Chemical Symposium series 534. New York: American Chemical Society, pp. 229-241.

Cunha, G.M., Maia, A.A., Rabelo, Neri E.D., Nogueira, N.A., and Matos, F.J. (1995). Antimicrobial activity of medicinal plants. *Rev. Bras. Farm.* 76: 5-6.

Eckerman, S.J. and Graham, K.J. (2000). Using chemical ecology to locate new antifungal products. In *Studies in Natural Products Chemistry* (Ed., Atta-ur-Rahman). Bioactive Natural Products, Volume 22. Amsterdam: Elsevier, pp. 55-92.

Farnsworth, N.R. (1994). Ethnopharmacology and drug development. In *Ethnobotany and the Search for New Drugs* (Eds., Ciba Foundation, Symposium 185). Chichester: J. Wiley and Sons, pp. 42-59.

Gentry, A. (1992). Tropical forest biodiversity and the potential for new medicinal plants. In *Human Medicinal Agents from Plants* (Eds., Kinghorn, A. and Balandrín, M.). American Chemical Symposium series 534. New York: American Chemical Society, pp. 229-241.

Gutkind, G.O., Martino, V., Graña, N., Coussio, J.D., and de Torres, R.A. (1981). Screening of South American plants for biological activities 1. Antibacterial and antifungal activity. *Fitoterapia* 52: 213-218.

Lahitte, H.B., Hurrel, J.A., Belgrano, M.J., Jankowski, L., Haloua, P., and Mehltreter, K. (1998). *Plantas Medicinales Rioplatenses.* Buenos Aires: L.O.L.A.

Lima, E.O., Morais, V.M.F., Gomes, S.T.A., Filho, V.C, Miguel, O.G., and Yunes, R.A. (1995). Preliminary evaluation of antifungal activity of xanthoxyline. *Acta Farm. Bonaerense* 14: 213-216.

Lorenti, A.S., Viale, A.A., Buschi, C.A., Gonzáles, M.D., Schteingart, C.D., Iribarren, A.M., and Pomilio, A.B. (1981). Antimicrobial activity of some Argentine higher plants. *Fitoterapia* 51: 81-85.

Penna, C.A., Radice, M., Gutkind, G.O., Van Baren, C., Broussalis, S., Muschietti, L., Martino, V., and Ferraro, G. (1994). Antibacterial and antifungal activities of some Argentinean medicinal plants. *Fitoterapia* 65: 172-174.

Phillipson, J. D. (1999). New drugs from nature. It could be yew. *Phytother. Res.* 13: 2-8.

Pomilio, A.B., Buschi, A.A., Tomes, C.N., and Viale, A.A. (1992). Antimicrobial constituens of *Gomphrena martiana* and *Gomphrena boliviana*. *J. Ethnopharmacol.* 36: 155-161.

Prance, G. T. (1977). Floristic inventory of the tropics: Where do we stand? *Ann. Mo. Bot. Gard.* 65: 659-684.

Sanabria, A. and Mantilla, J.R. (1986). Antifungal activity of higher Colombian plants. *Rev. Colomb. Cienc. Quim. Farm.* 15: 17-22.

Schultes, R.E. and Raffauf, R.F. (1990). Symptom, disease and treatment index. *The healing forest. Medicinal and toxic plants of the Northwest Amazonia* (Ed., Dudley, T.) Historical, ethno-and economic botanical series, Volume 2. Portland, Oregon: Disocordes Press, pp. 477-480.

Selitrennikoff, C.P. (1992). Screening for antifungal drugs. In *Biotechnology of Filamentous Fungi* (Eds., Firkeftein, D. and Ball, C.). London: Butterworth, pp. 189-217.

Tirillini, B., Rojas Velásquez, E., and Pellegrino, R. (1996). Chemical composition and antimicrobial activity of essential oil of *Piper angustifolium*. *Planta Med.* 62: 372-373.

Ugarte, R., Quilhot, W., Díaz, B., Vera, A., and Fiedler, P. (1987). Actividad antimicrobiana in vitro del ácido epiforelico 1 (sustancia liquénica). *Acta Farm. Bonaerense* 6: 65-69.

Zacchino, S.A., López, S.N., Pezzenati, G.D., Furlán, R.L., Santechia, C.B., Muñoz, L., Giannini, F.A., Rodríguez, A.M., and Enriz, R.D. (1999). In vitro evaluation of antifungal properties of phenylpropanoids and related compounds acting against dermatophytes. *J. Nat. Prod.* 62: 1353-1357.

Zacchino, S., Rodríguez, G., Pezzenati, G., Orellana, G., Enriz, R., and González Sierra, M. (1997). In vitro evaluation of antifungal properties of 8.O.4'-neolignans. *J. Nat. Prod.* 60: 599-662.

Zacchino, S., Rodríguez, G., Santecchia, C., Pezzenati, G., Giannini, F., and Enriz, R. (1998). In vitro studies of the mode of action of antifungal 8.O.4'-neolignans occurring in certain species of *Virola* and related genera of Myristicaceae. *J. Ethnopharmacol.* 62: 35-41.

Chapter 8

Current State and Future Directions in Plant-Derived Antimycotics

P. G. Latha
Acham Abraham
S. R. Suja
S. Rajasekharan

INTRODUCTION

Numerous diseases of serious concern in humans are caused by fungi (Bulmer, 1979). In addition, fungi produce toxins in food and can cause poisoning without even being physically present. Another problem caused by fungi is allergy to their spores. Most of these diseases are difficult to treat, and this problem is compounded by the increasing use of immunosuppressants generally in medicine, and more specifically in diabetes and AIDS, all of which increase human susceptibility to these stubborn fungal infections. Evaluation of the antimicrobial potential of natural products is important for the development of drugs from indigenous sources that may be more effective against microbial infections and less toxic to the host. The Centre for Agricultural and Biosciences International, United Kingdom, has published an annotated bibliography of plants with antimicrobial properties (Cousins, 1995).

The pharmacological principles of antifungal therapy are only partially understood. Fungal infections are traditionally divided into two distinct classes: superficial and systemic. Accordingly, the major antifungal agents are systemic antifungal agents and topical (superficial) antifungal agents (Bennet, 1991). Topical treatment is useful in many superficial fungal infections, that is, those confined to stratum corneum, squamous mucosa, or cornea, such as ringworm, candidiasis, fungal keratitis, etc.

Fungal infections of the hair, skin, and nails are a major source of morbidity throughout the world. It has been estimated that fungal infections account for 5 percent of new outpatient referrals to dermatologists in temperate climates and up to 20 percent in tropical climates. Most of the infections are caused by dermatophytes or by yeasts, most commonly *Candida* species.

Communal biochemical processes in eukaryotes make the prevention of growth of a fungal eukaryote within the plant, insect, or animal eukaryote difficult. These communal processes also make the development of appropriate bioassays more difficult. A major difficulty is that in vitro testing of fungi for either primary or secondary resistance to chemotherapeutic agents does not provide clinically useful information. Without a firm grasp on the concentration of drugs necessary to inhibit fungal growth in vitro, it is obviously difficult to estimate what concentration of drug might be therapeutic.

The development of resistance by pathogens to many of the commonly used antibiotics provides an impetus in the search for new antimicrobial agents to combat infections and overcome the problem of resistance and side effects of the currently available antimicrobial agents. The predominant factor that has stimulated the search for safer and more effective antifungal agents has been the increasing incidence of systemic mycoses in immunologically compromised patients and the unfortunate resistance of certain strains of *Candida albicans* to azole antifungals.

Chemicals may not be equally active against closely related fungi, or they may be equally active against unrelated fungi. Pathogens differ appreciably in their sensitivity to particular chemicals, and each disease reveals a number of chemicals that are not active against any other pathogens. It is difficult, therefore, to ensure that the organism selected for the primary test adequately covers all the targets regarded as important.

In the early stages of screening, large numbers of chemicals are tested against numerous diseases. The objective is to identify as quickly and as cheaply as possible the few that merit further study.

EXISTING ASSAYS FOR DETECTING ANTIFUNGAL COMPOUNDS

A major problem is controlling fungi that cause problems but cause no negative effect on man. Many fungi can be especially difficult to control because of their ability to grow on almost anything. Existing methods for detecting fungicides can be categorized roughly into (1) the inhibition of radial growth on an agar medium in a Petri plate, and (2) growth in liquid culture (which can be measured as increase in dry weight or increase in optical density at a given wavelength).

The radial growth method is the most popular, being very easy to perform. The change in optical density of a solution is presumed to occur through the increased scattering of light at wavelengths longer than 600 nm, and by increased number or size of cells in the solution. Dilution methods look for MIC (minimum inhibitory concentration) and are usually set up in twofold steps of fungicide concentration—in agar or liquid media. It is important to measure the response compared to standards under identical conditions for comparison of relative toxicity. Diverse biological samples, such as blood, present unique problems. Diffusion assay is the method of choice because only a small sample is required. Methods are detailed for the estimation of sensitivity of pathogenic fungi to therapeutic agents and for the assay of these agents in body fluids.

Paxton (1991) discussed induced experimental *Microsporum lanosum* infections in dogs, cats, and rabbits in order to screen antifungal compounds. *Trichophyton mentagrophytes* in infected epidermal scales obtained from guinea pigs also proved useful for testing antifungal compounds. Other in vitro assays for antifungal compounds used stratum corneum stripped off with translucent adhesive tape. Humans have been used to test control of *T. mentagrophytes* on arms. One arm was used to test the fungicide while the other arm served as control.

A simple bioautographic technique has also been devised for detection of fungitoxic sunstances. Chromatograms on Whatman 3 mm paper are developed with propanol-water (85:15) and after drying are sprayed with a conidial suspension of *Glomerella cingulata*. After incubation, clearly visible inhibition zones indicated the presence of fungitoxic substances per se. It was found that directly spraying the thin-layer chromatograms with a spore suspension of the test fungus in a glucose-mineral salt medium gave the most reliable results. Chromatograms were sprayed with a conidial suspension of *Cladosporium cucumerinum* in a mineral salt medium, containing 4.3 percent glucose. The plates were then incubated for two to three days at 25°C in a moist atmosphere. *Aspergillus niger, Botrytis cinerea, Fusarium culmorum,* and *Penicillium expansum* proved excellent test organisms for the detection of fungitoxic compounds.

Zacchino et al. (1998) devised a method for detecting antifungal activity using whole cell *Neurospora crarssa,* hyphal growth inhibition.

At the Central Drug Research Institute, Lucknow, India, a systematic program for antifungal testing is being conducted (Srivastava, 1984). Detailed methodologies for in vitro and in vivo antifungal testing have been described. Determination of viable counts, dry weights, optical density/transmission, the use of filter paper disc or cup assay, and poison food techniques are also applicable in the case of yeastlike fungi for the detailed quantitative evaluation of natural products. For mycelial fungi, the techniques do not prove accurate unless special methods are employed for their homogenization. Measurements of the diameter of the growing colony after regular intervals provide data for a growth curve, which in the great majority of cases proves to be a straight line. The evaluation of natural products can be easily accomplished with this technique and precise quantitative assessments obtained with the help of dose response curves.

Paxton (1991) reported á micromethod for fungicide bioassay on a TLC plate to evaluate antifungal compounds. This test was more sensitive than the paper disc impregnated bioassay test.

Only recently have methods for the mass screening of chemicals to control fungi been developed. Variation among tests is a problem and the relative potency is a useful concept in dealing with this variability. The total expenditure involved in research needed to discover and develop new fungicides, the factors critical to the design of screens, and the procedures used govern possible future developments. It currently costs nearly $30 million to research, discover, and develop a new fungicide. The country and the company dictate which organisms are used in the bioassay.

Anyone working with antibiotics needs to understand the principles of several kinds of bioassays, their applications, and their deficiencies. The principles of several kinds of assays for antibiotics, problems of application, reasons for assaying, and relation between structure and activity of the phytocompounds are discussed next.

Practical Aspects of the Bioassays

Practical aspects of bioassays (Paxton, 1991) that can have important influences on these tests include the following factors.

Chemical Reactivity

Many biologically active compounds owe at least part of their activity to their chemical reactivity. Therefore, they can react with many things in their environment, including, in some cases, themselves. The chemical purity of the candidate compound in many bioassays is questionable. Biologists often use archaic chemical methods to establish the identity and purity of these compounds.

Purity of Test Compound(s)

The purity of the test compound is especially important since it is well known that synergism occurs between many compounds, and isomers of compounds can vary widely in their biological activity. Many fungicides show similar activities dictated by stereochemical properties.

Time

Time can markedly affect the observed fungicidal activity of compounds and even the susceptibility of organisms to these fungicides. This may occur through the chemical change from inactive compounds to active compounds or through the chemical change from active compounds to inactive compounds. This can be a purely chemical change in the test compound, having nothing to do with action on the compound by the test organism. Many toxic compounds are stored in plants as glycosides, which are slowly hydrolyzed in aqueous in vitro environments to their often more toxic aglycones. Since all bioassays occur in aqueous environments by necessity and at room temperature in most cases, this is a phenomenon to consider seriously.

Stage of Growth

Organisms attack the plant or animal at quite different stages of the host's growth and their own growth (spore germination versus mycelial growth). This common knowledge is not always taken into account in designing bioassays. Mutants of the test organism in bioassays can arise over time and confuse the interpretation of results. Different strains of the same organism can differ greatly in their sensitivity to a given compound, and often unusual organisms are used in bioassays simply for convenience.

Change in pH

The change in pH of media with the growth of fungi is virtually inevitable, even in well-buffered media. This change may be inconsequential to the bioassay, or it may be a serious consideration in determining the effectiveness of the fungicide. The activity of many fungicides is affected by pH. It is probable that fungicides seldom encounter the same pH environment in application as those under which they were bioassayed.

Oxidation

Another chemical phenomenon that can occur with most compounds is oxidation. This takes place with different compounds at different rates depending on the temperature, oxygen partial pressure, and the presence of catalysts.

Air is 20 percent oxygen, but the partial pressure of oxygen can vary with altitude, as can the consumption of oxygen in liquids and other media (often in closed containers in which diffusion of oxygen is limited), and the composition of media. All can affect oxygen solubility. A microorganism growing in a medium can have quite a different oxygen supply than the same organism growing on top of the medium or, more important, in another organism, such as a plant or animal.

The catalysts for chemical reactions can be metal salts, metals, glass, and especially enzymes, such as peroxidases, from the test organisms. Higher temperatures and oxygen concentrations will accelerate this process. Other activators are the input of energy into the system through light—especially ultraviolet light. This can have an effect on the media and test organism as well as on the fungicide.

Composition of Medium

Unexpected effects on bioassays can occur because of the composition of the medium used in the bioassay. The nature and concentration of the carbohydrate source could strongly influence the formation of certain substances in culture which in turn could influence the toxicity of the fungitoxic compound.

Nature of Environment

In bioassays for potential fungicides, a major problem is understanding the environment in which the fungicide is supposed to function. It is rarely the same as the environment of the bioassay. In fact, the plant environment in which the fungicide functions is largely unknown. This creates obvious problems for generating realistic bioassays.

Scale of Bioassay

For various reasons, industry is switching to bioassaying potential fungicides directly in the systems in which they are expected to function. This means caring for extensive animal colonies or plant populations. The final results are, however, more directly applicable to end use of the compound in question. The serious problems of creating controlled repeatable disease epidemics in order to test the candidate compounds effectively are not minor ones.

Unfortunately, more and more screening of compounds occurs in an industry that is reluctant, for proprietary reasons, to discuss their methods. This slows progress in developing more effective and cost-efficient bioassays for fungicides.

THE MOST IMPORTANT ANTIFUNGAL COMPOUNDS FROM PLANTS

A group of natural products that has promise as fungicides but that has not been tested well is the phytoalexins. These phytoalexins are low-molecular-weight antimicrobial compounds, that are both synthesized and accumulated in plants after exposure to microorganisms (Paxton, 1991). Phytoalexins are of relatively low toxicity for mammalian cells and possess a broad spectrum of antifungal antibiotics useful in chemotherapy of human mycosis. It would be desirable to modify existing compounds or produce analogs that might show increased activity. The plasma membrane, or some process necessary for membrane function, is affected by phytoalexins.

The antibiotic activity of diterpenes and phenanthrene derivatives of diterpene origin have been studied (Paxton, 1991). They suggest antimicrobial agents from plants are of interest since they represent a natural defense against infection. A study of these agents may lead to agents of medical or veterinary significance.

A search for naturally occuring drugs with antifungal activity led to santolina oil, a volatile oil distillate of *Santolina chamaecyparissus* (Suresh et al., 1997). Studies revealed that the oil was effective in controlling experimental candidiasis in vitro and in vivo. It had a synergistic effect in controlling *Candida albicans* in vitro. It significantly controlled experimental vaginal candidiasis and experimental systemic candidosis. It was able to control superficial cutaneous mycoses. It is recommended as a potential candidate for further clinical studies.

The influence of a series of related isoflavonoids on growth of *Candida albicans* in liquid media was studied (Paxton, 1991). The order of effectiveness was identical with that previously found for a range of phytopathogenic fungi. Phaseollinisoflavan was the most active test compound, and the MIC was 250 μM for three days. It is reported that about 60 percent of the essential oils obtained from plants possess antifungal activity (Suresh et al., 1997). A whole gamut of other phytochemicals are reported to show potent antifungal property.

Table 8.1 provides information on some of the most important antifungal agents isolated from plants.

Mechanism of Action of Antifungal Agents

The major mechanism of action of antifungal agents is their direct effect on the cell membrane, affecting permeability and leading to leakage of intracellular compounds.

TABLE 8.1. Antifungal agents isolated from plants

Plants	Family	Plant Part	Phytocompounds	Reference
Aegle marmelos (L.) Corr.	Rutaceae	Leaves	Essential oils	Rana et al., 1997
Alpinia galanga Sw.	Zingiberaceae	Seeds	Diterpenes	Morita and Itokawa, 1988
Artemisia giraldii Pamp.	Asteraceae	Whole plant, aerial parts	Flavones	Zheng et al., 1996
Artemisia giraldii Lam.	Asteraceae	Whole plant	Sesquiterpene lactones	Tan et al., 1999
Bauhinia rutescens	Caesalpineaceae	Root bark	Stilbenes	Maillard et al., 1991
Bidens paludosa L.	Asteraceae	Whole plant	Polyacetylenes	Alvarez et al., 1996
Boswellia carterii Birdw.	Burseraceae	Bark	Essential oils	Wahab et al., 1987
Brosimopsis oblongifolia Ducke.	Moraceae	Roots	Isoprenylated flavones	Messana et al., 1987
Buddleja madagascarensis Lam.	Loganiaceae	Leaves	Triterpenoid saponins	Emam et al., 1996
Calycodendron milnei (A. Gray) A. C. Smith	Rubiaceae	Whole plant	Alkaloids	Saad et al., 1995
Clerodendron wildñ Moldenke.	Verbenaceae	Roots	Triterpenoid saponins	Toyota et al., 1990
Coccoloba dugandiana Fernandez-Perez	Polygonaceae	Leaves, twigs	Flavonoids	Li et al., 1999
Croton laciferus Blanco	Euphorbiaceae	Roots	Benzoquinones	Bandara and Wimalasiri, 1988
Curcuma longa L.	Zingiberaceae	Tubers	Essential oils	Apisariyakul et al., 1995
Dioscorea rotunda Poir.	Dioscoreaceae	Tubers	Dihydrostilbenes	Fagboun et al., 1987
Erythrina berteroana Urb.	Fabaceae	Stem bark	Prenylated flavone	Maillard et al., 1987
Ficus septica Lour.	Moraceae	Leaves	Alkaloids	Baumgartner et al., 1989
Garcinia gerrardii Harv.	Guttiferae	Root bark	Prenylated xanthones	Diserens et al., 1989
Gladiolus dalenii Van Geel.	Iridaceae	Leaves, corms	Saponins	Botha et al., 1989
Glehnia littoralis F. Schmidt.	Apiaceae	Roots	Polyenic alcohols	Matsura et al., 1996
Glycopetalum sclerocarpum Laws.	Celastraceae	Stem bark	Sesquiterpenes	Sotonaphun et al., 1999
Glycyrhiza glabra L.	Fabaceae	Roots	Flavonoids	Li et al., 1998
Hebe cupressoides (Hook. f.) Ckn. & Allan	Scrophulariaceae	Aerial parts	Flavonoids	Perry and Foster, 1994

Plants	Family	Plant Part	Phytocompounds	Reference
Helichrysum aureonitens Sch. Bip.	Asteraceae	Shoots	Trihydroxy flavones	Afolayan and Meyer, 1997
Heliotropium ellipticum R. Br.	Boraginaceae	Aerial parts	Pyrrolizidone alkaloids	Jain and Sharma, 1987
Heteromorpha trifoliata Eckl. & Zeyh.	Umbelliferae	Leaves	Falcarindiol	Villegas et al., 1988
Hypericum calcinum Linn.	Guttiferae	Aerial parts	Phloroglucinol derivatives	Decosterd et al., 1989
Indigofera oblongifolia Forsk.	Fabaceae	Leaves	Peptides	Dahot, 1999
Inula viscosa (Dryand.) Ait.	Asteraceae	Leaves	Sesquiterpene lactones	Maoz et al., 1999
Kigelia pinnata DC.	Bignoniaceal	Fruits	Naphthoquinones	Binutu et al., 1996
Limonia acidissima Linn.	Rutaceae	Stem and root bark	Psoralen	Bandara et al., 1988
Myrica gale Linn.	Myriaceae	Leaves	Essential oils	Stuart, 1998
Myristica fragrahs Houtt.	Myristicaceae	Mace	Resorcinols	Orabi et al., 1996
Parthenium hysterophorus Linn.	Asteraceae	Whole plant	Sesquiterpene lactones	Ganeshan and Ganeshan, 1993
Piper angustifolium Ruiz & Pav..	Piperaceae	Seed	Camphene	Tirillini et al., 1996
Piper betle Linn.	Piperaceae	Leaves	Essential oils	Garg and Jain, 1992
Pneumus boldus Molina	Monimiaceae	Leaves	Essential Oils	Vila et al., 1999
Polygonum hydropiper Neck.	Polygonaceal	Sesquiterpene dialdehydes	Whole plant	Lee et al., 1999
Psidium acetangulum DC.	Myrtaceae	Leaves	Chalcones	Miles et al., 1990
Raphanus sativus Linn.	Brassicaceae	Seeds	Proteins	Bolle et al., 1992
Rhinacanthus nasutus (Linn.) Kurz.	Acanthaceae	Leaves, stems	Naphthopyran derivatives	Kodoma et al., 1993
Serjania salzmanniana Schlecht.	Sapindaceae	Whole plant	Saponins	Ekabo et al., 1996
Sium nodiflorum Linn.	Apiaceae	Whole plant	Heterosides	Larshini et al., 1996
Sophora angustifolia Sieb. & Zucc.	Fabaceae	Root	Flavonoids	Honda and Tabata, 1982
Strychnos usambarensis Gilg.	Loganiaceae	Seeds	Alkaloids	Leclereq et al., 1995
Tabernaemontana divaricata (Linn.) R. Br. ex Roem. & Schult.	Apocynaceae	Leaves, flowers	Alkaloids	Beek et al., 1984
Terminalia bellerica (Gaertn.) Roxb.	Combretaceae	Fruit	Flavans, lignans	Valsaraj et al., 1997
Virola surinamensis Warb.	Myristicaceae	Leaves, bark	Neolignans	Zacchino et al., 1998

Zacchino et al. (1998) suggested the compounds could act by inhibiting cell wall polymer synthesis or assembly. Since fungal cells are encased in a carbohydrate-containing cell wall which serves as a protective barrier and is necessary for growth and viability of fungi, its inhibition represents a useful mode of action for antifungal agents. The first bioassay of Zacchino et al. (1998) used whole cell *Neurospora crassa* hyphal growth inhibition (Fukuda et al., 1991), an agar diffusion method, which is useful to detect agents whose mode of action is associated with cell wall polymer synthesis or assembly (Gunji et al., 1983). Since (1,3)-β-glucan is the most abundant carbohydrate polymer of fungal walls (Selitrennikoff, 1995), the inhibitory activity of (1,3)-β-glucan synthase, an enzyme that catalyses polymerization in the fungal wall, was measured.

Bolle et al. (1992), studying the mechanism of action of *Parthenium hysterophorus* sesquiterpene lactones as antifungal compounds, found the treatment caused lobulation of hyphal cell wall thickening and restrained mycelial growth. The lobulations resembled those produced by polyene antibiotics. In view of the importance of controlling filamentous fungi such as *Candida albicans,* an attempt was made to study combination effects of antifungal phenyl propanoids and nagilactones isolated from *Podocarpus nagi* root against *C. albicans* and two other fungi. The approach to enhancement of total biological activity by combining two or more substances seems to be a most promising strategy for efficient utilization of renewable natural resources.

Currently, antifungal drugs against systemic fungal pathogens are represented by amphotericin B, ketoconazole, fluconazole, and itraconazole. Their antifungal action is known to be accomplished through ergosterol, an important fungal membrane component. Amphotericin B is known to take advantage of ergosterol in the cellular membrane at its binding site. Polygodial, the active principle from *Polygonum hydropiper,* was found to exhibit antifungal properties, and it does not use ergosterol as its binding site on the cell. Thus, it is inferred that polygodial has a different mode of action than existing antifungal drugs of the azole and polyene classes (Lee et al., 1999).

Generally, it is accepted that antimicrobial therapy is dependant on both the drug's growth inhibition and the host's immune system. However, since most systemic fungal diseases occur in patients with seriously impaired immune systems, fungicidal properties of antifungal agents are considered to be very important.

CONCLUSION

The predominant factor that has stimulated the search for safe and more effective antifungal agents has been the increasing incidence of systemic mycoses in immunologically compromised patients and the unfortunate incidence of certain strains of clinical fungi becoming resistant to the existing antifungals, hence the need for developing a wider variety of antifungal agents from plants.

Research to elucidate the mode of action of the important plant-derived antifungal agents can be expected to suggest new antifungal targets. Further investigations on the antifungal action of these phytochemicals is warranted.

The need for economic, rapid, sensitive, and accurate bioassays for fungitoxic compounds is apparent from the problems caused by them. This chapter has concentrated on the types of in vivo and in vitro assays that are available for testing antifungal activity. Recent developments with antifungal drugs have pointed once more to the plant kingdom. Large numbers of plants continue to be used in traditional medicine for antifungal diseases, and such plants need to be tested for their efficacy. In vivo and in vitro antifungal assays can be used for assessment of activity of crude extracts and to aid fractionation procedures of plant extracts.

REFERENCES

Afolayan, A.J., and Meyer, J.J.M. (1997). The antimicrobial activity of 3, 5, 7 trihydroxyflavones isolated from shoots of *Helichrysum aureonitens*. *Journal of Ethnopharmacology* 57: 177-181.

Alvarez, L., Marguina, S., Villard, M.L., Alonso, D., Aranar, E., and Delgado, G. (1996). Bioactive polyacetylenes from *Bidens paludosa*. *Planta Medica* 62: 355-356.

Apisariyakul, A., Vanittanakom, N., and Buddhasukh, D. (1995). Antifungal activity of turmeric oil isolated from *Curcuma longa* (Zingiberaceae). *Journal of Ethnopharmacology* 49: 163-169.

Bandara, B.M.R., Gunatilaka, A.A.L., Wijerethne, E.M.K., and Adikaram N.K.B. (1988). Antifungal constituents of *Limonia acidissima*. *Planta Medica* 54: 374-375.

Bandara, B.M.R., and Wimalasiri, W.R. (1988). Diterpene alcohols from *Croton laciferus*. *Phytochemistry* 27: 225-226.

Baumgartner, B., Erdelmeier, C.A.J., Wright, A.D., Rali, T., and Sticher, O. (1989). Autofine, a strong antifungal alkaloid from *Ficus septica* leaves. *Planta Medica* 55: 652-653.

Beek, T.A.V., Deelder, A.M., and Svendson, A.B. (1984). Antimicrobial, anti-amoebic, antiviral screening of some *Tabernaemontana* species. *Planta Medica* 50: 181.

Bennet, J. (1991). Antifungal agents. In *The pharmacological basis of therapeutics,* Volume II. Gillmann, A.G., Rall, T.W., Nies, A.S., and Taylor, P. (Eds.). New York: Pergamon Press, pp. 1165-1198.

Binutu, O.A., Kayode, G., and Okugum, J.I. (1996). Antibacterial and antifungal compounds from *Kigelia pinnata. Planta Medica* 62: 352-354.

Bolle, M.F.C., Terras, F.R.G., Cammue, B.P.A., Rees, S.B., Broekaert, W.F., and Fritig, B. (1992). Mechanism of plant defence responses. In *Proceedings of the Second International Conference of European Foundation for Plant Pathology.* Legrand, M. (Ed.). The Netherlands: Kluwer Academic Publishers, pp. 433-466.

Botha, S., Gerritsma, L.M., and Vander Nest, D.G. (1989). Primary biological investigations of *Gladiolus dalenii. Planta Medica* 55: 115.

Bulmer, G.S. (1979). *Introduction to medical mycology.* Chicago, Illinois: Year Book, Medical Publishers.

Cousins, D.J. (Ed.) (1995). *Plants with antimicrobial properties.* Volume 2: *Antifungal properties.* Wallingford, UK: CAB International.

Dahot, M. (1999). Antibacterial and antifungal activity of small proteins of *Indigofera oblongifolia. Journal of Ethnopharmacology* 64: 277-282.

Decosterd, L.A., Stoeckli, H.E., Chapnis, J.C., Sordat, B., and Hostettman, K. (1989). New constituents with growth inhibitory activity against CO-115 human colon carcinoma cell line and a new antifungal phloroglucinol derivative from *Hypericum calycinum. Planta Medica* 55: 629.

Diserens, I.S., Marston, A., Hamburger, M., Rogers, B.C., and Hostettmann, K. (1989). Novel prenylated xanthones from *Garcinia gerardii. Planta Medica* 55: 584-589.

Ekabo, D.A., Farnsworth, N.R., Henderson, T.O., Guohua, M., Mukherjee, R., and Mao, G.H. (1996). Antifungal and molluscicidal saponins from *Serjania salzmanniana. Journal of Natural Products* 59: 431-435.

Emam, A.M., Moussa, A.M., Faure, R., Farvel, A., Delmas, F., Elias, R., and Balansard, G. (1996). Isolation and biological study of triterpenoid saponin from leaves of *Buddleja madagascariensis. Planta Medica* 62: 92-93.

Fagboun, D.E., Ogundana, S.K., Adesanya, S.A., and Roberts, M.F. (1987). Dihydrostilbene phytoalexins from *Dioscorea rotundata. Phytochemistry* 26: 3187-3189.

Fukuda, D., Nakadsukasa, W., Yao, R., Hunt, A., Gordee, R., Zeckner, D., and Myduse, J. (1991). A 45507 a complex of fungal cell wall inhibitors produced by a mould. 31 ICAAC meetings, poster No. 211.

Ganeshan, G., and Ganeshan, J. (1993). Antifungal activity of parthenin. Indian. *Phytopathology* 46: 193-194.

Garg, S.C., and Jain, R. (1992). Biological acivity of the essential oil of *Piper betle* L. *Journal of Essential Oil Research* 4: 601-606.

Gungi, S., Arima, K., and Beppu, T. (1983). Screening of antifungal antibiotics according to activities inducing morphological abnormalities. *Agriculture and Biological Chemistry* 47: 2061-2069.

Honda, G., and Tabata, M. (1982). Antidermatophytic substances from *Sophora angustifolia. Planta Medica* 46: 122-133.

Jain, S.C., and Sharma, K. (1987). Antimicrobial activity of pyrrolidone alkaloids of *Heliotropium ellipticum. Chemical and Pharmaceutical Bulletin* 35: 3487-3489.

Kodoma, O., Ichikawa, H., and Akatsuka, T. (1993). Isolation and identification of an antifungal naphthopyran derivative from *Rhinacanthus nasutus. Journal of Natural Products* 56: 292-294.

Larshini, M., Lazrek, H.B., Amarouch, H., and Jana, M. (1996). Investigations of antifungal and analgesise activities of extracts of *Sium nodiflorum. Journal of Ethnopharmacology* 53: 105-110.

Leclereq, J.Q., Fowel, A., Balansard, G., Regdi, P., and Angenot, L. (1995). Screening of in vitro antifungal activities of indole alkaloids. *Planta Medica* 61: 475-480.

Lee, S.H., Lee, J.R., Lunde, C.S., and Kubo, I. (1999). In vitro antifungal susceptibility of *Candida albicans* and other fungal pathogens to polygodial, a sesquiterpene dialdehyde. *Planta Medica* 65: 204-208.

Li, W., Asada, Y., Yoshikawa, T. (1998). Antimicribial flavonoids from *Glycyrrhiza glabra* hairy root cultures. *Planta Medica* 64: 746-748.

Li, X.C., Elsohy, H.N., Nimrod, A.C., and Clark, A.M. (1999). Antifungal activity of epigallocatechic gallate from *Coccoloba dugandiana. Planta Medica* 65: 780-782.

Maillard, M., Gupta, M.P., and Hostettmann, K. (1987). A new antifungal prenylated flavone from *Erythrina berteroana. Planta Medica* 53: 563-565.

Maillard, M.P., Iglesias, R.M.C., Saadou, M., Stoeckli-Evano, H., and Hostettman, K. (1991). Novel antifungal tetracyclic compounds from *Bauhinia rufescens* Lam. *Helvetica Chimica Acta* 74: 791-799.

Maoz, M., Kashmau, Y., and Neeman, I. (1999). Isolation and identification of a new antifungal sesquiterpene lactone from *Inula viscosa. Planta Medica* 65: 281-282.

Matsura, H., Saxena, G., Farmer, G.W., Hancock, R.E.W., and Towers, G.H.N. (1996). Antibacterial and antifungal polyine compounds from *Glehnia littoralis* ssp. *leiocarpa. Planta Medica* 62: 256-259.

Messana, J., Ferrari, F., and Aranjo, M.C.M. (1987). A new isoprenylated flavone from *Brosimopsis oblongifolia. Planta Medica* 53: 541-542.

Miles, D.H., Madeiros, D.J.M.R., Chittawong, V., Swithenbank, C., Lidert Z., weeks J.A., Atwood, J.L., and Hedin, P.A. (1990). 3'-formyl-2'4'6'-trihydroxy-5' Methyl dihydrochalcone, a prospective new agrochemical from *Psidium acutangulum. Journal of Natural Products* 53: 1548-1551.

Morita, H., and Itokawa, H. (1988). Cytotoxic and antifungal diterpenes from seeds of *Alpinia galanga. Plant Medica* 54 : 117-119.

Orabi, K.Y., Mossa, J.S., and El Feraly, F.S. (1996). Isolation and characterisation of two antimicrobial agents from mace *(M. fragrans). Journal of Natural Products* 54: 856-859.

Paxton, J.D. (1991). Assays for antifungal activity. In *Methods in plant biochemistry.* Hostettmann, K. (Ed.). London: Academic Press, pp. 33-46.

Perry, N.B., and Foster, L.M. (1994). Antiviral and antifungal flavonoids plus a triterpene from *Hebe cupressoides. Planta Medica* 60: 491-494.

Rana, B.K., Singh, U.P., and Tanja, V. (1997). Antifungal activity and kinetics of inhibition by essential oil from leaves of *Aegle marmelos. Journal of Ethnopharmacology* 57: 29-34. .

Saad, H.E.A., Sharkaroy, S.H.E., and Shier, W.T. (1995). Biological activity of pyrrolidinoindolin alkaloids from *Calycodendron milnei. Planta Medica* 61: 313-316.

Selitrennikoff, C. (1995). *Antifungal drugs. 1, 3, β-glucan synthase inhibitors.* Germany: Springer-Verlag, pp. 91-132.

Sotonaphun, U., Lipipun, V., Suttisri, R., and Bavovada, R. (1999). A new antiviral and antimicrobial sesquiterpene from *Glycopetalum sclerocarpum. Planta Medica* 65: 257:258.

Srivastava, O.P. (1984). Techniques for evaluation of antimicrobial properties of natural products. In *The use of pharmacological techniques for evaluation of natural products.* Dhawan, B.N., and Srimal, R.C. (Eds.). New Delhi: UNESCO, pp. 72-79.

Stuart, A.E. (1998). Antifungal effect of oil from leaves of *Myrica gale. Planta Medica* 55: 584.

Suresh, B., Sriram, S., Dhanaraj, S.A., Elango, K., and Chinnaswamy, K. (1997). Anticandidial activity of *Santolina chamaecyparissus* volatile oil. *Journal of Ethnopharmacology* 55: 151-159.

Tan, R.X., Lu, H., Wolfender, J.L., Yu, T.T., Yang, L., and Hostettman, K . (1999). Mono- and sesquiterpenes and antifungal constituents from *Artemisia* species. *Planta Medica* 65: 64-67.

Tirillini, B., Velasquez, E.R., and Pellegrim, R. (1996). Chemical composition and antimicrobial activity of essential oil from *Piper angustifolium. Planta Medica* 62: 372-374.

Toyota, M., Msonthi, J.D., and Hostettmann, K. (1990). A molluscidal and antifungal triterpenoid saponin from the roots of *Clerodendron wildii. Phytochemistry* 29: 2849-2851.

Valsaraj, R., Pushpangadan, P., Smitt , U.W., Adsersen, A., and Nyman, U. (1997). Antimicrobial screening of selected medicinal plants from India. *Journal of Ethnopharmacology* 58: 75-83.

Vila, R., Valenzuela, L., Bello, H., Canignerai, S., Montes, M., and Adset, T. (1999). Composition of antimicrobial activity of essential oil of *Pneumus boldus* leaves. *Planta Medica* 65: 178-181.

Villegas, M., Vargas, D., Msonti, J.D., and Hostettmann, K. (1988). Isolation of antifungal compounds falcarindiol and sarisan from *Heteromorpha trifoliata. Planta Medica* 54: 36-37.

Wahab, S.M.A., Aboutabe, E.A., Zalabani, S.M.E., Fouad, H.A., Pooter, H.L.D., and Fallaha, B.E. (1987). The essential oil of *Olibanum. Planta Medica* 53: 382-384.

Zacchino, S., Rodriguez, G., Santecchia, C., Pezzanati, G., Giannini, F., and Euriz R. (1998). In vitro studies on mode of action of antifungal 8-0-4'-neolignans occurring in certain species of *Virola* and related genera of Myristicaceae. *Journal of Ethnopharamacology* 62: 35-41.

Zheng, W.F., Tan, R.X., Yang, L., and Liu, Z.L. (1996). Two flavones from *Artemisia giraldii* and their antimicrobial activity. *Planta Medica* 62: 160-162.

Chapter 9

Plant Screening for Light-Activated Antifungal Activity

Olívia de Matos
Cândido P. Pinto Ricardo

INTRODUCTION

Economic factors and environmental concerns, together with the increasing resistance of pests and microorganisms to conventional pesticides, have led to the search for new sources and types of biologically active substances. When dealing with new pharmaceutical products, toxicity tests should give indications about the minimum level of effective compound, the exposure times, and the procedures to be adopted (Frazier, 1992; Ma et al., 1995).

Natural substances produced either by plants or by microorganisms have been self-imposing substitutes to synthetic compounds, due to their reduced environmental impact (Harborne, 1987; Clought et al., 1992; Osabe et al., 1992; Becker et al., 1993). Several microorganisms have been studied for their ability of producing bioactive compounds. Substances extracted from plants have also displayed herbicidal, insecticidal, antiviral, bactericidal, and antifungal activities (Takasugi et al., 1987; Bouguerra, 1990; Nakanishi, 1994).

Photopesticides are compounds that, in the presence of UVA (A ultraviolet or near UV) or visible light, cause damage to biological systems. These molecules belong to a larger group of compounds, the photosensitizers, which by absorbing light become reactive and trigger the chemical modification of certain biological molecules through specific processes (Becker and Maçanita, 1995; Heitz, 1995). These compounds are potentially very important for agriculture because the same light mechanism that activates

The authors wish to acknowledge the facilities provided by Professor António Maçanita in his laboratory and his advice in photochemistry. They are also very grateful to Dr. Maria João Marcelo Curto for chemical advice and lab support, and to Mrs. Isabel Pais for technical assistance.

them leads to their degradation. Therefore, a great decrease in the toxic residues introduced into the environment is expected by using such compounds.

The main purpose of the work in this chapter has been to detect antifungal activity in the extracts of several plants; to develop qualitative and quantitative methods for evaluation of their inhibitory effects, either in the dark or upon light irradiation; and to characterize the compounds responsible for the photoinhibition effect.

ANTIFUNGAL ACTIVITY IN PLANT EXTRACTS

Traditional Use of Products Extracted from Plants

Empirical knowledge about the properties of plants is very wide, and it has been maintained, deepened, and diversified over thousands of years. The discovery of the effects of opium—obtained from *Papaver somniferum* 4,000 years before the extraction of morphine—is an example of the antiquity and importance of this empirical knowledge. Despite the great development of chemotherapy, phytotherapy is still widely used today; in fact, it is reacquiring importance and credibility, as many damaging consequences of the use of synthetic compounds are observed (Font Quer, 1973).

Plants and their products have a determinant role also in the fields of agriculture and food preservation. Although scarce written information on past utilization of natural substances from plant origin in the combat of pests and diseases has been found, indications of their use exist in India and in the Himalayan region (Banerji et al., 1985).

The advent of agriculture, some 10,000 to 15,000 years ago, led to considerable transformations in the systems of human subsistence. The seeds and fruits of wild plants were quite resistant to insects and other predators. However, when cultivation of these plants was implemented and their products stored, catastrophic destruction occurred. This turn of events led to human intervention to find ways to combat pests. References from Egypt, Greece, and China more than 4,000 years ago note the protective effects of inorganic materials such as iodine, arsenic, and ashes, along with the use of organic products, such as tannins and vinegar. Reports from the fourth century B.C. to the sixteenth century about regions in Asia and Europe enumerate many plant species with properties useful for combating pests. Some examples include *Chrysanthemum marschalli* and *C. roseum,* which are used as insecticides (Banerji et al., 1985; Hostettman and Lea, 1987; Mendes, 1989).

Importance of Natural Products in Crop Protection

Agriculture and forestry have increased in productivity in order to cope with demographic growth. The decrease in cultivated areas (due to either migratory movements or the use of agricultural land for other purposes) also led to intensification of production. Thus, the use of chemical compounds, such as fertilizers, herbicides, and pesticides has risen considerably since the end of the nineteenth century (Heitz and Downum, 1986).

During World War II, progress in chemical syntheses led to a generalized use of synthetic pesticides, such as chlorinated hydrocarbons, organophosphorus compounds, and carbamates. As a result, toxicity problems, water contamination, persistence in the environment, residues in food, accumulation in animal and human fat tissues, and the resistance of parasites to pesticides have been serious concerns (Banerji et al., 1985; Kips, 1985). The need to continually increase pesticide doses and the high number of treatments led to dependence on these polluting agents, which are difficult to interrupt and from which no living being can be protected.

Despite this, synthesis of new compounds for the control of pests continues in several countries. In Germany, Sauter et al. (1999) reviewed the strobilurins, a new class of biologically active substances; in Russia (Moscow), Zakharychev et al. (1999) synthesized some derivatives of the 2-hydroxymino- and 2-alkoxyimino-3-(acylhydrazono) butanoic acids in order to prepare strobilurin analogs; in China, Wang et al. (1999) studied the relationship between the structure and the fungicidal activity of the beta-methoxyl acrylate esters; in India, Allimony et al. (1999) synthesized some new bactericidal and fungicidal derivatives of several imidazoles.

Conversely, natural pesticides could assume an important role because they combat pests and diseases while they maintain the ecological balance. These substances have a selective action; they are usually toxic only to the target organisms, act in quite small doses, and are easily degraded.

Natural Fungitoxins of Plant Origin

The fungi involved in plant interactions show a great heterogeneity of structures, life cycles, and metabolic activities. Evidence for the presence of fungitoxic compounds in plants may be obtained while studying the mechanisms of plant resistance to diseases. These studies may be a starting point for the evaluation of antifungal activity of plant extracts during the various phases of the extraction procedure (Harborne, 1987).

It can be assumed that natural fungitoxins may play important roles in the resistance to diseases considering that the secondary metabolites accumulated in the plants serve purposes in natural selection. Even if they do not prevent an infection, they may have a protective action in a certain stage of the plant's life (Harborne, 1987; Mayer, 1989).

Several studies on the detection of fungicide action have been done using either plants or in vitro cultures. Phenols, terpenoids, and alkaloids are the most frequent antifungal groups present in plants (Harborne, 1987). Table 9.1 presents examples of such compounds in the whole plant, while Table 9.2 shows some produced in vitro, and extracted from calli, cells in suspension, or culture media.

Natural Photosensitizers

The toxic action of several bioactive substances is triggered by absorption of solar light. After being irradiated, some photoactive molecules have been shown to promote toxic effects in living cells. This fact assumes special relevance when the photoactive chemical is taken up or ingested by agricultural or medical pests (Heitz and Downum, 1995).

According to Heitz and Downum (1995) the use of light to catalyze toxic reactions began in 1928 with studies performed on insects. Since then, a considerable amount of light-activated compounds have been tested to evaluate their reaction capacity against biological targets or, more specifically,

TABLE 9.1. Compounds of plant origin with fungicidal activity

Compounds	Plant species	Fungus	References
Triterpenoids	*Ophiobolus graminis*	*Gaeumannomyces graminis* *Ceratocystis ulmi*	Crombie et al., 1985
	Bellis perennis		Desevedavy et al., 1989
Carboxilic acids	*Chamaecyparis pisifera*	*Pyricularia oryzae*	Kobayashi et al., 1987
Insoluble phenols	*Solanum tuberosum*	*Phytophthora infestans* *Phoma exigua*	Ampomah and Friend, 1988
Polyphenols	*Stemonoporus* spp.	*C. cladosporioides*	Bokel et al., 1988
Flavonoids	*Helichrysum nitens*	Several	Tomás-Berberán et al., 1988
		Cladosporium herbarum	Tomás-Lorente et al., 1989
Alkaloids	*Chelidonium majus*	*Fusarium* spp.	Matos, Baeta, et al., 1999

TABLE 9.2. In vitro plant cultures producing compounds with fungicidal activity

Plant species	Type of culture	References
Scopolia japonica	Cells in suspension	Tabata et al., 1972
S. parviflora	Cells in suspension	Tabata et al., 1972
Papaver somniferum	Calli	Furuya et al., 1972; Ikuta et al., 1974
Macleaya cordata	Calli	Takao et al., 1983
Catharanthus roseus	Culture media	Balagué and Wilson, 1982; Bourgogne et al., 1989
Chelidonium majus	Culture media	Matos et al.,1988
Biscutella lusitanica *Iris taitii* *Picris spinifera*	Culture media	Matos, Gata-Gonçalves, et al., 1999

targets which act as photosensitizers (Spikes and Straight, 1987; Heitz and Downum, 1995). In the past four decades, an increasing number of studies have been carried out that make use of this kind of activity in diverse fields, mainly biomedicine (Parrish et al., 1982).

Agriculture is a relevant field toward the application of photoactive compounds. Several studies have been developed aiming to protect plants against pests and diseases through the use of photopesticides (Heitz, 1988; Osabe et al., 1992; Borges et al., 1995). In the past fifteen years, a great interest has emerged in the detection and development of a new type of pesticides (Becker and Maçanita, 1995). Several types of parasites are affected by compounds acting as photopesticides, such as arthropods, caterpillars, fruit fly larvae, yellow mealworms, black imported fire ants, mosquitoes and mosquito larvae, boll weevil, cockroaches, butterflies, house flies, cabbage butterfly larvae, fruit fly eggs, black cut worm larvae, alfalfa caterpillar, tobacco hornworm, and potato beetle (Becker and Maçanita, 1995; Heitz and Downum, 1995). Furthermore, many photosensitizers can act as potent herbicides and can control viruses, bacteria, algae, nematodes, yeasts, and fungi (Downum and Rodrigues, 1986; Smith, 1989; Becker and Maçanita, 1995).

Compounds capable of causing photodynamic action are produced by the natural defense mechanisms of plants and microorganisms against parasites. Several plants have been subjected to extraction procedures in the search for new natural photosensitizers (Table 9.3), such as acetophenones (encecalin), acetylenes (phenylheptatriyne), benzo- phenanthrenes (sanguinarine), coumarins (5,7-dimethoxycoumarin), β-carbolines (harmane),

TABLE 9.3. Photoactive compounds produced by plants

Class of compounds	Substance	Bioactivity	Plant family
Alkaloids	Dictamine	Antifungal	Rutaceae
Furanochromones	Khelin	Antiviral	Rutaceae
		Bactericidal	Umbelliferae
Coumarins	Angelicin	Larvicidal	Asteraceae, Leguminosae
	8-methoxypsoralen	Antifungal	Solanaceae, Umbeliferae
Quinones	α-Hipericin	Larvicidal Antiviral	Hypericaceae
Thiophenes	α-Terthienyl	Insecticidal	Asteraceae

extended quinones (cercosporin), furanochromones (khelin), furanocoumarins (8-methoxypsoralen), furanoquinolines (dictamine), isoquinolines (berberine), lignants (nordihydroquaiaretic acid), pterocarpans (pisatin), quinolines (camptothecin), sesquiterpenes (2,7-dihydroxycadalene), or thiophenes (alpha-terthienyl) (Heitz, 1988, 1995; Osabe et al., 1992).

In recent years, a diversity of natural photosensitizers with in vitro antifungal activity has been described. Simple and economic procedures for a rapid screening of plants and plant extracts for phototoxic activity have been developed (Towers and Champagne, 1986).

For the utilization of photoactive compounds, it is necessary to consider that a large part of the sun's radiation is emitted in the visible region of the spectrum. The capacity to deviate the absorption spectra to the region of visible light could highly increase the photoactive effectiveness of the compounds (Becker et al., 1993). Studies on the biological activity of known compounds and their thione derivatives, having absorption spectra toward the visible light region, were developed (Borges et al., 1995; Romão et al., 1997).

The elucidation of the phototoxic mechanisms of action is of great significance (Towers and Hudson, 1987). Lopes et al. (1995) studied the correlation between photophysical and photobiological properties of oligothiophenes.

Portuguese native plants were screened for their antifungal properties. Plant extracts were tested for their effectiveness in inhibiting crop pathogenic fungi. Special attention was given to the detection of compounds showing light-activated antifungal activity.

TECHNIQUES

Plant Material

Seeds of *Coriandrum sativum* L. and *Cuminum cyminum* L. were purchased; those of *Origanum vulgare* L., *Calamintha baetica* Boiss. and Reuter, and *Picris spinifera* Franco, and the leaves of *Ephedra fragilis* Desf., were collected from wild fields. The seeds were germinated in a greenhouse and the plants harvested, dried (under dark, with an air flow, at room temperature), and powdered. *Alethea officinalis* L. leaves and roots were supplied by industry (Aditiva-Fármacos e Suplementos, SA, Queluz, Portugal), dried, and powdered.

The *C. sativum* (Umbelliferae) is a Western European native plant that grows wildly in many countries of the Mediterranean region. It has been cultivated since antiquity for its aromatic and medicinal uses. Veterinary and perfumery uses are reported. Leaves and seeds are used as spices in soups, meats, and yogurts (Francisco and Herwing, 1986; Delgado, 1993; Font Quer, 1995). *C. cyminum,* also belonging to Umbelliferae, has been used for a long time as a spice of importance in food flavoring and preservation. It is a very important flavor for the food industry, largely used in meat, cheese, and liquors (Delgado, 1993; Font Quer, 1995; Bruneton, 1995).

A. officinalis belongs to the Malvaceae family. Both roots and leaves were used in folk medicine for their anti-inflammatory properties (Feijão, 1960; Bruneton, 1995; Amorim, 1998).

Ephedra fragilis Desf. (family Ephedraceae) grows wildly in Portugal and Spain (Coutinho, 1913; Franco, 1971; Castroviejo et al., 1986). Similar to other species of the same genus, it is rich in alkaloids, mainly ephedrine, and other substances, such as tannins, saponins, choline, and flavonolic dyes (Reti, 1953; Costa, 1978; Bruneton, 1995).

Calaminthe baetica, which belongs to the family Labiatae, has been used in folk medicine as an antispasmodic, a tonic, and a stimulant—and also as a spice (Muñoz, 1987).

Origanum vulgare is a species of family Labiatae occurring spontaneously throughout Europe to the Himalayas. Rich in essential oils with antimicrobial action, it is a widespread plant used as a spice and as a food preservative (Vasconcelos, 1949; Carmo et al., 1989; Kokkini and Vokou, 1989).

Picris spinifera (family Asteraceae) is a wild plant of Europe, occurring in barren, sandy land. Scarce information exists about its uses in folk medicine and about indications of its utilization for antimicrobial purposes (Vasconcelos, 1949; Font Quer, 1973, 1995).

Extraction Procedure

Extracts were prepared from leaves or roots by macerating with 500 ml either of methanol (MeOH) or ethyl acetate (EtOAc). For *P. spinifera, A. officinalis,* and *E. fragilis,* 100 g (dry weight) were macerated for 24 hours at room temperature. For aromatic plants *Coriandrum sativum, Cuminum cyminum, Origanum vulgare,* and *Calamintha baetica,* 500 g (dry weight) were macerated for five days, at 10°C, with continuous agitation. The macerates were filtered through a Buckner funnel with filter paper (Whatman 4, 90 mm in diameter) under vacuum, and centrifuged at 15,000 rpm. The supernatant was brought to dryness in a rotary evaporator under vacuum.

The crude extracts were fractionated with organic solvents: n-hexane (n-Hex), chloroform ($CHCl_3$), ethyl acetate (EtOAc), acetone (CO), ethanol (EtOH), and methanol (MeOH), either by successive direct washings or by medium pressure column chromatography (Silicagel Merck, 230 mesh). The fractions obtained were combined into calibrated flasks, evaporated to dryness, and weighted in order to determine the extraction's efficiency, and then redissolved in the corresponding solvent to a final concentration of 20 mg/ml.

All extracts were screened for their antifungal activity. The active fractions were reanalyzed by thin-layer chromatography (Silica-gel plates Merck 60 F_{254}, 0.2 μm) by development with the mixtures: $CHCl_3$, *O. vulgaris;* n-Hex:EtOH (50:50), *C. sativum* and *C. cyminum;* $CHCl_3$:MeOH (80:20), *P. spinifera;* $CHCl_3$:MeOH (95:5), *E. fragilis;* $CHCl_3$:MeOH (97.5:2.5), *A. officinalis*.

Fungus Selection and Growth

The following fungi were used as biological targets: *Cladosporium cucumerinum* (supplied by Departamento de Fitopatologia, Estação Agronómica Nacional, Oeiras, Portugal), *Fusarium culmorum, Aspergillus flavus, A. niger,* and *Botrytis cinerea* (supplied by Centro de Ferrugens do cafeeiro, Oeiras, Portugal).

Cladosporium cucumerinum Ellis & Arthur (order: Moniliales, family: Dematiaceae, genus: *Cladosporium* Link) is commonly used as a standard test fungus. This fungus is responsible for important losses in crops, affecting dried fruits and pods. Plants of the genera *Cucumis, Citrullus, Cucurbita,* and *Helianthus* are considered the main hosts (Vakalounakis and Williams, 1989). Spores of *C. cucumerinum* were observed in strawberry plant and tomato, and the symptoms of the disease were enhanced by high humidity conditions. This fungus grows in a temperature range of 0°C to 32°C, with optimum growth at 25°C (Alexopoulos and Mims, 1979).

Fusarium culmorum (W. G. Sm.) Sacc. (order: Moniliales, family: Tuberculariaceae, genus: *Fusarium* Link) frequently affects cereals of the genera *Zea, Triticum,* and *Avena.* This fungus causes crops root rot, being responsible for economically important losses of *Beta, Allium, Cucumis, Solanum, Melilotus, Phaseolus,* and *Vicia.* The fungus was also reported on *Agropirum, Bromus, Elymus,* and *Poa* (Stevens, 1925; Nelson et al., 1983).

Species of *Aspergillus* can be found in a large number of foods and products because of their good development under conditions of high temperature and low humidity. *Aspergillus flavus* L. is an ubiquitous fungus affecting seeds of many oil plants and causing serious economic damage. Maize, peanuts, and cotton are three of the main crops reported as susceptible to contamination during the drying process and storage (Petit et al., 1971; Klich et al., 1984).

Aspergillus niger L. has been well studied and is frequently found in food. In hot climates, this specie affects either fresh or dried and stored crops. Dry fruits, such as peanuts, nuts, almonds, and cereals, are reported as products highly affected by *A. niger* (Doupnick and Bell, 1971; Schindler et al., 1974; Pitt and Hocking, 1985). This fungus was also reported as being responsible for postharvest fruit decay of apples, pears, peaches, grapes, strawberries, tomatoes, and melons (Barkai-Golan, 1980). Due to its resistance to high temperature, *A. niger* is frequently isolated from dried grapes, dried fish, and spices (King et al., 1981; Pitt and Hocking, 1985).

Botrytis cinerea Pers. is a fungus well adapted to temperate regions, and affects a large range of crops—mainly small fruits, grapes being the most susceptible (Coley-Smith et al., 1980). This fungus was also isolated from apples, pears, strawberries, and tomatoes either fresh, postharvested, or stored (Hall and Scott, 1977; Dennis et al., 1979; Harwing et al., 1979).

The fungi were grown on potato dextrose agar media (PDA) in the dark at 23 ± 2°C for 5 to 10 days, depending on the fungus, and then used to collect the inocula necessary for the antifungal tests.

Aqueous suspensions of culture were prepared by disintegrating the mycelium previously grown in PDA media for five *(F. culmorum),* seven (*A. flavus, A. niger,* and *B. cinerea*), or ten days *(C. cucumerinum).* The liquid suspensions were filtered through a Buckner funnel with nylon nets (pore size 100 μm), under vacuum, in order to ensure their homogeneity. These suspensions were used as fungal inocula for the in vitro antifungal tests performed either in Petri dishes, on TLC plates, or in liquid media.

Assay Procedure

Biological Tests

Petri dishes (9 cm in diameter) containing 30 ml of solidified PDA medium were used for the antifungal assays. A fungus liquid inoculum (1 ml) was applied by pipette on the surface of the agar. Four holes (7 mm in diameter) were pierced in the agar to receive 100 μl of crude or fractionated extracts. Only one extract was tested per dish. Six dishes were used for each extract; of these, three dishes remained in the dark, while the other three were subjected to UVA light irradiation for 48 hours. Results were evaluated by the diameter of the halos of fungus nongrowth, observed three days after inoculation. Three replicates of each fungus growing without inhibitors, and controls for each one of the solvents, were maintained.

TLC plates were prepared which contained whole extracts, or the organic fractions applied by pipette directly into spots of 7 mm, or chromatographed. A standard liquid medium (Allen and Kuc, 1969) containing the fungus mycelium was sprayed on the surface of the plates. Two replicates of each plate were prepared and incubated in humid chambers at 23°C. One of the chambers remained in the dark for five days; the other, after a period of 24 hours in the dark, was irradiated with visible light, for eight to ten hours/day for three days, remaining in the dark until the end of the assay. The areas of the white spots without fungus growth gave a measure of the antifungal action. Photoinhibitory effects were detected by comparing each plate incubated under light with the corresponding one incubated in the dark.

Quantitative phototests were developed in liquid media. Samples (30 ml) of standard liquid media containing fungus spores and the extract (500 μl) were submitted to orbital shaking (100 rev min^{-1}) to increase the oxygenation, improve contact between fungus and extract, and ensure the best activation of the extract components. Ten replicates for each sample were tested, half of them being kept in the dark and the remaining irradiated with visible light. Photoinhibitions were evaluated by comparing the dry weight of mycelium produced in the irradiated assays with that obtained in the assays performed in the dark. Replicates for evaluation of the solvents and blank samples without extract or solvent were also tested both in the dark and under irradiation for the evaluation of the standard growth of each fungus.

Vanillin, MFA, KOH/EtOH, SO_4H_2/EtOH, and Dragendorff reagents were used to characterize the main chemical groups of active compounds.

Natural compounds commercially available or supplied by Departamento de Fotoquimica of Instituto de Tecnologia Química e Biológica,

Oeiras (flavone, coumarin, 7-methoxycoumarin, 7-methoxy- 4-methylcoumarin, cumaric acid, furanocoumarins, psoralen), were used as standards. 8-Methoxypsoralen (8-MOP) and 8-methoxythionepsoralen (8-MOTP) were used as references for the biological assays due to its well-known phototoxic behavior under UV and visible light, respectively. The concentration required was determined by estimating the overlap integral between the absorption spectrum of the standard and the lamp emission spectrum. The final concentrations were in the range of 10^{-4} to 2.10^{-6}.

Instrumentation

To perform the photobiological assays it was necessary to build an irradiation apparatus, schematically represented in Figure 9.1, which allowed the sample irradiation with either UVA light or visible light. The light sources were five fluorescent lamps of UVA light (Philips 40W/09N; emission range ~310-440 nm; λ_{max} = 354 nm; 73 µmol $m^{-2}s^{-1}$) or visible light (Osram L18w/10 daylight; emission range ~350-730 nm; λ_{max} = 485 nm; 48 µmol $m^{-2}s^{-1}$). Absorption spectra were determined in an OLIS 15 spectrophotometer.

Assays in Solid Media

The efficiency of the solvents EtOAc and MeOH for the extraction from the whole plant of substances active against *C. cucumerinum* and *F. cul-*

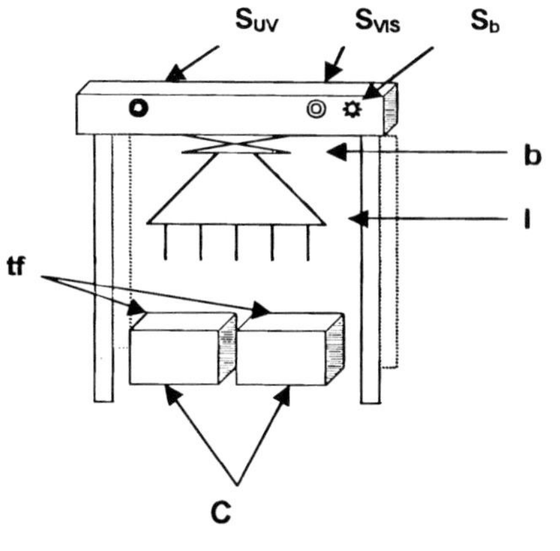

FIGURE 9.1. Schematic representation of a light irradiation apparatus. Light source (l), humid chambers or orbital shakers (C), transparent film (tf), UVA light switcher (S_{UV}), visible light switcher (S_{VIS}), cooling blade switcher (S_b), and cooling blades (b).

morum was evaluated. The plants analyzed (*A. officinalis, C. baetica, C. sativum, C. cyminum, E. fragilis, O. vulgare,* and *P. spinifera*) were potentially good sources of antifungal substances in spite of the differences of bioactivity observed and irrespective of the extraction solvent. The assays were performed as exemplified in Figure 9.2 and took place either in the dark or under UV irradiation (to detect photoactivity). Table 9.4 summarizes all the results obtained with the several plant extracts.

As a whole, *C. cucumerinum* was more inhibited by these extracts than was *F. culmorum,* especially under light-irradiated conditions. However, differences were detected in fungus inhibition, which were found to be dependent on the interaction solvent/plant. For instance, the EtOAc extracts of *A. officinalis, C. baetica, O. vulgare,* and *P. spinifera,* and the MeOH extract of *C. cyminum* were the most active against the two fungi. For *C. sativum,* ethyl acetate and methanolic extracts had equivalent antifungal activity. *Ephedra fragilis* extracts and methanolic extract of *A. officinalis* were inactive in the dark and showed just a slight photoactivity against *F. culmorum* *(A. officinalis)* or both fungi *(E. fragilis).*

We further analyzed the biological activity of some plant organs. Methanolic *C. sativum* and *P. spinifera* leaf and root extracts gave similar results to those obtained with the whole plants (results not shown). The results obtained with ethyl acetate leaf and root extracts are presented in Table 9.5.

The highest fungus inhibition was observed with the ethyl acetate *A. officinalis* root extracts. *Picris spinifera* leaf extracts were more active than those from roots, especially against *C. cucumerinum.* For *C. sativum,* no

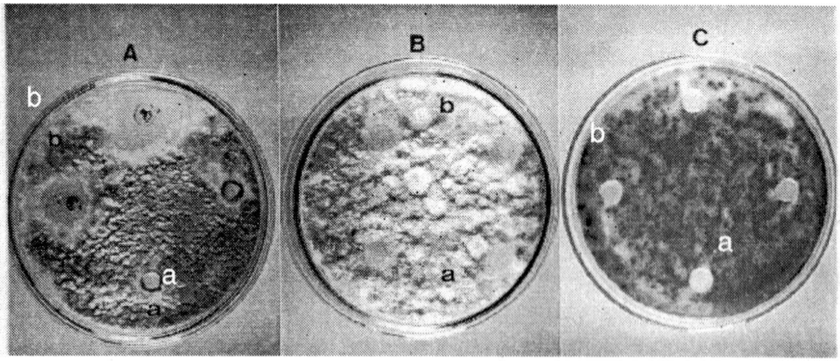

FIGURE 9.2. Petri dishes experiment showing the effect of methanolic (a) and ethyl acetate (b) extracts of *O. vulgare* on the growth of *F. culmorum,* for three days, either under irradiation with visible light (A) or in the dark (B). Growth observed with the solvents MeOH (a) and EtOAc (b) used as controls is shown in C.

TABLE 9.4. Effect of methanolic and ethyl acetate whole-plant crude extracts of *A. officinalis, C. baetica, C. sativum, C. cyminum, E. fragilis, O. vulgare,* and *P. spinifera* on the growth of *C. cucumerinum* and *F. culmorum.* The fungi were cultivated on solid media irradiated with UV light for a period of zero or eight hours/day, and growth inhibition was determined three days after inoculation. Inactive (–); slightly active (+); active (++); very active (+++).

Plants	Solvents	*C. cucumerinum*		*F. culmorum*	
		Light	Dark	Light	Dark
A. officinalis	EtOAc	+++	+	+++	++
	MeOH	–	–	+	–
C. baetica	EtOAc	+++	++	+	+
	MeOH	++	+	+	+
C. sativum	EtOAc	+++	+	+++	+
	MeOH	+++	+	+++	+
C. cyminum	EtOAc	++	+	++	+
	MeOH	+++	++	+++	+
E. fragilis	EtOAc	+	–	–	–
	MeOH	++	–	+	–
O. vulgare	EtOAc	+++	++	+++	+
	MeOH	+++	+	++	+
P. spinifera	EtOAc	+++	++	+++	+
	MeOH	++	+	++	+

significant differences were detected for the antifungal effects of leaf and root extracts. As a further step, it was attempted to use this procedure on solid media to analyze the bioactivity of the extract constituents separated by fractionation with a range of solvents of increasing polarity. However, the high toxicity of the solvents (mainly chloroform and acetone) did not allow any conclusions to be drawn (Amorim, Albuquerque, et al., 1996; Amorim, Matos, et al., 1986; Matos et al., 1998). Therefore, it was decided to use other assay procedures.

Assays on TLC Plates

To avoid the toxic effects of the partitioning solvents that hampered the solid media tests, assays were performed on TLC plates, which allowed the removal of the solvents by evaporation. The tests were performed either in

TABLE 9.5. Effect of ethyl acetate crude extracts of leaves and roots of *A. officinalis*, *C. sativum*, and *P. spinifera* on the growth of *C. cucumerinum* and *F. culmorum*. Growth inhibition of the fungi cultivated on solid media irradiated with UV light for a period of zero or eight hours/day was determined three days after inoculation: inactive (–); slightly active (+); active (++); very active (+++).

Plants	Plant Organs	*C. cucumerinum*		*F. culmorum*	
		Light	Dark	Light	Dark
A. officinalis	Leaves	++	+	+	+
	Roots	+++	++	+++	++
C. sativum	Leaves	+++	++	+++	++
	Roots	+++	++	++	+
P. spinifera	Leaves	+++	+	++	+
	Roots	++	+	+	–

the dark or under visible light irradiation (to detect photoactivity). Considering the large number of samples to be screened, these tests were performed on predetermined spots of each extract (20-100 μl) over the surface of the TLC plates.

For *A. officinalis*, *E. fragilis*, *P. spinifera*, and *O. vulgaris*, crude extracts of leaf and root as well as the samples obtained by fractionation with n-hexane, chloroform, ethyl acetate, acetone, ethanol, and methanol were analyzed against *C. cucumerinum* and *F. culmorum*. Table 9.6 summarizes these results.

As already observed in solid media assays, in general, crude extracts affected more *C. cucumerinum* than *F. culmorum*. However, after fractionation, higher inhibitory activities were observed in some of the fractions than in the corresponding crude extracts, e.g., *A. officinalis* leaves and roots (for *F. culmorum*), and *O. vulgaris* leaves (for both fungi). These observations apply to both dark- and light-stimulated effects, but the higher photoactivity of the less polar fractions of *A. officinalis* (roots) and *P. spinifera* (leaves and roots) and the more polar fractions of *A. officinalis* (leaves) and *E. fragilis* (leaves) should be noted.

CHARACTERIZATION OF THE BIOACTIVE COMPOUNDS

In an attempt to characterize the plant constituents responsible for the antifungal activities, the extracts or the extract fractions were chromatographed by liquid chromatography and/or by preparative thin-layer chro-

TABLE 9.6. Effect of ethyl acetate crude extracts from leaves and roots of *A. officinalis, E. fragilis, P. spinifera,* and *O. vulgare* and of the successive extract fractions obtained with n-hexane, chloroform, ethyl acetate, acetone, ethanol, and methanol on the growth of *C. cucumerinum* and *F. culmorum.* The fungi were developed on TLC plates irradiated with visible light for a period of 0 or 12 hours/day and growth inhibition was determined three days after inoculation. Inactive (–); slightly active (+); active (++); very active (+++); not assayed (·).

Fungi	Plants	Organs	Complete Extract		n-Hexane		Chloroform		Ethyl Acetate		Acetone		Ethanol		Methanol	
			Light	Dark	Light	Dark	Light	Dark	Light	Dark	Light	Dark	Light	Dark	Light	Dark
C. cucumerinum	*A. officinalis*	Leaves	++	+	++	+	–	–	–	–	–	–	–	–	–	–
		Roots	+++	+	++	++	–	–	++	+	–	–	–	–	–	–
	E. fragilis	Leaves	++	–	++	–	++	–	++	–	++	+	+++	+	+++	+
		Roots	·	·	·	·	·	·	·	·	·	·	·	·	·	·
	P. spinifera	Leaves	+++	+	+++	+	+++	–	++	–	·	·	·	·	+	–
		Roots	++	+	++	+	+	–	++	–	·	·	·	·	–	–
	O. vulgare	Leaves	++	+	+++	+++	+++	+	+	+	++	+	++	+	++	+
		Roots	·	·	·	·	·	·	·	·	·	·	·	·	·	·
F. culmorum	*A. officinalis*	Leaves	+	–	+	+	++	+	++	++	+++	++	–	–	–	–
		Roots	+	++	+++	+++	++	++	++	+	++	+	–	–	–	–
	E. fragilis	Leaves	+	+	–	–	–	–	–	–	–	–	–	–	–	–
		Roots	·	·	·	·	·	·	·	·	·	·	·	·	·	·
	O. vulgare	Leaves	–	+	+++	+++	++	+++	++	+	+++	++	–	–	–	–
		Roots	·	·	·	·	·	·	·	·	·	·	·	·	·	·

matography. Antifungal activity was detected on the TLC plates by white spots due to fungal growth inhibition. Phototoxicity was detected by comparing the plates kept in the dark with those irradiated with visible light.

For the identification of the main chemical groups of the active compounds, vanillin, MFA, KOH/EtOH, H_2SO_4/EtOH, and Dragendorff reagents were sprayed on equivalent TLC plates. Another means used to characterize the bioactive compounds was by subjecting chemical standards to similar TLC tests. After elution with the same solvent mixtures and testing against the same fungi, the R_f of the standards and those of the active compounds were compared. Figure 9.3 exemplifies the detection of compounds active against *C. cucumerinum* either in the dark or after light activation.

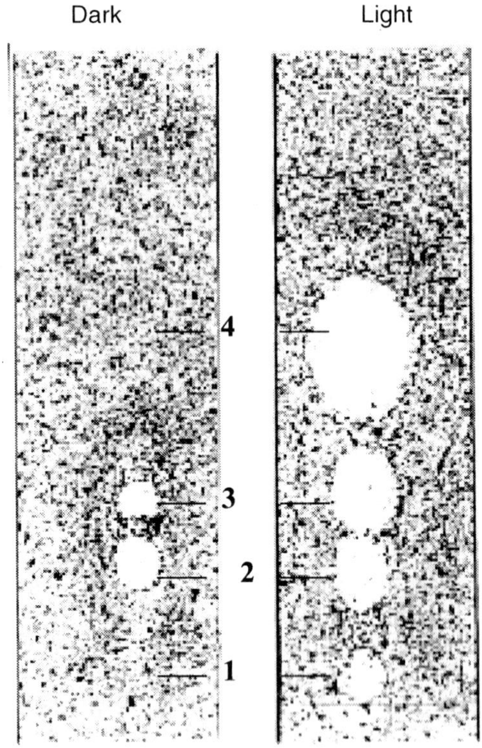

FIGURE 9.3. Effect of the chloroform leaf fraction of *O. vulgare* on the growth inhibition of *C. cucumerinum.* Tests performed in TLC plates in the dark or irradiated with visible light. Photoactive compounds: completely dependent on irradiation (1 and 4), slightly activated by light (3). Compound 2 shows antifungal activity independent of light.

In spite of good antifungal effects observed with *A. officinalis* fractions, no chemical characterization was possible. For *E. fragilis,* the dyeing reagents indicated that alkaloids were involved in the antifungal effects.

For *Origanum vulgare* several compounds were visualized under UV light (254 and 366 nm). However, only two photoactive compounds were isolated either from the whole extract or from the lower polar fractions (n-hexane and chloroform). Fractions treated with dyeing reagents revealed the presence of alkaloids, steroids, and flavonoids. Photoactivity was due to coumarin and cumaric acid. A flavone was identified as being active in the dark.

For *P. spinifera,* the dyeing reagents revealed high contents of coumarins in the leaves and alcohol derivatives, amides, and alkaloids in the roots. An unknown compound was detected as being active in the dark, and at least one coumarin was identified as being photoactive.

Two other plants, *C. sativum* and *C. cyminum*, showed very promising antifungal activity not only against *C. cucumerinum* and *F. culmorum* but also against *Aspergillus flavus*, *A. niger,* and *Botrytis cinerea* (Table 9.7).

Compounds of *C. sativum* were characterized by their R_f, fluorescence under UV light (366 and 254 nm), and ^{1}H NMR and ^{13}C RMN spectroscopies, which allowed identification of the main photoactive compound as the isocoumarin coriandrin (Delgado, 1993). For *C. cyminum,* a furanocoumarin similar to coriandrin was detected in the leaves (Carvalho, 1993), and cumaric acid was found both in leaves and roots (Matos, unpublished).

Liquid Media Assays

The persistance of antifungal effects was studied for the main active compound of *C. sativum* and *C. cyminum* leaves (furanocoumarin). These assays were performed in liquid media with *C. cucumerinum, F. culmorum,*

TABLE 9.7. Effects of *Coriandrum sativum* and *Cuminum cyminum* active fraction, isolated by a bioguided chromatographic separation, on the growth of *C. cucumerinum* (F1), *F. culmorum* (F2), *Aspergillus flavus* (F3), *A. niger* (F4), and *Botrytis cinerea* (F5). The fungi were cultured on TLC plates irradiated with visible light for a period of 0 or 12 hours/day and growth inhibitions observed three days from inoculation. Inactive (–); slightly active (+); active (++); very active (+++); strongly active (++++).

Plants	F1		F2		F3		F4		F5	
	Light	Dark	Light	Dark	Light	Dark	Light	Dark	Light	Dark
C. sativum	+++	++	+++	++	++	–	++	–	++	–
C. cyminum	++++	+++	+++	++	++	–	++	–	+	–

and *A. flavus* as targets. After 24 hours in the dark, the compounds were irradiated for 48 hours, remaining in the dark until the end of the assay. The inhibitions reached after three, five, and seven days in the dark or under irradiation with visible light are presented in Figure 9.4. Photoinhibitions were observed for both plants. In spite of the fact that the compound was irradiated only during hours 24 to 72 of culture, the photoactivated inhibition persisted during the whole incubation period (Matos et al., 1992; Carvalho, 1993; Delgado, 1993).

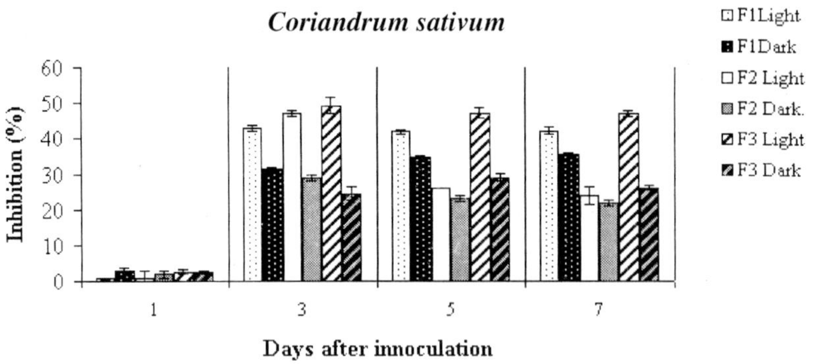

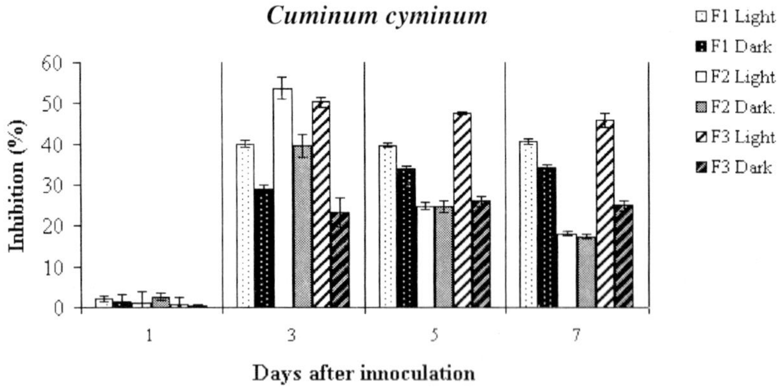

FIGURE 9.4. Effect of the main active compounds of *Coriandrum sativum* and *Cuminum cyminum,* on the growth of *Cladosporium cucumerinum* (F1), *Fusarium culmorum* (F2), and *Aspergillus flavus* (F3) grown in the dark or irradiated with visible light for 24 to 72 hours. The tests were performed in liquid medium, and the growth inhibitions (affected by standard deviation) were evaluated on the first, third, fifth, and seventh days after inoculation. Results are expressed as a percentage of the growth observed, with the solvents used as control.

Mycosis: A Great Challenge

For a natural compound to be considered a fungicide, it is necessary that in the biological tests a 50 percent inhibition (ED_{50}) of mycelium growth is produced for concentrations between 5 and 50 μg ml^{-1} (Harborne, 1987). The concentration of the extracts used in our assays was within this range.

In order to avoid or reduce to a minimum the problems of toxicity to the biological systems to be protected (diseased plant, beneficial microorganism, or insect) and also to reduce the impact on the environment, the highest efficacy and the lowest persistence must be achieved. Photopesticides may play an important role because they are toxic upon light exposure only and will be degraded by the light effect itself.

Phototoxic components occur in at least 30 plant families. The importance of these natural phototoxins in ecological interactions is stressed by the fact that the major plant families synthesizing such light-activated compounds are widely distributed in nature (Downum and Rodrigues, 1986).

Extracts of certain medicinal plants, namely of the Asteraceae family, have been known for centuries to be more active when exposed to light despite the fact that the mechanisms responsible for these activities were not understood. An example is the treatment of skin diseases, such as psoriasis and vitiligo, with extracts of *Ammi majus* and related species, which contain psoralens and furanochromones, now known common photoactive substances (Hudson and Towers, 1991).

Downum et al. (1989) verified that crude extracts of 12 species of the genus *Pectis* (family Asteraceae) were phototoxic to *Escherichia coli* and the yeast *Saccaromyces cerevisiae,* but the chemicals responsible for such phototoxicity were not identified. According to Downum and Rodrigues (1986), those compounds do not belong to currently recognized photosensitizer classes from Asteraceae. Acetylenic and thiophenic polyenes are two such common potent photosensitizers of Asteraceae, known to be responsible for elicitation reactions in species of this family. None of these compounds were detected in the extracts of *Picris spinifera* that expressed photoactivity.

For *A. officinalis* and *E. fragilis,* no references were found containing antifungal properties. The fact that antifungal effects of *A. officinalis* were found mainly in n-hexane, $CHCl_3$, and acetone fractions indicates that they could result from monoterpenic phenols, known for their effectiveness against fungi, and in which this plant is rich (Deans and Svoboda, 1990). Vanillic acid and p-hydroxybenzoic acid were identified in the roots of *A. officinalis,* which have been reported as active against fungi (Gudej, 1991;

Nagaoka et al., 1995). It is, therefore, possible that the antifungal effects of *A. officinalis* root extracts are due to the presence of these compounds.

In wild *O. vulgare*, a large number of bioactive substances have already been detected. Carvacrol and rosmarinic acid are referred to as the major *O. vulgare* compounds (Lamaison et al., 1991), but Carmo et al. (1989) refer linalool (≈37 percent) and thymol (≈32 percent) as the major constituents for the two Portuguese varieties identified. Carvacrol is believed to have a wide spectrum of antimicrobial action (Karapinar and Aktug, 1987; Beuchat and Golden, 1989), and antifungal properties were observed for thymol (Muller-Riebau et al., 1995).

Several authors reported antibacterial and antifungal activities of essential oil constituents of *O. vulgare* (Paster et al., 1990; Aureli et al., 1992; Aydin and Ozturk, 1995; Amorim, 1998). Fungistatic and fungicidal effects were reported against *Ascophaera apis*, and inhibition of spore germination and mycelium growth was described on *Aspergillus niger*, *A. ochraceus*, and *A. flavus* (Colin et al., 1989; Paster et al., 1990). Antifungal effects of *O. vulgare* methanol/water extracts against *A. fumigatus*, *Botrytis cinerea*, *Candida albicans*, *Fusarium oxysporum*, *Penicillium digitatum*, *Rhizopus nigricans*, *Saccharomyces pastorians*, and *Trichophyton mentagrophytes* were also reported (Guérin and Réveillère, 1985).

Most of the earlier studies on the detection of photopesticide used UV light to enhance the toxic reaction. In this way, carbolines, furanoquinolines, and related compounds have been evaluated for phototoxicity to bacteria and fungi. The studies with β-carbolines showed that some of them had strong UVA-dependent toxicity, while others had weaker and more selective activities. Hudson and Towers (1991) pointed out that UVA light could possibly have caused some photodamage. The knowledge that most of the sun's radiation is emitted in the visible light range encourages broadening of these studies, by stressing the importance in detecting naturally occurring compounds whose bioactivity is enhanced by visible light (Becker et al., 1993).

In previous work, we studied natural compounds and their thione derivatives for their antifungal activity after being irradiated with UV light or visible light (Borges et al., 1995). Thione compounds when irradiated by visible light displayed photoactivity levels comparable to those of their parent compounds under UV. Good antifungal efficiency was, thus, observed against *F. culmorum*, which has been reported as difficult to control. Present studies reveal that many plant extracts analyzed have antifungal properties similar to the inhibitory levels reached with thione compounds. This fact enhanced the importance of the studies performed with crude extracts, such as those referred in the present chapter. These could supply good information for further research and be the basis for the detection of additional antifungal compounds of plant origin.

CONCLUSIONS AND PERSPECTIVES

Chemical control of fungal diseases significantly reduces the losses of agricultural crops. Many of the traditional fungicides, in spite of their efficacy, can, as a result of intensive use, cause adverse effects on useful antagonistic organisms and predators or parasites of noxious insects (De Waard, 1993). The development of fungicide-resistant strains, leading to failure of disease control, is an additional serious risk.

In future research, the main goal should be the discovery of new compounds of plant origin, with novel ways of action showing low persistence in the environment and being active at low concentration. Most of these substances are synthesized during the natural defense mechanisms of plants as a response to the attack of parasites (De Waard, 1993; Downum and Wen, 1995).

The diversity of natural photoactive compounds isolated in recent years from bacteria, fungi, and plants suggests that such compounds are quite common in the biological systems. According to Towers and Champagne (1986), hundreds of compounds with this kind of action remain to be identified, and many interesting interactions could be expected (Heitz and Downum, 1986). It is important to study the new compounds evidencing photopesticide activity in relation to molecular structure and phototoxicity, elucidating their photophysical and photochemical properties. Nucleic acids, amino acids, proteins, steroids, unsaturated lipids, heterocyclic compounds, and some vitamins are included in the biological targets identified as being sensitive to the photodynamic action (Becker and Maçanita, 1995).

Despite notable success obtained with several classes of compounds tested as phototoxins, such as the thione derivatives of natural and synthetic coumarins, psoralens, chromones, and furanochromones (Becker et al., 1993; Borges et al., 1995), improvements should be done for the optimization of sunlight absorption, photoactivity, target selectivity, and biodegradation (Becker and Maçanita, 1995).

REFERENCES

Alexopoulos, C.J. and Mims, C.W. (1979). *Introductory Mycology*. New York: John Wiley and Sons. Inc.

Allen, E.H. and Kuc, J. (1969). α-Solanine and α-chaconine as fungitoxic compounds in extracts of Irish potato tubers. *Phytopathology* 58: 776-781.

Allimony, H.A., Saad, H.A., and El-Mariah, F.A.A. (1999). Synthesis and antimicrobial activity of some nitrogen heterobicyclic systems: Part 1. *Indian J. Chem. Section B* 38: 445-451.

Amorim, L.R. (1998). Actividade biológica de constituintes de espécies da flora portuguesa. Dissertação de mestrado em fisiologia e bioquímica de plantas (Dissertation for master's degree in plant physiology and biochemistry). Faculdade Ciências de Lisboa and Instituto Superior Agronomia (Eds.). Lisboa, 95 pp.

Amorim, L.R., Albuquerque, T., Amado, A., Baptista, S., Matos, O.C., and Ricardo, C.P. (1996). *X Congresso Nacional de Bioquimica*. Braga. Abstracts: 10.

Amorim, L.R., Matos, O.C., and Ricardo (1996). Utilização de extractos de plantas medicinais e condimentares no controlo de fungos patogénicos. *1° Colóquio Nacional de Plantas Aromáticas e Medicinais*. Vila Moura. Abstracts: 32.

Ampomah, Y.A. and Friend, J. (1988). Insoluble phenolic compounds and resistance of potato tuber disc to *Phytophthora* and *Phoma*. *Phytochem.* 27: 2533-2541.

Aureli, P., Constantini, A., and Zolea, S. (1992). Antimicrobial activity of some essential oils against *Listeria monocytogenes*. *J. Fd. Prot.* 55: 344-348.

Aydin, S. and Ozturk, Y. (1995). Investigation of *Origanum onites, Sideritis congesta* and *Satureja cuneifolia*. Essential oils for analgesic activity. *Phytother. Res.* 10: 342-344.

Balagué, C. and Wilson, G. (1982). Growth and alkaloid biosynthesis by cell suspension of *Catharanthus roseus* in a chemostat under sucrose and phosphate limiting conditions. *Physiologie Veg.* 20: 515-522.

Banerji, R., Misra, G., and Nigam, S.K. (1985). Role of indigenous plant material in pest control. *Pestic.* 19: 32-38.

Barkai-Golan, R. (1980). Species of *Aspergillus* causing post-harvest fruit decay in Israel. *Mycopath* 71: 13-16.

Becker, S.R., Chacravorti, S., Gartner, C.A., and Miguel, M.G. (1993). Photosensitizers: Comprehensive photophysics/chemistry and theory of coumarins, chromones, their homologs and thione analogs. *J. Chem. Soc. Far. Trans.* 89: 1007-1019.

Becker, S.R. and Maçanita, A. (1995). Biological photosensitizers: Phototoxic agents of the future? *Rev. Port. Quím.* 2: 30-44.

Beuchat, L.R. and Golden, D.A. (1989). Antimicrobials occurring naturally in foods. In Antimicrobial activity of some essential oils against *Listeria monocytogenes*. *J. Fd Prot.* 55: 344-348.

Bokel, M., Champika-Diyasena, M.N., Gunatikala, A.A.L., Kraus, W., and Sotheeswaran, S. (1988). Gunaliculalol, an antifungal resveratrol trimer from *Stemonoporous canaliculatus*. *Phytochem.* 27: 377-380.

Borges, M.L., Matos, O.C., Pais, I., Seixas de Melo, J., Ricardo, C.P.P., Maçanita, A., and Becker, R.S. (1995). Evaluation of a broad variety of coumarins, chromones, their furano homologues and thione analogues as phototoxins activated by UVA and visible light. *Pestic. Sc.* 44: 155-162.

Bouguerra, M.L. (1990). Pesticides phototoxiques: Des armes lumineuse contre les nuisibles. *Recherche* 271: 93-94.

Bourgogne, V., Labidalle, S., Galons, H., Foulquier, M., Jacquin-Dubreil, A., and Cosson, L. (1989). Biotransformation of a synthetic compound, 1,5-diphenyl sulphinyl-3-methyl-3-nitropentane by cell suspension of *Catharanthus roseus*. *Phytochem.* 28: 2345-2347.

Bruneton, J. (1995). *Pharmacognosy, phytochemistry, madicinal plants.* Paris: Lavoisier TEC and DOC; London: Intersect Ltd.

Carmo, M.M., Frazão, S., and Venâncio, F. (1989). The chemical composition of Portuguese *Origanum vulgarea* oils. *J. Essent. Oils Res.* 2: 69-71.

Carvalho, I.M.M.S. (1993). Acção de compostos naturais na inibição de microorganismos contaminantes de alimentos. Mestrado em ciência e tecnologia dos alimentos (master's degree in science and technology of food). Faculdade Ciências Lisboa. Lisboa, 146 pp.

Castroviejo, S., Laínz, M., González, G.L., Monserrat, P., Garmendia, F.M., Paiva, J., and Villar, L. (1986). *Flora Ibérica. Plantas vasculares de la Península Ibérica e Islas Baleares.* I (*Lycopodiaceae–Papaveraceae*). Madrid, 575 pp.

Clought, J.M, Evans, D.A., de Frain, P.J., Frazier, T.E.M., Godfrey, C.R.A, and Youle, D. (1992). Role of natural products in pesticide discovery. *Natural and Engineered Pest Management Agents. A.C.S. Symp. Ser.* 551: 37-53.

Coley-Smith, J.R., Verhoeff, K., and Jarvis, W.R. (Eds.) (1980). *The Biology of Botrytis.* London: Academic Press.

Colin, M.E., Ducos de Lahitte, J., Larribau, E., and Boué, T. (1989). Activité des huiles essentielles de Labiées sur *Ascophaera apis* et traitement d'un rucher. *Apidologie* 20: 221-228.

Costa, A.F. (1978). *Farmacognosia.* II. Fundação Calouste Gulbenkian (Eds.). Lisboa, 1117 pp.

Coutinho, A.X.P. (1913). *Flora de Portugal.* Aillaud, Alves e Companhia (Eds.). Paris, 710 pp.

Crombie, L., Crombie, W.M.L., and Whiting, D.A. (1985). The avenacins. Natural fungicides from oat roots active against "take-all" attack. *Coun. Fungic. Crop Prot.* 2: 267-271.

Deans, S.G. and Svoboda, K.P. (1990). The Antimicrobial properties of marjoram (*Origanum majorana* L.) volatile oil. *Flav. Frag. J.* 5: 187-190.

Delgado, F. (1993). Aspectos culturais e prospecção de compostos com actividade biológica do coentro (*Coriandrum sativum* L.). Mestrado em produção vegetal (master's degree in plant production). Instituto Superior Agronomia, Lisboa, 93 pp.

Dennis, C., Davis, R.P., Harris, J.E., Calcutt, L.W., and Cross, D. (1979). The relative importance of fungi in the breakdown of commercial samples and sulphited strawberries. *J. Sci. Fd. Agric.* 30: 959-973.

Desevedavy, C., Amoros, M., and Guirre, L. (1989). Antifungal agents: In vitro and in vivo antifungal extracts from the common daisy *Bellis perennis. J. Nat. Prod.* 52: 184-185.

De Waard, M.A. (1993). Chemical control of plant diseases: Problems and prospects. *Ann. Rev. Phytopathol.* 31: 403-421.

Doupnick, B. and Bell, D.K (1971). Toxicity to chicks of *Aspergillus* and *Penicillium* species isolated from moldy pecans. *Appl. Microbiol.* 21: 1104-1106.

Downum, K.R. and Rodrigues, E. (1986). Toxicological action and ecological importance of plant photosensitizers. *J. Chem. Ecology* 12: 823-833.

Downum, K.R., Villegas, S., Rodrigues, E., and Leil, D.J. (1989). Plant photosensitizers: A survey of their occurrence in arid and semiarid plants from North America. *J. Chem. Ecology* 15: 345-355.

Downum, K.R. and Wen, J. (1995). The occurrence of photosensitizers among higher plants. In Light-activated pest control. Heitz, J.R. and Downum, K.R. (Eds.). *ACS Symp. Sr.* 616: 135-143.

Feijão, R.O. (1960). *Elucidário Fitológico—Plantas Vulgares de Portugal Continental, Insular e Ultramarino* I. A-H. Instituto Botânico de Lisboa (Eds.), 473 pp.

Font Quer, P. (1973). *Plantas Medicinales.* Editorial Labor. S.A. Madrid (Eds.), 1033 pp.

Font Quer, P. (1995). *Plantas Medicinales.* El dioscorides renovado. Fifteenth edition. Editorial Labor. S.A. (Eds.). Madrid.

Francisco, I.E. and Herwing, V. (1986). *Plantas aromáticas e medicinais.* Incone (Eds.), São Paulo, 550 pp.

Franco, J.A. (1971). *Nova flora de Portugal (Continente e Açores).* I. *Lycopodiacea—Umbelliferae.* Escolar Editora (Eds.), Lisboa, 648 pp.

Frazier, J.M. (Eds.) 1992). *In vitro toxicity testing.* New York: Marcel Dekker, Inc.

Furuya, T., Ikuta, A., and Syono, K. (1972). Alkaloids from callus tissue of *Papaver somniferum. Phytochem.* 11: 3041-3044.

Gudej, J. (1991). Flavonoids, phenolic acids and coumarins from the roots of *Althaea officinalis. Planta Med.* 57: 284-285.

Guérin, J.C. and Réveillère, H.P. (1985). Activité antifongique d'extraits végétaux à usage thérapeutique. II. Étude de 40 extraits sur 9 souches fongiques. *Ann. Pharm. Fr.* 43: 77-81.

Hall, E.G. and Scott, K.J. (1977). *Storage and market diseases of fruits.* Commonwealth Scientific and Industrial Research Organization (Eds.), Melbourne, Australia, 52 pp.

Harborne, B.J. (1987). Natural fungitoxins. Biologically active natural products. *Proc. Phytochem. Soc. Eur.* 27: 195-211.

Harwing, J., Scott, P.M., Stoltz, D.R., and Blanchfield, B.J. (1979). Toxins of molds from decaying tomato fruit. *Appl. Envir. Microbiology* 38: 267-274.

Heitz, J.R. (1988). Photoactivated pesticides. In Evaluation of a broad variety of coumarins chromones, their furano homologues, and their analogues as phototoxins activated by UVA and visible light. *Pestic. Sc.* 44: 155-162.

Heitz, J.R. (1995). Pesticidal applications of pothotoactivated molecules. In Light-activated pest control. Heitz, J.R. and Downum, K.R., (Eds.), *A.C.S. Symp. Ser.* 616: 1-15. Washington, DC.

Heitz, J.R. and Downum, K.R. (1986). Light-activated pesticides. *A.C.S. Symp. Ser.* 339. Washington, DC.

Heitz, J.R. and Downum, K.R. (1995). Light-activated pest control. *A.C.S. Symp. Ser.* 616. Washington, DC.

Hostettman, K. and Lea, J.P. (1987). Biologically active natural products. *Ann. Proc. Phytochem. Soc. Eur.* Oxford: Clarendon Press.

Hudson, J.B. and Towers, G.H.N. (1991). Therapeutic potential of plants Photosensitizers. *Pharmac. Therapy* 49: 181-222.

Ikuta, A. Syono, K., and Furuya, T. (1974). Alkaloids of callus tissue and redifferenciated plantlets in the Papaveraceae. *Phytochem.* 13: 2175-2179.

Karapinar, M. and Aktug, S.E. (1987). Inhibition of food-borne pathogens by thymol, eugenol, and anethole. *Inter. J. Food Microbiol.* 4: 161-166.

King, A.D., Hocking, A.D., and Pitt, J.I. (1981). The mycoflora of some Australian foods. *Aust. Fd. Technology* 33: 55-60.

Kips, R.H. (1985). Environmental aspects. In *Pesticides Application: Principles and Practice.* P. T. Haskell (Ed.), Oxford, pp. 190-200.

Klich, M.A., Thomas, S.H., and Mellon, J.E. (1984). Field studies on the mode of entry of *Aspergillus flavus* into cotton seeds. *Mycologia* 76: 665-669.

Kobayashi, K., Nishini, C., Tomita, H., and Furushima, M. (1987). Antifungal activity of pisiferic acid derivatives against the rice blast fungus. *Phytochem.* 26: 3175-3179.

Kokkini, S. and Vokou, D. (1989). Carvacrol-rich plants in Greece. *Flav. Frag. J.* 4: 1-7.

Lamaison, J.L., PetitJean-Freytet, C., and Carnat, A. (1991). Lamiacées médicinales à propriétés antioxydantes, sources potentielles d'acide rosmarinique. *Pharm. Acta Helv.* 66: 185-188.

Lopes, A., Borges, M., Melo, J., Pais, I., Matos, O., Elisei, F., and Becker, R. (1995). Oligotiophenes as photosensitizers: A correlation between photophysical and photobiological properties. *III Congresso de Fotoquimica. San Lorenzo de el Escorial.* Abstract: C.39.

Ma, W.G., Fuzzati, N., Li, Q.S. Yang, C.R., Stoecki, E.H., and Hostettman, K. (1995). Polyphenols from *Eriosema tuberosum. Phytochem.* 39: 1049-1061.

Matos, O.C., Baeta, J., Silva, M.J.S., and Ricardo, C.P.P. (1999). Sensitivity of *Fusarium* strains to *Chelidonium majus* L. extracts. *J. Ethnopharmac.* 66: 151-158.

Matos, O.C., Baptista, S., Passarinho, J.A., Delgado, F., Ricardo, C.P.P., and Vilas-Boas, L. (1998). Watering and mineral supply effects on plant production, on hydrosoluble metabolites content and on the antifungal properties of three aromatic plants used as spices. *First International Meeting of Aromatic and Medicinal Mediterranean Plants.* Ansião-Conimbriga. Proceedings, pp. 158-163.

Matos, O.C., Delgado, F., and Carvalho, (1992). Inhibitory action of plant extracts on the growth of fungi affecting plants and food. *Biotec-92.* Santiago de Compostela. Proceedings. BV-6.

Matos, O.C., Gata-Gonçalves, L., Baptista, S., Pais, I., Azinheira, H., Bruno-de-Sousa, R., and Ricardo, C.P.P. (1999). Estudos in vitro de germinação, Indução de calli, regeneração de plantas e bioactividade de meios de cultura. *2° Workshop Biotecnologia de Plantas.* Faro. Abstracts, 11.

Matos, O.C., Melo, E.M., Ricardo, C.P.P., and Silva, M.J.S. (1988). Acção de meios de cultura de células e de extractos de *Chelidonium majus* L. no crescimento de fungos fitopatogénicos. *IV Congresso Nacional de Biotecnologia.* Coimbra. Abstracts, BV-6.

Mayer, A.M. (1989). Plant-fungal interactions. A plant physiologist's view point. *Phytochem.* 28: 311-317.

Mendes, M.E.E. (1989). Uso de substâncias no combate a pragas (monograph). LNETI-DTIQ, 4. Lisboa, pp 68.

Muller-Riebau, F., Berger, B., and Yegen, O. (1995). Chemical composition and fungitoxic properties to phytopathogenic fungi of essential oils of selected aromatic plants growing wild in Turkey. *J. Agric. Fd. Chem.* 43: 2262-2266.

Muñoz, F.L.B. (1987). Plantas medicinales y aromáticas: Estudio, cultivo y procesado. In *Aspectos culturais e prospecção de compostos com actividade biológica do coentro* (*Coriandrum sativum* L.). Lisboa: Instituto Superior de Agronomia.

Nagaoka, T., Ohira, J., Yoshimara, T., And Sakamura, S. (1995). Fungitoxic compounds from the roots of tomato stock. *Ann. Phytopathological Soc. Japan* 61: 103-108.

Nakanishi, K. (1994). Natural products as sources of pest management agents. In *Natural and engineered pest management agents.* Heitz, J.R. and Downum, K.R. (Eds.), *A.C.S. Symp. Sr.* 551: 11-24.

Nelson, P.E.; Toussoun, T.A., and Cook, R. J (Ed) (1983). Fusarium *diseases, biology and taxonomy.* University Park, PA: The Pennsylvania State University Press.

Osabe, H., Morishima, Y., Goto, Y., Masamoto, K., Nakagawa, Y., and Fujita, T. (1992). Quantitative structure-activity relationship of light-dependent herbicidal 4-pyridone-3-carboxanilides. I. Effect of benzene ring substituents at anilide moiety. *Pest. Sc.* 34: 17-25.

Parrish, J.A., Stern, R.S., Pathak, M.A., and Fitzpatrick, T.B. (1982). Photochemotherapy of skin diseases. In J.D. Regan and J.A. Parrish (Eds.), *The science of photomedicine.* New York: Plenium Press.

Paster, N., Juven, B.J., Shaaya, E., Menasherov, M., Nitzan, R., Weisslowicz, H., and Ravid, U. (1990). Inhibitory effect of oregano and thyme essential oil on moulds and foodborne bacteria. *Letters Appl. Microbiology* 11: 33-37.

Petit, R.E., Taber, R.A., Schroeder, H.W., and Harrison, A.L. (1971). Influence of fungicides and irrigation practice on aflatoxin in peanuts before digging. *Appl. Microbiology* 22: 629-634.

Pitt, J.I. and Hocking, A.D. (1985). *Fungi and food spoilage.* London: Academic Press.

Reti, L. (1953). Ephedra bases. In *Actividade biológica de constituintes de espécies da Flora Portuguesa.* L. R. Amorim. Faculdade de Ciências and Instituto Superior de Agronomia. Lisboa, 95 pp.

Romão, A.C., Bonifácio, V., Leitão, E., Borges, M., Matos, O.C., Duarte, M., Medeiros, J., Dias, A., and Maçanita, A. (1997). Photodegradation and photobiological activity of flavothiones. *V Encontro Nacional de Fotoquímica.* Braga. Abstracts, 17.

Sauter, H, Steglich, W., and Anke, T. (1999). Strobilurins: Evolution of a new class of active substances (review). *Angew. Chem. Int.* 38: 1329-1349.

Schindler, A.F., Abadie, A.N., Gecan, J.S., Mislivec, P.B., and Brickey, P.M. (1974). Mycotoxins produced by fungi isolated from inshell pecans. *J. Fd. Sc.* 39: 213-214.

Smith, K.C. (1989). *The Science of Photobiology.* New York: Plenum Press.

Spikes, J.D. and Straight, R.C. (1987). Biochemistry of photodynamic action. In J.R. Heitz and K.R. Downum (Eds.), *Light-activated pesticides. A.C.S. Symp. Sr.* 339: 98-108. Washington, DC: American Chemical Society.

Stevens, F.L. (1925). *Plant diseases fungi.* New York: The MacMillan Company.

Tabata, M., Yamamoto, H., Hiraoka, N., and Konoshima, H. (1972). Organization and alkaloid production in tissue cultures of *Scopolia parviflora. Phytochem.* 11: 949-955.

Takao, N., Kamigauchi, M., and Okada, M. (1983). Biosynthesis of benzo(c)phenanthridine alkaloids sanguinarine, chelirubine and macarpine. *Helv. Chim. Acta* 66: 473-484.

Takasugi, M., Kawashima, S., Katsui, N., and Shirata, A. (1987). Two polyacetylenic phytoalexins from *Arctium lappa. Phytochem.* 26: 2957-2958.

Tomás-Berberán, F.A., Msonthi, J.D., and Hostettmann, K. (1988). Antifungal epicuticular methylated flavonoids from *Helichrysum nitens. Phytochem.* 27: 753-755.

Tomás-Lorente, F., Iniesta-Sanmartin, E., Tomás-Berberán, F.A., Trowitzsch-Kienast, W., and Wray, V (1989). Antifungal phloroglucinol derivatives and lipophilic flavonoids from *Helichrysum decumbens. Phytochem.* 28: 1613-1615.

Towers, G.H.N. and Champagne, D.E. (1986). Fungicidal activity of naturally occurring photosensitizers. In Heitz, J.R. and Downum, K.R. (Eds.), *Light-activated pest control, A.C.S. Symp. Sr.* 339: 231-240. Washington, DC: American Chemical Society.

Towers, G.H.N. and Hudson, J.B. (1987). Potentially useful antimicrobial and antiviral phototoxins from plants. *Photochem. Photobiology* 46: 61-66.

Vakalounakis, D.J.E. and Williams, P.H. (1989). A cotyledon screen for resistance to scab *(Cladosporium cucumerinum)* in cucumber *(Cucumis sativus)* seedlings. *Ann. Appl. Biology* 115: 443-450.

Vasconcelos, J.C. (1949). Plantas medicinais e aromáticas. *Direcção Geral dos Serviços Agrícolas. Série: Estudos e Informação Técnica* 34: 112-128.

Wang, Z.W., Li, Z.M., Chen, H.S., and Li, S.Z. (1999). Studies on synthesis and fungicidal activities of beta-substituted phenyl (Phenoxyl) actylate esters (amides). *Chem. J. Chin. Univ.* 20: 1248-1253.

Zakharychev, V.V., Golubtsova, M.D., and Kovalenko, L.V. (1999). Synthesis of derivatives of 2-hydroxymino- and 2-alkoxymino-3-(acylhydrazono) butanoic acids. *Russ. Chem. Bull.* 48: 495-500.

Chapter 10

Antifungal Saponins from White Asparagus (*Asparagus officinalis* L.) Bottoms and Their Physiological Role in the Plant Defense System

Makoto Shimoyamada
Shigeki Nakashima
Kyoko Nakashima
Kazuyoshi Okubo
Kenji Watanabe

INTRODUCTION

Asparagus is a familiar food as well as medicinal plant in India and in other countries. White asparagus is most often processed into canned food. During this processing, the "bottom cut" is generally discarded because of its bitter taste, which is attributed to saponin constituents. Kawano et al. (1975, 1977) investigated the bitter components in asparagus and reported their structures to be furostanol saponins. Nine other steroidal saponins were independently isolated and characterized as asparagosides A to I (Goryanu et al., 1976a,b; Goryanu, Krokhmalyuk, Kintya, and Gltzin, 1976; Goryanu and Kintya, 1976, 1977) from *Asparagus officinalis,* whose structures were summarized in a review (Price et al., 1987). Recently, some new saponins also have been isolated from *A. officinalis* (Pant et al., 1988; Shao et al., 1997). Further, many saponins were isolated from other species in the genus *Asparagus,* namely *A. cochinchinensis* (Konishi and Shoji, 1979), *A. racemosus* (Sharma et al., 1981), *A. gonocladus* (Mandloi and Sant, 1981), *A. adscendens* (Sharma et al., 1982), *A. plumosus* (Sati, 1985), *A. stipularis* (Halim et al., 1989), *A. curillus* (Sharma and Sharma, 1993), *A. filicinus* (Sharma et al., 1996), *A. dumosus* (Ahmad et al., 1998), and *A. africanus* (Debella et al., 1999).

Many of the biological activities of saponins have been reviewed to date (Price et al., 1987; Mimaki and Sashida, 1996; Lacaille-dubois and Wagner,

1996; Mahato and Garai, 1998). Many biological activities of *Asparagus* saponins have been reported, including antioxytocic action (Gaitonde and Jetmalani, 1969), bitter taste (Kawano et al., 1975, 1977), molluscicidal acitivity (Sati et al., 1984), cytotoxic activity (Sati et al., 1984; Shao et al., 1997), spermicidal activity (Pant et al., 1988), replant problem (Lake et al., 1993), and antitumor activity (Shao et al., 1996). Although antifungal activity is one of the important biological activities, there is no report on antifungal activity of *Asparagus* saponins. We have thus attempted to determine the antifungal activity of the waste products from asparagus processing and to isolate and characterize the active antifungal principle (Shimoyamada et al., 1990, 1996). We review in this chapter the antifungal saponins from *Asparagus* and also discuss their role in the plant defense system of *Asparagus*.

TECHNIQUES

A crude saponin fraction was prepared from bottom-cut white asparagus (*Asparagus officinalis* L. cv Merry Washington 500W). This bottom-cut asparagus was obtained from factory waste.

Preparation and Fractionation of Crude Saponin Fraction from Asparagus

Dried bottom-cut asparagus (3 kg) was extracted twice with 60 percent ethanol (50 liters) at room temperature for 24 hours. The two extracts were combined, and the solvent was removed under reduced pressure. The product was dispersed in 1-butanol-water (1:1, v/v), the mixture was centrifuged, and the 1-butanol layer was collected and concentrated under reduced pressure. The concentrated extract was dispersed in benzene-water (1:1, v/v). The water phase was extracted again with 1-butanol, and the butanol extract was dissolved in methanol (300 mL). Ether (8 liters) was added dropwise with stirring, and the precipitate was collected as the crude saponin fraction (11.6 g). The fraction thus obtained was dispersed in methanol-water (60:40), subjected to octadecyl silyl (ODS) column chromatography (YMC GEL ODS-A; 230 × 18 mm id), equilibrated with methanol-water (60:40), and eluted with methanol-water (60:40), (70:30) and methanol, in that order. Each eluted fraction was monitored with TLC and antifungal activity.

Thin-Layer Chromatography (TLC)

The thin-layer chromatography plate was a Kieselgel 60F-254 (0.25 mm thickness, Merck). It was developed with chloroform-methanol-water (7:3:1; upper layer). The developed plate was dried and sprayed with a detecting reagent and heated as necessary. Detecting reagents were 10 percent sulphuric acid and Ehrlich's reagent, prepared by dissolving the 4-dimethylaminobenzaldehyde (1 g) in a concentrated hydrogen chloride-methanol (10:75) mixture (85 mL).

Measurement of Antifungal Activity of Saponins

Antifungal activity was determined by using the paper disc method. The fungi used in this study are shown in Table 10.1, and were preinoculated for one or two weeks on an agar culture broth (yeast extract, 3 g; malt extract, 3 g; peptone, 5 g; and glucose, 10 g; with agar, 10 g; in water, 1 liter). Mycelia were collected, suspended in NaCl solution (0.9 percent with 0.05 percent Tween), and inoculated into new agar plates (5×10^4 cells/mL).

Pieces of paper disc (Whatman AA discs, 6 mm diameter) were soaked in an ethanol solution (10 mg/mL) of a sample. These pieces, corresponding to 25 μL of a sample, were placed on the surface of the agar culture broth. A piece of control disc containing only ethanol was also placed on the same agar plate. The whole system was kept at 30°C, and the antifungal activity was measured after one week.

Measurement of Minimum Inhibitory Concentration (MIC)

The fungi used were preinoculated for one or two weeks on an agar culture. Fungal spores were suspended (5×10^4 cells/mL) in tubes containing broth (5 mL) without agar, the saponins being added at appropriate levels. The tubes were incubated for one week at 30°C and MIC was assessed.

High Performance Liquid Chromatography (HPLC)

High performance liquid chromatography was performed using a Waters 510 pump or a Jasco BIP-1 HPLC pump with Rheodyne model 7125 injection valve. The columns were a YMC Pack ODS AM-303 (5 μm, 250 × 4.6 mm i.d.) and a YMC Pack ODS AM-323-7 (7 μm, 250 × 10 mm i.d.). Detection was carried out with a Hitachi variable wavelength UV monitor (210 nm) or a Jasco 830-RI detector.

TABLE 10.1. Antifungal activities of the crude saponin fraction from *Asparagus officinalis* L. bottom cut

Fungus		Activity
Aspergillus candidus	AHU7017	–
Aspergillus chevalieri	AHU7443	–
Aspergillus fumigatus	IFO9733	–
Aspergillus luchuensis	AHU7092	±
Aspergillus niger	IFO4091	–
Aspergillus oryzae	S-01	–
Aspergillus repens	AHU7458	±
Aspergillus sulphureus	AHU7182	–
Aspergillus tamarii	AHU7367	–
Aspergillus terreus	FERM S-3	–
Aspergillus versioclor	AHU7200	–
Aspergillus wentii	AHU7207	–
Candida albicans	IFO1061	+
Chaetomium globosum	FERM S-11	–
Cryptococcus albidus	IFO1420	+
Epidermophyton floccosum	IFO9045	+
Fusarium oxysporum	IFO31213	–
Microsporum gypseum	IFO8307	+
Mucor racemosus	IFO4581	±
Penicillium funiculosum	FERM S-6	–
Penicillium italicum	IFO9419	+
Rhizopus stolonifer	IFO30816	–
Sporothrix schenckii	IFO8158	–
Trichophyton mentagrophytes	IFO5809	+
Trichophyton rubrum	IFO5808	+
Trichophyton tonsurans	IFO5945	+
Trichophyton violaceum	IFO31064	+

Note: Antifungal activity was measured by paper disc method. "–" = No antifungal activity; "+" = Positive antifungal activity; "±" = Unclear inhibition circle.

Acid Hydrolysis of Isolated Saponin and Separation of Aglycone and Constituted Sugars

A saponin (20 mg for analyzing aglycone and 2 mg for sugars) was previously dissolved in 0.5 mL of methanol and mixed with 2.5 mL of 3 M trifluoroacetic acid solution, then heated at 100°C for six hours. Hydrolyzate was evaporated. The dried sample was dispersed in 2 mL of water and then washed three times with an equal volume of diethyl ether. The ether-soluble fraction was identified through TLC and ^{13}C-NMR experiments.

The water-soluble fraction obtained from the acid hydrolysis of a saponin was transferred to aminopyridyl sugars following Kondo et al. (1990) with a slight modification and applied to HPLC. Sugar analyses were carried out with a Waters LC Module 1 (Millipore Co.) and a Waters M470 scanning fluorescence detector (e.g., 310 nm, emission, 380 nm, Millipore Co.). The column was a Palpak Type A (8 μm, 150 × 4.6 mm i.d.; Takara Co.), and the mobile phase was 0.4 M potassium borate buffer (pH 9.0)-acetonitrile (90:10).

Spectroscopy

NMR spectra were recorded on a JEOL GSX-400 (^{1}H at 400 MHz, ^{13}C at 100 MHz) spectrophotometer. Mass spectra were obtained with a JEOL JMS HX-105. IR spectra were recorded with system 2000 FT-IR spectroscopy (Perkin Elmer) by Kbr methods.

ANTIFUNGAL ACTIVITY OF SAPONIN FRACTION FROM WHITE ASPARAGUS BOTTOM CUT

Crude saponin fraction was prepared from white asparagus bottom cut, which had been discarded by a food factory. Antifungal activities of saponin fraction against several fungi were measured by paper disc method and the results were shown in Table 10.1. This fraction had clear antifungal activity against certain fungi, namely, *Candida albicans, Cryptococcus albidus, Epidermophyton floccosum, Microsporum gypseum,* and *Trichophyton* spp. Yet this fraction was ineffective against other fungi, e.g., *Rhizopus* and *Chaetomium.* Within the genera *Aspergillus* and *Penicillium,* some species were sensitive, e.g., *A. luchuensis, A. repens,* and *P. italicum,* but the others were insensitive. To examine this further, an attempt was made to separate and isolate the active saponins.

Isolation and Structural Elucidation of Antifungal Saponins

The crude saponin fraction prepared from the bottom cut of white asparagus was first applied to an ODS column (YMC GEL ODS-A; 230 × 18 mm i.d.), and three fractions (F1, F2, and F3) were separated. Each fraction was then concentrated under reduced pressure and analyzed with TLC (Figure 10.1). F1 (methanol-water = 60:40) consisted of nonsaponin components and F2 (methanol-water = 70:30) consisted of saponins that showed low R_f values on TLC. These constituents were considered furostanol saponins because they were positive to Ehrlich's reagent (Kawano et al., 1975). F3 (methanol fraction) consisted of saponins that showed high R_f values and were negative to Ehrlich's reagent. Regarding the antifungal activity of each

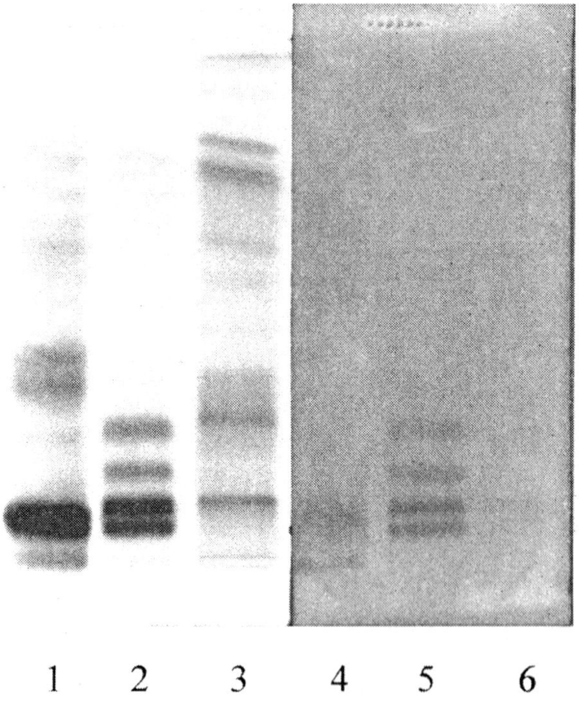

FIGURE 10.1. TLC patterns of eluents from ODS column chromatography of saponin fraction 1, 4, F1; 2, 5, F2; 3, 6, F3. 1, 2, 3, visualized by 10 percent sulfuric acid; 4, 5, 6, visualized by Ehrlich's reagent. Developing solvent used was chloroform-methanol-water (7:3:1; upper layer).

fraction, only F3 showed high antifungal activity. This active fraction was successively applied to the reversed-phase HPLC (YMC Pack ODS AM-323-7; 7 μm 250 × 10.0 mm i.d.). The mobile phase used was methanol-water (88:12) and eluents were monitored with RI detector. Four main fractions designated AS-1, AS-2, AS-3, and AS-4 were obtained and their anti-*Candida* activities were measured (Table 10.2).

AS-1 and AS-2 showed inhibitory effects against *Candida albicans,* but AS-3 and AS-4 did not. AS-1 and AS-2 fractions were collected, concentrated under reduced pressure, and lyophilized. AS-1 fraction obtained from HPLC contained only a single constituent, but AS-2 fraction was shown to consist of more than two components by NMR spectroscopy. Then this fraction was rechromatographed on HPLC with reconstituted elution systems (acetonitrile-water = 44:56). AS-2 fraction appeared to consist of three components, tentatively named AS-2-I, II, and III. In these constituents, AS-2-I showed strong anti-*Candida* activity, so AS-2-I was isolated. AS-1 was subjected to FAB-MS spectra and showed ion peaks at MHz 873 ($[M+H]^+$) and 895 ($[M+Na]^+$), indicating a molecular weight of 872. IR spectra showed bands at 3400(OH), 980, 915, 890, and 850 cm^{-1}. As the band at 915 cm^{-1} was stronger than at 890 cm^{-1}, the aglycone was considered to be 25S-spirostanol (Brain et al., 1968). Acid hydrolysis of AS-1 showed xylose and glucose (about 1:2; molar ratio). ^{13}C-NMR spectra of isolated saponins are shown in Tables 10.3 and 10.4. ^{13}C-NMR data of aglycone fraction from hydrolyzate showed the identical pattern to that of sarsasapogenin (Eggert and Djerassi, 1975), thus confirming the assignment

TABLE 10.2. Anti-*Candida* activities of separated asparagus saponin fractions

Saponin	MIC (μg/mL)
Crude saponin	32
AS-1	32
AS-2	8
AS-3	<128
AS-4	<128

Note: MIC (minimum inhibitory concentration) was measured as follows. The fungi used were preincubated for one or two weeks on an agar culture broth (yeast extract [3 g], malt extract [3 g], peptone [5 g], and glucose [10 g] with agar [10 g] in water [1 liter]). Fungal spores were suspended (5 × 10^4 cells/mL) in tubes containing broth (5 mL) without agar, the saponins being added at appropriate levels. The tubes were incubated for one week at 30°C and MIC was determined.

TABLE 10.3. ^{13}C-NMR data of *Asparagus* saponin aglycone moieties

	AS-1[a]	AS-1 Ag.[b]	Sars.[c]	AS-2-I[a]	AS-2-I Ag.[b]	Yamog.[c]	Dios.[c]	AS-P2-I[a]
1	30.8	29.9	29.9	37.6	37.3	37.3	37.2	37.5
2	26.8	27.8	27.8	30.3	31.5*	31.4	31.6	30.1
3	80.6	67.1	67.0	78.2	71.8	71.6	71.5	78.3
4	30.7	33.5	33.6	39.0	42.3	42.3	42.2	38.9
5	36.6	36.5	36.5	140.9	140.9	140.9	140.8	140.8
6	27.5	26.5	26.6	122.0	121.5	120.3	121.3	121.9
7	26.8	26.5	26.6	32.4	32.1	32.0	32.0	32.5
8	35.5	35.3	35.5	31.8	31.4*	31.4	31.4	31.7
9	40.3	40.3	40.3	50.4	50.9	50.1	50.1	50.3
10	35.2	35.3	35.3	37.2	36.7	36.6	36.6	37.1
11	21.1	20.9	20.9	21.2	20.9	20.9	20.9	21.1
12	40.2	39.8	39.9	39.9	39.8	39.8	39.8	39.9
13	40.9	40.7	40.6	40.5	40.3	40.3	40.3	40.6
14	56.4	56.5	56.4	56.7	56.6	56.5	56.5	56.6
15	32.1	31.7	31.7	32.3	31.9	31.8	31.8	32.3
16	81.3	81.0	80.9	81.3	80.9 (80.7)	80.8	80.7	81.1
17	61.7	62.3	62.1	62.7	62.2 (62.0)	62.1	62.1	62.6
18	16.6	16.5	16.5	16.4	16.3	16.3	16.3	16.4
19	24.0	23.9	23.9	19.5	19.5	19.4	19.4	19.4
20	42.5	42.1	42.1	42.6	42.2 (41.3)	42.2	41.6	40.8
21	17.9	14.3	14.3	15.0	14.6 (14.4)	14.3	14.5	16.4
22	109.7	109.7	109.5	110.0	109.4	109.7	109.1	110.7
23	27.0	27.1	27.1	27.6	26.0 (31.7)	26.0	31.4	37.0
24	26.4	25.9	25.8*	26.3	25.8 (28.9)	25.8	28.8	28.3
25	26.2	25.8	26.0*	26.5	27.1 (30.4)	27.1	30.3	34.4
26	65.1	65.1	65.0	65.2	65.2 (66.9)	65.2	66.7	75.4
27	16.3	16.0	16.1	16.5	16.1 (17.2)	16.1	17.1	17.4

Note: AS-1 Ag, AS-1 aglycone; Sars., sarsasapogenin; AS-2-I Ag., AS-2-I aglycone; Yamog., yamogenin; Dios., disogenin. In parentheses, additional signals.
*May be interchangeable.
[a]Solvent: d_5-pyridine
[b]Solvent: $CDCl_3$
[c]Solvent: $CDCl_3$ (Agrawal, 1992)

TABLE 10.4. ^{13}C-NMR data of sugar moieties of *Asparagus* saponins

	AS-1		**AS-2-I**		**AS-P2-I**
Glc-1'	101.7	Glc-1'	100.3	Glc-1'	100.2
2'	81.7	2'	78.1	2'	78.2
3'	75.3	3'	77.8	3'	77.9
4'	76.2	4'	78.7	4'	78.6
5'	76.3	5'	76.9	5'	76.8
6'	61.7	6'	61.3	6'	61.2
Glc-1''	105.4	Rha-1''	102.9	Rha-1''	102.8
2''	77.1	2''	72.4	2''	72.3
3''	77.9	3''	72.7	3''	72.6
4''	71.9	4''	74.0	4''	73.9
5''	78.6	5''	70.5	5''	70.4
6''	63.0	6''	18.7	6''	18.6
Xyl-1'''	105.5	Rha-1'''	102.2	Rha-1'''	102.0
2'''	75.0	2'''	72.4	2'''	72.3
3'''	78.4	3'''	72.6	3'''	72.5
4'''	70.8	4'''	73.8	4'''	73.7
5'''	67.4	5'''	69.7	5'''	69.5
		6'''	18.6	6'''	18.5
				Glc-1''''	105.0
				2''''	75.0
				3''''	78.2
				4''''	71.5
				5''''	77.7
				6''''	62.7

Note: Solvent: d_5-pyridine.

of sarsasapogenin. Proton assignments (Table 10.5) of sugar moieties of AS-1 were made by ^1H-NMR and by ^1H-^1H COSY and ^{13}C-^1H COSY techniques referring to the previous data (Agrawal, 1992). These data also showed that one glucose was the terminal of sugar chain and the other glu-

TABLE 10.5. ^{1}H-NMR spectral data of sugar moieties of AS-1 and AS-2-I

	AS-1[a]		AS-2-I[b]	
Glc-1'	4.47	(d, J=8)	Glc-1'	(d, J=8)
2'	3.62	(dd, J=8,9)	2'	(dd, J=8,9)
3'	3.71	(dd, J=9,9)	3'	(dd, J=9,9)
4'	3.52	(dd, J=9,9)	4'	(dd, J=9,9)
5'	3.38	(m)	5'	(m)
6'	3.81	(dd, J=4,12)	6'	(dd, J=3,9)
6'	3.87	(dd, J=2,12)	6'	(dd, J=4,9)
Glc-1"	4.70	(d, J=8)	Rha-1"	(d, J=2)
2"	3.17	(dd, J=8,9)	2"	(dd, J=2,3)
3"	3.36	(dd, J=9,9)	3"	(dd, J=3,9)
4"	3.21	(dd, J=9,9)	4"	(dd, J=9,9)
5"	3.25	(m)	5"	(m)
6"	3.66	(dd, J=5,12)	6"	(d, J=6)
6"	3.84	(dd, J=2,12)		
Xyl-1'''	4.34	(d, J=8)	Rha-1'''	(d, J=2)
2'''	3.19	(dd, J=8,9)	2'''	(dd, J=2,4)
3'''	3.31	(dd, J=9,9)	3'''	(dd, J=4,9)
4'''	3.49	(m)	4'''	(dd, J=9,9)
5'''	3.25	(dd, J=10,12)	5'''	(m)
5'''	3.90	(dd, J=5,10)	6'''	(d, J=6)

[a]Solvent: d_4-methanol
[b]Solvent: d_5-pyridine

cose linked two other sugars at C-2 and C-4 positions (Table 10.3). Three anomeric protons at δH 4.47 (d, J = 8 Hz), δH 4.70 (d, J = 8 Hz), δH 4.34 (d, J = 8 Hz) were identified, consistent with two β-linked glucopyranose moieties and one β-linked xylopyranose moiety, respectively. The linkages of the carbohydrate moieties were estimated by Nuclear Overhauser Effect (NOE) experiments, which were carried out at – δH 4.10 (intensity, 7.2 percent; H-3 of aglycone). From this data, AS-1 was estimated to be 3-*O*-[β-D-glucopyranosyl(1→2)} {β-D-xylopyranosyl(1→4)}- β-D-glucopyranosyl]-(25S), 5β-spirostan-3β-ol (Figure 10.2).

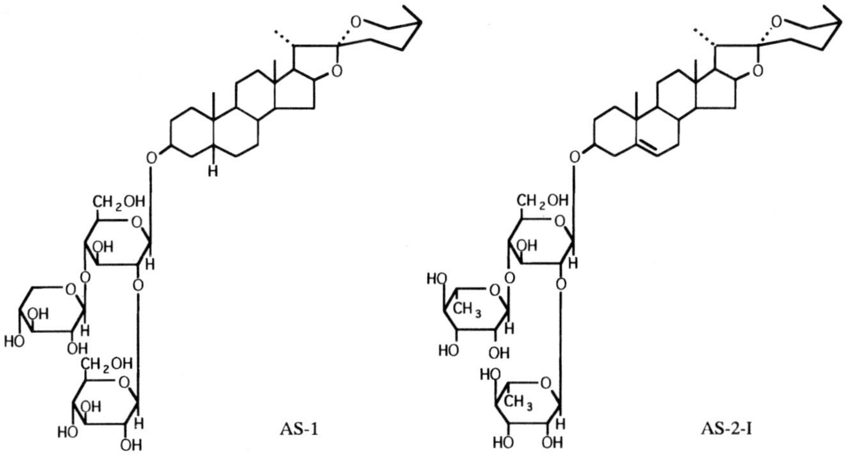

AS-1 AS-2-I

FIGURE 10.2. Structures of antifungal saponins isolated from white asparagus.

Then the structure of AS-2-I was estimated. The FAB-MS spectra of AS-2-I showed ion peaks at MHz 869 ($[M+H]^+$) and 891 ($[M+Na]^+$), indicating a molecular weight of 868. IR spectra showed bands at 3412(OH), 988, 920, 897, and 838 cm^{-1}. Similar to AS-1, the band at 920 cm^{-1} was stronger than at 897 cm^{-1}, and it showed the presence of 25S-spirostanol (Brain et al., 1968). Acid hydrolysis of AS-1 gave glucose and rhamnose (about 1:2; molar ratio). ^{13}C-NMR spectra (Table 10.2) of aglycone fraction from hydrolyzate showed the mixture of yamogenin and diosgenin (Eggert and Djerassi, 1975). From IR spectra of AS-I, the genuine aglycone was considered to be a yamogenin, and a diosgenin was considered to be an artifact formed during hydrolysis of the saponin. Proton assignments (Table 10.5) of sugar moieties of AS-2-I were made by ^{1}H-NMR and by ^{1}H-^{1}H COSY and ^{13}C-^{1}H COSY techniques. Three anomeric protons at δH 4.94 (d, J = 7 Hz), δH 5.84, and δH 6.38 (both singlet-like) were identified, consistent with one β-linked glucopyranose moiety and two α-linked rhamnoopyranose moieties, respectively. Similar to AS-1, AS-2-I has one 2,4-linked glucopyranose determined by processes such as ^{13}C-NMR (Table 10.4) and other data. AS-2-I was elucidated to be 3-*O*-[{α-L-rhamnopyranosyl(1→2)}{α-L-rhamnopyranosyl(1→4)}-(-D-glucopyranosyl]-(25S) spirost-5-en-3β-ol (Figure 10.2). This structure coincides with that of colletinside III from *Dioscorea collettii* (Liu et al., 1983). This constituent was first reported as yamoscin from the partial hydrolyzate of saponin fraction from *Trigonella foenum-graecum* (Bogacheva et al., 1976) and as 25S-isomer of dioscin from *Liriope platyphylla* (Watanabe et al., 1983).

ANTIFUNGAL ACTIVITIES OF ASPARAGUS *SAPONINS AND SAPONIN FRACTION OBTAINED FROM FRESH WHITE ASPARAGUS PLANT*

Antifungal activities of AS-1 and AS-2-I are summarized in Table 10.6. The MICs of AS-1 and AS-2-I were estimated against certain fungi. Both saponins inhibited the growth of some fungi, and especially strong inhibition was observed in the case of *Epidermophyton floccosum*. MICs of AS-2-I against *Candida albicans, Microsporum gypseum, Trichophyton rubrum,* and *Trichophyton mentagrophytes* were lower than those of AS-1. AS-2-I showed very low MICs against *Microsporum gypseum, Trichophyton rubrum,* and *Epidermophyton floccosum,* and slightly higher MICs against *Candida albicans* and *Trichophyton mentagrophytes*. On the other hand, neither AS-1 nor AS-2-I suppressed *Aspergillus oryzae*. Further, some of our data suggests that the crude saponin fraction from *Asparagus* did not inhibit the growth of *Saccharomyces cerevisiae* (unpublished data). These saponins may selectively suppress certain fungi.

In order to analyze native saponins and their antifungal activity, saponin fraction was prepared from fresh white asparagus. Harvested white asparagus was immediately dried and milled. The milled sample was extracted by a method similar to those discussed previously. Saponin fraction obtained was subjected to TLC analyses and measurement of MICs. As the result, the TLC pattern showed mainly furostanol saponins having low R_f values, which were visualized by Ehrlich's reagent (Kawano et al., 1975). This fraction also contained a small amount of other constituents, possibly spirostanol saponins. Antifungal activities of the saponin fraction were much lower than those from factory waste (Table 10.7), therefore these native saponins were considered to be furostanol saponins and antifungal inactive.

TABLE 10.6. MICs (μg/ml) of isolated saponins

Fungi	Crude	AS-1	AS-2-I
Candida albicans	5-10	30	10
Microsporum gypseum	5-10	2	0.5
Trichophyton rubrum	5-10	3	0.5
Trichophyton mentagrophytes	5-10	9	5
Epidermophyton floccosum	5-10	1	1
Aspergillus oryzae	>200	–	–

TABLE 10.7. MICs (μg/ml) of saponin fraction from fresh asparagus and effect of incubation

Fungi		Fresh	After incubation*
Candida albicans	IFO1061	>1,000	5~10
Cryptococcus albidus	IFO1420	>1,000	5~10
Trichophyton rubrum	IFO5808	100~150	5~10
Trichophyton tonsurans	IFO5945	50~100	10~20
Trichophyton violaceum	IFO31064	50~100	10~20
Epidermophyton floccosum	IFO9045	100~200	5~10
Penicillium italicum	IFO9419	>200	10~20
Aspergillus luchuensis	AHU7092	>200	>200
Aspergillus repens	AHU7458	>200	100~200

*Saponin fraction was extracted from the fresh asparagus mill which had been incubated at 25°C overnight with 0.1 percent NaN_3 solution.

This fraction was subjected to ODS column, silica gel column, and ODS HPLC. As a result, AS-P2-I, which was tentatively named, was obtained. AS-P2-I was identified as 3-*O*-[{α-rhamnopyranosyl(1→2)} {α-L-rhamnopyranosyl(1→4)}- β-D-glucopyranosyl] 26-*O*-{β-D-glucopyranosyl} (25S) furost-5-en-3β,5β, 26-triol (26-glucosyl AS-2-I) by NMR (Tables 10.3 and 10.4), IR spectra and sugar composition (Figure 10.6, see left). AS-P2 fraction, which mainly contains AS-P2-I, showed negligible antifungal activity against only *Penicillium italicum* (Table 10.8). Because the difference between AS-2-I and AS-P2-I was one glucose moiety linked to a side chain, the enzymatic removal of the glucose moiety from saponin was attempted. This saponin was incubated with commercial β-glucosidase from almonds. The resulting hydrolyzate showed Ehrlich's reagent negative constituent on TLC (Figure 10.3) and this constituent was identified as AS-2-I by HPLC (Figure 10.4). Then, the fresh asparagus mill (1 g) was mixed with 0.6 g of water containing 0.1 percent NaN3, which was expected to suppress microbes, and incubated overnight at 25°C. The saponin fraction was prepared from the resulting mixture, analyzed by TLC, and subjected to MIC measurement. From TLC patterns (Figure 10.5), antifungal saponins were generated through an incubation process; the resulting mixture showed higher antifungal activities (Table 10.7).

TABLE 10.8. Antifungal activities of AS-P2 fraction

Fungi		Activity
Candida albicans	IFO1061	
Trichophyton rubrum	IFO5808	
Epidermophyton floccosum	IFO9045	
Penicillium italicum	IFO9419	±

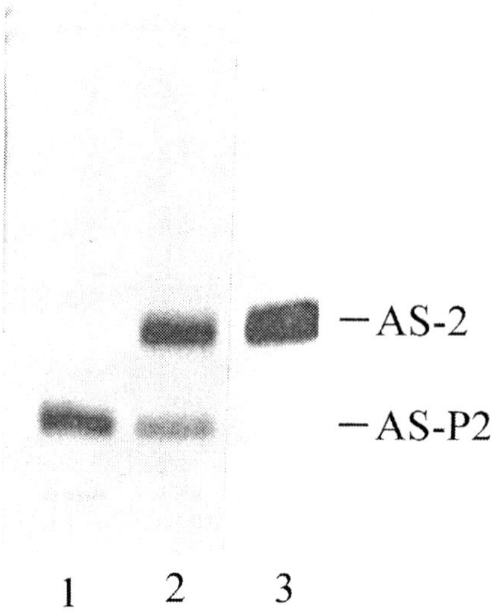

FIGURE 10.3. TLC patterns of AS-P2 after treatment with β-glucosidase. 1, AS-P2; 2, Hydrolyzate of AS-P2 by β-glucosidase; 3, AS-2.

RELATIONSHIP BETWEEN STRUCTURES AND ANTIFUNGAL ACTIVITIES OF ASPARAGUS *SAPONINS*

In this study, we isolated two antifungal saponins from white asparagus bottom cut. These saponins were spirostanol saponins and had 25S configurations. Takechi et al. (1991, 1992) and Takechi and Tanaka (1993) estimated the concentration causing 50 percent growth inhibition (GID_{50}) of natural or synthesized steroidal saponins. Dioscin, 25R-isomer of AS-2-I, showed the GID_{50} of 0.9 M against *Trichophyton mentagrophytes* and may be slightly more active than AS-2-I (approximately 6 M), although there are

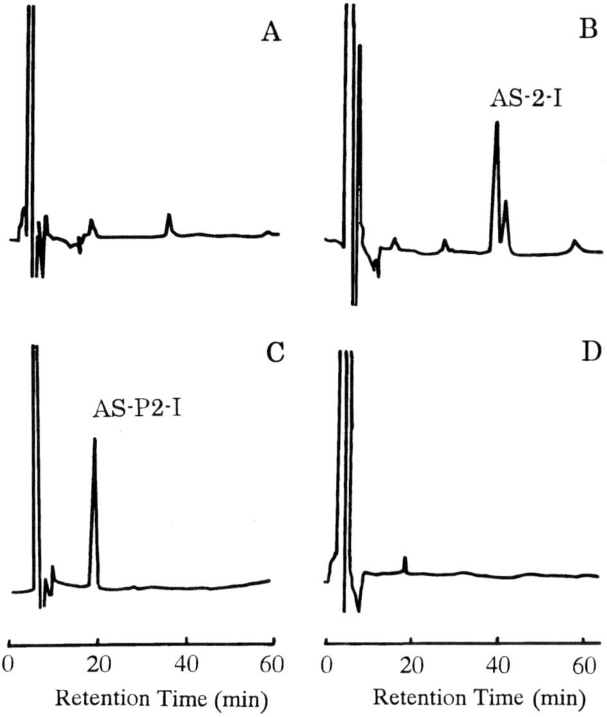

FIGURE 10.4. Liquid chromatograms of saponin fraction. A, B saponin fraction extracted from fresh asparagus mill after incubation; C, D saponin fraction from fresh asparagus mill with no treatment. Mobile phases; A, C, acetonotrile:water = 30:70; B, D, acetonitrile:water = 44:56.

some differences in the determination and expression of antifungal activities. The configuration of 25-methyl group may influence the activity of saponins to some extent. Antifungal *Asparagus* saponins and dioscin had three sugar residues. Takechi et al. (1991) also reported that the antifungal activity of triglycosyl saponin (dioscin) was higher than diglycosyl derivatives and that saponins having a branched sugar chain inhibited fungi growth more strongly. In fact, AS-3 and AS-4, which were expected to be diglycosyl saponins by preliminary NMR data (data not shown), showed very low antifungal activities (Table 10.2).

Comparing two *Asparagus* saponins, antifungal activity of AS-2-I was higher than AS-1 against some fungi evaluated, except *Epidermophyton floccosum*. Structural differences between AS-1 and AS-2-I were C5-C6 double bond and terminal rhamnose moieties. As for C5-C6 double bond,

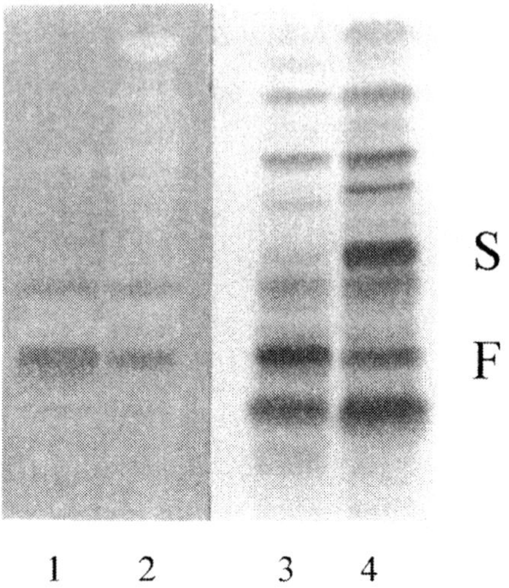

FIGURE 10.5. TLC patterns of crude saponin fraction from fresh asparagus mill. 1, 3, Saponin fraction from fresh asparagus mill with no treatment; 2, 4, saponin fraction from fresh asparagus after incubation whose conditions were referred to in the text. 1, 2, Visualized by Ehrlich's reagent; 2, 4, visualized by sulfuric acid. S, spirostanol saponins; F, furostanol saponins.

diosgenyl glycosides (with double bond) and tigogenyl glycosides (without double bond) have been reported to be almost equal antifungal activities (Takechi et al., 1992; Takechi and Tanaka, 1993). So the difference between AS-1 and AS-2-I was considered to be responsible for the difference in sugar composition and linkage, rather than the double bond in aglycones. The furostanol saponin isolated here showed hardly any antifungal activity. A glucose moiety linked to C-26 may inactivate saponins.

ANTIFUNGAL ACTIVITIES OF WHITE ASPARAGUS AND PHYSIOLOGICAL ROLE OF SAPONIN IN THE PLANT DEFENSE SYSTEM

A crude saponin fraction from white asparagus bottom cut strongly inhibited the growth of some kinds of fungi (Table 10.1). Interestingly, *Candida albicans,* which is responsible for contamination in food processing and candidiasis, and some species of *Trichophyton, Microsporum,* and *Epidermophyton,* which are responsible for so called ringworm, were suppressed effectively. On the other hand, the saponin fraction did not inhibit the growth of *Chaetomium globosum, Fusarium oxysporum, Rhizopus stolonifer, Sporothrix schenckii, Penicillium funiculosum,* and most species of the genus *Aspergillus*. Further, we have some data that the crude saponin fraction from *Asparagus* did not inhibit the growth of *Saccharomyces cerevisiae* (unpublished data). These saponins may selectively suppress only some fungi. The saponin inhibits some of the pathogenic fungi but not some useful fungi utilized in fermentation. This is important for further utilization of saponins in food processing.

This antifungal saponin fraction was prepared from white asparagus bottom cut, which had been discarded by a food factory. In a similar way, a saponin fraction was prepared from fresh white asparagus bottom cut and its antifungal activity was compared with those from factory waste. From TLC patterns (Figure 10.5), the saponin fraction from fresh asparagus contained mainly furostanol saponins that showed low R_f values and were visualized by Ehrlich's reagent, differing from those originating from factory waste, which also contained spirostanol saponins (Figure 10.1). Antifungal activities of the saponin fraction from fresh asparagus were very low against most fungi (Table 10.7). Isolated furostanol saponin mixture (AS-P2) also showed little antifungal activity against some fungi (Table 10.8), so furostanol saponins were thought to have no substantial antifungal activity.

In order to clarify the meanings of the antifungal activity of saponin fraction from factory waste asparagus, AS-P2 fraction, which contained AS-P2-I and also had little antifungal activity, was incubated with commercial β-glucosidase and formed an antifungal saponin which was identified as AS-2 (Figures 10.3 and 10.4). By enzymatic hydrolysis, inactive antifungal saponins were converted to active saponins. It was shown that *Asparagus* saponins were activated by glucosidase. Previously, oat leaf saponin avenacosides A and B were activated by endogenous β-glucosidase after invasion of pathogenic fungi (Nisius, 1988). Per Nisius (1988), avenacosides are reported to be stored in the vacuoles of plant cells in inactive forms, and saponin-activating β-glucosidase is in the plastid. Fungi that infect the plant cell damage membranes of cells and organelles to allow saponins to mix with the enzyme, resulting to activate avenacosides. The activation of the saponin by endogenous enzymes was then estimated in the asparagus plant. Asparagus plant mill was able to convert furostanol saponins to spirostanol ones even the presence of NaN_3, which suppressed microbes, so there were considered to be β-glucosidase(s) to activate saponins in asparagus plant tissues.

From this data, the role of an *Asparagus* saponin was supposed. Asparagus synthesizes and stores furostanol saponins, which is inactive against fungi. By invasion of fungi, membranes of organelles are broken and endogenous glucosidases, which are present in some organelles, are mixed with furostanol saponins of other organelles. Due to hydrolysis, strong antifungal saponins are produced (Figure 10.6). The *Asparagus* saponin was thought to be one of several important compounds responsible for plant defense systems.

CONCLUSION AND FUTURE DIRECTIONS

Saponins are widely distributed in the plant kingdom, and many biological activities were reported. However, there are limited reports for their physiological role in plants. From white asparagus, many kinds of saponins have been isolated which show a lot of biological activities, for example, molluscicidal, cytotoxic, antitumor activities. The saponins from factory waste of white asparagus have strong antifungal activities to some specific species of fungi, namely, *Trichophyton* spp., *Microsporum gypseum,* and *Epidermophyton floccosum*. The saponins effectively suppress the fungal growth at 0.5 to 10 μg/ml level. Some of them also inhibit the growth of *Candida albicans,* which would cause candidiasis, even if the MICs are slightly higher than those to other fungi, e.g., *Trichophyton, Microsporum,* etc. On the other hand, they hardly inhibit the growth of *Aspergillus* and some species of *Penicillium,* even if these fungi belong to same order, i.e., Moniliales to other sensitive fungi. The reasons for these phenomena are not clear. However, the saponin fraction from fresh white asparagus plant shows only negligible activity to the sensitive fungi. The data obtained from in vitro experiments show that *Asparagus* saponins are stored in the inactive

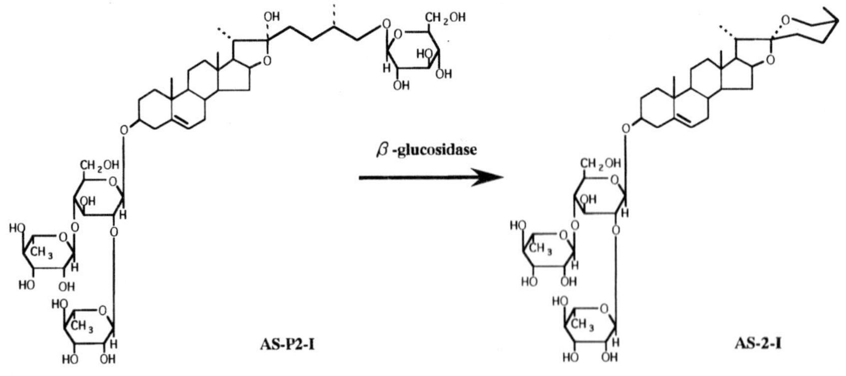

FIGURE 10.6. Enzymatic conversion and activation of *Asparagus* saponin.

furostanol saponins in the plant cell and that they are activated to the antifungal spirostanol by the enzmymatic removal of one sugar residue. This activation of saponins is very similar to the oat saponin activation process reported by Nisius (1988). These data allow us to suppose that the saponins have important relations to the defense system of the plant. Similar approaches may clarify the properties and significances of some constituents in other plants.

Biological activities of the saponins have some possibilities for application to food industry, even if their activities are generally weaker than chemically synthesized compounds. These natural substances such as saponins are considered to be essentially biodegradable and have only small effects on our ecological system. The antifungal activity of the *Asparagus* saponins is very specific to certain fungi. The saponins inhibit *Trichophyton* spp., *Microsporum gypseum,* and *Epidermophyton floccosum* strongly, and *Candida albicans* to a lesser extent, which are considered to be pathogenic fungi, but the saponins do not suppress other fungi, e.g., *Aspergillus, Rhizopus,* or *Saccharomyces cerevisiae,* etc., which are widely utilized in food processing. So this research is also interested in some applications, for example, selective incubations of some fungi where other fungi's growth must be suppressed.

Further investigation will clarify the mechanisms of antifungal activities of saponins and their very narrow antifungal spectra. These data would aid the development of more effective synthetic antifungicidal chemicals, being more specific against particular fungi and safer for humans. Finally, Nisius (1988) also referred to some *Dreschlera* species, which are capable of hydrolyzing the antifungal saponins and protecting themselves. These kinds of interactions between different organisms, namely pathogens and hosts, are very confusing and intriguing in terms of ecological aspects.

REFERENCES

Agrawal, P.K. (1992). NMR spectroscopy in the structural elucidation of oligosaccharides and glycosides. *Phytochemistry* 31: 3307-3330.

Ahmad, V.U., Khaliq-uz-Zaman, S.M., Shameel, S., Perveen, S., and Ali, Z. (1998). Steroidal saponins from *Asparagus dumosus*. *Phytochemistry* 50: 481-484.

Bogacheva, N.G., Kiselev, V.P., and Kogan, L.M. (1976). Isolation of 3,26-bisglycoside of yamogenin from *Trigonella foenum-graecum*. *Kim. Prir. Soedin.* 2: 268-269; *Chem. Abstr.,* 85: 106634e, 1976.

Brain, K.R., Fazli, F.R.Y., Hardman, R., and Wood, A.B. (1968). The rapid quantitative determination of C25 epimeric steroidal sapogenins in plants. *Phytochemistry* 7: 1815-1823.

Debella, A., Haslinger, E., Kunert, O., Michl, G., and Abebe, D. (1999). Steroidal saponins from *Asparagus africanus*. *Phytochemistry* 51: 1069-1075.

Eggert, H., and Djerassi, C. (1975). C-13 NMR spectra of sapogenins. *Tetrahedron Lett.* 42: 3635-3638.

Gaitonde, B.B., and Jetmalani, M.H. (1969). Antioxytocic action of saponin isolated from *Asparagus racemosus* on uterine muscle. *Arch. Int. Pharmacodyn. Ther.* 179: 121-129.

Goryanu, G.M., and Kintya, P.K. (1976). Steroidal saponins. XVII. Structures of asparagosides D and G. *Khim. Prir. Soedin.* 762-765; *Chem. Abstr.* 87:53516a, 1977.

Goryanu, G.M., and Kintya, P.K. (1977). Steroidal glycosides from *Asparagus officinalis* asparagosides F and H. *Khim. Prir. Soedin.* 810-813; *Chem. Abstr.* 88:136901u, 1978.

Goryanu, G.M., Krokhamalyuk, V.V., and Kintya, P.K. (1976a). Steroid glycosides. XVI. Structure of asparagosides C and E from *Asparagus officinalis*. *Khim. Prir. Soedin.* 823-824; *Chem. Abstr.* 86:103064d, 1977.

Goryanu, G.M., Krokhmalyuk, V.V., and Kintya, P.K. (1976b). Structure of glycosides of *Asparagus officinalis*. Structure of asparagosides A and B. *Khim. Prir. Soedin.* 400-401; *Chem. Abstr.* 85:90178w, 1976.

Goryanu, G.M., Krokhmalyuk, V.V., Kintya, P.K., and Gltzin, V.I. (1976). Medicinal *Asparagus* (*Asparagus officinalis* L.) as a source of steroidal glycosides. *Farmatsiya* 25: 66; *Chem. Abstr.* 85:156482z, 1976.

Halim, A.F., Mansour, E.S., and Waight, E.S. (1989). Steroidal saponin from *Asparagus stipularis* L. root. Mansoura *J. Pharm. Sci.* 5: 44-57.

Kawano, K., Sakai, K., Sato, H., and Sakamura, S. (1975). A bitter principle of *Asparagus:* Isolation and structure of furostanol saponin in *Asparagus* storage root. *Agric. Biol. Chem.* 39: 1999-2002.

Kawano, K., Sato, H., and Sakamura, S. (1977). Isolation and structure of furostanol saponin in *Asparagus* edible shoots. *Agric. Biol. Chem.* 41: 1-8.

Kondo, A., Suzuki, J., Kuraya, N., Hase, S., Kato, I., and Ikenaka, T. (1990). Improved method for fluorescence labeling of sugar chains with sialic acid residues. *Agric. Biol. Chem.* 54: 2169-2170.

Konishi, T., and Shoji, J. (1979). Studies on the constituents of Asparagi radix. I. On the structures of furostanol oligosides of *Asparagus* cochinchinensis (LOUREIO) Merrill. *Chem. Pharm. Bull.* (Tokyo) 27: 3086-3094.

Lacaille-dubois, M.A., and Wagner, H. (1996). A review of the biological and pharmacological activities of saponins. *Phytomedicine* 2: 363-386.

Lake, R.J., Falloon, P.G., and Cook, D.W.M. (1993). Replant problem and chemical components of *Asparagus* roots. *N. Z. J. Crop Hortic. Sci.* 21: 53-58.

Liu, C.-L., Chen, Y.-Y., Ge, S.-B., and Li, B.G. (1983). Studies on the constituents from *Dioscorea* plants. II. Isolation and identification of steroidal saponins from *Dioscorea collettii* Hook. F. Yaoxue Xuebao. 18: 597-606; *Chem. Abstr.*, 100: 9875g, 1984.

Mahato, S.B., and Garai, S. (1998). Triterpenoid saponins. *Prog. Chem. Org. Nat. Prod.* 74: 1-196.

Mandloi, D., and Sant, P.G. (1981). A saponin from *Asparagus* gonocladus. *Phytochemistry* 20: 1687-1688.

Mimaki, Y., and Sashida, Y. (1996). Steroidal saponins from the Liliaceae plants and their biological activities. *Adv. Exp. Med. Biol.* 404: 101-110.

Nisius, A. (1988). The stromacentre in *Avena* plastids: An aggregation of β-glucosidase responsible for the activation of oat-leaf saponins. *Planta* 177: 474-481.

Pant, G., Panwar, M.S., Negi, D.S., and Rawat, M.S.M. (1988). Spermicidal activity of steroidal and triterpenic glycosides. *Current Science* 57: 661-663.

Price, K.R., Johnson, I.T., and Fenwick, G.R. (1987). The chemistry and biological significance of saponins in food and feeding stuffs. *CRC Crit. Rev. Food Sci. Nutr.* 26: 27-135.

Sati, O.P. (1985). Spirostanol glycosides from *Asparagus plumosus*. *Phytochemistry* 24: 123-126.

Sati, O.P., Pant, G., and Hostettmann, K. (1984). Potent molluscicides from *Asparagus*. *Pharmazie* 39: 581.

Sati, O.P., Pant, G., Nohara, T., and Sato, A. (1985). Cytotoxic saponins from *Asparagus* and *Agave*. *Pharmazie* 40: 586.

Shao, Y., Chin, C.-K., Ho, C.-T., Ma, W., Garrison, S.A., and Huang, M.-T. (1996). Antitumor activity of the crude saponins obtained from *Asparagus*. *Cancer Lett.* 104: 31-36.

Shao, Y., Poobrasert, O., Kennelly, E.J., Chin, C.K., Ho, C.T., Huang, M.T., Garrison, S.A., and Geoffrey, A. (1997). Steroidal saponins from *Asparagus officinalis* and their cytotoxic activity. *Planta Med.* 63: 258-262.

Sharma, S.C., Chand, R., Bhatti, B.S., and Sati, O.P. (1982). New oligospirostanosides and oligofurostanosides from *Asparagus adscendens* roots. *Planta Med.* 46: 48-51.

Sharma, S.C., Sati, O.P., and Chand, R. (1981). Constituents of the fruits of *Asparagus racemosus* Willd. *Pharmazie* 36: 709.

Sharma, S.C., and Sharma, H.C. (1993). Oligofurostnosides from *Asparagus curillus* leaves. *Phytochemistry* 33: 683-686.

Sharma, S.C., and Thakur, N.K. (1996). Oligofurostanosides and oligospirostanosides from roots of *Asparagus filicinus*. *Phytochemistry* 41: 599-603.

Shimoyamada, M., Suzuki, M., Maruyama, M., and Watanabe, K. (1996). An antifungal saponin from white *Asparagus* (*Asparagus officinalis* L) bottoms. *J. Sci. Food Agric.* 72: 430-434.

Shimoyamada, M., Suzuki, M., Sonta, H., Maruyama, M., and Okubo, K. (1990). Antifungal activity of the saponin fraction obtained from *Asparagus officinalis* L. and its active principle. *Agric. Biol. Chem.* 54: 2553-2557.

Takechi, M., Shimada, S., and Tanaka, Y. (1991). Structure-activity relationships of the saponins dioscin and dioscinin. *Phytochemistry* 30: 3943-3944.

Takechi, M., Shimada, S., and Tanaka, Y. (1992). Structure-activity relationships of synthetic diosgenyl diglycosides. *Phytochemistry* 31: 3280-3281.

Takechi, M., and Tanaka, Y. (1993). Structure-activity relationships of synthetic tigogenyl glycosides. *Phytochemistry* 34: 1241-1243.

Watanabe, Y., Sanada, S., Ida, Y., and Shoji, J. (1983). Comparative studies on the constituents of *ophiopogonis tuber* and its congeners. I. Studies of the constituents of the subterranean part of *Liriope platyphylla* Wang & Tang (1). *Chem. Pharm. Bull.* (Tokyo) 31: 1980-1990.

Chapter 11

Activity of Plant Extracts, Essential Oils, and Pure Compounds Against Fungi Contaminating Foodstuffs and Causing Infections in Human Beings and Animals: A Six-Year Experience (1995-2000)

Guido Flamini
Pier Luigi Cioni

INTRODUCTION

Fungi can attack cultivated plants either during growth or when they are harvested and stored. These infections, besides causing economic damage due to heavy loss of production, are a very serious problem for food health. Many contain spores and mycelia, but their amounts, generally, are too low to cause acute poisoning. Sometimes, they can produce slow damage, and can be difficult to diagnose.

To control these infections, many synthetic products have been used, often in too large amounts. This has caused the induction of resistance phenomena in mycotic pathogens and the accumulation of residues in plants, soil, water, and herbivorous animals. The same is also true for zoophilic pathogens. The application of antifungal substances of natural origin is less studied, even if it represents a very interesting alternative.

In collaboration with the veterinary faculty and the Microbiological Institute of Livorno ASL, many plant extracts, essential oils, and pure compounds against some strains of human, animal, and plant fungal pathogens have been studied. The studies are based mainly from indications of folk medicine, a heritage that, unfortunately, is disappearing.

ACTIVITY ON FOODSTUFFS

One of the major problems of the food industry is to prevent the development of microorganisms in foodstuffs, which reduce the storage period or irremediably damage food. Many fungi isolated from plants belong to *Aspergillus, Fusarium,* and *Penicillium* genera, which produce very toxic metabolites for humans and animals. In particular, these fungi can colonize in many conserved products (Marth, 1972), comprising a number of foodstuffs; cereal and oleiferous seeds, legumes and ensiled hay are often infected (FAO-WHO-UNEP, 1977; Bottalico 1981). *Aspergillus* and *Penicillium* are also the main agents of moldiness of many products of animal origin (Tiecco, 1987). The mycotoxins cause liver and kidney damage (Lillehoj et al., 1970; Frank, 1984) and other very serious diseases especially to people who use contaminated foodstuffs as main nutritional sources. In countries where maize is the major food, for instance, a high frequency of esophageal cancer related to the presence of *Fusarium moniliforme* on this cereal has been found (Marasas et al., 1979). The toxicity of *A. niger* and *F. moniliforme* metabolites has been experimentally confirmed using laboratory animals; the treated animals showed gastrointestinal haemorrhages and liver and kidney lesions (Lillehoj et al., 1970; Frank, 1984). Because mycotoxins are secondary metabolites, they are synthesized by fungi only after the logarithmic phase of growth; for example, the aflatoxins of some *Aspergillus* and *Penicillium* species are synthesized only after 48 hours of spore germination.

The activity of some extracts, essential oils, and pure compounds against four mycetes that corrupt foodstuffs, *Fusarium moniliforme* var. *subglutinosus, Aspergillus niger, Pyricularia oryzae, Botritis cinerea* were studied. The examinations were performed both against hyphal and conidial-mycetes. The experiments on the mycelial growth were performed with the agar hole (6 mm) test on Sabouraud Dextrose agar. The petri dishes, containing 20 ml of the culture medium, were inoculated by inclusion of 0.1 ml of 48-hours-old fungal culture. The growth inhibition halo diameter around the wells, expressed in mm, was used as an index of the activity of each tested substance. For the evaluation of the inhibition of the conidial germination, these were taken from the aerial mycelium of aged cultures of each fungus in a standardized amount, than they were suspended in Sabouraud Broth containing different concentrations of the substances under examination. The inhibitory activity of the substances was evaluated on the basis of their ability to prevent, fully or in part, the germination of the conidia suspended in the cultural medium.

In the first study (Panizzi et al., 1995), the antimycotic properties of the essential oils of laurel (*Laurus nobilis* L.) and cedar (*Cedrus atlantica*

Man.) were evaluated in vitro against *Fusarium moniliforme* var. *subglutinosus* and *Pyricularia oryzae* (isolated from seeds), *Aspergillus niger* and *Botrytis cinerea* (both obtained from fresh fruit). The essential oils composition, obtained by GC-MS, is reported in Table 11.1.

The laurel essential oil showed antimycotic activity on all the studied molds, with wide and quite constant growth inhibition halos, while the ce-

TABLE 11.1. Chemical composition of laurel and cedar essential oils

Constituents	**cedar (%)**	**laurel (%)**
α-pinene	0.2	6.0
Camphene	0.9	0.8
β-pinene	0.2	3.2
Myrcene	–	0.9
Limonene	–	1.4
1,8-cineole	0.1	35.9
γ-terpinene	–	1.4
p-cymene	–	1.1
Linalool	–	10.6
α-cedrene	0.2	–
Terpinen-4-ol	–	2.6
Geraniol	–	0.1
α-terpineol	–	7.8
Borneol	–	5.1
Bornyl acetate	–	2.8
β-caryophyllene	–	1.1
α-gurjunene	1.3	–
α-himachalene	14.5	–
γ-himachalene	10.3	–
β-himachalene	42.1	–
δ-cadinene	2.1	–
(E)-α–atlantone	2.3	–
Eugenol	–	2.0
Metyl eugenol	–	1.4
α-terpinyl acetate	–	3.8

dar was completely ineffective against all the fungi (Table 11.2). Data showed that the essential oil of laurel also inhibited the germination of the conidia of all the tested fungi, however, for *A. niger* and *P. oryzae,* at the dose of 30 µl/ml the inhibition was full, at 10 µl/ml it was only partial (i.e., all the conidia contained in the nutrient medium were not inhibited). Conidia of *F. moniliforme* var. *subglutinosus* were the most sensitive, being completely inhibited at 10 µl/ml. Only in the case of *Botrytis cinerea,* although the two doses inhibited the germination of the conidia, when withdrawn and resuspended in a new medium without the essential oil, *B. cinerea* germinated. The cedar essential oil was less effective and it revealed low ability of conidial germination.

Another essential oil, which was tested against the same fungal species, was obtained from *Calamintha nepeta* (L.) Savi *sensu latu.* (Flamini, Cioni, et al., 1999). Although this plant is mainly known as spice, it is also used in Italian folk medicine for the treatment of respiratory and gastroenteric diseases (Bandini and Pacchiani, 1981); in parts of Sicily it is used as a wound disinfectant and cicatrizant (Aversa et al., 1976).

Few data exist about the antimicrobial effectiveness of the volatile oil of *Calamintha nepeta* (Panizzi et al., 1993). The essential oil of *C. nepeta* (Table 11.3) showed a wide antimicrobial spectrum of action (Table 11.4) including on *Aspergillus niger* also, which can be pathogenic to humans both by its spores and by the production of mycotoxins.

TABLE 11.2. Inhibitory activity of the essential oils on the hyphes at 30 µl/plate and on conidia germination at 30 and 10 µl/plate (inhibition halo diameters, mm)

Fungal species	Hyphes		Conidia			
	Laurel	Cedar	Laurel		Cedar	
	30 µl	30 µl	30 µl	10 µl	30 µl	10 µl
Fusarium moniliforme	22	–	+	+	±	±
Aspergillus niger	22	–	+	±	±	–
Pyricularia oryzae	24	–	+	±	±	±
Botrytis cinerea	23	–	[+]	[+]	–	–

– No significant difference with respect to untreated conidia
± Partial inhibition of conidia germination with respect to untreated conidia
+ Total inhibition of conidia germination with respect to untreated conidia
[+] Reversible total inhibition of conidia germination with respect to untreated conidia

TABLE 11.3. Composition (percent) of the essential oil of *Calamintha nepeta*

Constituents	Percent
α-pinene	0.5
Camphene	0.2
β-pinene	0.7
Myrcene	0.6
Limonene	7.0
1,8-cineole	0.3
p-cymene	0.7
Menthone	9.4
Isomenthone	0.3
β-caryophylene	2.2
Menthol	4.6
Pulegone	48.1
α-terpineol	0.9
Piperitone oxide	3.9
Piperitenone	3.4
Piperitenone oxide	4.6

TABLE 11.4. Inhibitory activity of the essential oil of *Calamintha nepeta* (inhibition halo diameters, mm)

Mycetes	mm
Fusarium moniliforme	10
Botrytis cinerea	12
Aspergillus niger	12
Pyricularia oryzae	15

All these data open interesting prospects for the research on the effectiveness of essential oil, at low doses, both against the vegetative forms and the spores of molds that corrupt foodstuffs.

The volatile fraction of the essential oils could also exert biocidal action as already demonstrated against arthropods, mycetes (Kurita et al., 1981;

Watanabe et al., 1989; Perrucci et al., 1996) and phytopathogenic fungi that attack stored grain (Paster et al., 1995). These volatile fractions, diffused in foodstuffs' storehouses, could prevent mold growth or the germination of spores, without the direct contact of the essential oils with the food.

Many species of *Mutisia* (Asteraceae) are used in South American folk medicine for the treatment of various diseases (Daily et al., 1988). On the basis of the medicinal use of its water extract (De Feo, 1992), the methanol and water extracts of the aerial parts of *Mutisia acuminata* var. *acuminata* against hyphal and conidia of the aforementioned pathogenic fungi (Catalano et al., 1998) have been evaluated. The dry extracts, dissolved in distilled water (1:10 w/v) and sterilized by filtration, were diluted in the culture broth to concentrations of 1:20 and 1:50. Only conidia of *B. cinerea* were inhibited by the methanol extract, and, when these were removed and incubated in fresh media, they were definitively inactivated. The water extract was inactive against both the hyphae or conidia of all the fungal species.

Mangiferin is a xanthone glucoside produced by many angiosperms and ferns. Phenolic compounds are often responsible for a generic resistance of plants toward bacteria and fungi (Wood, 1967). Particularly, mangiferin seems able to protect plants from pathogen tissue penetration (Ghosal et al., 1977; Ghosal et al., 1978; Chakrabarti and Ghosal, 1989). Because of these properties, the activity of mangiferin on the hyphae and conidia of the phytopathogenic fungi (Panizzi et al., 1997) have been assayed in vitro.

Solutions of 1:10 mangiferin were prepared and 0.1 ml were placed in a 6 mm well in the agar medium. In conidial germination tests these were suspended in broth containing three different doses of mangiferin, 50, 100, and 200 μl/ml.

Mangiferin was completely ineffective against the hyphae of the four mycetes. On the contrary, on *Botrytis cinerea* conidia, at 100 μl/ml, it was able to partially prevent conidia germination, and at 200 μl/ml conidia were permanently inactivated. Also, conidia of *A. niger* were partially inhibited at 200 μl/ml. At 200 μl/ml, *F. moniliforme* and *P. oryzae* were reversibly inhibited.

ACTIVITY AGAINST YEASTS (CANDIDA ALBICANS *AND* ASPERGILLUS NIGER)

Candida spp. are frequently present in the normal flora of the mouth, throat, large intestine, vagina, and skin. In patients whose immune defenses have been compromised by diseases or by secondary effects of drugs, normal flora may invade deeper tissues, producing severe infections. *Candida albicans,* the principal pathogenic species, causes mild to severe chronic superficial infections of skin, nails, and mucous membranes in individuals with normal immune defenses, as well as serious systemic infections in debilitated patients (Lennette et al., 1985).

Aspergilli are ubiquitous fungi, can be found growing on most organic materials, and are commonly present on grain and decaying vegetation, as well as in soil. Their conidia often become airborne in large numbers; no area open to air is free from *Aspergilli.*

The spectrum of diseases caused by pathogenic strains among the about 200 *Aspergillus* species is broad. They vary from allergic reactions due to the development of hypersensitivity to *Aspergilli,* to different types of clinical problems subsequent to colonization of various body sites, especially within the respiratory tract. Other colonized areas include ear canals and nails. Serious illness can result from the damage done by these molds as they invade tissues. In addition, some *Aspergillus* strains produce toxins which, when ingested, cause severe pathologic conditions ranging from acute or chronic toxicoses (i.e., aflatoxicosis) to hepatocarcinoma. Of the many species of *Aspergillus,* only four are commonly encountered as a cause of disease: *A. fumigatus, A. flavus, A. niger,* and *A. terreus.* Infrequently, other species classified in this genus may incite illness, particularly in immunocompromised or hypersensitive individuals. The effectiveness of plant extracts, essential oils, and pure compounds have been tested against these yeasts.

The plants were all extracted in a Soxhlet apparatus successively with hexane, chloroform, chloroform/methanol (9:1), and then at room temperature with methanol. Moreover, the decoction in water (1:5 w/v) was obtained. All the extracts were evaporated under reduced pressure and dissolved 1:2 in DMSO. The antimicrobial assays were performed using the agar diffusion technique; the tested substances were placed on 6 mm sterile paper disks.

Caulerpa taxifolia is a seaweed that was accidentally introduced into the Mediterranean basin, and has obtained a notable expansion. Its high degree of propagation, low-temperature resistance, and adaptability to every substrate has caused substantial modifications to the environment (Meinesz and Hesse, 1991). Moreover, *C. taxifolia* produces secondary metabolites that are very toxic for sea urchin eggs, cultivated mammal cells, and marine bacteria (Guerriero et al., 1993; Lemee et al., 1993). The fungicidal activity of some extracts of this alga against *C. albicans* (Della Pietà et al., 1996) have been tested. The hexane and chloroform extracts showed similar good activity on *C. albicans* (Table 11.5), while the chloroform/methanol 9:1 extract was the most active one. The more polar extracts (methanol and decoction)

TABLE 11.5. Fungicidal activity on *Candida albicans* of *Caulerpa taxifolia* extracts (inhibition halo diameters, mm)

Extract	mm
Hexane	16
Chloroform	15
Chloroform/methanol 9:1	21
Methanol	–
Decoction	–

were completely ineffective. It seems that the active principles of *C. taxifolia* could be due to medium polarity substances.

Achillea ligustica All. (Asteraceae) is a shrub that grows in Italy on arid slopes between 0 and 800 m. In Sicilian folk medicine, it is used as a hemostatic or against stomachache. Ethnopharmacological studies report the use of the sap obtained from the fresh plant as an anthelmintic (Ilardi and Raimondo, 1992); the infusion is used as a cataplasm against rheumatism and skin disorders (Bruni et al., 1997).

The composition of the essential oils obtained from the leaves and from the flowers (Table 11.6) has been studied and, considering that other *Achillea* species are endowed with good antimicrobial activity (Kedzia et al., 1990; Barel et al., 1991; De la Puerta et al., 1996), the two essential oils and some pure constituents against *C. albicans* (Bader et al., 1998) have also been tested. Both the oils showed a strong activity against this yeast (Table 11.7), with a inhibition halo of 24 mm. Considering this high susceptibility, the pure constituents carvone, 4-terpineol, and linalool were assayed. The first two, at 10 μg, were more effective than the essential oils, with inhibition halos of 30 and 26 mm, respectively; on the contrary, linalool was less effective. At lower doses, carvone maintained a good activity up to 4 μg, while 4-terpineol and linalool were active also at 2 μg. When comparing the activity of the pure compounds with some synthetic antimycotic drugs (Table 11.8), the natural compounds showed a higher inhibiting activity at about five times lesser doses.

Crithmum maritimum L. (Apiaceae), sea fennel, grows wild on Italian coastline rocks. The leaves are used in local folk medicine for their depurative properties. The composition (Table 11.9) and the activity (Table 11.10) of the essential oil and of some pure constituents against *C. albicans* and *A. niger* (Flamini et al., 1999) were evaluated. The former yeast was quite susceptible to the essential oil, with an inhibition halo of 24 mm. Among the main pure compounds screened along with the oil, only sabinene and

TABLE 11.6. Composition of the essential oil from leaves and flowers of *Achillea ligustica*

Compounds	Leaves (%)	Flowers (%)
α-pinene	1.7	1.2
β-pinene	5.8	3.3
Sabinene	3.3	1.4
α-terpinene	4.5	2.3
Limonene	1.0	1.0
β-phellandrene	6.8	5.4
γ-terpinene	7.2	4.0
p-cymene	3.0	2.9
Yomogi alcohol	2.0	–
α-thujone	2.1	2.7
Camphor	0.8	2.4
Linalool	1.6	20.4
Fenchol	–	2.1
4-terpineol	19.3	12.0
Myrcenol	1.6	1.2
Methyl cavichol	0.7	–
α-terpineol	2.4	1.6
Borneol	0.9	2.5
Germacrene D	2.9	–
Verbenone	0.9	–
Carvone	8.9	10.0
Bicyclogermacrene	1.0	–
Caryophyllene oxide	1.0	–
Cedrol	1.7	4.3
Thymol	–	0.7

methyl thymol showed positive inhibition toward *C. albicans,* while γ-terpinene and *p*-cymene were ineffective. *Aspergillus niger* was unaffected by the essential oil and the pure compounds.

Flavonoids are widely distributed among higher plants. They are often present as conjugated compounds, usually linked to sugars, inorganic sulphate, or organic acids. Free flavonoids, as inhibiting agents of many enzy-

TABLE 11.7. Antimycotic activity against *Candida albicans* of the essential oils of *Achillea ligustica* and of some pure constituents (inhibition halo diameters, mm)

Substances	10 μg	8 μg	6 μg	4 μg	2 μg
Leaves, essential oil	24	nt	nt	nt	nt
Flowers, essential oil	24	nt	nt	nt	nt
Carvone	30	24	18	12	0
4-terpineol	26	24	18	12	10
Linalool	17	14	15	11	8

nt = not tested

TABLE 11.8. Activity on *Candida albicans* of some synthetic antimycotics (inhibition halo diameters, mm)

Substance	Dose	mm
5-fluorocytosine	10 μg	0
Myconazole	50 μg	17
Econazole	50 μg	20
Nistatine	100 UI	15
Amphotericine	100 μg	18
Ketoconazole	50 μg	16
Clotrimazole	50 μg	13

matic reactions, are often toxic for living cells; therefore the linked forms seem to serve to transport the aglycones to the vacuoles, preventing cytoplasmic damage. Some less-polar aglycones are deposited within leaf cuticles to protect plants from insects, fungi, or bacteria. The antimicrobial activity of flavonoids could be related to the phenolic groups present in their molecules. These groups react with enzymatic proteins, altering biological processes.

The antimicrobial properties of two flavonoid aglycones and the two corresponding glycosides were evaluated to verify if the sugar moiety could interfere with the activity of the free aglycone (Cioni et al., 1998). Two flavonols, quercetin and rutin, and two flavanones, naringenin and naringin were used. *Aspergillus niger* was completely not affected by all the tested substances, while the flavanone aglycone naringenin was toxic for *C. albicans* (Table 11.11). It seems that glycosides are always ineffective while, among aglycones, the lack of a double bond at positions 2-3 of the flavonoid nucleus permits the inhibitory activity of the molecules.

TABLE 11.9. Composition of the essential oil of *Crithmum maritimum*

Constituents	**%**
α-pinene	1.2
Sabinene	30.0
Myrcene	1.4
Limonene	0.4
1,8-cineole	0.9
γ-terpinene	41.1
p-cymene	5.7
Methyl thymol	12.0
Dill apiole	1.6

VETERINARY PARASITES

Parasitic diseases due to fungal pathogens are an unresolved problem for veterinary medicine. The therapeutic control of parasitosis, in spite of the production of many synthetic drugs, is not yet solved. On the contrary, many antiparasitic therapies cause, at present, many new and severe problems. The features that an ideal drug should possess can include a wide spectrum of action, low cost, low toxicity for the animal, no resistance, lack of residues in products of animal origin, and low environmental impact. Unfortunately, contemporary synthetic drugs do not fully possess these characteristics. In fact, many drugs are effective only against one or few species, and most work on biological structures common to both parasite and host (Link, 1968; McKellar and Scott, 1990). Moreover, many synthetic drugs are not biodegradable and can cause environmental pollution, and changes in fauna (Robertson, 1988). The repeated and inconsiderate use of drugs to reduce parasitosis incidence have caused resistance to drugs and genetic mutations in parasites (Robertson, 1988; McKellar and Scott, 1990).

Although several studies confirm the therapeutic effectiveness of many plants, nowadays, synthetic active molecules, education and medicine changes, and gradual separation from tradition, have made their use obsolete. Some extracts, essential oils, and pure compounds of plant origin have been tested to try to eradicate these problems.

Saprolegniosis is a mycotic disease of animals living in fresh and mesohaline water (Shaperclaus, 1979). This infection is very dangerous for intensive freshwater fish farms and is widespread in all stages of the lifecycle,

TABLE 11.10. Antimycotic activity against *Candida albicans* and *Aspergillus niger* of the essential oil of *Crithmum maritimum* and of some pure constituents (inhibition halo diameters, mm)

Fungi	**Oil**	**Sabinene**					**Methyl thymol**					**γ-terpinene**					***p*-cymene**				
	10	10	5	3	2	1	10	5	3	2	1	10	5	3	2	1	10	5	3	2	1
Candida albicans	24	16	14	13	10	0	12	10	8	0	0	0	0	0	0	0	0	0	0	0	0
Aspergillus niger	0	0	0	0	0	0	0	0	0	0	0	0	0	0	0	0	0	0	0	0	0

TABLE 11.11. Antimycotic activity against *Candida albicans* and *Aspergillus niger* of some flavonoids (1 mg; inhibition halo diameters, mm)

Flavonoids	***Candida albicans***	***Aspergillus niger***
Quercetin	0	0
Rutin	0	0
Naringenin	14	0
Naringin	0	0

from egg to adult; the disease mainly affects the skin and the gills of fish, but internal organs can also be involved. The use of effective drugs (malachite green and formalin) is now restricted by law because they have been found to be toxic for animals and human consumers, and, furthermore, for their high environmental impact (Amadei et al., 1996). These reasons justify the research of new antimycotics of natural origin.

The effectiveness of four essential oils, *Satureja montana* L., *Calamintha nepeta* (L.) Savi, *Rosmarinus officinalis* L., and *Thymus vulgaris* L. were tested, in vitro, against *Saprolegnia ferax* (Table 11.12). Moreover, the antifungal potential of the pure compounds hydroquinone, caffeine, nicotinic acid, nicotine, thymol, and carvacrol (Perrucci et al., 1995) were tested.

Each tested substance was solubilized into the melted medium (Cornmeal Agar Difco) in 9 cm plates and left to solidify. Agar disks (10 mm diameter) taken from *Saprolegnia ferax* cultures were placed in the middle of the plates. Several dilutions for each extract were made to determine the minimum inhibitory concentration (MIC). The plates were incubated at 22°C and exposed to a light of 15 W and 400 lux. The fungal growth was checked after 24, 48, 72, and 96 hours and one week, and measured from the margin of agar disk (Table 11.13).

The composition of the four essential oils was determined by GC (Table 11.12).Thyme and savory essential oils, thymol, and carvacrol showed the best antimycotic activity. Particularly, thyme oil and carvacrol showed a MIC of 40 ppm, while formalin, a common antimycotic used in aquaculture, had a MIC of 150 ppm. Thymol completely inhibited the fungal growth at 100 ppm, and savory essential oil showed the same action at 300 ppm. Calamint and rosemary oils, hydroquinone, and caffeine were able to completely inhibit growth of *Saprolegnia* only at higher doses (2000 ppm); nicotine and nicotinic acid, at the same dose, showed only partial fungal growth inhibition. Malachite green and formalin, commonly used against saprolegniosis, showed in vitro MIC similar to those used in vivo (Ghittino,

TABLE 11.12. Composition (percent) of the essential oil of *Rosmarinus officinalis, Thymus vulgaris, Satureja montana*, and *Calamintha nepeta* tested against *Saprolegnia ferax*

Constituents	*Rosmarinus officinalis*	*Thymus vulgaris*	*Satureja montana*	*Calamintha nepeta*
α-pinene	25.30	2.66	0.93	0.71
Camphene	5.73	0.35	0.74	–
β-pinene	2.31	0.37	0.10	0.64
Myrcene	2.40	3.48	1.33	0.73
Limonene	3.71	0.55	0.15	6.42
1,8-cineole	17.31	0.85	0.59	0.43
γ-terpinene	2.07	4.11	10.50	–
p-cymene	1.01	16.15	13.48	1.17
Camphor	8.25	0.11	0.21	–
Linalool	4.27	12.68	1.12	–
Bornyl acetate	1.75	2.52	0.48	1.03
β-caryophyllene	1.03	0.87	0.92	2.26
α-terpineol	1.75	–	2.05	0.51
Borneol	8.87	–	1.57	–
Verbenone	5.77	–	0.41	–
Geraniol	1.13	0.30	0.53	–
Thymol	–	13.83	0.29	–
Carvacrol	–	28.98	54.32	–
Carvone	–	–	0.09	–
α-phellandrene	–	–	0.07	–
α-terpinene	–	–	1.43	–
Menthone	–	–	–	9.82
Isomenthone	–	–	–	0.20
Menthofuran	–	–	–	3.89
Menthol	–	–	–	4.85
Pulegone	–	–	–	46.00
Piperitone oxide	–	–	–	2.33
Piperitenone	–	–	–	2.07
Piperitenone oxide	–	–	–	2.56

TABLE 11.13. Antifungal activity against *Saprolegnia ferax* of essential oils and pure substances (fungal growth, mm; mean of seven trials ± SD)

Substances	Doses (ppm)	24 h	48 h	72 h	96 h	1 week
Thyme e.o.[a]	2000-40	–[b]	–	–	–	–
	30	4.2±0.75	11±2.61	25.8±2.71	35±2.10	f.p.[c]
Carvacrol[a]	2000-40	–	–	–	–	–
	30	1.33±1.24	2.83±2.67	6±6.85	11±10.83	19.16±17.88
Thymol[a]	2000-100	–	–	–	–	–
	75	–	–	–	–	3±2.53
	50	1.8±1.47	6.8±1.6	11.8±1.94	17.8±2.64	f.p.
Savory e.o.[a]	2000-300	–	–	–	–	–
	150	0.77±1.13	1.22±1.87	1.88±3.10	3.77±6.06	10±13.32
Hydroquinone	2000	–	–	–	–	–
	1000	1.12±0.78	3.5±1.41	5.37±2.12	7.12±3.06	9.62±3.24
Caffeine	2000	–	–	–	–	–
	1000	0.83±0.37	1.16±0.68	3±0.57	4.33±0.94	6.33±0.74
Nicotine	2000	1.6±1.36	5.4±3.5	8.6±5.28	11.6±7.74	f.p.
Nicotinic acid	2000	–	1.8±1.47	5±2.1	7.8±2.79	20.2±1.72
Rosemary e.o.[a]	2000-1000	–	–	–	–	–
	500	1.2±1.46	6±5.47	13.6±10.63	24.8±9.34	f.p.
Calamint e.o.[a]	2000-1000	–	–	–	–	–
	500	6.75±1.78	19.5±2.18	35±1.58	f.p.	f.p.
Malachite green	10-5-3	–	–	–	–	–
	1	1.2±0.98	1.4±0.8	2±0.6	2.8±1.17	7.2±1.17
Formalin	2000-150	–	–	–	–	–
	100	6±1.09	22.2±3.87	33.2±3.49	f.p.	f.p.

[a] Water-insoluble substances
[b] No fungal growth
[c] f.p. full fungal growth

1985; Roberts, 1990). However, in addition to the acute environmental impact on aquatic ecosystems, these substances can exhibit an immuno-depressive effect on repeatedly treated fish. The properties of hydroquinone (Srivastava, 1980) and caffeine (Prabhuji et al., 1983) against *S. ferax* have been assayed and the effective dosages obtained in other studies as well.

Because many species of *Santolina* and *Artemisia* have been found to have biological activity against several parasites (Lewis and Elvis-Lewis, 1977; Akila et al., 1987; Klayman, 1993; Ballero and Fresu, 1991), the activities of the *Artemisia verlotorum* and *Santolina etrusca* aqueous and methanol extracts and of two fractions were evaluated. Both were obtained by Amberlite IRC50 column chromatography, from the aqueous extracts (Macchioni et al., 1999).

Methanol extracts were prepared using 150 g of each dried and ground plant material, separately soaked in 500 ml of solvent for 24 h; aqueous extracts were obtained by boiling 100 g of plant material in 1000 ml of distilled water for ten minutes. The fractions were obtained from 3 g of water extract by column chromatography on Amberlite IRC50, eluting first with water and then with 90 percent ethanol to obtain aqueous and ethanol fraction, respectively. All the solvents were removed from extracts and fractions by rotary vacuum evaporation.

The results are summarized in Table 11.14. Aqueous extract of *A. verlotorum* at 1 percent concentration corresponded to its MIC against *Saprolegnia ferax,* while the methanol extract and the ethanol and aqueous fractions were more active (MIC = 0.25 percent). The aqueous extract of *S. etrusca* was less effective than that of *A. verlotorum,* showing a MIC of 2 percent; also the methanol extract was less effective, with a MIC of 1 percent. The ethanol and aqueous fractions had the same effectiveness as the corresponding fractions of *A. verlotorum*. The increased effectiveness of the chromatographic fractions, four times for *A. verlotorum* and eight times for *S. etrusca* with respect to the crude extracts, could depend on a higher concentration of the active principle(s). Even though the active doses were not very low compared with the previously used synthetic drugs (Amadei et al., 1996), it must be remembered that the substances used were crude materials and not pure principles. In the future, it may be possible to isolate one or more pure active principles so that a lower dose could be used with less toxic effects. At present, experimental work is still in progress.

In addition to an increased frequency of human cutaneous mycoses spread by domestic animals (Mantovani, 1978), some studies show an increased number of therapeutic failures of antimycotic drugs (Georgii and Korting, 1991; Macura, 1991), especially against *Microsporum canis* (Puccini et al., 1992). This led to the study of the activity of the essential oils of some spontaneous Lamiaceae of the Mediterranean area, thyme, savory, calamint, lavender, and rosemary, against two dermatophytes, *Microsporum canis* and *M. gypseum*. This kind of activity was previously unknown (Perrucci et al., 1994).

The essential oils were obtained from flowering wild plants, and their compositions are reported in Table 11.15. The dermatophytes were col-

TABLE 11.14. Antifungal activity against *Saprolegnia ferax* of extracts and fractions of *Artemisia verlotorum* and *Santolina etrusca* (fungal growth, cm; mean of five trials ± SD)

Concentration (%)	Aqueous extract				Ethanol fraction				Aqueous fraction				Methanol extract			
	24h	48h	72h	1w	24h	48h	72h	1w	24h	48h	72h	1w	24h	48h	72h	1w
A. verlotorum																
from 20 to 1	0	0	0	0	0	0	0	0	0	0	0	0	0	0	0	0
0.5	0.10 ±0	0.33 ±0.05	0.50 ±0.10	0.90 ±0	0	0	0	0	0	0	0	0	0	0	0	0
0.25					0	0	0	0	0	0	0	0	0	0	0	0
0.1					0.65 ±0.05	1.60 ±0.08	2.47 ±0.09	3.50	0.40 ±0.05	1.13 ±0.09	1.80 ±0.18	3.50	0.30 ±0	0.96 ±0.18	1.36 ±0.58	2.90 ±0.35
S. etrusca																
from 20 to 2	0	0	0	0	0	0	0	0	0	0	0	0	0	0	0	0
1	0	0.24 ±0.07	0.60 ±0.13	2.70 ±0.12	0	0	0	0	0	0	0	0	0	0	0	0
0.5					0	0	0	0	0	0	0	0	0	0	0	2.46 ±0
0.25					0	0	0	0	0	0	0	0	0	0.33 ±0.05	0.93 ±0.05	3.50
0.1					1.10 ±0	2.45 ±0	3.50	3.50	0.30 ±0.14	1.10 ±0.14	2.15 ±0.13	3.50	0.33 ±0.06	1.26 ±0.06	2.30 ±0	3.50

TABLE 11.15. Composition (percent) of the essential oils tested against *Microsporum canis* and *M. gypseum*

Constituents	*Lavandula angustifolia*	*Calamintha nepeta*	*Satureja montana*	*Thymus vulgaris*	*Rosmarinus officinalis*
α-pinene	0.1	0.6	0.9	1.2	30.7
Camphene	0.1	0.1	0.7	2.2	7.0
β-pinene	0.1	0.4	tr	tr	3.6
Myrcene	0.6	0.4	1.3	1.4	3.2
Limonene	0.2	2.2	tr	tr	3.8
1,8-cineole	0.5	0.3	0.6	2.6	11.0
γ-terpinene	–	0.5	8.7	9.2	0.7
p-cymene	9.1	0.1	22.5	31.2	tr
Menthone	–	15.3	1.7	–	–
Isomenthone	–	0.3	–	–	–
Menthofuran	–	2.2	–	–	–
Camphor	0.2	0.1	tr	tr	6.4
Linalool	19.7	0.2	1.1	tr	2.7
Linalyl acetate	21.1	–	–	–	–
Bornyl acetate	–	–	1.7	tr	4.4
β-caryophyllene	2.7	6.8	0.9	1.5	1.0
Menthol	–	2.3	–	–	–
Pulegone	–	23.6	–	–	–
α-terpineol	0.3	0.3	2.1	tr	1.0
Borneol	0.3	0.8	1.6	4.6	2.0
Verbenone	–	–	–	–	5.7
Carvone	–	1.4	–	–	–
Piperitone oxide	–	1.4	–	–	–
Piperitenone	–	15.8	–	–	–
Piperitenone oxide	–	14.9	–	–	–
Geraniol	0.4	0.1	0.5	tr	1.0
Thymol	–	–	1.7	7.3	–
Carvacrol	–	–	48.6	28.8	–

lected by brushing cats, dogs, and horses. The animals had never been treated with antimycotic drugs. The specimens were cultured on mycobiotic agar and incubated at 30°C. Phosphate buffered saline (PBS) suspensions were spread onto culture media containing the essential oils. The method used was the macrodilution test, which allows the incorporation of the drug into the medium at several dilutions to determine the MIC. To obtain a better distribution, the required amounts were dissolved in a 1 percent solution of Tween 80 in physiological saline. The inhibition of fungal growth indicated the effectiveness of the essences. To ascertain a fungicidal or fungistatic activity, parts of the media from plates without mycotic growth were transferred into new Sabouraud dextrose agar plates; no fungal growth after ten days incubation was indicative of fungicidal activity. Ten isolates of *M. canis* and thirteen of *M. gypseum* were also tested in vitro to detect their susceptibility to various common antifungal agents used in Italy: griseofulvin (25 μg/disk), ketoconazole (15 μg/disk), econazole (10 μg/disk), clotrimazole (10 μg/disk), tioconazole (10 μg/disk). The disks were placed on the surface of the cultures of dermatophytes and, after five days, the diameters of the inhibited growth areas were measured. The fungal susceptibility was classified according the following parameters: susceptible (imidazoles ≥20 mm; griseofulvin ≥10 mm), intermediate (imidazoles 12-19 mm), resistant (imidazoles ≤11 mm; griseofulvin ≤9 mm). The results are summarized in Tables 11.16 and 11.17. The essential oils of thyme and savory, for frequencies and repetitiveness of their results, showed an effective antimycotic action against the two fungial species. Their activity was constant also at the higher dilutions; moreover, these oils proved to be fungistatic at low doses and fungicidal at higher concentrations. When testing the in vitro susceptibility of *M. canis* and *M. gypseum* to synthetic antifungal drugs, isolates simultaneously resistant to two or more antimycotics were observed, while the same strains showed a great susceptibility to the essential oils of thyme and savory. The same effectiveness cannot be assigned to the lavender and rosemary oils. In particular, these essences showed a variable efficacy to inhibit fungal growth. Their antimycotic activity, almost exclusively fungistatic, was weak and became significant only at higher doses. The results obtained with calamint oil were positive enough to be accepted, but this essence was not available in sufficient amounts for further testing.

CONCLUSIONS AND FUTURE DIRECTIONS

Systemic fungal infections have increased significantly during the past decade. This increase is due to a greater use of broad-spectrum antibiotics,

TABLE 11.16. Fungistatic (S) and fungicidal (C) activity of the essential oils against isolates of *Microsporum canis* (in parenthesis no. of fungal isolates from different animals)

Ppm	12.5	25	50	75	100	300	Antimycograms		
Host							**r**	**i**	**s**
Thymus vulgaris									
Cat (4)	–	S	S	S	C	C			
Cat	–	S	S	S	S	S			
Cat*	–	S	S	S	S	S	1,2,3,4,5	–	–
Cat*	–	–	S	C	C	C	1,2,3	–	4,5
Cat*	C	C	C	C	C	C	2,3,4	–	1,5
Cat*	–	S	C	C	C	C	2,3,4,5	–	1
Dog (2)	–	S	S	S	C	C			
Dog*	S	C	C	C	C	C	1,2,3	–	4,5
Dog*	S	S	C	C	C	C	1,2,5	–	3,4
Horse	–	S	S	S	S	C			
Satureja montana									
Cat	–	S	S	S	C	C			
Cat (2)	–	S	S	S	S	S			
Cat*	–	S	S	S	C	C	1,2,3,4,5		
Cat*	S	C	C	C	C	C	1,2,5	–	3,4
Cat*	C	C	C	C	C	C	2,3,4	–	1,5
Cat*	–	–	S	S	C	C	1,2,5	–	3,4
Dog	–	S	S	C	C	C			
Dog*	S	S	C	C	C	C	2,3,4,5	–	1
Dog*	–	S	S	C	C	C	2,3,4	–	1,5
Horse	–	S	C	C	C	C			
Lavadula angustifolia									
Cat (2)	–	–	–	–	–	S			
Cat	–	–	S	S	C	C			
Cat*	–	–	S	S	S	S	1,2,3,4,5	–	–
Cat*	–	–	S	S	S	S	1,2,3	–	4,5
Cat*	–	–	S	S	S	S	2,3,4	–	1,5

Ppm	12.5	25	50	75	100	300	Antimycograms		
Host							**r**	**i**	**s**
Dog	–	–	–	–	S	S			
Dog	–	–	S	S	C	C			
Dog*	–	–	–	–	–	S	2,3,4	–	1,5
Horse	–	–	S	C	C	C			
Rosmarinus officinalis									
Cat (2)	–	–	S	S	S	S			
Cat	–	–	–	–	–	S			
Cat*	–	–	–	S	S	S	1,2,3,4,5	–	–
Cat*	–	–	–	–	S	C	2,3,4	–	1,5
Dog (3)	–	–	–	–	–	S			
Dog*	–	–	–	–	–	S	1,2,3	–	4,5
Horse	–	–	–	–	S	S			
Calamintha nepeta									
Cat (2)	–	–	–	–	–	S			
Cat (2)	–	–	S	S	C	C			
Cat*	–	–	–	S	S	C	2,3,4,5	–	1
Dog	–	–	–	–	–	S			
Dog	–	–	S	S	C	C			
Dog*	–	–	S	S	S	C	1,2,5	–	3,4
Horse	–	–	C	C	C	C			

Note: – = no activity; r = resistant; s = susceptible; i = intermediate; 1 = griseofulvin; 2 = ketoconazole; 3 = econazole; 4 = clotrimazole; 5 = tioconazole.
*Fungal isolates tested also with the antimycogram

immunosuppressive agents, organ transplantation, and AIDS. The first synthetic antifungal agents were nystatin and amphotericin B, synthesized during the 1950s (Espinel-Ingroff and Shadomy, 1989).

Until the early 1990s, little progress was made with the development of the azole drugs. During the past ten years there has been an expansion in the development of synthetic antifungal drugs, but there are still weaknesses in the current antifungal chemotherapy. In fact, many molecules are quite toxic for humans and animals and the use of these substances could result in economic damage for animal breeders and in health hazards for the final customers if some residues persist in the animal-derived foods.

TABLE 11.17. Fungistatic (S) and fungicidal (C) activity of the essential oils against isolates of *Microsporum gypseum*

Ppm	12.5	25	50	75	100	300	Antimycograms		
Host							r	i	s
Thymus vulgaris									
Cat	–	–	S	S	S	C	1,2	–	3,4,5
Cat	–	S	S	S	C	C	1,2	–	3,4,5
Cat	–	S	S	S	S	S	2,5	–	1,3,4
Cat	–	S	S	S	S	S	2	5	1,3,4
Dog	–	C	C	C	C	C	1,2	–	3,4,5
Dog	–	S	S	S	C	C	2,3	–	1,4,5
Horse	–	S	S	S	S	S	5	–	1,2,3,4
Horse	–	S	C	C	C	C	2,3,5	–	1,4
Satureja montana									
Cat	–	S	S	S	S	S	1,2	–	3,4,5
Cat	–	S	S	S	C	C	1,2	–	3,4,5
Cat	–	S	C	C	C	C	2,5	–	1,3,4
Cat	–	–	S	S	S	S	2	5	1,3,4
Dog	–	–	S	S	S	S	1,2	–	3,4,5
Dog	–	S	S	S	C	C	2,3		1,4,5
Horse	–	S	S	S	C	C	5	–	1,2,3,4
Horse	S	S	C	C	C	C	2,3,5	–	1,4
Calamintha nepeta									
Cat	–	–	S	S	S	S	1,2	–	3,4,5
Dog	–	–	–	–	–	S	2,3	–	1,4,5
Horse	–	–	C	C	C	C	5	–	1,2,3,4
Lavandula angustifolia									
Cat	–	–	–	–	S	S	2,5	–	1,3,4
Cat	–	–	–	–	–	S	2	5	1,3,4
Cat	–	–	S	S	S	S	2,3,5	–	1,4
Dog	–	–	–	–	S	S	2,3,5	–	1,4
Horse	–	–	C	C	C	C	5	–	1,2,3,4

Ppm	**12.5**	**25**	**50**	**75**	**100**	**300**	**Antimycograms**		
Host							**r**	**i**	**s**
Rosmarinus officinalis									
Cat	–	–	–	–	S	S	1,2	3,4,5	–
Cat	–	–	–	–	–	S	1,2	–	3,4,5
Cat	–	–	–	–	–	S	2,5	–	1,3,4
Dog	–	–	–	–	S	S	1,2	3,4,5	–
Dog	–	–	–	–	–	S	2,3	–	1,4,5
Dog	–	–	–	–	–	S	2	5	1,3,4
Horse	–	–	–	–	–	S	5	–	1,2,3,4
Horse	–	–	–	–	–	S	2,3,5	–	1,4

Note: – = no activity; r = resistant; s = susceptible; i = intermediate; 1 = griseofulvin; 2 = ketoconazole; 3 = econazole; 4 = clotrimazole; 5 = tioconazole.

Moreover, the use of fungicides is important in agriculture and in the food industry. In fact, plant pathogenic fungi are responsible for a significant loss of potential plant production for food and nonfood use. Synthetic antifungals, however, can be toxic for humans and animals that feed on treated crops. To reduce the use of potential hazardous fungicidal products in agriculture, the development and implementation of natural antifungal agents, the so-called green chemicals, should be increased in the future. Analytical methods for rapid and reliable evaluation of the presence of mycotoxins in foodstuffs should be developed.

Study results, supported also by many analogous literature data, demonstrate that it is possible to conduct in vivo experiments of new antifungal, natural molecules. The investigations have also pointed out that the effectiveness of many natural substances, at least in in vitro tests, is not lower than that of synthetic drugs. These findings support continuing the studies using a greater number of natural substances. It could be interesting to evaluate different chemical structures to obtain structure-activity correlations and to discover more effective principles, which could be employed at lower doses. These findings could decrease the environmental impact and the residues of toxins in foods. Moreover, fungi are becoming more and more resistant to common antifungal agents. Although the resistance phenomenon is also possible with natural substances, it is not always full and persisting; this is especially true for the essential oils.

Finally, it is necessary to establish adequate quality control criteria to reveal fungal contaminations.

REFERENCES

Akila, A., Raghunath, S.T., and Satya, P.P. (1987). Biosynthesis of artemisinin in *Artemisia annua. Phytochemistry* 26:1927-1930.

Amadei, A., Giorgetti, G., and Malvisi, J. (1996). Possibili alternative al verde malachite nella terapia di saprolegniosi, ictioftiriasi e malattie proliferative. *Riv. Ital. Acquacoltura* 31:119-126.

Aversa, M.C., Giannetto, P., and Romeo, G. (1976). Constituents of *Satureia calamintha.* Note I. Isolation of someursolicand 3-epiursolic acids *Chim. Ind. (Milan)* 58:448.

Bader, A., Cioni, P.L., Flamini, G., Morelli, I., and Panizzi, L. (1998). *Achillea ligustica:* Composizione e Attività Antimicrobica dell'Olio Essenziale e dei suoi Costituenti Principali. I Convegno estratti vegetali in cosmetica- Cosmetici, salute e qualitàdella vita. Istituto superiore di Sanità , Roma.

Ballero, M. and Fresu, I. (1991). Pianteofficinali impiegate in fitoterapia nel territorio del Marganai (SardegnaSud-Occidentale). *Fitoterapia* 62:524-531.

Bandini, P. and Pacchiani, M. (1981). Costituenti, proprietà e usi di *Calamimtha nepeta. Essenze Deriv. Agrum.* 51:325.

Barel, S., Segal, R., and Yashphe, J.(1991). The antimicrobial activity of the essential oil from *Achillea fragrantissima. J. Ethnopharm.* 33:187-191.

Bottalico, A. (1981). Problemi micotossicologicidei semi. *Informatore Fitopatologico* 12:41-46.

Bruni, A., Ballero, M., and Poli, F.(1997). Quantitative ethnopharmacological study of the Campidano valley and Urzulei district, Sardinia, Italy. *J. Ethnopharmacol.* 57:97-124.

Catalano, S., Cioni, P.L., Panizzi, L., and Morelli, I. (1998). Antimicrobial activity of extracts of *Mutisiaacuminata* var. *acuminata. J. Ethnopharm.* 59:207-209.

Chakrabarti, D.K. and Ghosal, S. (1989). The disease cycle of mango malformation induced by *Fusarium moniliforme* var. *subglutinans* and the curative effects of mangiferin-metalchelates. *J. Phytopathol.* 125:238-246.

Cioni, P.L., Flamini, G., Panizzi, L., and Morelli, I. (1998). Valutazione dell'attività antimicrobica, *in vitro,* di Quercetina e naringenina e dei corrispondentidiglicosidi. Acta VI Convegno sulla Chimica dei Carboidrati, Tirrenia(Pisa), 8-10 Ottobre, pp.118-119.

Daily, A., Seligmann, O., Nonnenmacher, G., Fessler, B., Wong, S.M., and Wagner, H. (1988). New chromone, coumarin, and coumestan derivatives from *Mutisia acuminata* var. *hirsuta. Planta Med.* 54:50-52.

De Feo, V. (1992). Medicinal and magical plants in the Northern Peruvian Andes. *Fitoterapia* 63:417-440.

De la Puerta, R., Saenz, M.T., and Garcia, M.D.(1996). Antibacterial activity and composition of the volatile oil from *Achillea ageratum* L. *Phytother. Res.* 10:248-250.

Della Pietà, F., Panizzi, L., Cioni, P.L.,Cinelli, F., and Morelli, I. (1996). Attivit^ antimicrobica di estratti di *Caulerpa taxifolia* su Batteri e *Miceti terrestri. Acta of 2nd Internat. Workshop on Caulerpa taxifolia.* In Ribera M.A., Ballesteros E.,

Boudoresque C.F., Gomez A., Gravez V. (Eds.). Barcelona: Publicacions Universitat, pp. 261-264.

Espinel-Ingroff, A. and Shadomy, S. (1989). *Eur. J. Clin. Micorobiol.* 8:352.

FAO-WHO-UNEP (1977). *Global perspective on mycotoxin.* Document Myc, 4th Joint Congress about Mycotoxins, Nairobi, Kenya, September 19-20, p. 87.

Flamini, G., Cioni, P.L., Puleio, R., Morelli, I., and Panizzi, L. (1999). Antimicrobial activity of the essential oil of *Calamintha nepeta* and its constituent pulegone against bacteria and fungi. *Phytother. Res.* 13:349-351.

Flamini, G., Mastrorilli, E., Cioni, P.L., Morelli, I., and Panizzi, L. (1999). Essential oil from *Crithmum maritimum* grown in Liguria (Italy):Seasonal variation and antimicrobial activity. *J. Essent. Oil Res.* 11:788-792.

Frank, B. (1984). Mycotoxins from mold fungi-weapons of uninvited fellow-boarders of man and animal: Structures, biological activity, biosynthesis, and precautions. *Angew. Chem.* 23:493-505.

Georgii, A. and Korting, H.C. (1991). Antifungal susceptibility testing with dermatophytes. *Mycoses* 34:193-199.

Ghittino, P. (1985). *Tecnologia e Patologia in Acquacoltura*, Volume 2. (pp. 151-155). Torino, Italy: Tipografia Bono E.

Ghosal, S., Biswas, K., Chakrabarti, D.K., and Chaudhari, K.C. (1977). Control of *Fusarium* wilt of safflower by Mangiferin. *Am. Phytopathol. Soc.* 67:548.

Ghosal, S., Biswas, K., and Chattopadhyay, B.K. (1978). Differences in the chemical constituents of *Mangifera indica,* infected with *Aspergillus niger* and *Fusarium moniliformae. Phytochemistry* 17:689-694.

Guerriero, A., Marchetti, F., D'Ambrosio, M., Senesi, S., Dini,F., and Pietra, F. (1993). New ecotoxicologically and biogenetically relevant terpenes of the tropical green seaweed *Caulerpa taxifolia* which is invading the Mediterranean. *Helv. Chem. Acta* 76:855-864.

Ilardi, V. and Raimondo, F.M. (1992). *Quad. Bot. Ambientale Appl.* 3:41-51.

Kedzia, B., Krzyzaniak, M., and Holderna, E. (1990). Effect of yarrow essential oil *(Oleum millefolii)* and its components on pathogenic microorganisms. *Herba Pol.* 36:117-125.

Klayman, D.L. (1993). Artemisia annua: From weed to respectable antimalarial plant. In A.D. Kinghorn and M.F. Balanotrin (Eds.), Human Medicinal Agents from Plants (pp. 242-255). Washington, DC: Am. Chem. Soc. Symp. Series.

Kurita, N., Miyaji, M., Kurane, R., and Takahara, Y. (1981). Antifungal activity of components of essential oils. *Agric. Biol. Chem.* 45:945-952.

Lemee, R., Pesando, D., Durant-Clement, M., Dubreuik, A., Meinesz A., Guerriero A., and Pietra F. (1993). Preliminary survey of toxicity of the green alga *Caulerpa taxifolia* introduced into the Mediterranean. *J. Appl. Phyc.* 5:485-493.

Lennette, E.H., Balows, A., Hausler, W.J. Jr., Shadomy, H.J. (1985). *Manual of Clinical Microbiology,* edited by American Society for Microbiology, Washington, DC.

Lewis, N.H. and Elvis-Lewis, P.F. (1977). *Medical Botany Plants Affecting Man's Health.* Lawana, MA: Quartermen Publ. Inc.

Lillehoj, E.B., Ciegler, A. and Detry, R.W. (1970). In F.R. Blood (Ed.), *Essays in Toxicology,* Volume 2. New York: Academic Press, pp. 1-136.

Link, R.P. (1968). *Veterinary Pharmacology and Therapeutics*. Ames: Iowa State University Press.

Macchioni, F., Perrucci, S., Flamini, G., Cioni, P.L., and Morelli, I. (1999). Antimycotic activity against *Saprolegnia ferax* of extracts of *Artemisia verlotorum* and *Santolina etrusca*. *Phytother. Res.* 13:242-244.

Macura, A.B. (1991). Fungal resistance to antimycotic drugs. A growing problem. *Int. J. Derm.* 30:181-183.

Mantovani, A. (1978). *Mycopahtolgia* 65:61-66.

Marasas, W.F.O., Kriek, N.P.J.,Wiggins, V.H., Steyn, P.S., Towers D.K., and Hastie T.J. (1979). Incidence, geographic distribution, and toxigenicity of *Fusarium* species in South African corn. *Phytopathology* 69:1181-1185.

Marth, E.H. (1972). *Mycotoxins, A review. Fungi and Food. Proceedings of 7th Annual Symposium.* October 19, New York.

McKellar, Q.A. and Scott, E.W. (1990). The benzimidazole anthelmintic agents—A review. *J. Vet. Pharmacol. Ther.* 13:223-247.

Meinesz, A. and Hesse, B. (1991). Introduction et invasion de l'alguetropicale *Caulerpa taxifolia* en mediterranee nord-occidentale. *Oceanol. Acta* 14:415-426.

Panizzi, L., Catalano, S., Cioni, P.L., Flamini, G., and Morelli, I. (1997). Mangiferina: Prove di inibizione dello sviluppodel micelio e della germinabilità delle spore di alcuni miceti responsabili dei marciumi. *Acta Phytotherapeutica* 4:27-30.

Panizzi, L., Flamini, G., Cioni, P.L., and Morelli, I. (1993). Composition and antimicrobial properties of essential oils of four Mediterranean Lamiaceae. *J. Ethnoph.,* 39:167-170.

Panizzi, L., Flamini, G., Cioni, P.L., Perrucci, S., Morelli, I. (1995). Propriet^ antimicotiche *in vitro* di alcuni oliessenziali su ife e spore di muffe contaminanti gli alimenti. *Acta II Congresso Nazionale di Chimica degli Alimenti, Giardini Naxos,* 24-27 Maggio, 1143-1147.

Paster, N., Menasherov, M., Ravid, U., and Juven, B. (1995). *Journal of Food Protection* 58:81-85.

Perrucci, S., Cecchini, S., Pretti, C., Varriale Cognetti, A.M., Macchioni, G., Flamini, G., and Morelli, I. (1995). *In vitro* antimycotic of some natural products against *Saprolegnia ferax*. *Phytother. Res.* 9:147-149.

Perrucci, S., Cioni, P.L., Flamini, G., Taccini, F., Morelli, I., and Macchioni, G. (1996). The activity of volatile compounds from *Lavandula angustifolia* against *Psoroptes cuniculi*. *Phytotherapy Research* 10:5-8.

Perrucci, S., Mancianti, F., Cioni, P.L., Flamini, G., Morelli, I. and Macchioni, G. (1994). *In vitro* antifungal activity of essential olis against some isolates of *Microsporum canis* and *Microsporum gypseum*. *Planta Med.* 60:184-187.

Prabhuji, S.K., Srivastava, R.C., Rizvi, S.J.H., and Mathur, S.N. (1983). 1,3,7-trimethylxanthine (caffeine); a new natural fish fungicide. *Experentia* 39:177-179.

Prost, M. and Sopinska, A. (1989). Reaction of mucous cells of the carp (*Cyprinus carpio* L.) to an insecticide as a biological indicator of water pollution. *Med. Weter.* 40:2064-267.

Puccini, S., Valdre, A., Papini, R. and Mancianti, F. (1992). *In vitro* susceptibility to antimycotics of *Microsporum canis* isolates from cats. *J. Am. Vet. Med. Ass.* 201:1375-1377.

Roberts, R.J. (1990). Parassitologia. *Patologia dei Pesci,* Giorgetti G. ed. Roma, Italy: Edizioni Agricole Calderini, pp.178-182.

Robertson, E.L. (1988). *Veterinary Pharmacology and Therapeutics.* Ames: Iowa State University Press.

Shaperclaus, W. (1979). *Fischkrankheiten,* Volumes 1 and 2. Berlin: Academie-Verlag, pp. 510 and 1089, respectively.

Srivastava, R.C. (1980). Chemotherapeutic investigations on fish mycoses. *Monatshefte fŸr Veterinarmedizin* 35:267-269.

Tiecco, G. (1987). *Ispezione Degli Alimenti di Origine Animale Freschi Econservati,* Second Edition, Bologna: Edagricole.

Watanabe, F., Tadaki, S., Takaoka, M., Ishino, M., and Morimoto, I. (1989). Killing activities of the volatiles emitted from essential oils for *Dermatophagoides pteronyssimus, Dermatophagoides farinae* and *Tyrophagus putrescentiae. Shoyakugaku Zasshi* 43:163-168.

Wood, R.K.S. (1967). In W. Black (Ed.), *Physiological Plant Pathology.* Oxford: Scientific Publ., pp. 413-421.

Chapter 12

Antimycotic Activity of Essential Oils: The Possibility of Using New Bioactive Products Derived from Plants

K. P. Svoboda
S. K. Kyle
J. B. Hampson
G. Ruzickova
S. Brocklehurst

INTRODUCTION

At the beginning of the eighteenth century, there were no synthetic preparations used in any sphere of human activity. Today, in contrast, we enjoy the benefits of synthetic products, but we also suffer from their side effects as well as the pollution caused by their production and usage. There is a general growing trend among consumers for more natural rather than synthetic products in a whole range of industries, including food and drink, cosmetics, agricultural, and pharmaceuticals (Bauer et al., 1997; Svoboda and Deans, 1998; Glaser, 1999; Traffic International, 1999; Walton and Brown, 1999). Volatile oils may have an important role to play in the preservation of foodstuffs against fungi, in fungicidal applications against plant diseases, and in the fight against various human fungal infections. Recent literature has shown the biological activities of essential oils and their individual pure

The authors would like to acknowledge the following: Andrew Syred of Microscopix (Wales) for excellent microphotographs; Professor Y. Asakawa for GC-MS analysis; CABI Bioscience for fungi identification; Dr. Ruiradh Bain (SAC Auchincruive) for his work on potato storage; Dr. R. A. Evans (SAC Edinburgh) for his assistance on antifeedant activity; Lorna Pearce (SAC Auchincruive) for her work with *Arnica;* Murielle Letissier (SAC Auchincruive) for her work on *Hyssopus* and *Oregano* species; Dr. Neil McRoberts (SAC Auchincruive) for the regression model of *Fusarium* inhibition. SAC Auchincruive receives funding from the Scottish Office Agriculture and Fisheries Department.

components, and has documented the inhibitory activity of these substances against the growth of various fungi (Marotti et al., 1994; Pattnaik et al., 1996; Ezer and Abbasoglu, 1996; Thoppil et al., 1998; Baratta et al., 1998). A list of aromatic plant species used in these experiments is given in Table 12.1.

Volatile Oils and Their Constituent Compounds

Volatile oils are very complex mixtures of compounds. The constituents of the oils are mainly monoterpenes and sesquiterpenes with the general formula $(C_5H_8)_n$. Oxygenated compounds derived from these hydrocarbons include alcohols, aldehydes, esters, ethers, ketones, phenols, and oxides (Dewick, 1997). It is estimated that there are more than 1,000 monoterpene and 3,000 sesquiterpene structures. Other compounds include phenylpropenes and specific compounds containing sulphur or nitrogen. Hundreds of new natural substances are being isolated and identified every year, but data concerning their biological activities are known for only some (Ogura and Sankawa, 1997). In certain plants, one main constituent may predominate. In basil, for example, methyl chavicol can make up 90 percent of the volatile oil fraction (Grayer et al., 1996). In other species, there may be no single component which predominates, but a mixture of various components. In the oil of sweet marjoram, for example, the individual chemicals are represented by 0.1 to 10 percent of total oil volume (Lawrence, 1988-1991). The presence of trace components, often still unidentified, can also influence the odor, flavor, and biological activity of the oil to a significant degree.

Sources of Volatile Oils and Other Aromatic Compounds

Volatile oils are derived from herbs and spices, and have been known since antiquity to possess biological activities, namely antibacterial, antifungal, and antioxidant. In recent years, attempts have been made to isolate and identify the compound(s) responsible for such activities. In only a few cases have these compounds of interest been identified and fully characterized using analytical techniques, such as GLC, GC-MS, and NMR (Colgate and Molyneux, 1993). Lawrence (1993) lists the world's top twenty essential oils used in the flavor and fragrance industry, and points out that there are approximately 300 natural products used as raw materials in these industries. Synthetic or semisynthetic compounds are still preferred for many reasons, including property rights, economics, safety, patentability, and reproducibility of results.

TABLE 12.1. List of aromatic plants from temperate regions used in experiments to demonstrate bioactivity

Plant	Family
Abies sp.	Pinaceae
Achillea sp.	Compositae
Acorus calamus	Araceae
Allium cepa	Liliaceae
Anethum graveolens	Compositae
Angelica archangelica	Umbelliferae
Artemisia herba-alba	Compositae
Cedrus deodora	Coniferae
Chamaecyparis obtuse	Cupressaceae
Cryptomeria japonica	Pinaceae
Cupressus sempervirens	Cupressaceae
Daucus carota	Umbelliferae
Foeniculum vulgare	Umbelliferae
Helichrysum sp.	Compositae
Heracleum sp.	Umbelliferae
Houttuynia cordata	Saururaceae
Hyssopus sp.	Labiatae
Inula helenium	Compositae
Lamium garganicum	Labiatae
Lavandula sp.	Labiatae
Lippia multiflora	Verbenaceae
Melissa officinalis	Labiatae
Minthostachys verticillata	Labiatae
Mentha sp.	Labiatae
Monarda citriodora var. citriodora	Labiatae
Myrica gale	Myricaceae
Ocimum sp.	Labiatae
Origanum sp.	Labiatae
Pelargonium sp.	Geraniaceae
Petroselinum crispum	Umbelliferae
Pimpinella anisum	Umbelliferae

TABLE 12.1 *(continued)*

Plant	Family
Pinus sp.	Pinaceae
Rosmarinus sp.	Labiatae
Salvia sp.	Labiatae
Satureja sp.	Labiatae
Sideritis sp.	Labiatae
Tagetes erecta	Compositae
Thymus sp.	Labiatae
Verbena sp.	Verbenaceae

Source: Aromatic and Medicinal Plant Abstracts, CAB 1995-1999.

Mosses and liverworts (Bryophyta) represent another large group as a potential source of secondary metabolites. There are about 2,400 species of bryophytes in the world (Asakawa, 1998). A number of these have been used as medicinal plants in Europe, North America, China, and India to cure burns, bruises, external wounds, and inflammations. It has been proven by bioassays and experiments that they show biological activity, probably due to their sesqui- and diterpenoid components and other lipophilic aromatic compounds. Generally, they are not damaged by insects, snails, slugs, or other small animals. Some liverworts, however, can cause intense allergenic reactions by contact, including dermatitis and allelopathy.

BIOLOGICAL ACTIVITY OF VOLATILE OILS

Research is being done on the testing of a wide range of plants for antiviral, anticancer, antibacterial, and antifungal properties, and often to a very high standard. If the biological activity of the oils is comparable to that of synthetically produced pharmacological preparations, they should be investigated in a thorough and exhaustive manner (Colgate, 1993; Baratta et al., 1998; Svoboda, Hampson, et al., 1998; Svoboda, Inglis, et al., 1998; Svoboda et al., 2000). Generally, their action is the result of the combined effect of both their active and inactive compounds. These inactive components might influence resorption, rate of reactions, and bioavailability of the active compounds. Several active components may have a synergistic effect. To add to the complexity of volatile oils, there is evidence that the time of harvest also influences the oil composition, and consequently the potency of their biological activity (Deans and Svoboda, 1988; Lis-Balchin et al., 1992; Galambosi et al., 1993; Marotti et al., 1994). Other factors such as genotype, chemotype, geographical origin, and environmental and agronomic conditions, can all influence the composition of the final natural product (Collins et al., 1994; Svoboda et al., 1995; Galambosi et al., 1999). In addition, the enantiomeric composition of various monoterpenes in different species can further complicate the biological activity of a given oil (Ravid et al., 1987; Ravid, Putievsky, Katzir, Ikan, et al., 1992; Ravid, Putievsky, Katzir, Weinstein, et al., 1992; Ravid, Putievsky, and Katzir, 1994d, 1996). Examples of enantiomeres in essential oils are listed in Table 12.2.

ANTIMICROBIAL ACTIVITY OF VOLATILE OILS

Volatile oils of many plants are known to exhibit antimicrobial activity (Deans et al., 1992; Kandil et al., 1994; Prudent et al., 1995). This has evolved for the plants own protection as a chemical defense against plant pathogenic diseases. Pathogens can readily penetrate at wound sites caused, for example, by herbivores. Wounding of leaves which are covered with volatile oil glands results in the rupture of glands, which causes the oil to flow over the wound. The existence, therefore, of antimicrobial properties in the oil, would be of considerable benefit to plants. Indeed, a good majority of aromatic and medicinal plants do not succumb to many of the most common diseases, a high percentage of which are fungal. A complex oil presents a greater barrier to pathogen adaptation than a more simple mixture of monoterpenes. This theory is well-documented in a detailed study of *Myrica gale* volatile oil (see Photos 12.1, 12.2, and 12.3) and its inhibitory properties against a broad spectrum of fungal species (Carlton et al., 1992; Svoboda, Hampson, et al., 1998). Although the complicated mixtures of monoterpenes and sesquiterpenes present in the whole oil represented the most effective barrier to fungal infection, the antifungal activity of a newly discovered flavonoid glycoside kaempferol-3-(2,3-diacetoxy-4-p-coumaroyl) rhamnoside isolated from the leaves of *Myrica gale,* has been investigated for its inhibitory activity against five species of fungi which had been isolated from leaves of *Myrica gale* growing in the wild (Carlton et al., 1991). The inhibitory properties were tested in a liquid medium at concentrations of 50 and 100 μg/ml. This flavonoid glycoside had clear antifungal properties and significantly inhibited the growth of *Penicillium citrinum* Thom., *Epicoccum nigrum* Link, *Apiospora montagnei* Sacc., and *Fusarium sporotrichoides* Sherb.

TABLE 12.2. Enantiomeres in essential oils

Enantiomere
Alpha terpineol
(1R)-(+)-borneol
(1S)-(–)-borneol
Camphor
(S)-(+)-carvone
(R)-(–)-carvone
(–)-citronellol
(+)-citronellol
Citronellyl acetate
Fenchone
Iso-menthol
(1R,4R)-(+)-isomenthone
(1S,4S)-(–)-isomenthone
Linalol
R-(–)-linalyl acetate
Menthol
(1S,4R)-(+)-menthone
(1R, 4S)-(–)-menthone
Methyl acetate
(4S)-(+)-piperitone
(4R)-(–)-piperitone
(1R)-(+)-pulegone
(1S)-(–)-pulegone
Terpinen-4-ol
Verbenone

Source: Adapted from Ravid et al., 1987; Ravid, Putievsky, Katzier, Ikan, et al., 1992; Ravid, Putievsky, Katzir, Weinstein, et al., 1992; Ravid, Putievsky, and Katzir, 1994a,b,c,d, 1996.

The potential role of the volatile resin exudate from the stem bark of *Commiphora rostrata* (Burseraceae) from Ethiopia in plant defense was investigated by McDowell et al. (1988). These trees do not suffer from herbivore damage and pathogen attacks. The whole resin, as well as its three major components (2-decanone, 2-undecanone, and 2-dodecanone) at a dose level of 5,000 ppm, were screened for antifungal activity against *Aspergillus*

PHOTO 12.1. *Myrica gale*—Bog myrtle growing in the wild

PHOTO 12.2. *Myrica gale*—Bog myrtle showing harvested shoot with rooting of stem

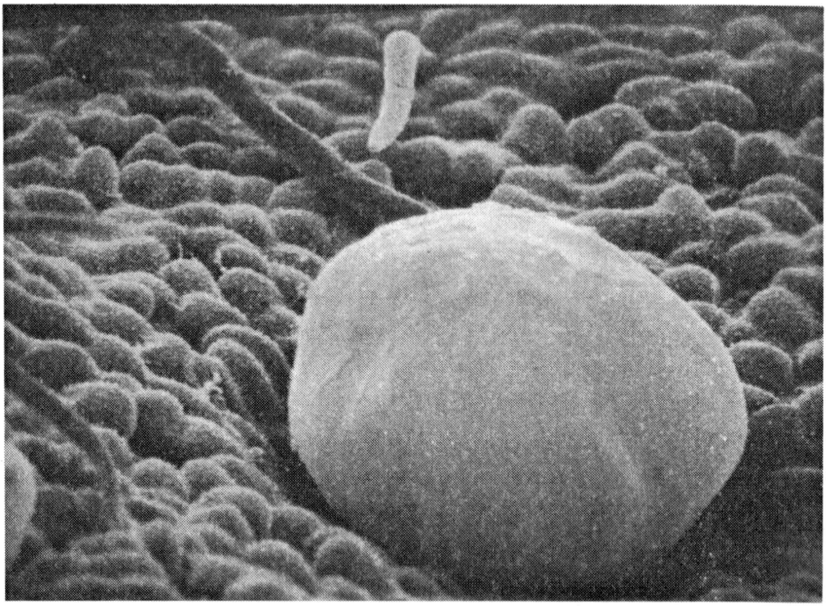

PHOTO 12.3. *Myrica gale*—Bog myrtle – secretory trichomes (× 607)

and *Penicillium* species. In addition to the inhibition of fungal growth, the resin also prevented the synthesis of mycotoxins by the *Aspergillus* species.

Deans et al. (1992, 1994) tested volatile oils of several temperate and tropical aromatic plants for their antimicrobial and antioxidant activity. Antifungal testing was carried out using filamentous fungi, wherein the inhibition of growth was monitored by the change in the dry weight of mycelium using the following formula:

$$\text{Percentage inhibition} = \frac{C - T}{C} \times 100$$

where T = mean weight of mycelium from test flasks; C = mean weight of mycelium from control flasks

Even at low concentrations (1 μl per 1 ml of growth media) there was marked antifungal activity recorded with several species. Volatile oil from *Heteromorpha trifoliata* (Wendl.) Eckl. & Zeyh. (Apiaceae) and *Heteropyxis natalensis* Harv. (Heteropyxidaceae) from Zimbabwe showed significant activity against mycotoxigenic filamentous fungi of the genus *Aspergillus* (Gundidza et al., 1993; Deans et al., 1994).

A list of major bacteria and fungi used in antimicrobial assays is presented in Table 12.3.

TABLE 12.3. List of bacteria and fungi used in antimicrobial assays

Bacteria	**Fungi**
Acinetobacter	Alternaria
Aeromonas	Apiospora
Alcaligenes	Aspergillus
Apiospora	Botrytis
Azotobacter	Candida
Bacillus	Cryptococcus
Beneckea	Epicoccum
Brevibacterium	Fusarium
Campylobacter	Helminthosporium
Citrobacter	Macrophoma
Clostridium	Microsporum
Enterobacter	Mucor
Escherichia	Ovularia
Erwinia	Penicillium
Flavobacterium	Ramularia
Klebsiella	Rhizopus
Lactobacillus	Saccharomyces
Leuconostoc	Septoria
Listeria	Sclerotium
Micrococcus	Sporotrichum
Moraxella	Trichoderma
Mycobacterium	Trichophyton
Proteus	
Pseudomonas	
Salmonella	
Serratia	
Staphylococcus	
Vibrio	
Yersinia	

Source: Aromatic and Medicinal Plant Abstracts CAB 1995-1999.

THE BRINE SHRIMP BIOASSAY—A SIMPLE METHOD FOR TESTING OF VOLATILE OIL TOXICITY

The use of volatile oils as fungicidal agents will require further testing in regard to their potential toxic effects on other organisms present in soil or on the plant surface, i.e., beneficial or harmless insects. The brine shrimp bioassay is a useful method in toxicology studies (Fatope et al., 1993; He et al., 1997).

The brine shrimp is a crustacean belonging to the subclass Branchiopoda, order Anostraca. It is found worldwide, in water ranging from the brackish to the ultrasaline. This high tolerance to a wide range of salinity (10-220 g/l) makes it a relatively easy animal to culture and study (Meyer et al., 1982). The eggs of the brine shrimp are about 0.2 mm in diameter (see Photo 12.4), and as long as they remain dehydrated, can be stored for long periods. When they are returned to the saline solution, the eggs absorb water and embryogenesis begins, and is completed between 16 and 36 hours. The embryo develops antennae and mandibiles, breaks away from the hatching membrane, and becomes an active, free-swimming nauplius (see Photo 12.5). The larva develops over about 15 stages and survives for 72 hours on its yolk resource alone. The availability of the eggs, the ease of hatching them into the larvae, the rapid growth of nauplii, and the relative ease of maintaining a population under laboratory conditions, make the brine shrimp bioassay an effective experimental tool. All stages in the life cycle of the brine shrimp have been used. The hatching rate of the eggs after exposure to petroleum oil, pesticides, carcinogens, and other environmental contaminants has been used as a criterion for their toxicity. The most common stage used, however, is that of 24 to 48 hours after hatching, although the older stages can also be used for screening. Identification of the lethal concentration for 50 percent mortality after six or eight hours of exposure (the acute LD_{50}) makes the test rapid and simple.

Test Methodology and Results

Five oils were used in preliminary tests (see Table 12.4). For each oil, four duplicate multiwell plates (30 wells per plate) were used to characterize the relationship between the mortality of the shrimps and the concentration of oil. In each plate, there were seven different concentrations of oil in each of four adjacent wells plus two controls where only water was added (see Photo 12.6). The test concentrations were 30, 60, 100, 150, 200, 240, and 480 ppm. These concentrations were chosen after estimates of the LD_{50} were obtained from a preliminary experiment using a wider range of concentrations—

PHOTO 12.4. *Artemia* sp.—Brine shrimp eggs (× 5)

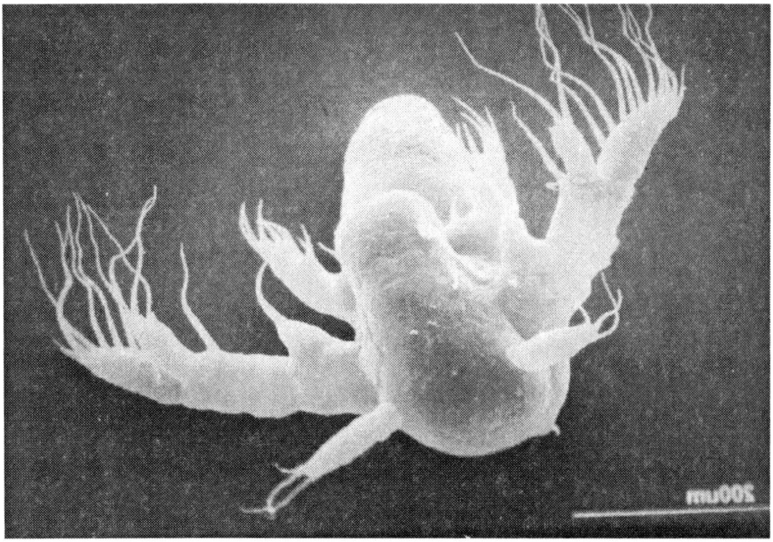

PHOTO 12.5. *Artemia* sp.—Brine shrimp nauplius

TABLE 12.4. Toxicity study of five volatile oils using the brine shrimp bioassay (seven concentrations)

Volatile oil	Estimates of LD_{50}	Model Parameters (Slope)	LD_{50} Transformed Scale (95% Confidence Interval)	LD_{50} Point Estimate	Original Scale (95% Confidence Interval)
German chamomile *Matricaria recutita*	2.445 ± .043	6.923 ± 1.12	2.358-2.532	279	228-340
Fennel *Foeniculum vulgare*	2.029 ± .043	8.715 ± 2.17	1.942-2.116	107	88-131
Thyme *Thymus vulgare*	2.093 ± .043	6.923 ± 1.12	2.006-2.180	124	101-151
Roman chamomile *Chamaemelum nobile*	2.284 ± 0.50	6.923 ± 1.12	2.184-2.384	192	153-242
Lemon balm *Melissa officinalis*	2.216 ± .043	6.923 ± 1.12	2.129-2.303	164	135-201

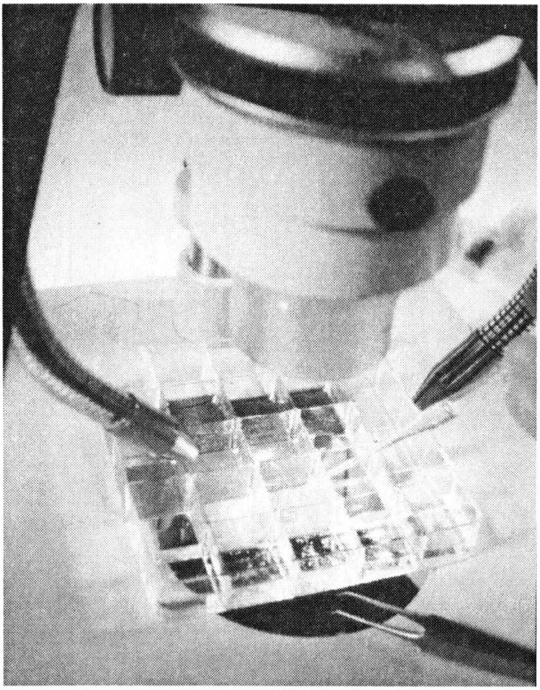

PHOTO 12.6. Brine shrimp bioassay using multiwell plate

33, 100, 500, and 1,000 ppm. The data for each well consisted of the total number of shrimps pipetted into the well, the number alive after six and eight hours, and, finally, the oil concentration. This gave 16 replicates of each oil concentration on which to assess the toxicity of the oil at varying dilutions measured against controls. Results are presented in Table 12.4.

To provide a wider choice of volatile oils to gauge suitability for further testing on fungicidal activity, an additional sixteen oils were then investigated. The same method as described previously was used, but with only three oil concentrations—30, 300, and 500 ppm. Most of the sixteen oils showed very few deaths at concentrations of 500 ppm. If any exhibited an LD_{50} value greater than 1,000 ppm, they were omitted from further statistical analysis. From these sixteen oils, only the volatile oils of dragonhead, Roman chamomile, lavender, basil, and mint were chosen for further analyses. Each plate was summarized by its slope and LD_{50} value. Further tests were carried out using individual components to establish a more accurate LD_{50}, especially for the oils which were not toxic at the given concentrations. However, the results indicated relatively low toxicity of these oils (see Table 12.5). No oil, apart from dragonhead, had an LD_{50} of 50 ppm, or indeed, of 1 ppm which is a recognized level for insect toxicity. There is a strong possibility, therefore, that these oils would not be harmful in the natural environment. In addition, it is necessary to consider the biodegradability of the oils in soil, as suggested and discussed by Noma and Asakawa (1996) and Nishimura and Yoshiaki (1996).

Although there are no references available regarding volatile oils, terpinen-4-ol, carvone, camphor, limonene, menthone, and citral have all been tested previously and showed relatively low toxicity if used between 500-1,800 ppm (Sam, 1993; Maganga et al., 1996; Southwell et al., 1997). The significance of the tests is mainly as an indicator of the possible antitumour activity of compounds (ideal LD_{50} being less than 40 ppm) or for use as an insecticide (ideal LD_{50} being around 1ppm) (Colgate and Molyneux, 1993).

Statistical Methods for the Analysis of Data

A generalized linear model was applied to relate the logic of the mortality to the logarithm (to base 10) of the concentration of oil. The plate to plate variation was incorporated into this model, and wells of identical concentration within a plate were aggregated prior to fitting the model. This resulted in estimates of the LD_{50} and the slope for each plate. The oil effect on these estimates was then investigated using an analysis of variance. The mean values and confidence intervals from the analyses of variance are shown in Tables 12.4 and 12.5. (Genstat 5 release 4.1, copyright 1997, Lawes Agricultural Trust, Rothamstead Experimental Station, was used for all statistical analyses.)

CASE STUDIES

The following are three case studies undertaken at SAC Auchincruive using both in vitro and in vivo experimental techniques.

Case Study 1: Phytochemical Investigation and Biological Activities of **Tanacetum vulgare** *(L.)*

Introduction

Tanacetum vulgare (L.) (tansy) is a herbaceous perennial native to Europe and Asia, belonging to the Compositae family (see Photo 12.7). Prepa-

TABLE 12.5. Toxicity study of five volatile oils using the brine shrimp bioassay (three concentrations)

	$Log_{10}LD_{50}$		LD_{50}		Log_eSlope		Slope	
Volatile Oil	**E ±SE**	**CI**	**E**	**CI**	**E ±SE**	**CI**	**E**	**CI**
Dragonhead *Dracocephalum modavica*	1.583±0.191	1.167-1.999	8	15-100	2.72±0.376	1.901-3.539	15.18	6.69-34.44
Roman chamomile *Chamaemelum nobile*	2.57±0.191	2.154-2.986	372	143-968	1.33±0.376	0.511-2.149	3.78	1.67-8.58
Lavender *Lavandula angustifolia*	2.68±0.191	2.264-3.096	479	184-1247	0.59±0.376	-0.229-1.409	1.80	0.80-4.09
Basil *Ocimum sanctum*	1.958±0.191	1.542-2.374	91	35-236	2.09±0.376	1.271-2.909	8.08	3.56-18.34
Peppermint *Mentha piperita*	2.299±0.191	1.883-2.715	199	76-518	1.53±0.376	0.711-2.349	4.62	2.04-10.48

E = Estimates
SE = Standard Errors
CI = 95 percent confidence intervals for LD_{50} and slopes

PHOTO 12.7. *Tanacetum vulgare*—Tansy growing at SAC Auchincruive

rations (powders, tinctures, infusions) from tansy leaves were used in folk medicines as anthelmintics, abortives, in the treatment of diarrhea, and as an insect repellent. Today, due to its high thujone content, the use of tansy oil is limited. In their finished form, all alcoholic beverages and other drinks and food preparations, must be thujone-free. Tansy oil is still added to fragrances. Several chemotypes of tansy have been described, including camphor, thujone, bornyl acetate, sabinene, germacrene and umbellulone types (Holopainen, Hiltunen, Lokki, et al., 1987; Holopainen, Hiltunen, and Von Schantz, 1987). The plant material used in these experiments was of a pure thujone chemotype (see Photo 12.8).

In this study, tansy plants were cultivated at Auchincruive in experimental plots, and regular observations of phenological phases and measurements for growth analysis (height, diameter at base, number of stems, date of budding, date of flowering, overwintering) were taken. Distillation and GC analysis of the oil was carried out as described previously (British Standard Method, 1985; Svoboda, Inglis, et al., 1998). Plants reached a height of 150 cm during the second year and 160 cm during the third year. The highest oil yield was obtained from the vegetative stage harvest (1.3%v/w), followed by flowering stage (1.2% v/w), and flower bud stage harvest (1.1% v/w), all using fresh plant material. During senescence, the oil content fell

PHOTO 12.8. *Tanacetum vulgare*—Tansy varieties growing at SAC Auchincruive

rapidly to 0.05 percent. Thujone was consistently high throughout the growth period representing between 87 and 93 percent of the total oil, with the highest percentage present during the seed stage.

In Vitro Testing

Fusarium coeruleum, Phoma exigua var. *foveata, Helminthosporium solani,* and *Polyscytalum pustulans,* commercially important potato pathogens, were used to test for any antifungal activity from the vapor given off by tansy oil.

Plugs of each fungi, two millimeters in diameter, were inverted onto plates of potato dextrose agar in petri dishes. Small plastic wells were placed onto the inside of the plate lids, into which either 80 microliters of tansy oil or distilled water (control) were dispensed. The base of the plate containing the fungal culture was then turned upside down onto the lid and the dish was sealed with plastic tape. The plates were randomized, then incubated at 2°C. The diameters of the colonies were measured every two days over a period of 14 days (see Table 12.6).

TABLE 12.6. Antifungal activity of volatile oil vapor of *Tanacetum vulgare* (tansy) on three fungal cultures in petri dishes

Day After Inoculation	*Fusarium solani* var. *coeruleum*		*Helminthosporium solani*		*Polyscytalum pustulans*	
	Tansy oil	Control	Tansy oil	Control	Tansy oil	Control
2	8	–	2	–	2	–
4	10	–	2	–	3	–
6	15	–	2	–	5	–
8	22	–	3	–	6	–
10	24	Trace	3	Trace	9	–
12	30	Trace	5	Trace	12	–
14	32	Trace	5	Trace	14	–

Note: Numbers are the average of ten numbers (diameter in mm).

In Vivo Testing

In vivo tests were also carried out on two species of potato (Cara and Pentland Hawk varieties). Ten tubers were each wounded four times and inoculated with *Phoma exigua* var. *foveata* and ten were wounded and inoculated with *Fusarium coeruleum.* Each set of ten was placed into separate, air-tight boxes with three replicates. Three control boxes were set up for each pathogen. Plastic wells were glued to the inside of two diagonal corners, into each of which was dispensed either 40 μl of oil (a total of 80 μl per box), or in the case of the controls, distilled water. The variety Cara was stored in a cold room at 5 to 6°C, and the variety Pentland Hawk was stored in an outside shed at approximately 8°C. Lesion diameters and depths were measured once a week over a period of ten weeks, and the means calculated. Disease development was generally greater for the control than the oil treatment, but differences were not statistically significant. A more active movement of air (ventilation) would promote better volatile oil distribution, and thus enhance its action.

In both the in vitro (agar plates) and in vivo tests (potato storage), the results were promising. It is important to find a noncontact (fumigant) method of preventing disease from spreading in potatoes. In the Netherlands, a program using volatile oil of caraway *(Carum carvi)* as a biological sprouting inhibitor for potatoes and as a biocide for insects and fungi, has been approved for large-scale implementation (Meijer, 1993).

Toxicity Against Fauna

If there is any likelihood of the volatile oils being applied in a field situation, it is also important to test the bioactivity of potentially active oils against fauna. For these preliminary tests, caterpillars were used *(Pieris brassicae).* Ten caterpillars were starved for 24 hours before being placed in a 9 cm diameter plastic petri dish containing a piece of cabbage leaf which had been dipped in a solution of tansy oil and distilled water. Tween 20 had also been added to act as a surfactant which allows the solution to adhere to the leaf surface. After two hours, the caterpillars were given untreated leaves. Their reactions from being exposed to tansy essential oil resulted in the following findings:

Percentage Oil Concentration	Results
10	All died within a few hours.
1	All died within a few hours.
0.1	All died within six hours.
0.01	One died within 24 hours; others alive but ate little and died before pupation.
0.001	One died within 24 hours; others alive but ate little and died before pupation.

It would appear that the oil at the higher concentrations had a fumigant or suffocating effect, which caused the caterpillars to squirm and wriggle and vomit green fluid. It was difficult to distinguish between the fumigant action and any feeding deterrent action. Tansy oil was a deterrent to the caterpillars down to a concentration of 0.001 percent, although this may have been due to the vapor having an effect on them. Further experiments should be carried out on this and other essential oils to see how low concentrations of the oil disrupt molting between caterpillar instars (of which there are five before pupation) as these molts are also controlled by insect hormones.

For comparison, two additional oils were tested [*Monarda citriodora* var. *citriodora* (L.) and *Hyssopus officinalis* (L.)]. The results were similar. At 0.1 percent all caterpillars died at the prepupal stage, but at 0.01 percent, pupated. Those caterpillars that did survive after eating the treated leaves exhibited deformities at pupation, and some apparently normal pupae failed to develop to adults. These results suggest, as did the brine shrimp experiments, that there is variability in the toxicity of different volatile oils at different concentrations. The toxicity of various natural extracts against fauna has not been sufficiently tested and there are not many references available (Suomi et al., 1986; Klingauf and Weil, 1988; Hough-Goldstein, 1989).

Case Study 2: **Fusarium** *Infection on* **Arnica montana** *(L.)*

Introduction

Arnica montana (L.) is a member of the Compositae family, one of the largest families of flowering plants whose distribution is almost worldwide. The genus *Arnica* consists of about 30 species of rhizomatous perennials which are found mainly in the subalpine zones of the Northern Hemisphere (Bown, 1995). *Arnica montana* (L.) is indigenous to central Europe but its range covers most of Europe and Western Asia. It is found in mountain pastures and open woodland on poorer, acid soils. *Arnica* is in decline in the wild, and may even have become extinct within some parts of its range. There have been several research projects undertaken to attempt to discover the reasons for this decline. Some of these include changes in their habitats caused by alterations in the management of grasslands, high fertilizer application, soil acidification, and alteration of the mineral balance. Overexploitation of the wild populations is also a problem in some areas. Field cultivation is needed to protect wild populations, and breeding programs are making seeds of high yielding genotypes available (Galambosi et al., 1998). In cultivation, *Arnica* suffers from various fungal diseases. These include *Entyloma arnicale* and *Sphaerotheca fuliginea,* which have occurred in trials in France but have not proved to be damaging or debilitating, and *Fusarium* wilt, the symptoms of which are black patches on the stem and leaves and stem rot.

Arnica has been a popular remedy for centuries. The flowers were used to make ointment or tinctures for external application to sprains, bruises, wounds, and chilblains, and internally to stimulate circulatory, nervous, and digestive systems. At present, *Arnica* is mainly used in homeopathic preparations, which are made from either the whole plant, the root, or the flowers. The remedies are used to minimize the immediate effects of bruising, shock, falls, and injuries caused by blunt objects (Arkopharma, 1999). The tincture is recommended to reduce the inflammation and pain of bruises, aches, and sprains (*British Herbal Pharmacopoeia,* 1979). It is also added to cosmetics such as bath preparations and shampoos, and it is listed as a natural food and beverage flavoring (Newall et al., 1996).

Isolation of Fungi

Arnica montana (L.) plants infected with an unknown vascular wilt fungus were studied. Segments of diseased plant tissues were cut, rinsed in three changes of sterile distilled water, blotted dry on filter paper, and then placed onto Oxoid potato dextrose agar (PDA) modified to contain 100 mg

ml^{-1} of both erythromycin and streptomycin (PDAES). Plates were incubated at 18 to 20°C. After five to seven days, hyphal tips of mycelia growing from the segments were transferred to single petri dishes of PDAES. Once the fungal hyphae were well established on the plates, samples were examined microscopically and identified as *Fusarium,* with its typical, dark-red coloring. Growth was very dense, and after replating, a pure culture was obtained. The fungus was identified by CABI Bioscience Identification Services, Egham, Surrey, as being *Fusarium avenaceum* (Fr.) Sacc. In addition to being a cereal pathogen, strains of this species may cause seedling disease of other plants. On mature plants, it may form symptomless infections, become invasive if host resistance declines, and become associated with root infections or die-back symptoms.

Experimental Design

In vitro bioassays to test the antifungal activity of five essential oils (Roman chamomile [*Chamamaelum nobile* (L.)], German chamomile [*Matricaria recutita* (L.)], sage [*Salvia officinalis* (L.)], swiss mint [*Mentha* spp.], and basil [*Ocimum basilicum* (L.)]) against *Fusarium* cultures were set up. Six flasks, each containing 100 ml of PDA, were autoclaved and cooled to 40 to 50°C. Five concentrations (2,000, 3,000, 4,000, 5,000, and 6,000 ppm) of each oil were then added to respective flasks, with one left as a control with nothing added. The contents of each flask were then distributed equally between four replicate plates. Seven-millimeter plugs taken from the fungal cultures were placed inverted on the centre of each plate. On days one through seven inclusive after the experiment was set up, three radial measurements of fungal growth were taken from each plate and the percentage inhibition relative to controls was calculated.

The growth responses of *Fusarium* sp. to different concentrations of plant essential oils was assessed by regression analysis (see Table 12.7). Linear regressions of fungal colony radius against $\log_{10}$ oil concentration, revealed a variety of responses. The regression parameters were obtained by fitting a model for the basil (A) data set and then testing whether significantly different parameter values are required for the other samples. Only the regression constants for the spearmint samples differed significantly from the value for the basil (A) data set. The regression coefficients for the chamomile and blue chamomile differed significantly from the estimate for the basil (A) data set. In each case, the estimated values were larger than that for basil (A) suggesting a less severe inhibition of colony growth in response to increasing oil concentration for these samples. In all the data sets, the highest oil concentration (1,000 ppm) induced an approximately tenfold decrease in colony growth.

TABLE 12.7. Linear regression models of inhibition of fungal colony growth in vitro by plant essential oils

Oil[a]	Regression Constant (a)[b] (mm)	*p* Value for Inclusion in Model	Regression Coefficient (b)	*p* Value for Inclusion in Model
Basil (A)	38.0	<0.001	–11.87	<0.001
Basil (B)	38.7	0.800	–11.99	0.930
Blue chamomile (A)	39.4	0.630	–8.34	0.010
Blue chamomile (B)	43.1	0.080	–8.07	0.001
Chamomile (A)	39.1	0.700	–9.24	0.054
Chamomile (B)	39.0	0.720	–7.89	0.004
Sage (A)	36.6	0.630	–9.82	0.133
Sage (B)	36.4	0.582	–9.34	0.260
Spearmint (A)	28.5	0.001	–9.33	0.259
Spearmint (B)	31.3	0.022	–11.21	0.627

[a]Letters in brackets refer to replicate experiments.
[b]Fitted regression models are of the form; colony radius (mm) = a + b*log (oil conc. + 0.5).

Case Study 3: Effect of Hyssop and Oregano Essential Oils on Fungal Growth and Fungal Infections

Introduction

The volatile oils of hyssop [*Hyssopus officinalis* (L.)] and oregano [*Origanum vulgare* (L.)] are used extensively by the flavor and fragrance industries (see Photo 12.9). These oils also have a broad range of medicinal applications. In vitro and in vivo experiments assessing the antimycotic activity of the oils and the synergistic effects of individual components were carried out in tests against six economically important fungal pathogens: *Erysiphe graminis* (barley powdery mildew); *Podosphaera leucotricha* Ell. Ev. Salm (apple powdery mildew); *Uromyces viciae-fabae* (broad bean rust); *Pyrenophora avenae* (rice blast); *Pyricularia oryzae* (oat leaf stripe and seedling blotch); and *Botrytis fabae* (broad bean chocolate spot). In vitro experiments included assays on the inhibition of mycelial growth and spore germination. In vivo experiments assessed the effects of whole oils and single components on the fungi when applied before and/or after inoculation of the plants.

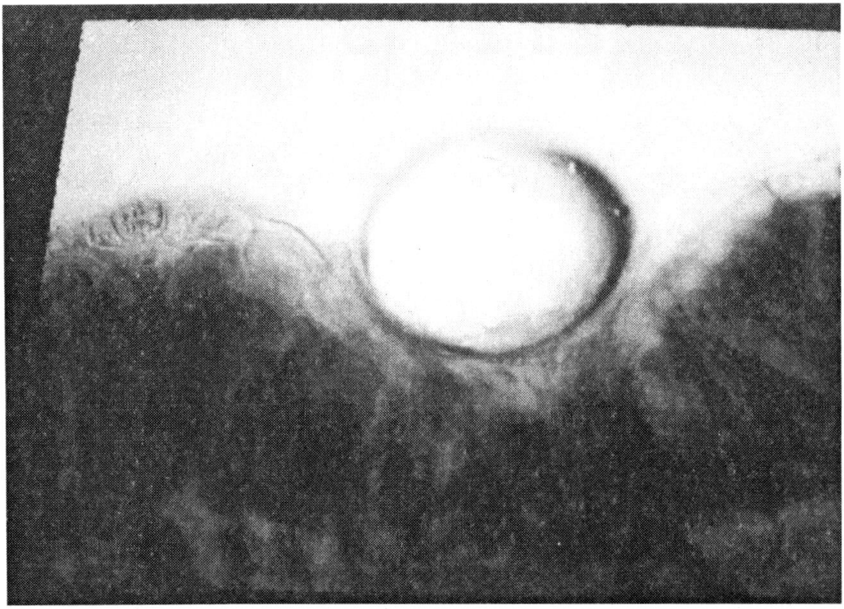

PHOTO 12.9. *Origanum heracleoticum*—Greek oregano sessile secretory gland (× 400)

Both hyssop and oregano plants were harvested from the herb garden at SAC Auchincruive, and dried at 35°C. Volatile oils were extracted by hydrodistillation using the British Pharmacopoeia distillation apparatus and procedure (British Standard Method, 1985) and analyzed by GC as described by Svoboda, Inglis, et al. (1998). The main components of hyssop were β-pinene (4.0-8.9 percent), iso-pinocamphene (47.2-57.7 percent), pinocamphene (11.1-22.1 percent), bornyl acetate (3.2-4.3 percent), and borneol (4.2 percent). The main components of oregano oil were carvacrol (72 percent), γ-terpinene (6.2 percent), and p-cymene (2.9 percent). The numbers in brackets represent percentage of the total oil.

In Vitro Experiments

Inhibition of mycelial growth. The oils or individual components of the oils were added to flasks of PDA at varying concentrations of between 0.01 to 0.4 percent for hyssop oil and 0.005 to 0.02 percent for oregano after they had been autoclaved and cooled to 40 to 50°C. The media were then dispensed into petri dishes to a depth of 3 mm. Plugs of mycelia from five-day-old cultures of *P. avenae* and *P. oryzae* were taken and placed inverted onto the center of each petri dish. Hyssop oil (0.01 percent) had no effect on *P. avenae* and *P. oryza.* At 0.05 percent and 0.1 percent, however, mycelial growth was significantly reduced, and at 0.4 percent it completely inhibited growth. Oregano oil had a more drastic effect. At 0.005 percent, growth was substantially reduced, and at 0.01 percent and 0.02 percent mycelial growth was completely inhibited (see Photo 12.10).

β-pinene, l-bornyl acetate, and iso-pinocampheol, components of hyssop oil, when tested individually at low concentrations, had either no effect or actually appeared to increase fungal growth slightly. Bornyl acetate at 0.002 and 0.004 percent reduced the growth significantly (see Photo 12.11). Combinations of these three individual components however, even at low concentrations, caused significant reductions in growth of both fungi. Of the individual components of oregano, γ-terpinene had no significant effect on growth, while *p*-cymene, thymol, and carvacrol caused substantial inhibition of mycelial growth of *P. avenae,* but had no significant effect on *P. oryzae.* The combination of all four components completely inhibited fungal growth of both *P. avenae* and *P. oryzae.*

Assays were also carried out using three-compartment petri dishes wherein one compartment contained PDA and *P. avenae,* a second one contained PDA and *P. oryzae,* and the third, PDA and varying concentrations of each oil. As there was no contact between the fungi and the oil, effects found could be attributed solely to the volatile components.

The volatile components of 0.4 percent hyssop oil stopped mycelial growth of both fungi after four days, and 0.1 percent reduced growth of *P. avenae* by 50 percent, but had less effect on *P. oryzae.* With the oregano oil, 0.005 percent and 0.01 percent substantially inhibited mycelial growth of both fungi, and after four days, 0.04 percent stopped growth completely.

Inhibition of spore germination. A stock plant infected with *U. v-fabae* provided a supply of spores, and the method described by Leach and Moore (1966) was used to induce in vitro sporulation of *B. fabae.* A modified De Cals method was used to assess the effects of varying oil concentrations on spore germination. Lactophenol cotton blue stain was applied after 20 hours and the percentage spore germination calculated (Bell and Daly, 1962). At concentrations of 0.1 percent and 0.4 percent, hyssop oil had a small but significant effect on *B. fabae* and caused a substantial reduction in spore germination of *U. v-fabae.* Oregano at a concentration of 0.04 percent caused a slight increase in spore germination in *U. v-fabae,* but 0.2 percent and 0.04 percent oregano oil significantly reduced spore germination of *B. fabae.*

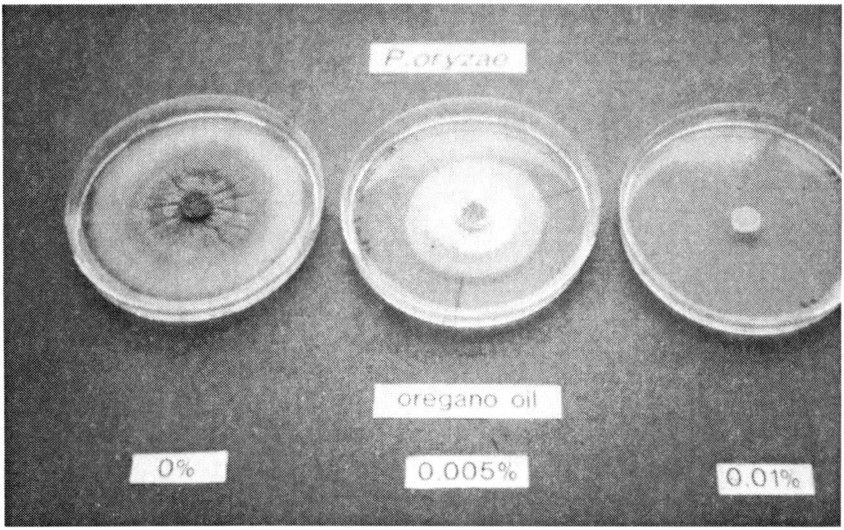

PHOTO 12.10. *Pyricularia oryzae*—Petri dishes showing zones of inhibition using two dilutions of oregano oil

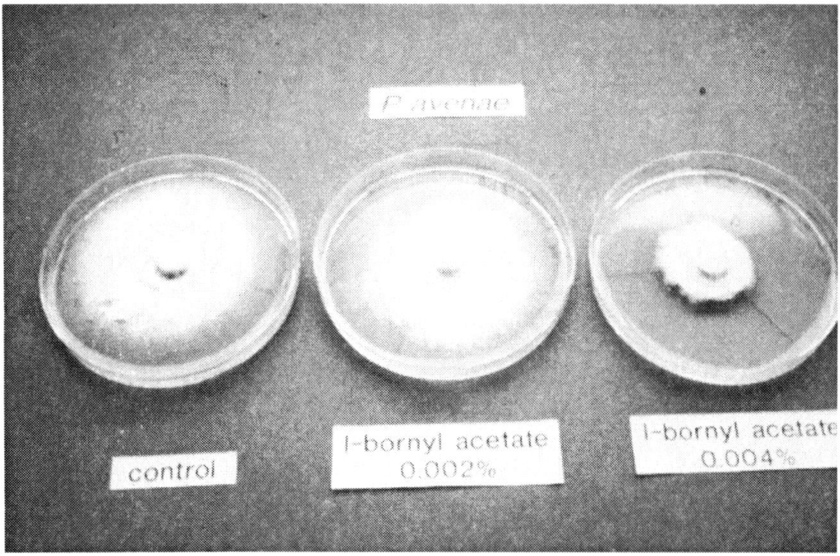

PHOTO 12.11. *Pyrenophora avenae*—Petri dishes showing zones of inhibition using two dilutions of bornyl acetate

In Vivo Experiments

Initial experiments were carried out using oil concentrations ranging from 0.1 percent to 3.2 percent. When toxicity was detected, subsequent experiments were done using oil concentrations below this level to determine the lowest effective concentration.

In the experiments with barley and broad bean plants, two separate trials were carried out, one with preinoculation spraying with oil and one with postinoculation spraying (see Photos 12.12 and 12.13). Experiments with apple seedlings involved one trial only, with both pre- and postinoculation application of oils.

Disease assessment for barley powdery mildew involved recording the percentage area of the second leaf affected by powdery mildew. This was done using a visual key on the sixth day after inoculation. A summary of results significantly different from the controls is given in Table 12.8.

Broad bean rust infection was expressed as number of rust pustules per cm^2 of the second pair of leaves. Results significantly different from controls are listed in Table 12.9.

Apple seeds were initially stratified at –10°C for 14 weeks to induce germination. Apple, barley, and broad bean seedlings were grown in a controlled environment for four weeks and then inoculated with spores and conidia. Apple powdery mildew infection was assessed, using a visual key, and infection was recorded on the third leaf 13, 15, and 17 days after inoculation. Results, listed in Table 12.10, showed significant differences between treated plants and controls.

CONCLUSIONS ON FUNGICIDAL BIOACTIVITY OF ESSENTIAL OILS

On the whole, in vitro assays showed both oils to be extremely effective at inhibiting fungal growth. Oregano oil, however, appeared to be considerably more potent than hyssop. The in vivo experiments showed variable results. With both oils, the greatest inhibitory effects could be attributed to their main components, a finding also reported by others (Zaika, 1988; Lis-Balchin and Hart, 1998a,b). However, a more complex oil is more likely to overcome a broader range of pathogens and pathogen adaptation. This may be due to individual components having complementary modes of action. Kurita et al. (1981), found the antifungal activity of certain phenolic compounds to be dependent upon the size of an added alkyl group, and others have noted a ranking order in the activity of aromatic chemical vapors.

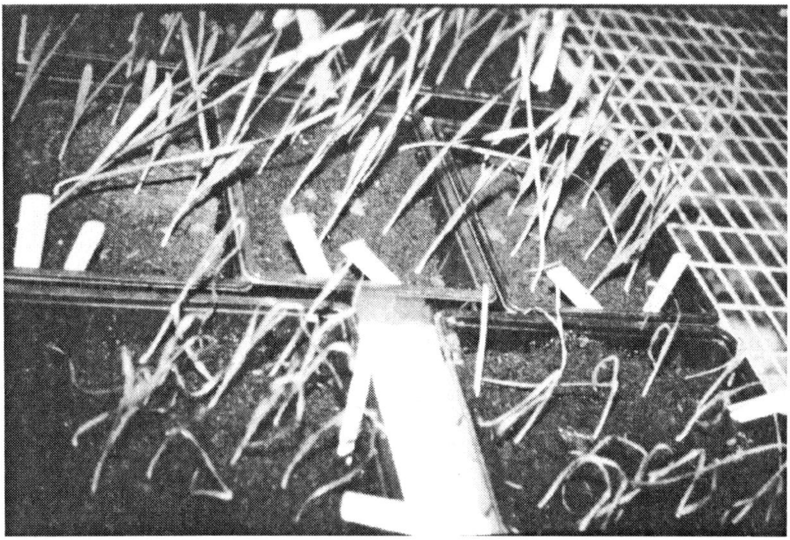

PHOTO 12.12. Barley plants grown in glasshouse showing effects of spraying with oils to induce inhibition of mycelial growth

PHOTO 12.13. Bean plants grown in glasshouse showing effects of spraying with oils to induce inhibition of mycelial growth

TABLE 12.8. Effects of hyssop and oregano oil on barley powdery mildew

Application	Oil Concentration (%)	Observed Effect	Phytotoxicity (% Oil)
Effects of hyssop oil on barley powdery mildew			
			0.2 on second day
Preinoculation			
1 day	0.1 and 0.01	++	
2 days	0.1	+	
2 days	0.01	– –	– – – –
3 days	All except 0.01	– –	– – – –
Postinoculation			
1 day	All concentrations	– –	– – – –
3 days	All concentrations	–	– – – –
Effects of oregano oil on barley powdery mildew			
			0.1 on second day
Preinoculation			
1 day	0.01	– –	– – – –
1 day	All except 0.01	++	– – – –
2 days	0.005	++	– – – –
2 days	All except 0.005	– –	– – – –
3 days	All concentrations	++	– – – –
Postinoculation			
1 day	0.005 and 0.01	++	– – – –
2 days	0.1	– – –	– – – –
2 days	0.01	++	– – – –
3 days	0.05 and 0.1	– –	– – – –

Note: Effects of individual components of hyssop on barley powdery mildew: Only β-pinene (0.008 percent) was found to significantly reduce infection, although others caused a slight reduction; iso-pinocampheol and pinocampheol (0.05 percent) reduced infection, and it was noted that the addition of ether, especially postinoculation, caused a reduction in infection. Effects of individual components of oregano on barley powdery mildew: Carvacrol at low (0.36 percent) and endogenous(0.72 percent) concentrations increased infection, and low concentrations of thymol (0.36 percent) decreased infection, but neither significantly.
+ = increased infection
– = decreased infection

TABLE 12.9. Effect of hyssop and oregano oil on broad bean rust

Application	**Oil Concentration (%)**	**Observed Effect**	**Phytotoxicity (% Oil)**
Effects of hyssop oil on broad bean rust			
			None till fifteenth day
Preinoculation			
1 day	0.05	– –	– – – –
1 day	0.2	+++	– – – –
2 days	All concentrations	– –	– – – –
3 days	0.05 and 0.2	– –	– – – –
3 days	0.1	++	– – – –
Postinoculation			
1 day	All concentrations	– –	– – – –
2 days	All concentrations	– –	– – – –
Effects of oregano oil on broad bean rust			
			0.2 on second day
Preinoculation			
1 day	All concentrations	++	– – – –
2 days	All concentrations	++	– – – –
3 days	0.01 and 0.1	– –	– – – –
3 days	0.05	++	– – – –
Postinoculation			
1 day	0.1	– –	– – – –
1 day	All except 0.1	++	– – – –
2 days	0.01 and 0.05	++	– – – –
2 days	0.005 and 0.1	– –	– – – –
3 days	All except 0.1	++	– – – –

\+ = Increased infection
– = decreased infection

More recent work has been carried out by Lis-Balchin and Hart (1999), on the mode of action of *Lavandula angustifolia* P. Millar, *Leptospermum scoparium* J.R. and G. Forst, *Kunzea ericoides* (A. Rich) J. Thompson, and a selection of oils from *Pelargonium* species. But, on the whole, little is

TABLE 12.10. Effect of hyssop and oregano oil on apple powdery mildew

Application	Oil concentration (%)	Observed Effect	Phytotoxicity (% Oil)
Effects of hyssop oil on apple powdery mildew			
			0.3
Overall:	2 day preinoculation treatments were less infected than controls		
	2 day postinoculation treatments were more infected than controls		
Effects of oregano oil on apple powdery mildew			
			0.2
Overall:	2 day preinoculation treatments increased infection		
	2 day postinoculation treatments decreased infection		

known of the mechanisms underlying the biological activity of essential oils and their individual components. Consequently, it is envisaged that molecular plant pathology will help in the elucidation of the mode of action of essential oils and also in the gene expression in fungal plant pathogens.

THE FUTURE OF MOLECULAR PLANT PATHOLOGY— GENE EXPRESSION

Molecular plant pathologists at SAC Auchincruive are presently using differential display and reverse transcription polymerase chain reaction (DDRT-PCR) assays (Liang and Pardee, 1992; Bauer et al., 1993) to elucidate the differential gene expression induced in fungal plant pathogens exposed to essential oils and their components. Initial experiments are being carried out using oils from *Ocimum* sp. and two plant-pathogen interactions; *Ocimum basilicum–Botrytis cinerea* and *Brassica napus–Pyrenopeziza brassicae*. *Botrytis cinerea* (grey mold) is both a pre- and postharvest problem in basil crops, and *P. brassicae* (light leaf spot) is one of the most economically important fungal pathogens of oilseed rape. Both fungal infections are encouraged by leaf wetness and so are particularly problematic in the United Kingdom. *Ocimum* oils were chosen because extensive research into the species and varieties of the genus had already been carried out at Auchincruive (Svoboda et al., 1999; Kyle, 1999, 2000) (see Photo 12.14). They have a characteristically high variation in oil constituents, thus giving a wide scope for the discovery of useful bioactive components. The technique involves culturing fungi on its own and also in the presence of inhibiting but

PHOTO 12.14. Field scale cultivation of basil *(Ocimum basilicum)* in the United Kingdom

nonlethal concentrations of the essential oil or its individual component(s). Ribonucleic acid (RNA) is then isolated from the fungi and an RT-PCR is performed in which the messenger RNA (mRNA) species are transcribed into complimentary DNA (cDNA), which is then amplified using DD-PCR. The products of this are screened for differentially displayed genes, i.e., gene expression induced by exposure to the essential oil components using polyacrilimide gel electrophoresis (PAGE) and silver staining. The bands containing cDNA of putative differentially expressed genes are cut from the gel, reamplified, and sequenced. Different techniques can be used to extend a cDNA fragment to obtain the full length cDNA sequence, which can then be converted to an amino acid sequence to enable the encoded protein and its biological function to be characterized. Similar work has been successfully carried out by Benito et al. (1996), Benito and van Kan (1998), Benito et al. (1998), and Melin et al. (1999), albeit not using volatile oils. The hope is, therefore, that this present work will uncover the molecular mechanisms of action of the oil and its components, and thus be a step toward allowing the full potential of essential oils as bioactive compounds to be realized. This could be a very important development in the area of novel fungicides for use in crop protection.

REFERENCES

Arkopharma (1999). CO_2 extract:Marigold flowers. France: Arkopharma Laboratoires Pharmaceutiques.

Asakawa, Y. (1998). Biologically active compounds from Bryophytes. *J. Hattori Bot. Lab.* No. 84: 91-114.

Baratta, M.T., Dorman, H.J.M., Deans, S.G., Figueiredo, A.C., Barroso, J.G., and Ruberyo, G. (1998). Antimicrobial and antioxidant properties of some commercial essential oils. *Flavour Frag. J.* 13: 235-244.

Bauer, D., Muller, H., Reich, J., Riedel, H., Ahrenkiel, V., Warthoe, P., and Strauss, M. (1993). Identification of differently expressed mRNA species by an improved display technique (DDRT-PCR). *Nucleic Acids Res.* 21: No.18.

Bauer, K., Garbe, D., and Surburg, H. (1997). Preparation, properties and uses. Third completely revised edition. *Common Fragrance and Flavor Materials* (p. 269). Weinheim, Wiley-VCH.

Bell, A.A. and Daly, J.M. (1962). Assay and partial purification of self-inhibitors of germination from uredospores of the bean rust fungus. *Phytopathology*. 52: 262-266.

Benito, E.P., Prins, T. and van Kan, J.A.L. (1996). Application of a differential display RT-PCR to the analysis of gene expression in a plant-fungus interaction. *Pl. Mole. Biol.* 32: 947-957.

Benito, E.P., ten Have, A., van't Klooster, J.W., and van Kan, J.A.L. (1998). Fungal and plant gene expression during synchronized infection of tomato leaves by *Botrytis cinerea*. *Eur. J. Pl. Path.* 104: 207-220.

Benito, E.P. and van Kan, J.A.L. (1998). DDRT-PCR for screening of fungal gene expression. In P.D. Bridge, D.K. Arora, C.A. Reddy, and R.P. Elander (Eds.), *Applications of PCR in mycology,* CAB International, pp. 47-62.

Bown, D. (1995). *Royal Horticultural Society encyclopedia of herbs and their uses.* London: Dorling Kindersley.

British Herbal Pharmacopoeia (1979). Part 1. West Yorkshire, England: British Herbal Medicine Association.

BSM (1985.) British standard method for testing of spices and condiments. Determination of volatile oil content. BS 4585 Part 15. ISO 6571-1984. London: British Standards Institute.

Carlton, R., Deans, S.G., Gray, A.I., and Waterman, P.G. (1991). Antifungal activity of a flavonol glycoside from the leaves of bog myrtle *(Myrica gale). Chemoecology* 2: 69-71.

Carlton, R.R., Waterman, P.G., Gray, A.I., and Deans, S.G. (1992). The antifungal activity of the leaf gland volatile oil of sweet gale *(Myrica gale)* (Myricaceae). *Chemoecology* 3: 55-59.

Colgate, S.M. (1993). Detection, isolation and structural determination. In *Bioactive natural products.* London: CRC Press, pp. 50-51.

Colgate, S.M. and Molyneux, R.J. (1993). Detection, isolation and structural determination. In *Bioactive natural products.* London: CRC Press, pp. 442-454.

Collins, J.E., Lepp, N.W., Svoboda, K.P., and Deans, S.G. (1994). Changes in the volatile oil from *Monarda citriodora* var. *citriodora* due to ontogeny. *Chromat. Anal.* 34: 5-7.

Deans, S.G., Kennedy, A.I., Gundidza, M.G., Mavi, S., Waterman, P.G., and Gray, A.I. (1994). Antimicrobial activities of the volatile oil of *Heteromorpha trifoliata* (Wendl.) Eckl. and Zeyh. (Apiaceae). *Flavour Frag. J.* 9: 245-248.

Deans, S.G. and Svoboda, K.P. (1988). Antibacterial activity of French tarragon (*Artemisia dracunculus* Linn.) essential oil and its constituents during ontogeny. *J. Hort. Sci.* 63: 503-508.

Deans, S.G., Svoboda, K.P., Gundidza, M., and Brechany, E.Y. (1992). Essential oil profiles of several temperate and tropical aromatic plants: Their antimicrobial and antioxidant activities. *Acta Hort. 306,* 229-232.

Dewick, P.M. (1997). *Medicinal natural products*. Chichester, UK: J. Wiley and Sons.

Ezer, N. and Abbasoglu, U. (1996). Antimicrobial activity of essential oils of some *Sideritis* species growing in Turkey. *Fitoterapia* 67: 474-475.

Fatope, M.O., Ibrahim, H., and Takeda, Y. (1993). Screening of higher plants reputed as pesticides using the brine shrimp lethality assay. *Int. J. Pharmacog.* 31: 250-254.

Galambosi, B., Svoboda, K.P., Deans, S.G., and Hethelyi, E. (1993). Agronomical and phytochemical investigation of *Hyssopus officinalis. Agric. Sci. Finl.* 2: 293-302.

Galambosi, B., Svoboda, K.P., Hampson, J.B., and Asakawa, Y. (1999). Agronomical and phytochemical investigation of *Picnanthemum* spp. in Finland. *Z. Arzn. Gew. Pfl.* 4: 19-23.

Galambosi, B., Sz-Galambosi, Z., Svoboda, K.P., and Deans, S.G. (1998). Flower yield and antioxidant properties of *Arnica montana* L. grown in Finland. *Drogenreport.* Jg. 11, 19: 10-13.

Glaser, V. (1999). Billion-dollar market blossoms as botanicals take root. *Nat. Biotechnol.* 17: 17.

Grayer, R.J., Kite, G.C., Goldstone, F.J., Bryan, S.E., Paton, A., and Putievsky, E. (1996). Infra specific taxonomy and essential oil chemotypes in sweet basil *O. basilicum. Phytochemistry* 43 (5): 1033-1039.

Gundidza, M.G., Deans, S.G., Kennedy, A.I., Mavi, S., Waterman, P.G., and Gray, A.I. (1993). The essential oil from *Heteropyxis natalensi*: Its antimicrobial activities and phytoconstituents. *J. Sci. Food Agric.* 63: 361-364.

He, K., Zeng, L., Shi, Guoen, Zhao, G-X., Kozlowski, J.F., and McLaughlin, J.L. (1997). Bioactive compounds from *Taiwania cryptomerioides. J. Nat. Prod.* 60: 38-40.

Holopainen, M., Hiltunen, R., Lokki, J., Forsen, K., and Von Schantz, M. (1987). Model for the genetic control of thujone, sabinene and umbellulone in tansy. *Hereditas* 106: 205-208.

Holopainen, M., Hiltunen, R., and Von Schantz, M. (1987). A study on tansy chemotypes. *Planta Med.* 53(3): 284-287.

Hough-Goldstein, J.A. (1989). Antifeedant effects of common herbs on the Colorado beetle *(Coleoptera: Chrysomelidae). Environ. Entomol.* 19(2): 234-238.

Kandil, O., Radwan, N.M., Hassan, A.B., Amer, A.M.M., El-Banna, H.A., and Amer, W.M.M. (1994). Extracts and fractions of *Thymus capitatus* ehxibit antimicrobial activities. *J. Ethnopharmacol.* 44: 19-24.

Klingauf, F. and Weil, B. (1988). Antifeedant activity of plant extracts against the Colorado potato beetle. *Gesunde-Pflanzen.* 40(6): 219-224.

Kurita, N., Miyaji, M., Kurane, R., and Takahara, Y. (1981). Antifungal activity of components of essential oils. *Agr. Biol. Chem. Tokyo.* 45(4): 945-952.

Kyle, S.K. (1999). Investigation of volatile oil composition and DNA fingerprinting of *Ocimum basilicum.* SAC and University of Glasgow Honours Thesis.

Kyle, S. K. (2000). *The effects of essential oils on fungal gene expression.* From: MBFP2000, University of East Anglia, July 26-28th, 2000.

Lawrence, B.M. (1988-1991). Progress in essential oils. *Perfumer and flavorist.* Wheaton, IL: Allured Publishing Corp, 226 pp.

Lawrence, B.M. (1993) In J. Janik and J.E. Simon, (Eds.), *New Crops,* John Wiley and Sons Inc., pp. 620-627.

Leach, R. and Moore, K.G. (1966). The effect of 6-benzylaminopurine (benzyladenine) on senescence and chocolate spot *(Botrytis fabae)* of winter beans *(Vicia faba). Ann. Appl. Biol.* 61: 65-76.

Liang, P. and Pardee, A.B. (1992) Differential display of eukaryatic messenger RNA by means of the polymerase chain reaction. *Science,* 257: 967–971.

Liang, P. and Pardee, A.B. (1997). Differential display methods and protocols. *Methods in Molecular Biology,* 85: 306.

Lis-Balchin, M. and Hart, S. (1998a). Studies on the mode of action of the essential oils of Manuka *(Leptospermum scoparium)* and Kanuka *(Kunzea reicoides),* Myrtaceae, on guinea pig smooth muscle. *J. Pharm. Pharmacol.* 50: 809-811.

Lis-Balchin, M. and Hart, S. (1998b). Studies on the mode of action of essential oils of scented-leaf *Pelargonium* (Geraniaceae). *Phytother. Res.* 12(3): 215-217.

Lis-Balchin, M. and Hart, S. (1999). Studies on the mode of action of the essential oil of lavender (*Lavandula angustifolia* P. Miller). *Phytother. Res.* 13(6): 540-542.

Lis-Balchin, M., Simmonds, M., Hart, S., and Deans, S.G. (1992). Agrochemical and medicinal usage of essential oil of representative species of the Geranaceae. *Proceedings of the 12th International Congress Flavor, Fragrance and Essential Oils,* Vienna, Austria.

Maganga, M.E., Gries, G., and Gries, R. (1996). Repellency of various oils and pine oil constitutents to house flies *(Diptera:Muscidae). Environ. Entomol.* 25(5): 1182-1187.

Marotti, M., Piccaglia, R., Giovanelli, E., Deans, S.G., and Eaglesham, E. (1994). Effects of variety and ontogenic stage on the essential oil composition and biological activity of fennel (*Foeniculum vulgare* Mill.). *J. Ess. Oil Res.* 6: 57-62.

McDowell, P.G., Lwande, W., Deans, S.G., and Waterman, P.G. (1988). Volatile resin exudate from stem bark of *Commiphora rostrata:* Potential role in plant defence. *Phytochemistry* 27(8): 2519-2521.

Meijer, W.J.M. (1993). Relevance of specialist crops for arable farming and industry in the Netherlands. 3-12. In *The production and impact of specialist minor crops in the rural community: Proceedings of an EC workshop,* Brussels, April 27-28.

Melin P., Schnurer, J., and Wagner, G.H. (1999). Changes in *Aspergillus nidulans* gene expression induced by bafilomycin, a Streptomyces-produced antibiotic. *Microbiology*. 145: 1115-1122.

Meyer, B.N., Ferrigni, J.E., Putnam, L.B., Jacobsen, D.E., Nichols, and McLaughlin, J.L. (1982). Brine shrimp: A convenient general bioassay for active plant constituents. *Planta Med.* 45: 31-34.

Newall, C.A., Anderson, L.A., and Phillipson, J.D. (1996). *Herbal medicines, a guide for health care professionals.* London: The Pharmaceutical Press.

Nishimura, H. and Yoshiaki, N. (1996). Microbial transformation of monoterpenes: Flavor and biological activity. In G.R. Takeoka, R. Teranishi, P.J. Williams, and A. Kobayashi (Eds.), *Biotechnology for improved foods and flavours.* American Chemical Society.

Noma, Y. and Asakawa, Y. (1996). Biotransformation of terpenoids and related compounds by micro-organisms—Production of biologically active substances. 27th International Symposium on Essential Oils. September 8-11. Vienna, Austria.

Ogura, K. and Sankawa, U. (1997). Molecular biological approaches. In K. Ogura and U. Sankawa, (Eds.), *Dynamic Aspects of Natural Products Chemistry.* Tokyo: Kodansha Publishers; and Amsterdam: Harwood Academic Publishers, 295 pp.

Pattnaik, S., Subramanyam, V.R., and Kole, C. (1996). Antibacterial and antifungal activity of ten essential oils *in vitro. Microbios.* 86: 237-246.

Prudent, D., Perineau, F., Bessiere, J.M., Michel, G.M., and Baccou, J.C. (1995). Analysis of the essential oil of wild oregano from Martinique (*Coleus aromaticus* Benth.)—Evaluation of its bacteriostatic and fungistatic properties. *J. Ess. Oil Res.* 7: 165-173.

Ravid, U., Bassat, M., Putievsky, E., Weinstein, V., and Ikan, R. (1987). Isolation and determination of optically pure carvone enantiomeres from caraway (*Carum carvi* L.), dill (*Anethum graveolens* L.), spearmint (*Mentha spicata* L.) and mint (*Mentha longifolia* L. Huds.). *Flavour Fragr. J.* 2: 95-97.

Ravid, U., Putievsky, E., and Katzir, I. (1994a). Chiral GC analysis of enantiomerically pure (R)(–)-linalyl acetate in some Lamiaceae, myrtle and petitgrain essential oils. *Flavour Fragr. J.* 9: 275-276.

Ravid, U., Putievsky, E., and Katzir, I.. (1994b). Chiral GC analysis of menthone and isomenthone with high enantiomeric purities in laboratory-made and commercial essential oils. *Flavour Fragr. J.* 9: 139-142.

Ravid, U., Putievsky, E., and Katzir, I. (1994c). Chiral GC analysis of (1R)(+)-pulegone with high enantiomeric purity in essential oils of some Lamiaceae aromatic plants. *Flavour Fragr.* 9: 205-207.

Ravid, U., Putievsky, E., and Katzir, I. (1994d). Enantiomeric distribution of piperitone in essential oils of some *Mentha* spp., *Calamintha incana* (Sm.) Heldr. and *Artemisia judaica* L. *Flavour Fragr. J.* 9: 85-87.

Ravid, U., Putievsky, E., and Katzir, I. (1996). Stereochemical analysis of borneol in essential oils using permethylated ß-cyclodextrin as a chiral secondary phase. *Flavour Fragr. J.* 11: 191-195.

Ravid. U., Putievsky, E., Katzir, I., Ikan, R., and Weinstein, V. (1992). Determination of the enantiomeric composition of citronellol in essential oils by chiral GC analysis on a modified ?-cyclodextrin phase. *Flavour Fragr. J.* 7: 235-238.

Ravid, U., Putievsky, E., Katzir, I., Weinstein, V., and Ikan, R. (1992). Chiral GC analysis of (S)(+)- and (R)(-)- carvone with high enantiomeric purity in caraway, dill and spearmint oils. *Flavour Fragr. J.* 7: 289-292.

Sam, T. W. (1993). Toxicity testing using the brine shrimp: *Artemia salina*. In S. M. Colgate and R.J. Molyneux, (Eds.), *Bioactive Natural Products,* London: CRC Press Inc., pp. 441-456.

Southwell, I.A., Freeman, S., and Rubel, D. (1997). Skin irritancy of tea tree oil. *J. Ess. Oil Res.* 9 (1): 47-52.

Suomi, D., Brown, J.J., and Akre, R.D. (1986). Responses to plant extracts of neonatal codling moth larvae, *Cydia pomonella* L. *J. Entomol. Soc. British Columbia.* 83: 12-18.

Svoboda, K.P. and Deans S.G. (1998). Presentation of the research and development data base. *Concerted action air 3 ct 94 2076 final report.* 3: 48-60. In *Towards a model of technical and economic optimization of specialist minor crops.* Brussels: Commison European, DG V1 F, 11. 3.

Svoboda, K.P., Gough, J., Hampson, J., and Galambosi, B. (1995). Analysis of the essential oils of some *Agastache* species grown in Scotland from various seed sources. *Flavour Fragr. J.* 10: 139-145.

Svoboda, K.P., Hampson, J.B., and Hunter, A. (1998). Secretory tissues—Storage and chemical variation of essential oils in secretory tissues of higher plants and their bioactivity. *Int. J. Aromathy.* 9(3): 124-131.

Svoboda, K.P., Inglis, A., Hampson, J.B., Galambosi, B., and Asakawa, Y. (1998). Biomass production, essential oil yield and composition of *Myrica gale* L. harvested from wild populations in Scotland and Finland. *Flavour Fragr. J.* 13: 367-372.

Svoboda, K.P., Juhasz, A., and Hampson, J.B. (2000). Bioactivity of essential oils and their components—Anti-inflammatory models in pharmacological studies (Part 1). *Aroma Res.* 1(1): 70-74.

Svoboda, K.P., Kyle, S.K., Sinclair, W., Hampson, J.B., and Asakawa, Y. (1999). Investigation of essential oil composition and DNA fingerprinting of *Ocimum basilicum*. 30th International Symposium on Essential Oils, Leipzig, September 5-8.

Traffic International (1999). Europe's medicinal and aromatic plants: Their use, trade and conservation. Goldaming, Surrey.

Walton, N.J. and Brown, D.E. (Eds.) (1999). *Chemicals from plants. Perspectives on plant secondary products.* London: Imperial College Press and World Scientific.

Zaika, L.L. (1988). Spices and herbs—Their antimicrobial activity and its determination. *J. Food Safety* 9(2): 97-118.

Chapter 13

Antimycotic Potential in Some Naturally Occurring Essential Oils

A. K. Pandey
Mahendra Rai

INTRODUCTION

In recent years, there has been a gradual revival of interest in the use of medicinal and aromatic plants in developed as well as in developing countries, because plant-derived drugs have been reported to be safe and without side effects. Thus, there is a great need for new plants and for better and cheaper substitutes. Recently, some products of higher plant origin have been shown to be effective sources of chemotherapeutic agents without undesirable side effects and with strong fungicidal activity (Fawcett and Spencer, 1970; Anonymous, 1977; Dubey et al., 2000; Aquil et al., 2000; Chatterjee and Parkashi, 1995; Singh, 1999). Very few antifungal substances are available in the market when compared to antibacterial substances, and they are also relatively unsatisfactory in controlling the diseases caused by fungal infections (Roxburgh and Borrie, 1973). In fact, there is a need for powerful and specific antimycotic agents on an increasing scale to combat fungal infections.

The presence of antimycotic principles in higher plants has long been recognized as an important factor in disease resistance (Mahadevan, 1982). Essential oils from different plant species are known to exhibit various kinds of biological activities. Their narrow or wide range of antifungal activity may prove useful in the development of effective antifungal substances. A survey of the literature reveals that many essential oils have earlier been reported to have antifungal activity against a number of pathogenic fungi (Maruzzella and Henry, 1958; Sbrajia, 1975; Sawhney et al., 1977; Misra and Dikshit, 1978; Sharma and Singh, 1979; Kubo and Nakanishi, 1979; Nene and Thapliyal, 1979; Chandra et al., 1982; Dikshit et al., 1982; Dubey et al., 1982; Singh, 1984; Maiti, et al., 1985; Singh, Dubey, et al., 1986; Tripathi, Singh, et al., 1986; Misra et al., 1990; Srivastava et al., 1993;

Samuel and Tripathi, 1994; Lis Balchin et al., 1996; Singh, 1996; Thoppil et al., 1998; Guddewar et al., 1999). However, there is a need to assess the practical potency of other essential oils, which have shown antifungal properties during initial screening. The knowledge of mechanism of antimicrobial activity of essential oils is insufficient. Knobloch et al. (1986) reported phenols as the most effective terpenoids, followed by aldehydes and ketones, alcohol, and hydrocarbons.

Immense prospects for the commercial exploitation of essential oils as antimycotic agents exist. Large-scale trials are needed before recommendations for formulation as antifungal ointments can be made. Such essential oils will be superior in action to the prevalent synthetic antifungal agents, most of which have severe side effects on the mammalian system as well as on the environment.

CYMBOPOGON WINTERIANUS *(CITRONELLA)*

Cymbopogon is an important genus of the family Poaceae and is represented by about 120 species and or its varieties. Out of this, 40 species are distributed in old world tropics and subtropics, and only eight species, viz., *C. martinii* (Roxb) Wats, C. *winterianus* Jowitt, *C. citratus* (DC ex Nees), *C. flexuosus* (Nees ex Steud), *C. pendulus* (Nees ex Steud) Wats, *C. coloratus, C. nardus,* and *C. jwarancusa* are commercially exploited for geraniol, citronellol, citral, and piperitone.

Cymbopogon martinii has two varieties, i.e., *motia* and *sofia.* These two varieties can be distinguished on the basis of morphology, cytology, kareotype, and amino acid pattern. *Cymbopogon martinii* var. *motia,* commonly known as Rosha grass or Tikhadi, is valued for its sweet-scented rosaceous aroma. It has dark-green leaves, which are leathery, prominently midribbed, roundish at the base, and form an obtuse to right angle with the stem. It has a diploid chromosome number of 2n = 20, and the oil is rich in geraniol, geranyl acetate, and linalool. On the other hand, variety *sofia,* known as ginger grass, has a curved leaf base attached at an acute angle to the stem, with a diploid chromosome number of 2n = 40. The oil contains less of geraniol and other components such as perillyl alcohol (14 percent), carnecol (13.5 percent), carvon (20 percent), diterpene (24.1 percent), etc. Although palmarosa oil (motia oil) is used in high-grade perfumes, sofia oil is mostly used in low-grade perfumes. The oil from palmarosa is a major source of high-grade geraniol (Guenther, 1950). It has antifungal and antibacterial properties. This oil is also used for curing skin diseases.

The essential oil has been used as a fungicidal agent (Singh, et al., 1980; Saikia, et al., 1999). Citronella oil of Java quality is obtained by steam distillation of leaves of *C. winterianus*. It is an important source of perfumery chemicals, such as citronellal, citronellol, geraniol, and hydroxy citronellal. Until 1973-1974, the oil was imported to meet the domestic demands of the country (India). *Citronella* and geranium oil have been reported to show antifungal activity against *Aspergillus flavus, A. niger, Candida albicans, Microsporum gypseum, Sporothrix schenckii,* and *Trichophyton rubrum* (Aggarwal et al., 2000). They found the oil components d- citronellol and geraniol were most active. Isomer d-citronellol was relatively more active than l-isomer against all the fungal strains at a twofold level. This probably explained the marginal high activity of citronella oil over geranium oil. Geraniol had higher activity against *A. niger* than the closely related species *A. flavus.*

Palmarosa oil is also known to possess antimicrobial activity (Gangrade et al., 1991; Jeyalakshmi and Seetharaman, 1998). Babu et al. (2000) conducted a study on medicinal growth and spore germination to test antifungal properties of palmarosa oil against *Alternaria solani,* the tomato leaf -blight pathogen. The palmarosa oil inhibited the mycelial growth and spore germination of *A. solani* at 0.3 percent concentrations. Mycelial growth and spore germination were observed at very low concentrations of the oil (below 0.2 percent). Similarly, Jain and Jain (2000) also conducted a study concerning essential oils of palmarosa, sacred basil, and six fungicides against *Alternaria tenius,* which causes seed spoilage in *Withania somnifera*. The objective was to explore the possible effectiveness of these essential oils as better substitutes for fungicides at 0.15 concentrations, against the normal doses of fungicides. Among the essential oils and fungicides, palmarosa oil was found most effective.

Essential oils of citronella (*Cymbopogon winterianus* Jowitt.), geranium (*Pelargonium graveolens* L.), lemon grass [*Cymbopogon citratus* (D.C.) Stapf.], palmarosa (*C. martinii* Roxb.), clove *(Syzygium aromaticum),* and patchouli (*Pogostemon cablin* Benth.) were tested in vitro for their antifungal efficacy against *Sclerotium rolfsii* which causes collar rot of mentha species. The results revealed that the oils were effective and that their effect was dependent on their concentrations. The oils of citronella, clove, and geranium were more effective in reducing the mycelial growth at low concentrations (100 to 500 ppm), appreciably and completely at 1,000 ppm (Singh et al., 2000). These results showed similarity to the work done on the antifungal efficacy of *Cymbopogon* spp. (Nidiry 1998). The essential oil of lemon grass was found effective against soil-borne plant pathogens (Handique and Singh, 1990).

TURMERIC OIL

Turmeric *(Curcuma longa)* is a well-known, indigenous herbal medicine used as a stomachic and blood purifier, as well as in treatment for the common cold, leprosy, intermittent fever, infection of the liver, dropsy, wound healing, and inflammation. A review of the literature reveals that turmeric is useful in treating a variety of ailments and metabolic disorders. The roots of this plant are known to be antiseptic and aromatic (Chopra et al., 1958; Patnaik, 1993). The paste is used in cleansing and disinfecting the skin and skin ulcers without drying out its natural oils. Its major constituents, curcumin, various curcuminoids, and dl-ar-turmerone exhibit a wide range of biological activities. The essential oil fractions from *C. longa* rhizomes exhibit fungistatic activity particularly against *Aspergillus niger, Physalospora tucumanesis, Ceratocystis paradoxa, Sclerotium rolfsii, Curvularia lunata, Helminthosporium sacchari,* and *Fusarium moniliforme.*

Turmeric rhizome and leaf contain 4.8 percent and 1.5-2 percent essential oils, respectively (Ramachandraiah et al., 1998). The turmeric rhizome parts as a whole and rhizome oil have antibacterial, antifungal, and antiviral properties (Rath et al., 1999; Khanna, 1999). Rath et al. (1999) reported antimycotic activity of the essential oils of turmeric against *Candida albicans,* and *Cryptococcus neoformans*. Turmeric leaf and rhizome oil were found to be fungicidal against *C. neofermans* MTCC-1437 strains. Leaf oil showed better activities in comparison to rhizome oil. Leaf oil killed the pathogens within 1 to 15 minutes. From the GC analysis studies, Rath et al. (1999) reported that turmeric leaf oil is rich in monoterpenes, and rhizome oil is rich in terpenoids and terpenoid-ketone compounds. The turmeric leaf is treated as a waste product during postharvest operations of turmeric rhizome.

EUCALYPTUS OIL

There are more than 700 species of the genus *Eucalyptus* which are native to Australia. A large number of them contain essential oils in their leaves. The oil of *Eucalyptus,* obtained by steam distillation of mature leaves, differs considerably in composition among species. Based on their chemical composition, the oil can be broadly divided into three groups: cineole-rich; citronellal-rich; and plyllandrene-rich oils. *Eucalyptus globules* produce cineole-rich oil, *Eucalyptus citriodora* gives citronellal-rich oil, and *Eucalyptus dives* yields an essential oil rich in phyllandrene.

Rai et al. (1999) and Pandey et al. (2000) studied the antimycotic activity of essential oils obtained from different species of *Eucalyptus.* Volatile oils from the leaves of *Eucalyptus* spp. growing in Nigeria exhibited considerable antifungal activity against *C. albicans* (Oyedeji et al., 1999). Shahi, Shukla, Bajaj, et al. (1999) also evaluated antifungal activities of oils of *Eucalyptus.* They found the oil of *E. citriodora* to be the most effective as an antifungal agent against *Microsporum nanum, Tricchophyton mentagrophytes,* and *T. rubrum,* while the pure oil (100 percent) killed *M. nanum* in just 15 seconds, at minimal fungicidal concentration. It required 5 h 30 min against *M. narum* and 4 h against *T. mentagrophytes* and *T. rubrum.* On comparing the minimum effective concentrations of the oil with those of prevalent synthetic antifungal drugs, the oil was found to be more effective. Moreover, it did not exhibit any side effects on mammalian skin in up to 5 percent concentrations.

The oil in the form of an ointment broad-spectrum antimycotic drug (BSAD) was subjected to topical testing on patients at the department of Moti Lal Narhul Medical College, Allahabad, India. At the end of treatment, while 55.5 percent of patients recovered completely, 44.5 percent showed significant improvement from the fungal disease. The ointment not only showed maximum efficacy, but was also found to be cost effective, had a long shelf life, and showed an absence of any adverse side effects. Shahi, Shukla, and Dikshit (1999) further studied antifungal activities of some stored essential oils of *Eucalyptus* spp. viz., *E. amygdalina, E. citriodora, E. dalrympleana,* and *E. laveopinea* against dermatophytes, viz. *Epidermophyton floccosum, Microsporum gypseum, M. nanum, Trichophyton mentagrophytes, T. rubrum,* and *T. violaceum.* The efficacy of the oil increased for inhibiting the mycelial growth of the test organism at its normal pH 5.6. The minimum fungistatic and fungicidal concentrations decreased in the range from 12.5 to 80 percent, and 11.1 to 75 percent, respectively, against all the test organisms, except the oil of *E. citriodora,* which showed an increase of 25 percent to the minimum fungistatic concentrations against *T. rubrum* and *T. mentagrophytes.* The oils did not show any side effects on mammalian skin at up to 5 percent concentrations at normal and alkaline. The variations in minimum fungistatic concentrations among various *Eucalyptus* spp. against the test organisms is probably due to variations in cineole contents (27-78 percent). Hence, the oil of *E. citriodora,* because of its strong antifungal activity, and with no irritations on human skin, can be used successfully in the form of broad-spectrum antimycotic drugs for the control of fungal infections in human beings.

Pandey et al. (2000) studied the effects of essential oils of six species of *Eucalyptus,* viz., *E. camaldulensis, E. citriodora, E. globulus, E. hybrid, E. robusta,* and *E. tereticornis* on the growth of three opportunistic pathogens, viz., *F. solani, R. arrhizus,* and *A. alternata*. All six essential oils were found to be more effective against *R. arrhizus* and *A. alternata* than against *F. solani*. The essential oil of *E. rostata* Schlecht. at a dose of 3,000 ppm proved significant to *F. moniliforme*. Significant reduction in fruit rot was observed if tomatoes were treated with oil prior to storage (Misra et al., 2000). The fungistatic property of the oil remained unaffected by temperature treatment, prolonged storage, and increase in inoculum density.

Eucalyptus leaves are reported to contain various antifungal substances (Egawa et al., 1977). The need to develop fumigants with sufficient vapor action for the control of postharvest losses of perishables has been stressed by Tarr (1972). Essential oils are volatile and have sufficient vapor action coupled with antimicrobial efficacy. Therefore, these may be used as fumigants. The leaves of *Eucalyptus* fall on the ground and are wasted because of their unpalatability as feed for animals. Earlier literature shows that the recovery of oil from dried leaves of *E. rostrata* is greater in comparison to fresh leaves (Singh et al. 1988).

SCREENING FOR ANTIFUNGAL ACTIVITY

Several reports exist on the antifungal screening of essential oils. Most of the work comprises laboratory trials under in vitro conditions. Several plant products have been reported effective against *A. solani* (Singh, Tripathi, et al., 1986). Dey and Choudhary (1989) report that eugenol and methyl eugenol of *Ocimum sanctum* oil inhibited the mycelial growth of *A. solani*. Tripathi, Banerjee, et al. (1986) studied the toxicity of essential oil of *Ocimum gratissimum* against *C. capsici, A. alternata,* and *S. rolfsii* and found eugenol to be more toxic.

Dubey et al. (1983) and Saxena et al. (1990) noticed fungal toxicity of essential oils of *Acorus calamus* against *Helminthosporium oryzae*. The essential oil extracted from *Trachyspermum ammi, Caryophyllus aromaticus, Piper nigrum, Sassurea lappa, Algelica archangelica,* and *Acorus calamus* was tested for evaluation of fungicidal properties against *Fusarium oxysporum* Schelcht. Essential oils of *T. ammi* and *C. aromaticus* were found most effective completely against fungal growth followed by *A. archangelica, P. nigrum,* and *A. calamus*.

Essential oils of palmarosa, *Eucalyptus globulus,* and *Ocimum canum* were found to be active against *Sclerotium rolfsii* inciting scletorial wilt in jasmine and barley (Singh and Dwivedi, 1987). The essential oils of peppermint, clove, and *Eucalyptus* have been evaluated for their antifungal activity against *A. niger, Alternaria alternata,* and *Fusarium* spp. Maximum antifungal activity was detected in clove oil followed by peppermint and eucalyptus (Aquil et al., 2000).

Nigam and Rao (1977) reported that the oil of *Cuminum cyminum* was toxic to *Aspergillus fumigatus, F. tenuis,* and *Trichoderma viride* but was ineffective against *A. flavus, A. vericolor, A. oryzae, Curvularia* sp., *Penicillium javanicum,* and *P. funiculosum*. Sharma and Singh (1979) reported that the oil was toxic to *Alternaria* sp., *A. candidus, A. flavus, A. nidulans, A. niger, Cladosparium herbarum, Cunninghamella echinulata, Fusarium moniliforme, F. oxysporum, Heliminthosporium sacchari, Microsporum cookei, M. gypseum, Mucor mucedo, P. distatum, Rhizopus* sp., *Trichophyton mentagrophytes, T. rubrum,* and *Trichothecium roseum*. Saxena (1984) found antifungal activity against *C. albicans, C. tropicalis, Keratinomyces ajelloe, M. gypseum, T. equinum,* and *T. rubrum*. The study conducted by Sarika et al. (1999) revealed that essential oil from the seeds of *Cuminum cyminum* may prove an ideal fungicide for the safe storage of food commodities. However, its use in practice must await results of in vitro studies.

Cyperus oil is extracted from *Cyperus scariosus* (Nagar motha), which belongs to the family Cyperaceae. It is extensively used in medicines, perfumery, and for the preparation of agarbatti. The oil is also known to possess antimicrobial activity (Lahariya and Rao, 1979).

Tree-borne essential oils of *Chloroxylon swietina* Roxb. Corom (family: Rutaceae) exhibit antifungal properties. Commonly known as bhirrha in Hindi, and satinwood in English, it gives an essential oil on distillation. The oil showed good antifungal activity (Garg and Oswal, 1981) against *A. tenuis, A. oryzae,* and *Candida albicans*. *Artemisia pallens* Wall is an aromatic, annual, small herb, which belongs to the family Asteraceae and yields an essential oil which is used as an antifungal (Nakhare and Garg, 1996).

The essential oils from the rhizome of *Luvunga scandens* Roxb. (family: Rutaceae) has been studied for in vitro antifungal activity against four keratinophillic fungi, viz., *Arthroderma benhamiae, Microsporum gypseum, Trichophyton mentagrophytes,* and *Ctenomyces*. The oil exhibited very good to moderate inhibitory effect against the fungi (Garg and Jain, 1999). The oil has already been found to exhibit inhibitory effect against 15 bacteria and eight fungi including *Aspergillus* sp. and *Microsporum cookei*.

Geranium oil, whose most abundant constituents are citronellol and geraniol (Lawrence, 1996a) showed highest activity against the mycelial growth of *C. gloeosporoides*. The most abundant constituent of peppermint oil is menthol (Lawrence, 1997), which is a monoterpene secondary to alcohol. Menthol has been identified as a chief antifungal principle in peppermint oil (Nidiry, 1998). The most abundant principles of patchouli oil are sesquiterpenoids (Lawrence, 1995a), while pogostone, sesquiterpene, and ketone were identified as active principles earlier (Anonymous, 1977).

Nidiry and Chandravadana (1994) reported that patchouli alcohol, a sesquiterpene alcohol present in the oil, also showed mycelial growth inhibition of *C. gloeosporoides*. Carvone, the main constituent of dill oil, which is a monoterpene ketone (Lawrence, 1996a,b), showed antifungal activity. The major constituents of davana oil are davanones, which are sesquiterpene ketones (Lawrence, 1995a,b). Antifungal activity of davana oil has been reported by others also (Alankara Rao and Prasad, 1981). There is no report on the antifungal activity of davanones or other compounds present in davana oil. However, davana oil contains about 8 percent geraniol, which has antifungal properties.

Many essential oils possess antimicrobial properties, which are invariably dependent on the chemical nature of the constituents present in them. Kurita et al. (1981) studied the antifungal activity of forty different constituents present in various essential oils. The compounds octanal, nonanal, citronellal, and decanal had lower activity than their corresponding alcohols, namely octanol, nonanol, citronellol, and decanol, respectively. Primary alcohols, in general, possess higher activity than secondary and tertiary alcohols. These findings were corroborated by recent studies on the structure and fungitoxicity relationship of the monoterpenoids of geranium and peppermint oil (Chandravadana and Nidiry, 1994; Nidiry, 1998). The fungitoxicity of the essential oils, viz., davana oil (*Artemisia* pallens), dill oil *(Anethium graveolens),* and patchouli oil *(Pogostemon patchouli)* to the mycelial growth of *Colletotrichum gloeosporoides* was evaluated and compared with the earlier reported fungitoxicity of geranium and peppermint oil (Nidiry, 1998). The antifungal activity of peel oil of *Citrus* spp. was fungitoxic against *A. cladosporium* and *Fusarium*.

The essential oils extracted from plant species, such as *Citrus sinensis, Cuminum cyminum, Eucalyptus lanceolatus, Hyptis suaveolens, Ocimum basilicum, Syzigium aromaticum,* and *Xanthoxylum alatum* have been tested against nine sugarcane pathogenic fungi. The oil of *C. cyminum* was the best in all tested oils (Singh and Pandey, 1998).

The essential oil from *Lantana aculeata* leaves was found to be active against *Penicillium digitatum, P. notatum, Rhizophus stolonifer,* and *M. gypseum* (Saxena and Sharma, 1999a). The essential oil of *Carum copticum* (syn. *T. amni*) inhibited the growth of *Phomopsis distrutum, A. niger,* and *A. flavus* (Srivastava et al., 1999). *Toddalia* leaf essential oil was found to be active against *A. fumigatus, A. niger, R. stolonifer,* and *M. gypseum* even at a dilution of 1:20 (Saxena and Sharma, 1999b). The essential oils of *Hyptis suaveolens* leaves showed antifungal activity against *Candida albicans* (Asekun et al., 1999). In vitro fungicidal properties of the essential oil of *Calmintha nepeta* were studied against *A. niger, A. parasiticus, R. oryzae,*

C. albicans, and *Colletotrichum musae*. The oil was found to be most effective against *C. albicans, R. oryzae,* and *C. musae* (Tajo et al.,1999).

Callistemon viminalis Cheel. (family: Myrtaceae), commonly known as bottlebrush, is cultivated in Indian gardens. The leaves of the plant yield an essential oil, which has exhibited good antifungal activity against *F. solani, R. nodosus, T. rubrum, T. viride, A. tenius, A. flavus,* and *C. albicans* (Garg and Kasera, 1998). The efficacy of the oil against *F. solani,* which causes disease in wheat, oats, and barley, is very significant. The antifungal activity of the oil against *T. rubrum* and *C. albicans* shows that the oil is of great therapeutic importance in fungal infections of the inner surface of thighs and nails, cutaneous and mesocutaneous and digital webs of toes, and sides of feet (athlete's foot). The essential oil of *Callistemon lanceolatus* L. was tested in vitro for antifungal efficacy against *Sclerotium rolfsii* which causes collar rot of mentha species. The oil showed fungistatic effect, not fungicidal (Singh et al., 2000).

The essential oils of *Tagetes,* palmarosa, clove, citrus peel, and *Chenopodium* were evaluated for their antifungal properties against *Colletotrichum falcatum*—a causal agent of red rot disease in sugarcane. Amongst the oils, tagetes oil was found to be the most effective (Srivastava et al.,1997). *Ocimum* species belonging to the family Lamiacae are cultivated for essential oils and have one or more economically important constituents such as estragole, geraniol, linalool, and other such compounds. Essential oils of different plants belonging to the family Lamiacae have been tested for their in vitro antifungal activity (Dubey et al., 1989; Gangrade et al., 1989, Thakur et al., 1989). The essential oil of *Ocimum gratissium* showed its applicability as an antidermatophytic agent.

CONCLUSION AND FUTURE PROSPECTS

Many essential oils have been found to have antifungal activities. Because of their notable antimycotic properties coupled with pleasing flavors, essential oils can be used to treat skin diseases caused by fungi. Essential oils also have advantages over chemical antimycotics due to their low mammalian toxicity, their environment-friendly status, target specificity, ability to overcome pest resistance, low cost, and easy accessibility. The commercial use of volatile oils for controlling various fungal diseases seems to be a neglected field, though a large number of oils are known to possess antimicrobial activities. Very little is known on the mechanism of antimicrobial activity of essential oils. More studies should be conducted on the mechanism of antifungal activity of essential oils. The possible relationship between fungitoxicity of essential oils and the chemical nature of the most abundant principles present in the oils as well as the active principles already identified should be examined. However, this kind of study has been undertaken for few species.

REFERENCES

Aggarwal, K.K., Ahmad A., Kumar Santha, T. R., Jain, N., Gupta, V.K., Sushil K., and Khanuja, S.P.S. (2000). Antimicrobial activity spectra of *Pelargonium graveolens* L. and *Cymbopogon winterianus* Jowitt oil constituents and acyl derivatives. *Journal of Medicinal and Aromatic Plant Sciences* 22: 544-548.

Alankara Rao, G.S.J.G. and Prasad, Y.R. (1981). Investigation on the antifungal activity of the essential oils from *Artemisia pallens* Well and *Artemisia vulgaris* Linn. *Ind. Perf.* 25(1): 112-113.

Anonymous (1977) . Isolation and structural analysis of pogostone—An antifungal compound from Chinese drug "Kwang Ho Hsiang" (*Pogostemon cablin* (Blanco) Benth. *Kohsurh Tung Rao,* 22 (7): 318-320; *Chem Abstr* 87: 172771q.

Aquil, F., Beg Arina, Z., and Ahmad, I. (2000). In vitro toxicity of plant essential oil against soil fungi. In *National Seminar on the Frontiers of Research and Development in Medicinal plants.* CIMAP, Lucknow, JMAPS, 22(Supplement): 177-181.

Asekun, O.T., Ekundayo, O., and Adeniyi, B. A. (1999). Antimicrobial activity of the essential oil of *Hyptis suaveolens* leaves. *Fitoterapia* 70(4): 440-442.

Babu, S., Seetharaman, K., Nand Kumar, R., and Johnson, I. (2000). *In vitro* antifungal activity of palamrosa (*Cymbopogon martinii* Roxb. Watson) oil against *Alternaria solani. Ind. Perf.* 44(4): 275-277.

Chandra, N., Pandey, D.K., Tripathi, N.N., and Dikshit, S.N. (1982). Toxicity of the essential oil of *Eupatorium capillifolium* against *Helminthosporium oryzae. Ind. Perf.* 26: 231-236.

Chandravadana, M.V. and Nidiry, E.S.J. (1994). Antifungal activity of the essential oil *Pelergonium graveolens* and its constituents against *Colletotrichum gloeosporioides. Indian J. Exp. Biol.* 32: 908-909.

Chatterjee, A. and Parkashi, S. (1995). *The Treatise of Indian Medicinal Plants,* Volume IV. New Delhi: Publication and Information Directorate CSIR.

Chopra, R.N., Chopra, I.C., Handa, K.L., and Kapoor, L.D. (1958). *Indigenous Drugs of India.* Calcutta: U.N. Dhar and Sons.

Dey, B.B. and Choudhary, M. A. (1989). Essential oil of *Ocimum sanctum* L. and its antimicrobial activity. *Ind. Perf.* 28: 82-87.

Dikshit, A., Dubey, N.K., Tripathi, N.N., Tripathi, G.D., and Dikshit, S.N. (1982). Control of fungal deterioration of some species during storage by the essential oil of *Cedrus deodara. Ann. App. Biol.* 100 (Suppl.): 56-57.

Dubey, G., Rao, T.A.S., and Nigam, S.S. (1983). Antifungal efficacy of some Indian essential oils. *Ind. Perf.* 27 (1): 5-8.

Dubey, N.K., Kishore, N., Tripathi, N.N., Tripathi, R.D., and Dikshit, S.N. (1982). Fungitoxicity of the essential oil of *Citrus medica* against storage fungi. *Ann. App. Biol.* 100: 58-59.

Dubey, N.K., Tripathi, P., and Singh, H.B. (2000). Prospects of some essential oils as antifungal agents. *Journal of Medicinal and Aromatic Plant Sciences* 22: 350-354.

Dubey, S., Upadhya, P.D., and Tripathi, S.C. (1989). Antifungal, physicochemical and insect repellent activity of essential oil of *Ocimum basilicum* L. *Can. J. Bot.* 67: 2085-2087.

Egawa, H., Tsutsui, O., Tatsuyama, K., and Hatta, Y. (1977). Antifungal substances found in leaves of *Eucalyptus* spp. *Experientia* 33: 889-890.

Fawcett, C.N. and Spencer, D. M. (1970). Plant chemotherapy with natural products. *A Rev. Phytopath.* 8: 403-408.

Gangrade, S.K., Srivastava, R. D., Sharma, O.P., Jain, N.K., and Trivedi, K.C. (1989). Evaluation of anti fungal properties of essential oils of *Ocimum* species. *Ind. Perf.* 33(2): 97-101.

Gangrade, S.K., Srivastava, R.D., Sharma, O.P., Jain, N.K., and Trivedi, K.C. (1991). *In vitro* antifungal effects of the essential oils. *Ind. Perf.* 35: 46-48.

Garg, S.C. and Jain, R. (1999). Antifungal activity of *Luvunga scandens* against some keratinophilic fungi. *Indian Drugs* 36(4): 265-266.

Garg, S.C. and Kasera M.K. (1998). Antimicrobial activity of the essential oil of *Callistemon viminalis*. *Ind. Perf.* 42(3): 138-141.

Garg, S.C. and Oswal, V.B. (1981). *In vitro* antifungal activity of the essential oil from the leaves of *Chloroxylon swietina*. *Indian Drugs* 19(5): 189-191.

Guddewar, M., Naik, S.N., and Durga Prasad (1999). Evaluation of fungicidal activity of certain essential oils against *Fusarium oxysporum* Sclecht. *Ind. Perf.* 43(1): 26-28.

Guenther, E. (1950). *The Essential Oils,* Volume IV. New Jersey: D Van Nostrand Co. Inc.

Handique, A.K. and Singh, H.B. (1990). Antifungal activity of lemon grass oil on soil borne plant pathogens. *Ind. Perf.* 34: 232-234.

Jain, N.K. and Jain, P.K. (2000). A comparative study of some essential oils and fungicides as seed dressers against *Alternaria tenuis* in *Withania somnifera*. In Sushil Kumar, S.A. Hasan, Samresh Dwivedi, A.K. Kukreja, Ashok Sharma, Anil Kumar Singh, Srikant Sharma, and Rakish Tiwari (Eds.), *National Seminar on the Frontiers of Research and Development in medicinal plant.* Lucknow: CIMAP, pp 56-57.

Jeyalakshmi, C. and Seetharaman, K. (1998). Biological control of root-rot and die-back of chilli with plant products and antagonistic microrgamisms. *Pl. Dis. Res.* 13: 46-48.

Khanna, N.M. (1999). *Turmeric:* Natures precious gift. *Curr. Sci.* 76(10): 1351-1356.

Knobloch, K., Weis, N., and Weigand, H. (1986). Mechanism of antimicrobial activity of essential oils. *Planta Medica* 52: 556.

Kubo, I. and Nakanishi, K. (1979). *Advances in Pesticides Science.* Oxford and New York: Pergamon Press.

Kurita, N., Miyaji, M., Kurana, R., and Takahara, Y. (1981). Antifungal activity of components of essential oils. *Agric. Biol. Chem.* 45(4): 945-952.
Laharya, A.K. and Rao, J.T. (1979). *In vitro* antimicrobial studies of the essential oil of the *Cyperus scariosus* and *Ocimum basilicum. Indian Drugs* 150-152.
Lawrence, B. M. (1995a). *Progress in essential oils. Perfum and Flav.* 20 (1): 54.
Lawrence, B.M. (1995b). *Progress in essential oils. Perfum and Flav.* 20 (3): 72-73.
Lawrence, B.M. (1996a). *Progress in essential oils. Perfum and Flav.* 21 (3): 64-65.
Lawrence, B.M. (1996b). *Progress in essential oils. Perfum and Flav.* 21 (6): 58-62.
Lawrence, B.M. (1997). *Progress in essential oils. Perfum and Flav.* 22 (4): 57-66.
Lis Balchin, M., Deans, S.G. and Hart, S. (1996). Bioactivity of geranium oils from different commercial sources. *J. Essen. Oil Res.* 8: 281-291.
Mahadevan, A. (1982). *Biochemical Aspects of Plant Disease Resistance.* I. New Delhi: Today's and Tomorrow's Printers and Publishers.
Maiti, D., Cole, C.R., and Sen, C. (1985). Antimicrobial efficacy of some essential oils. *J. Pl. Dis. Prot.* 92(1): 64-68.
Maruzzella, J.C. and Henry, P.A. (1958). The antifungal activity of perfume oils. *J. Amer. Phar. Ass.* 47: 471-476.
Misra, D., Jaiswal, P.K., Misra, D., Chaudhary, N., and Siddiqui, N. (2000). Control of Fusarium rot of tomato by essential oils of *Eucalyptus rostrata* Sclecht. *Ind. Perf.* 44(2): 79-82.
Misra, D.N., Tiwari, R., Mishra, A.K., and Upadhyay, P.S. (1990). Antifungal activity of *Anethum graveolens* L. against some storage fungi. *Vegetos* 3: 90-93.
Misra, S.B. and Dikshit, S.N. (1978). Fungitoxic activity of volatile fraction of three plants. *Ind. Perf.* 21: 61-62.
Nakhare, S. and Garg S.C. (1996). Antimicrobial activity of the essential oil of *Artemisia pallens* Wall. *Ind. Perf.* 40 (4): 118-120.
Nene, Y.L. and Thapliyal, A.N. (1979). *Fungicides in Plants Disease Control.* New Delhi: Universal Oxford and UYG Publishing Co.
Nidiry, E.S.J. (1998). Structure-fungitoxicity relationships of the monoterpenoids of the essential oils of peppermint *(Mentha piperita)* and scented geranium *(Pelargonium graveolens). J. Ess. Oil Res.* 10(6): 628-631.
Nidiry, E.S.J. and Chandravadana, M.V. (1994). Annual Report of the Indian Institute of Horticultural Research. Hessaraghatta, Bangalore: Indian Institute of Horticultural Research, pp. 89.
Nigam S.S. and Rao T.S.S. (1977). Antimicrobial efficacy of some Indian essential oils. In L.D. Kapoor and Ramakrishnan, (Eds.), *Advances in Essential Oil Industry*, New Delhi: Today's and Tomorrow's Printers and Publishers, pp. 177-180.
Oyedeji A.O., Ekundayo O., Olawore, O.N., Adeniyi, B.A., and Koenig, W.A. (1999). Antimicrobial activity of the essential oils of five Eucalyptus species growing in Nigeria. *Fitoterapia* 70(5): 526-528.
Pandey, A.K., Rai, M.K., and Qureshi S. (2000). *In vitro* antimycotic effect of volatile oils of *Eucalyptus* species. *Ind. Perf.* 44: 74-77.
Patnaik, N. (1993). *In the Garden of Life*. New York: Doubleday.
Rai, M.K., Qureshi, S., and Pandey, A.K. (1999). *In vitro* susceptibility of opportunistic *Fusarium* spp. to essential oils. *Mycoses* 42(1,2): 97-101.

Ramachandraiah, O.S., Azeemoddin, G., and Charyulu, K. (1998). Turmeric (*Curcuma longa* L.) leaf oil, a new essential oil for perfumery industry. *Ind. Perf.* 42(3): 124-127.

Rath, C.C., Dash , S.K., Mishra R.K., and Charyulu, J.K. (1999). In vitro evaluation of antimycotic activity of turmeric (Curcuma longa L.) essential oils against *Candida albicans* and *Cryptococcus neoformans. Ind. Perf.* 43(4): 172-178.

Roxburgh, A.C. and Borrie, P. (1973). The English Language Book Soc. and H.K. Lewis and Co. Ltd. XII Edition.

Saikia, D., Shantha Kumar, T.R., Kahol, A.P., and Khanuja, S.P.S. (1999). Comperative bioevaluation of essential oils of three species of *Cymbopogon* for their antimicrobial activities. *J. Med. and Arom. Plant Sciences* 21(1): 24.

Samuel, C.O. and Tripathi, S.C. (1994). Fungitoxic properties of essential oil of *Hedychium spicatum. Ind. Perf.* 38(4): 105-111.

Sarika, N., Singh, J., and Tripathi, S.C. (1999). Fungitoxicity of Cumin *(Cuminum Cyminum)* oil. *Ind. Perf.* 43(3): 142-146.

Sawhney, S.S., Suri, R.K., and Thind, T.S. (1977). Antimicrobial efficacy of some essential oil in-vitro. *Indian Drugs* 15: 30-32.

Saxena, D.B., Tomar, S.S., and Singh, R.P. (1990). Fungitoxicity of chemical components and some derivatives from *Anethium sowa* and *Acorus calamus. Ind. Perf.* 34(3): 199-203.

Saxena, N.K. (1984). Comparative evaluation of some essential oils for their antifungal activities against some dermatophytes. *Ind. Perf.* 28: 35-37.

Saxena, V.K. and Sharma, R.N. (1999a). Antimicrobial activity of essential of *Lantana aculeata. Fitoterapia.* 70(1): 67-70

Saxena, V.K. and Sharma, R.N. (1999b). Antimicrobial activity of the essential oil of *Toddalia asiatica. Fitoterapia* 70(1): 64-66.

Sbrajia, R. J. (1975). Chemical control of plant diseases: An exciting future. *A. Rev. Phytopath.* 13: 257-267.

Shahi, S.K. Shukla, A.C., Bajaj, A.K., Medgely, G. and Dikshit, A. (1999). Broad spectrum antimycotic drug for the control of fungal infections in human beings. *Curr Sci* 76: 836-839.

Shahi, S.K. , Shukla, A.C. and Dikshit, A. (1999). Antifungal studies of some essential oils at various pH levels for betterment of antifungal drug response. *Curr. Sci.* 77(5): 703-706.

Sharma, S.K. and Singh, V.P. (1979). The antifungal activities of some essential oils. *Indian Drugs Pharm.* 14: 3-6.

Singh, A., Chaturvedi, C., and Singh, H.B. (2000). Antifungal efficacy of some essential oils against *Sclerotium rolfsii. Ind. Perf.* 44: 71-73.

Singh, A.K., Dikshit, A., Sharma, M.L., and Dikshit S.N. (1980). Fungitoxic activity of some essential oils. *Econ. Bot.* 34: 184-190.

Singh, G. (1996). Studies on fungicidal activity of essential oils. *Euro. Cosmetic* 4: 27-32.

Singh, G. (1999). Studies on the biocidal activities of certain essential oils. *J. of Med. and Aro. Pl Sciences* 21: 1119-1130.

Singh, R.K. and Dwivedi, R.S. (1987). Effect of oil on *Sclerotium rolfsii* causing foot- root of barley. *Indian Phytopath.* 40: 531-533.

Singh, S., Dubey, N.K., Tripathi, S.C., and Singh, S.K. (1984). Fungitoxicity of some essential oils against *Aspergillus flavus*. *Ind. Perf.* 28: 164-166.

Singh, S., Singh, S.K., and Tripathi, S.C. (1988). Fungitoxic properties of essential oil of *Eucalyptus rostrata*. *Ind. Perf.* 32(3): 190-193.

Singh, S.P., Dubey, P., and Tripathi, S.C. (1986). Fungitoxic properties of essential oil of *Trachyspermum amni*. Sprague. *Mykosen* 29: 37-60.

Singh, S.P. and Pandey, S.K. (1998). Antifungal efficacy of volatile constituents of higher plants against sugarcane fungal pathogens. *Ind. Perf.* 42(2): 82-85.

Singh, Y.,Tripathi, R.D., and Tripathi, N.N. (1986). The isolation and properties of fungitoxic principle from *Zingiber officinale*. *Indian J. Pl. Path* 1: 89-96.

Srivastava, M., Saxena, A., and Baby, P. (1999). GC–MS investigation and antimicrobial activity of the essential oil of *Carum opticum* Benth and Hook. *Acta Alimentaria* 28(3): 291-295.

Srivastava, S., Naik, S.N., and Maheshwari, R.C. (1993). In-vitro studies on antifungal activity of palmarosa and Eucalyptus oils. *Ind. Perf.* 37(3): 277-279.

Srivastava S., Walia, D.S., and Naik, S.N. (1997). Antigungal activity of certain essential oils against *Colletotrichum falcatum* Went. *Ind. Perf.* 41(1): 30-32.

Tajo, A., Minija, M.J., Deena, M.J., and Thoppil, J.E. (1999). Antimicrobial properties of *Calamintha nepeta* (L.) essential oil. *Ind. Perf.* 43(4): 179-181.

Tarr, S.A.J. (1972). Physical and chemical methods of disease control. In *Principle of Plant Pathology*. London and Basingstoke: Macmillan, pp. 490-499.

Thakur, R.N., Singh, P., and Khosla, M.K. (1989). *In vitro* studies on antifungal activities of some aromatic oils. *Ind. Perf.* 33(4): 257-260.

Thoppil, J.E, Tajo, A., and Minija, J. (1998). Antibacterial and antifungal activity four varieties of *Ocimum basilicum*. *Fitoterapia* 69: 191-192.

Tripathi, R.D., Banerjee, R., Sharma, M.L., Balasubramanium, M.L., and Nigam, S.K. (1986). Toxicity of essential oil from new strain of *Ocimum gratissimum* (Clocimum) against betelwine pathogenic fungi. *Agric. Biol. Chem.* 49(8): 2277-2282.

Tripathi, S.C., Singh, S.P., and Dubey, S. (1986). Studies on antifungal properties of *Trachyspermum amni* Sprague. *Phytopath Z.* 113-120.

Chapter 14

Antimycotic Essential Oils: Impact on Skin Microflora

Jean-Pierre Chaumont

INTRODUCTION

All over the world, mycoses take a prominent place in dermatology (Rippon, 1987). In tropical countries, populations are largely affected by fungal infections (Omidynia et al., 1996; Vandemeulebroucke et al., 1999). For the past thirteen years, due to HIV infections, fungal diseases have increased (Dromer and Dupont, 1996; Hostettmann, 1997; Porro et al., 1997; Kaufman, 1998; Stevens et al., 1999). Immunocompromised patients have risks of mycoses. In temperate lands, ailments have also considerably progressed in the past thirty years. There are many factors responsible for skin infections: aberrant clothing or living customs; excessive chemotherapy; and excessive use of contraceptives (Daltio and Lunardi, 1991; Bodey et al., 1992; Puccini et al., 1992; Chabasse et al., 1999).

Synthetic antimycotics have been used for oral or local treatments and have shown strong efficacy. Nevertheless, some resistance and many relapses appear to be caused by genetic adaptation or insufficient treatment (Chabasse et al., 1999).

Synthetic antimycotics are often expensive and are not covered by insurance. Underprivileged peoples hesitate to have medical attention. Due to the various problems with synthetic antimycotics, many are in favor of alternative therapies based on the development of natural drugs (Ake Assi, 1982; Viollon and Chaumont, 1994; Franchomme et al., 1996; Hostettmann, 1997; Baratta et al., 1998; Hammer et al., 1999; Ozcan, 1999). The use of antimycotic essential oils or natural volatile compounds (NVC) are gaining ground (Ake Assi, 1982). Several essential oils of varied origin and chemical makeup have been tested for their antimicrobial activities against some bacteria and fungi of skin microflorae, mainly foot, armpit, or other natural folds exposed to intertrigo. On each organ, fungi and bacteria may constitute a typical population by their association.

THE TECHNIQUES

Pathogenic Fungi

Pathogenic fungi were recovered from patients in the hospital of Besançon (France).

Pure Essential Oils

Essential oils were obtained by hydrodistillation of plant raw materials and verified by GC-MS. Pure chemical compounds were obtained from Sigma-Aldrich Chimie, France.

Culture and MIC

Fungi were grown on Sabouraud dextrose agar medium and bacteria on nutrient agar. Essential oils were added at increasing concentrations in the sterilized culture medium at 40°C, and then distributed into petri plates before solidification. Fungi (yeast) were inoculated by blastospore suspension and dermatophytes mycelium disposition in center plates. Bacteria were seeded by suspension with a multipoint inoculator. Cultures were incubated for 24 hours at 37°C (yeast and bacteria) and for 7 days at 24°C (dermatophytes). The lowest concentration of compounds that completely inhibited strains was considered to be the minimum inhibitory concentration (MIC), expressed in μg/ml. The experiments were performed and compared to control tests.

Antimicrobial Spectrum of Essential Oils

Each essential oil is a complicated mixture that includes light constituents (100 to 600), some of which are often largely dominant, and quite all easily detectable by GC-MS.

Synthetic compounds present a large and strong antimicrobial spectrum. Essential oils and their constituents show less impressive MIC (more than ten times in vitro) but, above all, selective inhibitions. So, if the drug is not toxic, some concentrations can reach only undesirable targets. Therefore, the extension of useful saprophytes becomes easier. Many inhibition tests in vitro have shown essential oils to ameliorate the balance of skin microflorae.

IMPACT OF ESSENTIAL OILS ON SKIN MICROFLORA

Table 14.1 shows twelve essential oils with their antimicrobial activities on skin microflorae. The MIC vary largely depended on the kinds of targets and oil used. Essential oil extracted from *Cinnamom zeylanicum* exhibited strong antifungal activities on all the fungal strains. The antimycotic properties of cinnamom bark and cinnamaldehyde are well known (Viollon and Chaumont, 1994). Although essential oils are irritant and allergenic, they are prescribed in low dilutions for topical treatments on skin and even on mucous (Franchomme et al., 1996).

The other essential oils presented generally, a broad antifungal spectrum, except for *Acorus,* sandalwood or vetiver. But they all were efficient in vitro on dermatophytes.

The main components of essential oils are always phenols or aldehydes and oxygenated terpenoids, as indicated in Table 14.2. All these constituents present MICs closely related to the correspondent essential oils. On the contrary, in vitro, the hydrocarbons are inactive against the whole strains and are probably inefficacious when applied directly on the skin. That is the case, for example, of a majority of essential oils of conifers and members of the family Apiaceae.

Impact in Veterinary Medicine

Concerning pets, large applications of essential oils on the whole surface of the fur appear conceivable with difficulty. Nevertheless, the special microflora of the external ear may be modified and treated locally (Kiss et al., 1997). External otitis is a critical problem for veterinary surgeons because of the complex anatomy of the organ and antibioresistances (Lloyd and Noble, 1999). The main targets of this therapy are *Malassezia pachydermatis* and *Staphylococcus intermedius* associated in many cases with *Pseudomonas aeruginosa*. Some essential oils inhibit yeasts and *Staphylococcus intermedius,* whereas the very resistant *Pseudomonas aeruginosa* is not affected.

Antimycotic Effects and Associated Bacteria

Because skin microflorae at about each organ level constitute a whole community of microorganisms more or less associated or competitive, it is essential to consider the activities of the essential oils on the main bacteria representative of the outside of the organs (Marshall et al., 1987, 1988; Viollon and Chaumont, 1994; Lacoste et al., 1996; Yang et al., 1996).

TABLE 14.1. In vitro antimicrobial properties of 12 essential oils against human skin, pathogenic fungi, and commensal bacteria

	Minimal Concentration (MIC) (μg/ml)											
	A.C.	C.Z.	C.W.	P.G.	L.S.	O.O.	C.M.	P.C.	S.A.	S.H.	T.V.	V.Z.
Foot microflora												
Bacteria												
*Staphylococcus epidermidis**	> 1000	400	>1000	400	300	200	600	400	100	400	300	1000
*S. hominis**	> 1000	150	>1000	400	300	200	800	400	100	500	300	500
*S. cohnii**	> 1000	400	>1000	600	250	400	800	>500	500	600	300	>1000
Corynebacterium gr. C*	> 1000	300	700	400	100	200	400	400	50	200	400	300
Corynebacterium gr. B*	> 1000	100	600	500	150	400	300	150	50	500	400	400
Corynebacterium gr. D2*	1000	250	600	800	150	200	500	100	50	500	300	50
*Micrococcus luteus**	400	200	600	800	150	200	500	75	50	400	300	100
*Micrococcus sedentarius**	1000	200	600	800	150	200	250	200	50	500	300	100
Acinetobacter sp.	>1000	150	700	800	150	400	500	>500	>1000	500	400	>1000
Moraxella sp.	300	100	500	400	50	200	100	200	100	500	300	100
Alcaligenes sp.	>1000	300	>1000	>1000	400	200	>1000	>500	>1000	600	500	>1000
Fungi												
Candida albicans	>1000	60	500	400	175	200	200	>500	>1000	200	100	>1000
Cryptococcus neoformans	nt	100	300	300	200	150	200	250	100	150	150	200
Trichophyton rubrum	150	30	125	400	75	100	200	50	100	200	100	100
T. mentagrophytes	250	200	100	400	100	100	200	80	400	100	100	50
T. interdigitale	250	200	175	400	100	100	200	400	400	200	100	200
Microsporum canis	250	50	125	300	75	100	150	200	50	200	100	200

Armpit microflora												
Staphylococcus xylosus	>1000	400	1000	800	250	400	500	400	300	400	400	200
S. epidermidis	>1000	400	>1000	400	300	200	600	400	100	400	300	1000
S. haemolyticus	>1000	400	>1000	>1000	400	200	800	400	500	800	400	>1000
Micrococcus luteus	400	200	600	800	150	200	500	75	50	400	300	50
*Corynebacterium xerosis**	nt	150	700	600	150	400	500	150	50	500	300	100

Note: A.C. = *Acorus calamus* (rhizome), C.Z. = *Cinnamom zelyanicum* (bark), Citronelle: C.W. = *Cymbopogon winterianus* (leaves), "Bourbon" geranium: P.G. = *Pelargonium graveolens* var. *rosat* (leaves), L.S. = *Lippia sidoides*, oregano: O.O. = *Origanum officinale* (fresh flowers), *Palmarosa:* C.M. = *Cymbopogon martinii* var. *motia* (dried plant), patchouli: P.C. = *Pogostemon cablens 0* (dried plant), sandalwood: S.A. = *Santalinus albus* (wood), savory: S.H. = *Satureia hortensis* (fresh plant), thyme: T.V. = *Thymus vulgaris* (dried plant), vetiver: V.Z. = *Vetiveria zizanoides* (rhizome).
*Bacteria associated with smelly odors; nt = not tested

TABLE 14.2. Comparative evaluation of antimicrobial properties of six essential oils to their main components

	MIC (μg/ml)											
	Oxygenated Terpenoids				Phenols				Aldehydes			
	Palmarosa	Geraniol	Sandal-wood	Santalol	*Lippia sidodes*	Thymol	Clove	Eugenol	*Eucalyptus Citriodora*	Citronellal	Cinnamom	Cinnam-aldehyde
	Geraniol: 76%		Santalol: 72%		Thymol: 70%		Eugenol: 83%		citronellal: 82%		Cinnam-aldehyde	
Food microflora												
Bacteria												
Staphylococcus epidermidts	600	500	100	100	300	200	700	800	>1000	>1000	400	250
*S. hominis**	800	500	500	400	300	200	600	800	>1000	>1000	150	400
*S. cohnii**	800	500	500	300	250	200	600	600	>1000	>1000	400	400
Corynebacterium gr. C*	400	400	20	100	100	200	700	500	>1000	>1000	300	150
Corynebacterium gr. B*	300	400	20	100	150	200	700	500	>1000	>1000	100	200
Corynebacterium gr. D2	500	400	50	50	150	50	600	500	>1000	>1000	250	400
Micrococcus luteus	500	300	20	30	150	200	600	500	>1000	>1000	200	400
M. sedentarrius	250	400	20	100	150	50	600	400	>1000	>1000	200	400
Acinetobacter sp.	500	250	>1000	>1000	150	200	800	250	>1000	>1000	150	150
Moraxella sp.	100	200	20	100	50	200	700	200	>1000	>1000	100	50
Alcaligenes sp.	>1000	>1000	>1000	>1000	400	400	1000	800	>1000	>1000	300	250
Fungi												
Candida albicans	200	200	>1000	>1000	175	75	400	400	>1500	1500	60	
Cryptococcus neoformans	200	100	100	100	200	50	200	150	600	600	100	
Trichophyton rubrum	200	200	100	200	75	50	100	100	500	500	30	

T. mentagrophytes	200	200	400	200	100	50	100	100	500	300	200	
T.interdigitale	200	200	400	200	100	50	100	200	500	500	200	
Microsporum canis	150	175	50	200	75	50	100	100	200	300	50	
Armpit microflora												
S. xylosus	500	500	120	200	250	400	1000	600	>1000	>1000	400	150
S. epidermidis	600	500	100	100	300	200	700	800	>1000	>1000	400	250
S. haemolyticus	800	600	500	400	400	400	>1000	800	>1000	>1000	400	400
Micrococcus luteus	500	300	20	30	150	200	600	500	>1000	>1000	200	400
*Corynebacterium xerosis**	500	400	20	20	150	200	600	600	>1000	>1000	150	250

*Bacteria associated with smelly odor

Diversity in Activity of Essential Oils

The impact of these oils on skin microflorae appears very dissimilar: Essential oils of *Acorus calamus* are devoid of activities against yeasts and any bacteria. The essential oil of *Cymbopogon citratus* has only fungistatic properties. Sandalwood and patchouli essential oils inhibit many pathogenic fungi, but also *Corynebacterium xerosis* (Rennie et al., 1990). These oils are simultaneously antagonistic against some Gram-positive bacteria such as *Staphylococci* or *Micrococci* (linked to smelly feet) with fermentations by exoenzymes (proteases or lipases). On the contrary, they do not inhibit the suitable saprophytes, which are not producers of exoenzymes. So, these perfumes may be proposed as pleasant deodorants. It would be very desirable if such products could, at the same time, prevent or cure superficial mycoses.

Cryptococcus neoformans is a resistant, opportunistic fungus present in patients stricken with HIV (Stevens et al., 1999). Cutaneous cryptococcosis is an external manifestation of a systemic mycosis. All selected oils behaved well against this yeast. Perhaps these volatile compounds can penetrate the protective mucilage around cells. Many essential oils may be recommended as adjuvant treatments in primary cryptococcosis. Nevertheless, in no case would essential oils be substituted for classic antibiotherapy.

CONCLUSIONS

For the general public, and also for many doctors, essential oils may be considered strong antimicrobial agents. The results of studies have shown that essential oils should be considered as treatments against fungal and bacterial infections. First, some oils with very effective MICs are able, with suitable concentrations, to eliminate pathogenic fungi. Moreover, it is artful to utilize some specificities with a view to promote the development of saprophytes. The presence of such microorganisms covering the skin protects it against parasitic invaders.

In the future, investigations with essential oils might reveal a best approach to an ideal antimycotic able to conserve beneficial microflorae. It is possible to treat tineas, intertrigo, ringworm, athlete's foot, or many onychomycoses with cheap, perhaps long, external treatment founded on a specific aromatherapy. A network of laboratories localized in Brazil, Cameroon, China, France (including Antilles), India, Spain (Canaries), Switzerland, Togo, Turkey, and Uzbekistan is in place. All over the world, it is possible to find plants that produce antimycotic essential oils.

REFERENCES

Ake Assi, L . (1982). Santé et valorisation des plantes médicinales en Côte d'Ivoire. *Bothalia*. 14 (3-4): 603-605.

Baratta, M.T., Dorman, H.J.D., Deans, G., Biondi, D. and Ruberto, G. (1998). Chemical composition, antimicrobial and antioxydative activity of laurel, sage, rosemary, oregano and coriander essential oils. *J. Essent. Oil Res.* 10: 618-627.

Bodey, G., Bueltmann, B., Duguid, W., Gibbs, D., Hanak, H., Hotchi, M., Mall, G., Martino, P., Meunier, F., Milliken, S., et al. (1992). Fungal infections in cancer patients: An international autopsy survey. *Eur. J. Microbial Infect. Diseases* 11(2): 99-109.

Chabasse, D., Guigen, C. and Contet-Audonneau, N. (1999). *Mycologie Médicale*. Masson (Ed.). Paris.

Daltio, R. and Lunardi, M. (1991). Prevalence of superficial mycosis in the Aosta valley region of Italy from 1984 to1989. *Mycopathologia*. 116: 155-158.

Dromer, F. and Dupont, B. (1996). The increasing problem of fungal infections in the immunocompromised host. *J. Mycol. Med.,* 6(Suppl. 1): 1-7.

Franchomme, P., Jollois, R. and Penoel, D. (1996). *L'aromathérapie Exactement,* Jollois, R. (Ed.), Limoges, France.

Hammer, K.A., Carson, C. and Riley, T.V. (1999). Antimicrobial activity of essential oils and other plant extracts. *J. Appl. Microbiol.* 86: 985-990.

Hostettmann, K. (1997). *Tout Savoir sur le Pouvoir des Plantes Sources de Medicaments*, Favre S.A. (Ed.), Lausanne, Suisse.

Kaufman, L. (1998). Penicilliosis marneffei and pythiosis emerging tropical diseases. *Mycopathologia* 143: 3-7.

Kiss, G., Radvanyi, S.Z. and Szigeti, G. (1997). New combination for the therapy of canine otitis externa. I. Microbiology of otitis externa. *J. Small Animal Practice,* 38: 51-56.

Lacoste, E., Chaumont, J.P., Mandin, D., Plumel, M.M. and de Abreu-Matos F.J. (1996). Les propriétés antiseptiques de l'huile essentielle de *Lippia sidoides* Cham. Application à la microflore cutanée. *Ann. Pharm. Fr.* 54(5): 228-230.

Lloyd, D.H. and Noble, W.C. (1999). Use and abuse of antibiotics in veterinary dermatology. *Veterinary Dermatology* 10(3):161.

Marshall, J., Holland, K. and Gribbon, E.M. (1988). A comparative study of the cutaneous microflora of normal feet with low and high levels of odour. *J. Appl. Bacteriol.* 65: 61-68.

Marshall, J., Leeming, J.P. and Holland, K. (1987). The cutaneous microbiology of normal human feet. *J. Appl. Bacteriol.* 62: 139-146.

Omidynia, E., Farshchian, M., Sadjjadi, M., Zamanian, A. and Rashidpouarei, R. (1996). A study of dermaphytoses in Hamadan the government ships of West Iran. *Mycopathologia* 133: 9-13.

Ozcan, M. (1999). Antifungal properties of propolis. *Grasas y aceites* 50(5): 395-398.

Porro, A.M., Yoshioka, M.C.N., Kaminski, S.K., De Palmeira, M., Fishman, O. and Alchorne, M.M.A. (1997). Disseminated dermatophytosis caused by *Micro-*

sporum gypseum in two patients with the HIV syndrome. *Mycopathologia* 137: 9-12.

Puccini, S., Valdre, A., Papini, R. and Manciati, F. (1992). *In vitro* susceptibility to antimycotics of *Microsporum canis* isolates from cats. *Javma* 201(9): 1375-1377.

Rennie, P.J., Holland, K.T., Mallet, A.I. and Watkins, W.J. (1990). The skin microflora and the formation of human axillary odour. *Int. J. Cosmet. Science* 12: 197-207.

Rippon, J.W. (1987). Symposium on medical mycology. *Mycopathologia* 99: 143-146.

Stevens, D., Denning, D., Shatsky, S., Armstrong, R., Adler, J.D. and Bradley, H.L. (1999). Cryptococcal meningitis in the immunocompromised host: Intracranial hypertension and other complications. *Mycopathologia* 146: 1-8.

Van De MeuleBroucke, E., Mounkassa, B., De Loye, J., Jousserand, P., Poujade, F. and Petithory, J.C. (1999). Teigne du cuir chevelu en milieu scolaire rural au Mali. *J. Mycol. Med.* 9: 111-113.

Viollon, C. and Chaumont, J.P. (1994). Antifungal properties of essential oils and their main components upon *Cryptococcus neoformans*. *Mycopathologia* 128: 151-153.

Yang, D., Michel, D., Mandin, D., Andriamboavonjy, H., Poitry, P., Chaumont, J.P. and Millet-Clerc, J. (1996). Propriétés antifongiques et antibactériennes, *in vitro,* de trois huiles essentielles de Patchouly d'origines différentes. *Acta Bot. Gallica* 143(1): 29-35.

Chapter 15

Antimycotic Potential in Plants of Central and West Africa

Jean-Pierre Chaumont

INTRODUCTION

Since the second half of twentieth century, human mycoses have increased due to numerous factors, which are very different from one another, but all related, in fact, to some modern practices of life (Rippon, 1987). These practices include wearing tight, waterproof clothing that does not allow evaporation of sweat; frequent use of swimming pools; and use of more powerful cleaners that destroy the superficial, protective structure of the skin. Chemotherapy, which leaves the immune system vulnerable to infection, can foster the development of external or systemic mycoses. The growth of opportunistic fungi can occur on sites of catheters, and in some cases in the transplantation of organs. Finally, patients affected with HIV are more prone to invasion of opportunistic fungi which are able to develop the disease toward a fatal phase. This is due to weakening of the immune system.

In the tropics, there is a wide distribution of fungal pathogens. However, some fungi are limited to a particular region (Adou-Bryn et al., 1977; Badillet, 1991; Moalic et al., 1999; Van De MeuleBroucke et al., 1999; Chandenier et al., 2000). Fungal infections are increasing at a fast rate, for example, tinea capitis, generally caused by dermatophytes, such as *Microsporum audouinii* var. *langeronii* or *Trichophyton soudanense* (a suggestive name) is increasing in cosmopolitan cities (Chabasse et al., 1999). In Africa, with the explosive outbreak of AIDS, the situation has worsened quickly during the past few years. Particularly distressing are immunocompromised patients who are victims of extremely grave and often mortal systemic mycosis. Various reports exist of candidoses (Bodey and Anaissie, 1989; Di Silverio et al., 1991; Taelman et al., 1991; De Fatima-Costa-Pires et al., 1996; Nucci et al., 1998), invasive aspergilloses (Perraud et al., 1987; Denning, 1992; Gyanchandani et al., 1998) and especially cryptococcosis (Ehui et al., 2000), which decimate the populations of HIV-positive persons by developing into invasive meningitis against which it is impossible to be cured due to lack of effective and cheap medicines. Taking into consideration such catastrophic conditions and scarcity of means, new, effective, and inexpensive therapeutic alternatives must be developed. The situation is serious especially in case of opportunistic pathogens, which may develop resistance to available antimycotics.

Africa is an immense continent with well-variegated flora in mainly tropical climates (Walker and Sillans, 1961). The leaves and barks of trees, for example, represent a totally unexploited vegetable biomass, wherein pharmacologically active molecules are metabolized. It is hoped that this will be a source of untapped effective therapeutic reserve against pathogenic fungal infections because synthetic antiseptics and antibiotics have become gradually less effective due to the development of resistance (Viollon and Chaumont, 1994). Some products of vegetable origin also play additional roles in classic therapies.

Historically and culturally, Africans have used plant-derived treatments in traditional medicine. That is why researchers have begun to scientifically verify the benefits of plants of traditional medicine. A rather recent science, ethnopharmacology has increased in Africa. Periodical surveys and scientific screenings have assisted in the development of many new drugs. Today, the speedy progress in ethnopharmacology has generated the reliability, and has encouraged the screenings of plants (Hostettmann, 1997).

In regard to the search for plants with antimycotic properties, very active investigations have begun in Central Africa and in West Africa (Baba-Moussa et al., 1997; Baba-Moussa, 1999).

CENTRAL AFRICA: CAMEROON, REPUBLIC OF CONGO, AND ZAIRE

Walker and Sillans (1961) have reported on the customary use of plants of the Gaboon. Thirty plants are used for "cosmetic" purpose, and only two trees are clearly used for the treatment of "the tinea": *Daniella klainei* and *Pterocarpus soyauxii*. Clearly, it is necessary to search for more precise information.

Baba-Moussa et al. (1999) investigated volatile fractions of vegetables. Bouchet, Legin, et al. (1986) investigated padouk, or coralwood, *Pterocarpus soyauxii,* a very common tree in Black Africa. The colored wood contains isoflavonoids, type homopterocarpin, which are by-products of the pterostilbene and red pigments, santalin A and B (Cardon and du Chatenet,

1990). According to Bouchet, Masanes, et al. (1986), the antidermatophytic activity is attribued to the homopterocarpin. Triponney (2000) found activities in Gymnoascaceae with etheropetrolic, chloroformic, and methanolic extracts of *P. soyauxii* wood. On the other hand, the barks and the sheets of the same tree are divested of such properties.

Investigations were carried out by Laine et al. (1985) on some Congolese plants in which the aqueous extracts were rich in saponins and tannins. No significant inhibition was able to be discovered on *Candida albicans*. On the other hand, *Garcinia huilensis, Maesopsis eminii,* and *Pentaclethra eetveldeana* extracts reduced appreciably in vitro growth of *Microsporum canis*.

Bouchet, Masanes, et al. (1986) also investigated saponosidic extracts of 12 plants of Zaire to clarify the role of these compounds on nine strains of human pathogenic fungi common in Africa. Results were very dissimilar according to fungal target envisaged. In regard to yeasts, camelliagenine A and A1-borrigenol from *Harpullia cupanoides* presented a remarkable activity against *Candida albicans* and *C. glabrata* (see Table 15.1).

The jegosapogenol of *Majidea fosteri* seemed the most adapted to the local treatment of dermatophytes. Saponins supplied by this plant strongly af-

TABLE 15.1. Antimycotic activities of saponins extracted from some plants of Zaire

		Concentrations mg/mL^{-1}			
Saponins from	**Strains Tested**	**4**	**2**	**1**	**0(control)**
Olax obtusifolia	*Microsporum gypseum*	0	++	+++	++++
Blighia welwitschii	*Microsporum canis*	+++	+++	+++	++++
	M. gypseum	++	+++	+++	++++
Majidea fosteri	*Microsporum gypseum*	0	0	++	++++
	M. canis	0	++	++	++++
	M. nanum	+	++	++	++++
	Trichophyton tonsurans	0	0	++	++++
	T. mentagrophytes	+	++	++	++++
	T. rubrum	+	++	++	++++
	T. soudanense	0	+	+	++++
	T. ajelloi	++	++	++	++++
	Epidermophyton floccosum	++	++	++	++++

Source: Adapted from Bouchet, Masanes, et al., 1986.

fected the walls and the cytoplasm of *Microsporum gypseum* (Yaziji et al., 1991).

Current research is focused on Cameroonian Rutaceae belonging to the genus *Zanthoxylum* (Ngono Ngane et al., 2000). The various organs of two species, *Z. leprieuri* and *Z. xanthoxyloides,* were tested against some human pathogenic fungi. The best activities have been observed against some dermatophytes, and the yeasts *Candida albicans* and *Cryptococcus neoformans.*

Recently, Amvam-Zollo et al. (1998) reported on the essential oil of plants of Cameroon, obtained by hydrodistillation, and subjected to tests for antimycotic activities. Simultaneously, these oils were analyzed by GC-MS. Three essential oils from the family Lamiaceae showed strong antimycotic activities. They were from *Ocimum basilicum* (linalool chimiotype), *O. gratissimum* (thymol chimiotype), and *Thymus vulgaris* (rich in thymol, terpinene, and *p*-cymene). These results confirmed what had been noticed already for some plants geographically disseminated all over the world, but possessing the same active compounds.

WESTERN AFRICA: TOGO AND BENIN

Problems caused by mycosis in Western Africa are quite identical to those previously discussed. Dermatophytes, widespread all over Africa (Adou-Bryn et al., 1977), are found in great majority on other strain opportunists associated to immunodeficiencies and particularly to HIV. According to Baba-Moussa (1999) *Aspergillus fumigatus* is also a pathogenic agent dreaded in veterinary medicine.

In Togo and Benin, essential oils have been the object of attentive studies. Convincing results were first displayed by Baba-Moussa et al. (1997, 1999). In one study, an oriented in vitro screening of 12 different essential oils was tested against different human pathogenic strains. The most promising oils were from *Cymbopogon citratus* and *C. schoenanthus, Eucalyptus citriodora, Lippia multiflora,* and *Ocimum gratissimum* (Baba-Moussa, 1999). *Aspergillus flavus* (rarely met as a human parasite) was found to be very resistant to these volatile antifungals, whereas *Microsporum gypseum,* in the same conditions, was the most vulnerable. *Ocimum gratissimum* contains several chemotypes, some of which are rich in thymol (Pino et al., 1996), and others in eugenol or in ethyl cinnamate (Dubey et al., 2000). For a long time, these compounds have been known for antifungal properties used in medicine and agriculture.

Baba-Moussa (1999) evaluated the antimycotic properties of *Cymbopogon citratus* against *Candida albicans* and *Microsporum gypseum*. Recently, this work was supported by Koba (Chaumont et al., 2001) who targeted pathogenic fungi localized on skin and nails. The target fungi were the most significant representatives of cutaneous microflora. The results of the two studies were significant. The essential oils showed dominant concentrations of the mixture geranial-neral from 75 to 84 percent. "Citral plants" are powerful antimycotics in numerous circumstances (see Table 15.2). In some studies, *Cymbopogon schoenanthus* essential oil was an excellent anti-

TABLE 15.2. MIC ($\mu l \cdot ml^{-1}$, on Sabouraud agar) of essential oils of some plants growing in Togo against pathogenic fungi

Fungi	Cc	Cs	Ap	Pa	Og	Oc	Os	Ec
Candida albicans	100	750	150	600	300	1000	600	>1000
Cryptococcus neoformans	100	750	75	400	150	600	400	600
Aspergillus fumigatus	100	600	150	600	300	1000	800	>1000
Trichophyton mentagrophytes	100	600	60	250	200	>1000	250	500
T. mentagrophytes var *interdigitale*	100	300	75	300	150	600	250	500
T. rubrum	200	300	60	150	150	>1000	125	500
T. erinacei	150	450	70	250	150	600	200	–
T. soudanense	100	450	50	500	150	500	200	500
T. violaceum	100	400	40	250	200	500	200	400
Epidermophyton floccosum	100	600	70	350	200	650	200	–
Microsporum canis	100	600	70	300	200	700	300	200
M. gypseum	200	600	100	200	200	>1000	200	500
Scytalidium dimidiatum	600	>1000	500	>1000	600	>1000	>1000	>1000

Source: Adapted from Chaumont et al., in press.
Note: Cc: *Cymbopogon citratus,* Cs: *Cymbopogon schoenanthus,* Ap: *Aeolanthus pubescens,* Pa: *Platostoma africana,* Og: *Ocimum gratissimum* (thymol chemotype), Oc: *Ocimum canum,* Os: *Ocimum sanctum,* Ec: *Eucalyptus citriodora.*

mycotic in human medicine, but in a lesser degree in others. Moreover, piperitone, the main constituent in this oil is not considered a major antifungal. Nevertheless, the oil of three *Cymbopogon* spp. showed remarkable properties in vitro against some human pathogenic fungi in tropical regions: *Microsporum canis, Scytalidium dimidiatum, Trichophyton soudanense,* and *T. violaceum* (Koba, unpublished). *Cryptococcus neoformans* is sensitive in vitro to the essential oil of *Cymbopogon citratus* (MIC: 100 $\mu g \cdot ml^{-1}$). Koba and Chaumont (in press) investigated other, lesser known Togolese Lamiaceae, *Aeolanthus pubescens, Platostoma africana.*

Kpakote et al. (1998) reported that *Burkea africana, Parinari curatellifolia, Prosopis africana,* and *Terminalia glaucescens* extracts inhibited more or less the growth of *Candida albicans* in vitro. *Aspergillus fumigatus,* a relatively more resistant fungus, is inhibited only by *Garcinia ovalifolia.* The most important contribution in this field was made by Baba-Moussa (1999) on members of the family Combrétaceae belonging to genera *Combretum, Pteleopsis,* and *Terminalia.* In *Pteleopsis suberosa,* methanolic extracts of barks and leaves presented a significant MIC (Table 15.3). The extracts were subjected to column chromatography. Some fractions were found to be very active against pathogenic yeasts.

The potentialities of exploitation of the medicinal resources of African plants are still far from being totally investigated. Adequate measures to protect a maximum of plant taxa should be taken before it is too late.

TABLE 15.3. MIC ($\mu g\ ml^{-1}$ on Sabouraud agar) of methanolic extracts of *Ptelopsis suberosa* from Benin, against pathogenic yeast and dermatophytes

Fungi	**Leaves Extract**	**Bark Extract**
Candida albicans	500	250
C. glabrata	500	500
Cryptococcus neoformans	–	500
Epidermophyton floccosum	125	125
Microsporum canis	125	125
M. gypseum	500	250
Trichophyton rubrum	250	250
T. mentagrophytes	125	125

Source: Adapted from Baba-Moussa, 1999, p. 95.

CONCLUSION

Human mycoses are largely widespread in Central and West Africa including tinea, ringworm, athlete's foot, and onychomycosis, etc. These fungal diseases constitute a permanent problem because of many factors: difficulty of stocking antifungal medicines in remote villages; expensive cost of synthetic drugs for impoverished populations; and resistance of many strains to antimycotics.

Phytotherapy by means of healers is largely used in many African countries. This practice is debatable because its origins are found mainly in the empiricism of traditional and folk medicine.

The role of the ethnopharmacologist is to demonstrate the bioactivity of natural and well-known drugs with scientific tests and proofs. Several African plants have shown efficacious antimycotic properties, particularly plants of West and Central Africa. In addition, these natural extracts are cheap and easy to use. Cultures of more potential plant species all over the world in similar soils and climates should be undertaken. The yield of active compounds would have to be optimized and it would be essential to formulate topical drugs with very common and cheap plant extracts.

REFERENCES

Adou-Bryn, K. D., Yeo, N., Kassi, E. A., Ouhon, J., Assoumo, U. A. and Penali, K. N. (1977). Athlete's foot: Mycological etiology in marine soldiers in Abidjan (Ivory Coast). *J Mycol Med.* 3: 142-144.

Amvam-Zollo, P. H., Biyiti, L. C., Tchoumbougnang, F., Menut, G. and Bouchet, P. (1998). Aromatic plants of tropical central Africa. Part XXXII. Chemical composition and antifungal activities of thirteen essential oils from aromatic plants of Cameroon. *Flavor and Fragrance Journal* 13: 107-114.

Baba-Moussa, F. A. K. (1999). Recherche sur les propriétés antifongiques de plantes utilisées en médecine traditionnelle au Bénin et au Togo. Thèse Sciences *Pharmaceutiques n°* 208, Reims, France.

Baba-Moussa, F. A. K., Akpagana, K. and Bouchet, P. (1999). Antifungal activities of seven West African Combretaceae used in traditional medicine. *J of Ethnopharmacology* 66: 335-338.

Baba-Moussa, F. A. K., Koumaglo, K., Ayedoun, A., Akpagana, K., Moudachirou, M. and Bouchet, P. (1997). Activité antifongique d'huiles essentielles extraites au Bénin et au Togo. *Cryptogamie-Mycologie* 18(2): 165-168.

Badillet, G. (1991). *Dermatophyties et Dermatophytes. Atlas Clinique et Biologique.* Paris: Varia.

Bodey, G. P. and Anaissie, E. J. (1989). Chronic systemic candidasis. *Eur J Clin Microbiol Infect Dis,* 855-857.

Bouchet, P., Legin, H., Gassita, J. N., Issembre, S., Bobichon, H. and Quirin, F. (1986). Propriétés antifongiques d'une plante africaine d'usage traditionnel: *Pterocarpus soyauxii*. *Bull Soc Fr Mycol Med.* XV(2): 509-512.

Bouchet, P., Masanes, C. and Delaude, C. (1986). Activités antidermatophytes et antilevures d'une série de saponines extraites de végétaux récoltés au Zaïre. *Bull Soc Fr Mycol Med.* XV(2): 533-538.

Cardon, D. and du Chatenet, G. (1990). Guide des teintures naturelles, Delachaux et Niestlé, Lausanne.

Chabasse, D., Guigen, C. and Contet-Audonneau, N. (1999). *Mycologie Médicale.* Paris: Masson.

Chandenier, J., Renard, J. C., Goma, D., Nzounza, R., de Bievre C., Ravisse, P., Huerre, M., Moyen, G., Iyouangaporo, A., Gathse, A., et al. (2000). Tropical mycoses diagnosed at Brazzaville University Hospital Congo. Clinical and mycological appearance before, during and after treatment. *J Mycol Med.* 10(2): 67-78.

Chaumont, J. P., Mandin, D. and Koba, K. S. (in press). Activiés antimicrobiennes de cinq huiles essentielles de Lamiaceae togolaises vis-à-vis de germes représentatifs de la microflore cutanée. *Acta Botanica Gallica.*

De Fatima-Costa-Pires, M., Birman, E., Riera-Costa, C., Gambale, W. and Rodriguez, P. C. (1996). *Candida albicans* biotypes isolated from the oral cavity of HIV positive patients. *Rev Microbiol Sao Paulo* 27(1): 46-51.

Denning, D. W. (1992). Aspergillose invasive au cours du Sida. Revue générale. *J Mycol Med* 1(suppl.): 35-41.

Di Silverio, A., Brazzelli, V., Brandozzi, G., Barbarini, G., Maccabruni, A. and Sacchi, S. (1991). Prevalence of dermatophytes and yeasts (*Candida* sp and *Malassezia furfur*) in HIV patients. *Mycopathologia* 114: 103-107.

Dubey, N., Tiwari, N., Mandin, D., AmdrianBoavonjy, H. and Chaumont J. P. (2000). Antifungal properties of *Ocimum gratissimum* essential oil (ethylcinnamate chemotype). *Fitoterapia* 71: 567-569.

Ehui, E., Eholie, S. P., Kakou, A., Bery, I., Diomande, I. M., Adou-Bryn, K. D., Bissagnene, E. and Kadio, A. (2000). Disseminated Cryptococcosis revealed by cutaneous manifestations in the course of AIDS. *J Mycol Med* 10: 51-61.

Gyanchandani, A., Khanz, K. and Prasad, R. (1998). Fungal infections in immunocompromised patients with pulmonary tuberculosis as a secondary disease. *J Med Mycol* 3: 147-152.

Hostettmann, K. (1997). Tout savoir sur le pouvoir des plantes sources de médicaments. Ed Favre, Lausanne, Suisse.

Kpakote, K. G., Akpagana, K., De Sousa, C., Nenonene, A. Y., Djagba, T. D. and Bouchet, P. (1998). Les propriétés antimicrobiennes de quelques espèces de cure-dents du Togo. *Annales Pharmaceutiques Françaises* 56(4): 184-186.

Laine, C., Banikina, J., Vaquette, J., Chaumont, J. P. and Simeray, J. (1985). Activités antifongiques d'écorces de troncs de sept phanérogames congolaises. *Pl Med Phytother* XIX(2): 75-83.

Moalic, E., Zerilli, A. and Le Flohic, A. M. (1999). Biodiversity of oral fungal flora in Bénin children and relation with dental caries. *J Mycol Med* 9(4): 214-217.

Ngono Ngane, A., Biyiti, L., Amvam-Zollo, P. H. and Bouchet, P. (2000). Evaluation of antifungal activity of extracts of two cameroonian Rutaceae: *Zanthoxylum leprieuri* and *Zanthoxylum xanthoxyloides*. *J of Ethnopharmacology* 70: 335-342.

Nucci, M., Silveira, M. I., Spector, N., Silviera, F., Velasco, E., Martin, C. A., DeRossi, A., Colombo, A. and Pulcheri, W. (1998). Fungemia in cancer patients in Brazil: Predominance of non *C. albicans* species. *Mycopathologia* 141: 65-668.

Perraud, M., Piens, M. A., Nicoloyannis, N., Girard, P., Sepetjan, M. and Garing, J. P. (1987). Invasive nosocomial pulmonary aspergillosis: Risk factors and hospital building works. *Epidem Inf* 99: 407-411.

Pino, J. A., Rosado, A. and Fuentes, V. (1996). *Ocimum gratissimum* grown in Cuba. *J of Essential oil Research* 8(2): 139-141.

Rippon, S. W. (1987). Symposium on medical mycology. The new opportunistic fungal infection diagnosis, isolation, identification and impact on mycology. *Mycopathologia* 99: 143-146.

Taelman, H., Clerin X., Kafame A., Batungwanyo J., Nyirabareja A. and Bogaerts J. (1991). Oral candidasis in HIV infections. *The Lancet* 338: 761.

Triponney, F. (2000). *Pterocarpus soyauxii* Tauber: Études botanique et chimique; propriétés antimicrobiennes sur certaines microflores cutanées. These, pharmacie n° 250016, Besançon, France.

Van De MeuleBroucke, E., Mounkassa, B., de Loye, J., Jousserand, P., Poujade, F. and Petithory, J. C. (1999). Teigne du cuir chevelu en milieu scolaire rural au Mali. *J Mycol Med* 2: 111-114.

Viollon, C. and Chaumont, J. P. (1994). Antifungal properties of essential oils and their main components upon *Cryptococcus neoformans*. *Mycopathologia* 128: 151-153.

Walker, A. and Sillans, R. (1961) Les plantes utiles du Gabon, Lechevallier, Paris.

Yaziji, A. L., Al Awad, M. and Bouchet, P. (1991) Intérêt de la technique de Thiery (PAT Ag) pour l'étude des dermatophytes en microscopie électronique à transmission. Application à l'étude des saponines antifongiques de *Majidea fosteri* (Sapindaceae). *J Mycol Med* 1: 193-200.

Chapter 16

Antimycotic Principles of *Pentanema indicum* (L.) Ling

Saradha Vasanth
Mahendra Rai

INTRODUCTION

Pentanema indicum (L.) Ling occurs in China, Pakistan, Myanmar, Sri Lanka, Thailand, and Africa. In India, the plant is distributed up to 1,200 m in Madhya Pradesh, Uttar Pradesh, Bihar, Maharashtra, Jammu and Kashmir, and Himachal Pradesh. *Pentanema indicum,* also known as banjauri (Hindi), is used as an ethnomedicinal plant in the Bihar state of India, and is reported to produce sterility in women. Banjauri literally means, an agent which makes a woman sterile. In folkloric practice, the whole plant with some pappus heads are ground with water and administered for three consecutive days on an empty of stomach immediately after delivery. Phytochemical studies on the plant have resulted in the characterization of new sesquiterpene lactones, vicolides A-E (Purushothaman et al., 1981; Purushothaman and Vasanth, 1986; Vasanth and Kundu, 1995).

Monoterpenediol, vicodiol, triterpenoid glucoside, vicoside A and B (Vasanth et al., 1990, 1991), and 28-noroleananes, vicogenin, and vicorgenin are present in *P. indicum* (Vasanth et al., 1991; Balakrishna et al., 1995). A flavone, 6-hydroxy luteolin 7.3'-dimethylether has been reported (Vasanth and Ahmad, 1998). The vicolides were shown to exhibit antifertility (Susan et al., 1985), antiinflammatory (Alam et al., 1992), antimicrobial (Hamsaveni et al., 1992; Rai and Vasanth, 1995), antifeedant (Vasanth et al., in press a,b) and phytogrowth activities (Krishnaveni et al., 1997). The antimycotic activities of the vicolides, flavone, and essential oil of the plant would be discussed.

Saradha Vasanth wishes to thank the director, Central Council for Research in Ayurveda and Siddha, New Delhi, for financial assistance.

MORPHOLOGY

Petnanema indicum is an herb. It is erect, pubescent with a woody rootstock, simple or branched at the top, up to 75 cm tall; stems are terete, grooved, hairy, and glabrous with age. The leaves are oblong lanceolate, with a semiamplexicaul base, acuminate, entire, or serrate dentate with recurved margins, 2-10 × 0.1-1 cm, scabrous, dark-green above, pale beneath, and sessile. The inflorescence is yellow, radiate, solitary axillary or terminal 1-2 cm across, on 1-6 cm long peduncles. The receptacle is convex, pitted, involucre campanulate; involucral bracts 3 to 4-seriate, linear-lanceolate, 0.2-0.4 cm long, acute, glabrous, and erect. Ray florets are 1-seriate, females are 0.8-1.0 cm long, three-toothed; corolla yellow. Disc florets are yellow, bisexual, 0.5 cm long, and five-lobed. Achenes are brown, obovate, 1 mm long, obtuse, narrow at base, thinly hairy. The pappus is white, hairs filiform, up to 0.3 cm long, few. Flowering and fruiting: September to June (Hajra et al., 1995).

CHEMISTRY

Phytochemical examination of the plant resulted in the isolation and characterization of (1) sesquiterpene lactone, (2) triterpenoids, and (3) hydrocarbons/esters.

Isolation of Phytochemicals

Sesquiterpenoid Lactones

Four germacranolides and one guaianolide were isolated: vicolides B (Figure 16.1:2), C (Figure 16.1:4), D (Figure 16.1:3), E (16.1:5), and A (Figure 16.1:1). Plants from the Madras region yielded vicolide A, B, and D. Assays of vicolides by HPLC revealed that the leaves were a rich source of vicolide A,B, and D, followed by the flowers. Vicolide concentration was very low in the roots, and vicolide B was practically absent in the stems. This observation was in agreement with an earlier investigation in certain Compositae species (Rodriguez et al., 1976).

The vicolides A,B,C,D and flavone were isolated from the chloroform extract of *Pentanema indicum B* by column chromatography over silica gel and crystallization (Figure 16.1). The essential oil was obtained by steam distillation of fresh plant (2 g/15 kg).

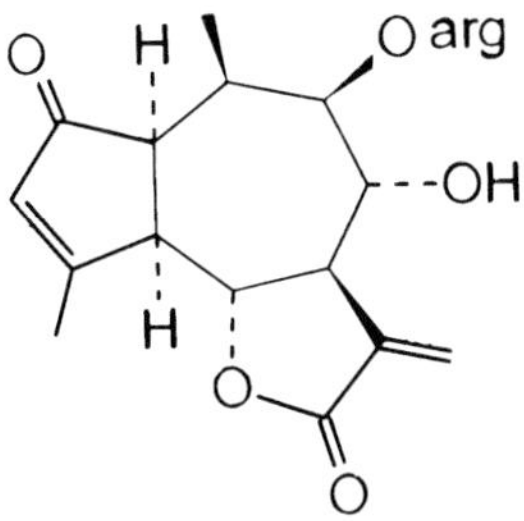

1.Vicolide A

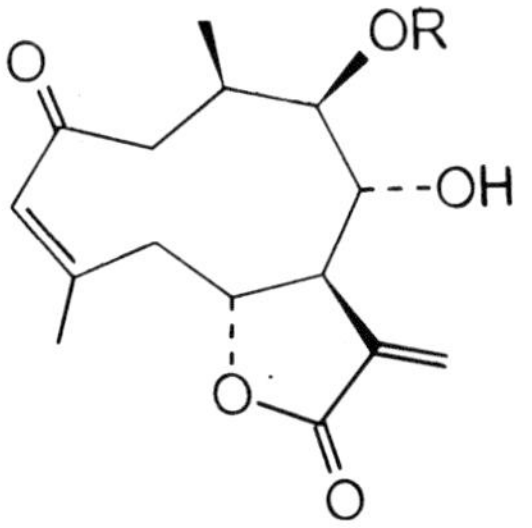

2.Vicolide B : R = angeloyl
3.Vicolide D = R = epoxy angeloyl

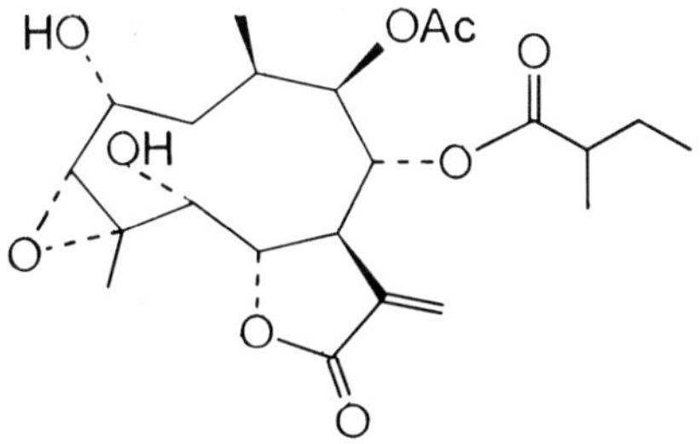

4.Vicolide C

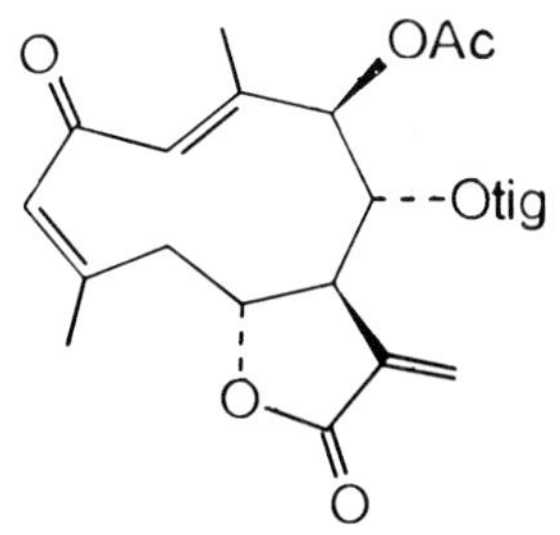

5.Vicolide E

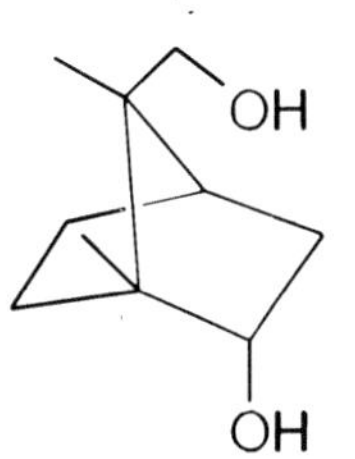

6.Vicodiol

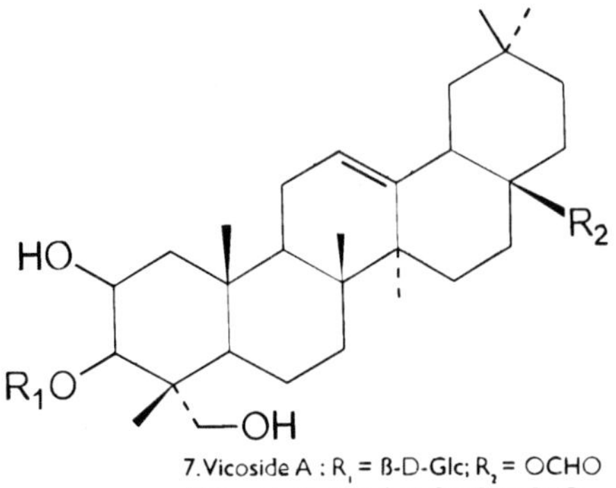

7.Vicoside A : R_1 = β-D-Glc; R_2 = OCHO
8.Vicoside B : R_1 = β-D-Glc; R_2 = CHO
9.Vicosigenin : R_1 = H; R_2 = OCHO
10.Vicogenin : R_1 = H; R_2 = OH

FIGURE 16.1. (1-5) Sesquiterpene lactones; (6) a monoterpene diol; (7-8) oleanane glucosides; (9-10) 28-noroleananes from *Pentanema indicum*

THE TEST ORGANISMS

Six fungi were used for antimycotic screening and maintained on Sabouraud dextrose agar and potato dextrose agar media at 28 ± 2°C. The fungi included *Candida albicans, C. tropicalis, C. krusei, Microsporum gypseum, Trichophyton terrestris*, and *Chrysosporium tropicum*.

Parent Suspension (PS)

For each test fungus, parent suspension was prepared by harvesting the test fungi grown in petri dishes for two weeks. The sterilized, washed conidia after filtration were diluted tenfold by serial dilution technique. The colony-forming units per ml were calculated and the test inocula were maintained between 1-5 × 10^5/ml.

Determination of Minimum Inhibitory Concentration (MIC)

MIC was determined by using a twofold serial dilution technique. The values were determined by visual observation. The assay tube had no growth and was recorded as MIC. The antifungal activity of essential oil and vicolides was conducted against pathogenic fungi.

VICOLIDES AS POTENT ANTIMYCOTICS

Antimycotic activity of vicolides and flavones against the studied fungi are given in Table 16.1. They were devoid of activity against all *Candida*

TABLE 16.1. Sensitivity of vicolides (A, B, C, D) to *M. gypseum, C. tropicum,* and *T. terrestris*

Drug	No.	MG (mg)	CT (mg)	TT (mg)	Average
Vicolide A	(1)	62.50	15.62	62.50	46.87
Vicolide B	(2)	125	31.25	31.25	62.50
Vicolide C	(3)	62.50	15.62	62.25	46.87
Vicolide D	(4)	125	31.25	31.25	62.50
Flavone	–	250	–	125	–
MN (Control)	–	7.81	15.62	15.62	13.01

Note: MG = *Microsporum gypseum,* TT = *Trichophyton terrestris,* CT = *Chrysosporium tropicum,* MN = *Miconazole nitrate.*

species even up to 1 mg/ml. Vicolides B and D showed similar MIC (31.25 μg/ml) for *M. gypseum, C. tropicum,* and *T. terrestris*. At a higher concentration (62.50 μg/ml) vicolide A and C showed similar inhibition of the test fungi. For *C. tropicum,* vicolides A and C exhibited almost equal MIC. However, these vicolides were active against *Bacillus subtilis, Staphylococcus aureus,* and *S. citreus*. The flavone showed activity at 125 μg/ml and 250 μg/ml against *T. terrestris* and *M. gypseum,* respectively. All were less active than miconazole nitrate.

Mares (1987) suggests that variation in the sensitivity may be due to varying permeability of the mycelial and conidial walls of these fungi. However, the ability of unsaturated lactones to act as inhibiting substances against several microbes is due to the ability of the molecule to penetrate the microbial cell. According to another school of thought, the activity of sesquiterpene lactones is due to their cytotoxic nature. The physiological processess of the test fungi are disturbed due to their cytotoxic nature.

The essential oil of the plant showed good activity against *C. albicans* (26 mm), *C. tropicalis* (18 mm), and *C. krusei* (36 mm) (Standard nystatin 10 mg/disc, 18 mm). The excellent inhibition against yeast is suggestive of its use as a potent drug for candidiasis and as a preservative when these yeasts are contaminants.

The antimycotic activity of the plant is due to the presence of several compounds, viz., vicolides, flavonoids, and also the essential oils of the plant.

REFERENCES

Alam, M.M., Susan, T., Joy, S., and Kundu, A.B. (1992). Antiinflammatory and antipyretic activity of vicolides of *Vicoa indica* DC. *Indian J. of Exptl. Biol.* 30:38-41.

Balakrishna, K., Vasanth, S., Kundu, A.B., Bhima Rao, R., and Connolly, J.D. (1995). Vicogenin, a 28-nor-12-oleananepent from *Vicoa indica* DC. *Phytochemistry* 40:335-336.

Hajra, P.K., Rao, R.R., Singh, D.K., and Uniyal, B.P. (1995). *Flora of India,* Volume 13, *Asteraceae.* Dehra Dun, India: Botanical Survey of India (BSI), Northern circle.

Hamsaveni Gopal, R., Vasanth, S., and Kundu, A.B. (1992). *In vitro* antimicrobial efficacy of *Vicoa indica, Fitoterapia* LXIII (6):546-549.

Krishnaveni, R., Shymala Devi, C.S., and Vasanth, S. (1997). Antiinflammatory and analgesic actions of 4', 5,6-trihydroxy 3', 7-dimethoxy flavone from *Vicoa indica. Indian J. Pharmacol.* 29:178-180.

Mares, D. (1987). Antimicrobial activity of protoanemonin, a lactone from Ranunculaceous plants. *Mycopathol.* 98:133-140.

Purushothaman, K.K. and Vasanth, S. (1986). Vicolide D, a new germacranolide from *Vicoa indica* DC. *Indian J. Chem.* 25 B:417.

Purushothaman, K.K., Vasanth, S., Cox, P.J., Akinniyi, J.A., Connolly, J.D., Rycroft, D.S., and Sim, G.A. (1981). Vicolide A, B and C—New sesquiterpenoid lactones from *Vicoa indica* DC., X-ray structure of vicolide B and C. *J. Chem. Res.* (Synop):374-375. *J. Chem. Res.* (Mini print):4371-4392.

Rai, M.K. and Vasanth, S. (1995). Laboratory evaluation of sensitivity of three keratinophilic fungi to some vicolides. *Hindustan Antibiotics Bulletin* 37(1-4): 48-50.

Rodriguez, E., Towers, G.H.N., and Mitchell, J.C. (1976). Biological activity of sesquiterpene lactones. *Phytochem.* 15:1573-1580.

Susan, T., Alam, M.M., and Purushothaman, K.K (1985). Antifertility activity of vicolide B in albino rats. *Indian Drugs* 22:512-515.

Vasanth, S. and Ahmad, A.A. (1998). 6-hydroxyluteolin 7,3' -dimethyl ether from *Vicoa indica* DC. *Fitoterapia* 69(2):179.

Vasanth, S. and Kundu, A.B. (1995). Vicolide E, a new sesquiterpene lactone from *Vicoa indica. Fitoterapia* 66:181-82.

Vasanth, S., Kundu, A.B., Panda, S.K., and Patra, A. (1991). Vicoside A, a 28-norteterpenoid glucoside from *Vicoa indica. Phytochemistry* 30:3053-3055.

Vasanth, S., Kundu, A.B., and Patra, A. (1992). Further oleanane triterpenoids from *Vicoa indica. J. Nat. Prod.* 55:1149-1151.

Vasanth, S., Kundu, A.B., Purushothaman, K.K., Patra, A., Vasantha, P., and Connolly, J.D. (1990). Isolation and characterization of vicodiol, a new monoterpenediol from *Vicoa indica* DC. *J. Nat. Prod.* 53:354-358.

Vasanth, S., Regina Mary, and Govinda Rajan, S. (in press, a). Antifeedant activity of vicolides from *Pentanema indicum. Fitoterapia.*

Vasanth, S., Regina Mary, and Govinda Rajan, S. (in press, b). Studies on vicolides of *Pentanema indicum* as phytogrowth promoters. *Journal of Medicinal and Aromatic Plant Abstracts.*

Chapter 17

Rapid Test Methods for Evaluation of Antifungal Activity

Lennart Gip
Christina Gip

INTRODUCTION

The assessment of topical antimycotic agents is frought with several difficulties. In vitro testing usually bears little resemblance to clinical usage. The growth of a dermatophyte on a nutrient culture medium is not comparable to its growth on stratum corneum in vivo or in vitro (Emmons, 1934; Klein, 1965). The testing of antimycotic compounds in experimentally infected animals, usually rabbits or guinea pigs, is expensive, complicated, time-consuming, and does not always give reproducable results. Clinical trials in humans can be difficult to control because of the lack of uniformity of naturally occurring lesions. The use of experimental infections in humans for the clinical evaluation of drugs may be ethically problematic. Also in experimental infections in animals and humans, there may be difficulties in detecting therapeutic effect because of the natural remission of the infections.

New ways to consider human models for in vivo and in vitro evaluation of antimycotic activity were suggested by Knight (1973) and Knudsen (1976). Knight (1973) assessed antimycotic activity within the stratum corneum by culturing conidia of *Trichophyton mentagrophytes* on stripped stratum corneum. Growth of the dermatophyte was compared on untreated and previously treated stratum corneum. By taking strippings over a period of time after one application of the drug, it was possible to follow the length of time that the drug activity remained in the stratum corneum. However, in other experiments using this model, it was difficult to obtain reproducable results, because of failing or poor growth of dermatophytes in the stripped stratum corneum from the untreated patients. It was even more difficult when *Candida* species were chosen as the test fungus.

Knudsen (1976) treated tape strippings already exposed to fungal infection of the horny layer and evaluated the efficacy. The infected tapes were either painted or suspended in different agents to be examined. After treatment, pieces of tape were placed, adhesive side down, on nutrient agar. The results were recorded after four weeks as growth or no growth of dermatophytes. No studies were run for other fungi, e.g., *Candida,* nor were micromorphological characteristics in cases of suppressed growth investigated. The advantage of both methods was the possibility to study the effect of fungi in their natural environment, the stratum corneum.

Assessing the retention time of drugs to avail maximum use of antimycotics is difficult. Gip (1988) has proposed a new technique of in vitro determination of lingering antimycotic effect of omoconazole cream.

The main goal of this chapter is to present a novel and fast screening test method for assessment of the effect of topical antimycotics, plant extracts, and essential oils. The method involves stripping the infected and untreated skin of patients suffering from tinea inguinalis or candidosis of the groin. After strippings, the infected tape is exposed to the test drug on a glass slide, whereafter it is pressed gently onto the agar surface of a microculture slide (Oricult DTM, Orion Diagnostica, Helsinki, Finland). Using this method, which combines Knudsen's method with microculture technique (Gip, 1977, 1981; Gip and Gip, 1984), quicker and more reliable results can be achieved by making perpetual studies of the possible micromorphological changes of the fungi of the treated infected scales.

The retention time of the drug can be evaluated by a simple method proposed by Gip (1988). The lingering effect of two 1 percent omoconazole nitrate cream formulations were studied following single topical applications. Tape strippings were taken at various intervals from the once-treated skin and applied onto to the surface of a substrate, previously inoculated with a dermophyte. The lingering effect of omoconazole nitrate in epidermis was determined by the degree of inhibition or by the noninhibition of growth of fungus under the tape.

THE TECHNIQUES

Rapid Test Method for In Vitro Evaluation of Antifungal Drugs

Collection of Infected Specimen from Patients

Dermatophyte infections of the groin were selected for tests because of the uniformity of such lesions. The patients were comprised of five men with tinea cruris (mean age 21 years), and five women with candidosis of the groin (mean age 64 years). None of the patients had been treated locally or systemically prior to the study and in none of the cases had the infections lasted for more than ten days. The infections were, in each case, confirmed by positive direct examination and conventional culture on Sabourauds's glucose agar.

Sampling Method

The strippings were taken with ultrathin, transparent, Scotch tape 850 (3M). Ten strippings (4 × 2 cm) were taken from each patient. Separate tape was used for each infected part(s) of the skin. Every second tape saturated with stratum corneum was exposed to the drugs to be tested. Proper controls (without exposure to drugs) were also maintained for comparison.

Treatment of Infected Tapes to Drugs

The test drug (0.5 ml) was placed on a sterile glass slide exposed to the agent. The infected tape was placed onto the glass slide, sticky side down, so that the whole stratum corneum area of the tape was exposed to the agent. Pilot experiments had earlier shown that exposure times of 1 and 30 minutes were the most suitable. After treatment, the tape was removed from the glass slide and rinsed in sterile water. Before inoculation, it was ensured that the various agents had not detached the stratum corneum from the tape. Every second tape was run the same way without exposure of the tape to the test drug as control. For each drug to be tested, and for each time of exposure and type of infection, there was one test run. Test substances are provided in Tables 17.1 and 17.2.

Inoculation and Incubation

The sticky side of each tape was pressed gently onto the agar surface of an Oricult DTM slide. These are transparent plastic slides, 7 × 2 cm, formed as shallow troughs and filled with DTM agar (Taplin et al., 1969) to a thickness of 4 mm. The slides were attached to the lid of a sterile plastic container and were easily removed from the lid for examination under a microscope. They were incubated at 28°C with the lids left loose to allow free passage of air.

TABLE 17.1. Growth on microculture slides (Oricult DTM) of fungi from infected tapes exposed to various drugs and chemicals

Test Drug	Grading of Growth* of Dermatophytes and *Candida albicans* (Within parentheses) Exposure Time 1 min		30 min	
Gentian violet, aqueous solution 0.1%	+	(+)	–	(–)
Potassium permanganate 0.1%	+++	(+++)	+	(+)
Potassium permanganate 1.0%	+	(+)	+	(+)
Alcoholic solution 70%	+	(–)	–	(–)
Chlorhexidine in alcohol 5%	+++	(–)	–	(–)
Benzoic acid in alcohol 5%	–	(–)	–	(–)
Tolnaftate solution 2%	–	(+++)	–	(+++)
Econazole solution 1%	–	(–)	–	(–)
Clotrimazole solution 1%	–	(–)	–	(–)
Bifonazole solution 1%	–	(–)	–	(–)
Oxiconazole solution 1%	–	(–)	–	(–)

*Normal growth = +++, suppressed growth = +, no growth = –

Grading of Growth

The microcultures were checked daily for growth with the final reading after one week. The grading of growth was assessed as follows:

- *Normal growth:* The same growth as in the control culture slide.
- *Suppressed growth:* Dermatophytes—short, fat, deformed, and/or curved hyphae seen outside the scales. *Candida*—reduced number of blastospores and torted hyphae.
- *No growth:* No fungal elements outside the infected scales.

Trichophyton rubrum was isolated in three, and *Epidermophyton floccosum* in two of the tinea cruris cases, whereas *Candida albicans* was recovered from all five patients suffering from candidosis. All of the microcultures used as controls were positive. Table 17.1 shows the gradings of growth of each test drug, exposure time, and fungus. The same results were recorded after two days and after one week of growth.

TABLE 17.2. The test chemicals/drugs and their concentrations

Chemical/Drug	Concentration
Gentian violet, aqueous solution	0.1%
Potassium permanganate, aqueous solution	0.1% and 1%
Alcoholic solution	70%
Chlorhexidine in alcohol	5%
Benzoic acid in alcohol (Spir. Mycocten, LEO)	5%
Tolnaftate solution (Focusan, Lundbeck)	2%
Econazole solution (Pevaryl, C@ag)	1%
Clotrimazole solution (Canesten, Bayer)	1%
Bifonazole solution (Mycospor, Bayer)	1%
Oxiconazole solution (Oceral, Roche)	1%

None of the tapes treated with imidazole derivatives showed any growth outside the scales. As expected, tolnaftate was shown to be effective only against dermatophytes. Differences in effect were noted for the two exposure times with some of the disinfectants. Suppressed, or even normal growth, was recorded after 1 min treatment. However, for all test drugs, except potassium permanganate, complete inhibition was noted after 30 min exposure. Therefore, a new series of experiments were run with three hours exposure to potassium permanganate 1 percent solution, however, without change in the test results.

In Vitro Determination of Lingering Antimycotic Effect

The Test Drug

Twenty-five gms of omoconazole nitrate cream (1 percent, batch RBT03 and 8608 E 12) was chosen as the test drug.

Volunteers

Eighteen healthy volunteers (13 women and 5 men, mean age 23 years) without a previous nor present history of dermatophyte infection were selected. Also, none of the volunteers were actually treated with antibiotics, corticosteroids, or other systemic drugs.

Application of the Drug

An area of skin, approximately 10 × 16 cm of the right arm of each volunteer, was treated once with the test cream. No previous treatment of the test area with soap, water, etc., was undertaken during the last three hours prior to the application of the drug.

Approximately 1 g of cream was applied to the test area of the right arm, which was then covered with a covering bandage. A plastic cover was used for the protection of the bandage when the volunteers took showers.

Strippings from the treated skin area were taken at 1, 2, 6, 12, 24, 48 hours, and 7 days after the single treatment of the skin. Transparent, ultra-thin Scotch tapes were used for the stripping. Each treated area was sampled only once with a 4 cm piece of the tape, which was then applied onto the surface of a microculture slide, previously inoculated with the aerial mycelium of a dermatophyte.

Controls

Strippings from the untreated skin area of the left upper arm of each volunteer were taken at the same time intervals and then applied onto the surface of a microculture slide.

The microculture slides were transparent culture slides, 2 × 7 cm, formed as shallow troughs and filled with DTM agar to a thickness of 4 mm (Gip, 1981). The slides were attached to the lids of sterile plastic containers and were easily removed for examination under a microscope.

Inoculation of each slide was carried out with the aerial mycelium of a four-week culture of a fresh isolate of *Trichophyton mentagrophytes* var. *granulosum* on Sabouraud glucose agar, whereby an unused tape (Scotch 850) first flattered the aerial mycelium of the culture plate and then the surface of the agar of the microculture slide. Pilot studies had earlier shown that this type of inoculation procedure gave a very even and uniform spread of round spores to the surface of the medium on the microculture slide (approximately 1,000 spores per mm^2). These slides were incubated at 28°C. Reading and registering of growth was carried out after two days.

Two strippings per time interval and per volunteer were taken—one from the treated area of the right arm and one as control from the untreated area of the left arm. Thus, there were a total of 144 samples from treated skin areas and the same number of samples as controls from the nontreated skin areas.

Grading of Growth

- +++ = *Normal growth:* As registered on the microculture slides with tape strippings from untreated skin, this corresponds to the production of slender septate hyphae emerging from the round spores after 2 days incubation time. In cases of tape strippings from treated skin areas, this indicated no lingering effect of the drug in the sampled skin, because the time period between the treatment and the sampling was too long.
- + = *Suppressed growth:* Short, fat, deformed, and/or curved hyphae. With tape strippings from treated skin areas, this indicates a reduced lingering effect of the tested antimycotic in sampled skin.
- – = *No growth:* After the 2 days incubation period, only round spores are present. This indicates lingering antimycotic effect of the drug in the sampled epidermis.

Table 17.3 shows that the lingering antimycotic effect of 1 percent omoconazole nitrate cream, following a single application, remained in all the cases for at least 48 hours.

All of the microcultures onto which tapes from normal, untreated skin were applied showed normal growth (the controls).

TABLE 17.3. The lingering antimycotic effect of two 1 percent omoconazole nitrate cream formulations following single topical application

Time Between the Treatment of the Skin and Sampling	Grading of Growth No. of Samples			Grading of Growth No. of Samples		
	+++	+	–	+++	+	–
	RBT 03			8608 E12		
Hours	0	0	9	0	0	9
1	0	0	9	0	0	9
2	0	0	9	0	0	9
6	0	0	9	0	0	9
12	0	0	9	0	0	9
48	0	0	9	0	0	9
Days						
3	4	4	1	2	4	3
7	9	0	0	9	0	0

BENEFITS

The method described allows the study of fungi in their natural growth medium, the stratum corneum. Therefore, this technique could be called a modified in vivo method. However, it is so only formally, because just a single layer of the stratum corneum is used, and this one layer must be better penetrated by the drug tested than would be the whole stratum corneum under pure clinical conditions (Knudsen, 1976). Some of these differences could be overcome by the use of repeated strippings with the same tape.

The suppressed growth of the dermatophytes caused by some of the disinfectants was observed in culture. This is in accordance with earlier observations by Paldrok (1953, 1955) and Gip (1964). Exposure of *Candida*-infected tapes to the same substances in subinhibitory concentrations revealed reduced growth of blastospores and distorted pseudomycelium. Observations by Plempel and Regel (1982) have shown that only growth of *Candida* budding cells without pseudomycelia occurred when *C. albicans* was grown in normal Eagle's medium with addition of bifonazole in subinhibitory doses.

The rapid test method technique produced consistent results not only with dermatophytes, but also with *Candida* as test organisms. Also, the method enabled material from the same patient to be tested by many different substances.

The rapid test method is easy and quick. The microculture technique allowed simple detection of fungal growth within a couple of days. Also, in the cases of no growth (no fungal elements outside the infected scales), the fungus inside the stratum corneum scale could be easily identified, thus serving as an extra control.

The described method is useful not only in making rapid and simple screening tests of the effects of new antimycotic drugs, but also in testing the activity of drugs against fungi in cases of therapeutic failure. Finally, this method may be used to check the possible increase of drug resistance of pathogenic fungi.

The rapid test method is a new method for assessment of the lingering effects of omoconazole cream. The basic tools necessary for this method are only microculture slides and transparent tape.

Both testing methods can be used for in vitro evaluation of plant extracts, essential oils, and other plant-derived antimycotics. This is an economic, time-saving, novel, and rapid method of assessment of antimycotics.

REFERENCES

Emmons, C. W. (1934). Dermatophytes—Natural grouping based on the form of the spores and accessory organs. *Archives of Dermatology* 30: 337.

Gip, L. (1964). Antimycotic effect *in vitro* of a detergent—Fungicidal and fungistatic effect of the detergent and its components on *Trichophyton mentagrophytes, Trichophyton rubrum* and *Epidermophyton floccosum. Acta Dermatol. Venereol.* 44: 211-218.

Gip, L. (1977). A new method for the rapid identification of pathogenic fungi on the skin. *Current Therapeutic Research* 22: 57-64.

Gip, L. (1981). A rapid screening method for the detection of pathogenic fungi on the skin. *Mykosen* 24: 17-26.

Gip, L. (1988). The *in vitro* determination of lingering antimycotic effect of two 1 percent omoconazole-nitrate cream formulations following single topical application. *Mycoses* 31(3): 155-158.

Gip, L. and Gip, Ch. (1984). Screening test method for the determination of the *in vitro* activity of topical antimycotics. *Mykosen* 27: 348.

Klein, D. T. (1965). Morphological response of *T. metagrophytes* to keratin. *Mycologia* 26: Ill.

Knight, A. G. (1973). Human models for in vivo and *in vitro* assessment of topical antifungal compounds. *Brit. J. of Dermat.* 89: 509-514.

Knudsen, E. A. (1976). Test for the in vitro activity of topical antimycotics. *Acta Dermatol. Venereol.* 56: 481-483.

Paldrok, H. (1953). On the variability and classification of dermatophytes. *Acta Dermatol. Venereol.* 33: 1.

Paldrok, H. (1955). The effect of temperature on the growth and development of dermatophytes. *Acta Dermatol. Venereol.* 35: 1.

Plempel, M. and Reget, E. (1982). Antimycotic properties of the topical azole bifonazole *in vitro* and *in vivo.* International Antifungal Symposium: Bifonazole. *Excepta Medica,* pp. 29-37.

Taplin, D. N., Zaias, N., Rebell, G. and Blank, H. (1969). Isolation and recognition of dermatophytes on a new medicum (DTM). *Acta Dermatol. Venereol.* 99: 203.

Chapter 18

A New Technique for the Evaluation of Antifungal Activity of an Alcohol Extract of *Eugenia caryophyllata* Thunberg on *Penicillium digitatum*

Carlo Romagnoli
Gianni Sacchetti

INTRODUCTION

Given the wide variability in growth rate and spread of test phytopathogenic fungi, their use in testing for new natural substances with antimycotic potential can require the application of different evaluation techniques. The aim of this chapter is to evaluate the antifungal activity of an alcohol extract of cloves of *Eugenia caryophyllata* Thunberg using the phytopathogen *Penicillium digitatum* as a test fungus.

Penicillium digitatum has a very rapid growth rate and produces abundant spores, therefore, the classic diameter-measurement method is useless. A new and more appropriate technique, that of counting the number of colonies formed within 48 hours, was applied. This technique was then applied using the alcoholic extract of cloves of *E. caryophyllata* Thunberg, a plant that belongs to the Myrtaceae family. This plant is well-known for having numerous components, i.e., furfurals, esters, ketones, and several sesquiterpenes, the most abundant of which are α and β-karyophyllene and eugenol (85 percent). Moreover, the cloves are well-known in folk medicine, and are used particularly in soaps, insect repellents, perfumes, mouthwashes, medicines, and as antiseptics.

The antibacterial and antifungal action of the extracts as a whole have been documented in the literature (Guérin and Révèillere, 1985; Perez and

---

Investigations supported by grants from "Mazzoni Vivai" Tresigallo, FE (Italy) and the Italian Ministero dell'Università e della Ricerca Scientifica (MURST) of Italy are thankfully acknowledged.

Anesini, 1994; Wilson et al., 1997). Greater data are available on the activity of a single component, eugenol, which is considered the main component involved in its multifarious pharmacological activity. In particular, the anticarcinogenic (Zheng et al., 1992), anti-inflammatory and antirheumatic (Sharma et al., 1994), antioxidant (Toda et al., 1994; Reddy and Lokesh, 1994), antibacterial and antifungal (Laekeman et al. 1990; Tawata et al., 1996) activity of this compound have been recognized.

In this chapter, the activity of an extract of *E. caryophyllata* and identification of which components are most likely responsible for its antifungal activity will be explored. To this purpose, both eugenol and karyophyllene were chosen as pure substances to be tested in parallel with the extract.

THE TECHNIQUE

To evaluate the antifungal activity of an alcohol extract of cloves, a new method was set up and adapted to the growth characteristics of the phytopathogenic fungus *Penicillium digitatum* (Whem) strain no. 48947 which was supplied by the American Type Culture Collection. The fungus was maintained at 4°C in agar slants on potato dextrose agar (PDA). To set up the experiment, it was necessary to establish how to obtain dishes containing a statistically valid number of colonies, repeatable in all dishes. To do so, two mother cultures were prepared by inoculating *P. digitatum* in 75 ml of a Sabouraud dextrose (SD) liquid medium and incubating them under constant agitation at 27°C: one was incubated for 24 h, the other for 48 h. From each of these, a 20 ml mother culture was taken aseptically and placed in a sterile Falcon centrifuge test tube and centrifuged at 2,000 rpm for 5 min. The supernatant was taken from each of these two samples and spectrophotometric analysis was performed using a Perkin-Elmer spectrophotometer to determine the wavelength to be used in the optical density (OD) readings. Scanning showed a maximum absorption at a wavelength of 315 nm. For this reason, 315 nm was chosen as the wavelength for optical density measurements, relating it to the number of spores present.

Petri dishes were prepared containing PDA (Difco). Ten sterile glass spheres or balls (5 mm in diameter) were placed on the solid medium and then the centrifuged cultures whose optical density (OD) was previously measured were added at 100, 200, and 500 μl of each. The spores were uniformly distributed over the entire surface of the spheres by rotating, and by applying linear back-and-forth and left-right movements using the plate. Finally, the petri capsules were turned over so that the spheres were no longer in contact with the culture medium and were then set in incubation for 48 h at 27°C.

The data from these preliminary tests showed that the optimal optical density for the inoculation was between 0.020 and 0.040. These values were obtained with a 24 h culture and the most suitable volume proved to be 200 μl. The combination of these parameters yielded 25 to 40 colonies. The mother culture at 48 h presented a high optical density and the resulting number of colonies was too high to be counted easily.

Evaluation of the Antifungal Activity

Evaluation of the antifungal activity of the alcoholic extract of cloves *(Eugenia caryophyllata)* and of the pure extracts of eugenol and karyophyllene was performed using the previous method.

Petri dishes containing PDA were prepared with and without the extract or standards to be tested. Initially, the pure eugenol and karyophyllene samples were used at concentrations of 5, 10, and 50 μg/ml. Because the eugenol proved highly active, it was also used at concentrations of 2.5 and 1 μg/ml. On the other hand, the amount of extract used was calculated so that its eugenol content was 1, 2.5, 5 μg/ml. The ethanol concentration was adjusted to 0.1 percent in the final solution. Controls were set up with equivalent quantities (0.1 percent) of the solvent (ethanol).

Ten sterile glass spheres of 5 mm diameter, were placed on the solid medium and 200 μl of the centrifuged culture was added (see Photo 18.1). The spores were then distributed over the surface of the plate. Finally, the petri dishes were turned over and incubated for 48 h at 27°C (see Photo 18.2).

The antifungal activity of the substances was evaluated by comparing the number of colonies which grew in the controls and in the treated samples. The values obtained were the average of three experiments performed in duplicate.

Chemistry

The ethanol extract of cloves was obtained from 1,000 g of dried flowers which were concentrated until 691 ml of the extract itself was collected. Quali-quantitative determination of the main components, eugenol and karyophyllene, was performed using high-performance thin-layer chromatography AMD/HPTLC (Camag, Switzerland). The resulting eugenol and karyophyllene concentrations in the extract were, respectively, 63.5 μl/ml and 3.77 mg/ml.

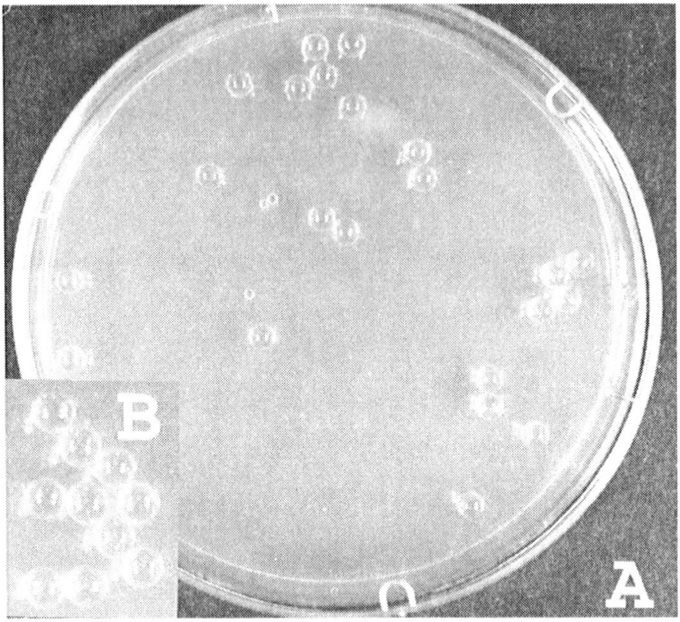

PHOTO 18.1. Aspect of a control culture before inoculum (A), showing the glass spheres, evident in the insert (B)

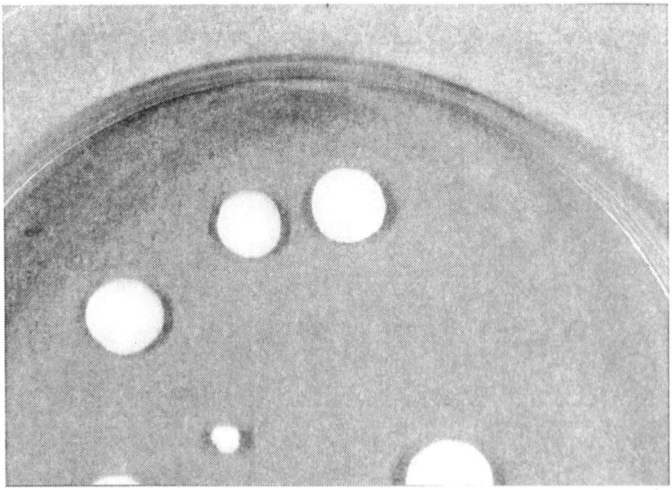

PHOTO 18.2. Particular of the same culture after 48 hours from inoculum, showing some colonies

DISCUSSION AND CONCLUSION

Table 18.1 shows that the pure eugenol was highly active, and that it totally prevented fungal growth, even at the lowest doses, thus confirming the data in the literature regarding its real effectiveness. On the other hand, karyophyllene did not show any effect whatsoever, and was, therefore, tested only at the highest doses. The extract was quantified on the basis of its eugenol concentration. In all cases, the extract totally inhibited fungal growth as in the plates treated with the pure eugenol standard. This experiment demonstrates the fungistatic activity and the intense colony-growth inhibition activity exerted by this substance. This leads to a hypothesis that eugenol is also fungicidal, although this must be confirmed. Further study is being performed to determine the ultrastructural modifications induced by this substance and its action mechanism.

These findings confirm the antimycotic activity of eugenol and *E. caryophyllata* using a new, rapid, and simple technique for the evaluation of the antifungal activity against *Penicillium digitatum.* This novel and rapid method can also be used for evaluation of sensitivity of several other fungi.

TABLE 18.1. Number of colonies in control petri dishes and treated with the extract and the pure standards

	Doses (μl/ml)	**No. of Colonies**
Controls		30
Pure eugenol	50	0
	10	0
	5	0
	2.5	0
	1	0
Pure karyophyllene	50	31
	10	33
	5	31
Extract, equal to:		
Eugenol	5	0
	2.5	0
	1	0

REFERENCES

Guérin, J.C. and Réveillère, H.P. (1985). Activitè antifongique d'extraits végétaux à usage thérapeutique. II. Étude de 40 extraits sur 9 sourches fongiques. *Ann. Pharmaceutiques Françaises* 43: 77-81.

Laekeman, G.M., Van Hoof L., Haemers, A., Vanden Berghe, D.A., Herman, A.G., and Vlietinck, A.J. (1990). Eugenol a valuable compound for *in vitro* experimental research and worthwhile for further *in vivo* investigation. *Phytother. Res.* 4: 90-96.

Perez, C. and Anesini, C. (1994). Antibacterial activity of alimentary plants against *Staphylococcus aureus* growth. *Am. J. Chin. Med.* 22: 169-174.

Reddy, A.C.P. and Lokesh, B.R. (1994). Studies on the inhibitory effects of curcumin and eugenol on the formation of reactive oxygen species and the oxidation of ferrous iron. *Mol. Cell. Biochem.* 137: 1-8.

Sharma, J.N., Srivastava, K.C. and Gan, E.K. (1994). Suppressive effects of eugenol and ginger oil on arthritic rats. *Pharmacology* 49: 314-318.

Tawata, S., Taira, S., Kobamoto, N., Ishihara, M. and Toyama, S. (1996). Synthesis and fungicidal activity of new thiophosphorylated monoterpenoids and related compounds. *J. Pest. Sci.* 21: 141-146.

Toda, S., Ohnishi, M., Kimura, M. and Toda, T. (1994). Inhibitory effects of eugenol and related compounds on lipid peroxidation induced by reactive oxygen. *Planta Medica* 60: 282.

Wilson, C.L., Solar, J.M., El Ghaouth, A. and Wisniewski, M.E. (1997). Rapid evaluation of plant extracts and essential oils for antifungal activity against *Botrytis cinerea*. *Plant Dis.* 81: 204-210.

Zheng, Guo-Qiang, Kennedy, P.M. and Lam, L.K.T. (1992). Sesquiterpenes as potential anticarginogenic agents. *J. Nat. Prod.* 55: 999-1003.

Chapter 19

Palestinian Plants As a Source of Antimycotics

M. S. Ali-Shtayeh
Rabee A. G. Zayed
Rana M. F. Jamous

INTRODUCTION

The use of plants in folk medicine still constitutes a significant part of the heritage of many Arab countries including the United Arab Emirates (UAE) (Tanira et al., 1994); Saudi Arabia (Mossa et al., 1983); Qatar (Rizk, 1982); Sudan (El-Sheikh et al., 1982); Morocco (Haiji et al., 1993); Egypt (Hanafy and Hatem, 1991); Iraq (Jawad et al., 1988); Jordan (Alkofahi et al., 1990); and Palestine (Ali-Shtayeh et al., 2000).

The abundance of species, condensed on a very small geographical area, is a major characteristic of Palestinian florae. About 2,600 species are found in this area, out of which more than 700 species are mentioned in ethnobotanic data (7,100,101). This richness is due to the diversity of habitats created by the soil and climatic conditions.

Folk remedies are prepared as powders, poultices, ointments, baths, decoctions, infusions, and teas. Decoctions are the most popular form of home remedy. Decoctions, infusions, and teas are usually prepared just before application and filtered through a cloth of cotton or wool. Most plants are stored for use in the dry state, which permits their utilization throughout the year; sometimes fresh plants are used.

Many plant species have been used in folkloric medicine in Palestine to treat various ailments including skin disorders (Ali-Shtayeh et al., 2000). Mycotic infections are common diseases in developing countries including the Palestinian area (Ali-Shtayeh and Arda, 1986; Gruseck et al., 1996; Bokhari, 1999; Costa et al., 1999). The use of medicinal herbs in the treatment of skin diseases, including mycotic infections, is an age-old practice in many parts of the world (Irobi and Daramola, 1993). This use has been supported by the isolation of active antifungal compounds from plant extracts (Fabry et al., 1996). These antimycotics represent secondary metabolites that serve as defense agents against invading microorganisms.

An antimycotic agent may be defined as "a chemical substance derived from a living source (plants, animals or microbes) that in dilute solutions has the capacity to inhibit the growth of or destroy fungi and yeast" (Grayer and Harborne, 1994, p. 20). Antimycotic agents have been slow in their development, compared to antibacterial agents, because mycotic infections are less common than bacterial infections, and because it is difficult to develop specific antimycotic agents without affecting host cells (Ali-Shtayeh et al., 1998).

Reports of antimycotic activity of indigenous plants have been published from a few regions of the world (Amoros et al., 1987; Bagchi et al., 1999; Caceres, Jauregui, et al., 1991; Caceres, Lopez, et al., 1991). Previous studies on antifungal activity of Palestinian plants (Ali-Shtayeh and Abu Ghdeib, 1999) indicate that the majority of the plants tested are an important source of antifungal compounds that may provide renewable sources of useful antifungal drugs against dermatophytic infections in humans. Among plant species tested, *Juglans regia, Pistacia lentiscus, Capparis spinosa, Inula viscosa,* and *Anagalis arvensis* were shown to have high antidermatophytic activity. In fact, MIC values for some of these plants, e.g., *J. regia,* were comparable to those of griseofulvin. This obviously justifies the use of many of these plants in traditional medicine to cure dermatophyte infections.

In the constant effort to improve the efficacy and ethics of modern medical practice, researchers are increasingly turning their attention to folk medicine as a source of new drugs (Haslam et al., 1989). One aspect of the modern scientific approach to natural products is to chemically isolate, identify, and screen the active principles from medicinal plants.

MAJOR GROUPS OF ANTIMYCOTIC COMPOUNDS FROM PLANTS

Antimycotic substances in the plants are either "constitutive," i.e., formed in the plant before infection, or they are "induced" (also called phytoalexins) after infection or in response to chemical or environmental factors (Grayer and Harborne, 1994). However, with some substances, it is difficult to determine whether they are phytoalexins or constitutive antimycotic compounds, as the distinction is not always clear.

Constitutive substances, also designated by Schmidt (1933) as prohibitins, can also be classified into three groups: "prohibitins," which include preinfectional plant metabolites normally present in concentrations high enough to inhibit most fungi; "inhibitins," which include antimycotic substances that may be present at low concentrations in the plant but may increase enormously following infection; and "postinhibitins," which include antimycotic substances produced in response to infection but whose formation does not involve the elaboration of a biosynthetic pathway within the tissues of the host (Grayer and Harborne, 1994). The antifungal compounds (250 new substances) present in the plant taxa, surveyed (1982-1993), belong to a very wide range of chemical classes. Even closely related species produce their own specific antimycotic substances.

Plant constitutive antifungal substances belong to several classes of secondary metabolites including: terpenoids, nitrogen- and/or sulphur-containing constituents, aliphatics, and aromatics (Grayer and Harborne, 1994). Of the 250 antifungal substances, half are constitutive, whereas the remainder are induced as phytoalexins. They are typically secondary metabolites, mainly being of terpenoid or phenolic biosynthetic origin.

In general, useful antimicrobial phytochemicals can be divided into several groups (Cowan, 1999) including phenolics and polyphenols (simple phenols, phenolic acids, quinones, flavonoids, flavones, flavonols, tannins, and coumarins), terpenoids and essential oils, alkaloids, lectins and polypeptides, and polyacetylenes.

PLANTS USED IN FOLK MEDICINE IN PALESTINE FOR THE TREATMENT OF SKIN DISEASES AND AILMENTS: AN ETHNOBOTANICAL STUDY

Information regarding medicinal plants was collected using interviews and questionnaires of 110 individuals representing different geographical regions of the Palestinian area (Jenin, Tulkarm, Qalqilia, Nablus, and Salfit in the north; Ramallah and East Jerusalem in the center; Bethlehem and Hebron in the south) during the period May 1998 and December 2000. People interviewed were either native-born or had been living in the region for more than 30 years. Some of the people interviewed were practitioners of herbal medicine. The majority of the informants were above the age of 50 and only very few were under age 40. Ethnobotanic data were mainly collected and analyzed following the techniques of Ali-Shtayeh et al. (2000). Voucher specimens of species recalled by interviewees were collected.

Ethnobotanical data on plants used for the treatment of skin ailments and diseases in the Palestinian area are presented in Table 19.1.

TABLE 19.1. Plants used in Palestinian folk medicine for the treatment of skin ailments*

Plant Species (Family)/ No. of Voucher Specimen	Common Name	Part Used **	Medicinal Use/ Properties	Forms of Use/Preparation/Administration	Active Constituents
Achillea fragrantissima (Forssk) Sch-Bip. (Compositae) A-S100	Yarrow	WP	Skin ailments	Applying the infusion on the ailing skin	Essential oil (terpinen- 4-o1)
Adiantum capillus-veneris L. (Adiantaceae) A-S201	Venus hair	L	Skin ailments, and hair loss	Applying the ground leaves on the infected skin Applying the infusion of the plant together with rue *(Ruta chalapensis)*, and aloe *(Aloe vera)* on the hair	Mucilaginous compounds, tannin, bitter substances
Ajuga orientaslis L. (Labiatae) A-S202	Eastern bugle	WP	Skin ailments, and wounds	Applying the ground fresh plant on wounds or the infected skin for 14 days	Tannin
Alcea acaulis (Cav.) Alef (Malvaceae) A-S203		R	Burns	Applying a boiled mixture of the plant with olive oil, soap, and bee wax on skin burns	
Alhagi maurorum Medik (Papilionaceae) A-S204	Alhagi	R	Skin ailments	Applying the infusion on the ailing skin	
Alkanna orientalis (Boraginaceae) A-S205	Alkanna	R	Burns	Applying a boiled mixture of the plant with olive oil, soap and bee wax on skin burns	
Allium cepa L. (Liliaceae) A-S206	Onions	F	Eczema	Applying a boiled mixture of the plant with olive oil on the infected skin	Flavones, steroidal saponins, sapogenins
Allium sativum L. (Liliaceae) A-S207	Garlic	F	Skin ailments (psoriasis)	Applying a mixture of the garlic and honey	Allicin, alliin, ajoene, diallyl disulfide
Ammi majus L (Umbelliferae) A-S208	Bishop's weed	S	Skin ailments (psoriasis)	Applying the infusion of seeds on the infected skin	Furanocoumarins
Amygdalus communis L. (Rosaceae) A-S209	Sweet al-monds	F	Skin ailments	Applying the almond oil on the infected skin	Amygdaline, fatty oil (amygdalarum), fixed oil, proteins, mucilage, emulsin
Anagallis arvensis L. (Primulaceae) 5408	Red pim-pernel	L, WP	Skin ailments	Applying the infusion on the ailing skin	Saponins, triterpenoids, flavonoids (narigenin, neoponcirin), arvenins, tan-nins
Anchusa strigosa Banks and Sol. (Boranginaceae) A-S5412	Prickly alkanet	WP, R, L	Skin ailments	Applying the macerated plant on the infected skin several times	Cynoglossine, consolidine, alkaloids, mucilage, acids, tannins

Artemisia inculta (Compositae) A-S210.	Wormwood	WP	Skin ailments (allergy)	Drinking as an infusion or decoction	Essential oil, thyjon, thyjol, proazulene, artabsin
Asphodeline lutea (L.) Reichenb, (Liliaceae) A-S5413	Jacob's rod	R, S, WP	Skin ailments	Applying the macerated plant on the infected skin several times	
Atropa belladonna (Solanaceae) A-S211	Deadly nightshade	L	Skin ailments	Applying the macerated plant on the infected skin with the use of bandage to hold the maceration in place	Alkaloids (e.g., hyoscyamine, scopolamine)
Capparis spinosa L. (Capparidaceae) A-S5402	Caper bush	R, FL, F	Skin ailments (psoriasis)	Applying the macerated root mixed with egg albumin on the infected skin	Choline, flavonoids, glucosinolates, rutoside, pectin, enzymes (myronases), saponin, *p*-methoxy benzoic acid, capparidis, capparenol
Carthamus tinctorius L. (Compositae) A-S212	Safflower	FL	Skin ailments	Applying dried flowers or infusion on the infected skin	b-sitosterol, palmitic acid, myristic acid, lauric acid, flavonoids, steroles, phenolic amide, nonacosane
Centaurea dumulosa Boiss (Compositae) A-S213	Shrubby centaury	WP	Skin ailments	Applying the ground plant as an ointment on the infected skin	Bitter constituents (centorine), glycosides (cayanidine)
Citrulus colocynthis L. (Schader) (Cucurbitaceae) A-S252	Colocynth	F	Skin ailments	Applying fruit juice on infected skin	Cucurbitacins, colocynthin, essential oils, sterols, alkaloides
Citrus limon L. (Rutaceae) A-S 214	Lime	F	Pimples, pustules, psoriasis	Applying lemon juice mixed with olive oil on skin infected with psoriasis, pimples, or pustules Applying the extract of boiled thorns on infected hand	Essential oils, hesperidin
Clematis cirrhosa L. (Rananculaceae) A-S5411	Clematis	ST	Skin ailments (warts)	Ironing warts using the plant twigs	Saponins, triterpenoids, oleanolic acid
Coriandrum sativum L. (Umbelliferae) A-S215	Coriander	WP	Skin ailments	Applying plant juice on the affected skin	Essential oil (coriadri), volatile oils, fats, proteins, linalol (oriandrol), limonene, α-cymeme, *p*-terpinene, α-pinene
Coridothymus capitatus (L.) Reichb. (Lameaceae) A-S216	Thyme	L	Hair loss	Washing the hair with plant infusion	Thymol, phenols, saponins, resins, flavonoids, fixed oils
Crocus sativus L. (Iridaceae) A-S217	Saffron	EO	Skin ailments (gecko and leprosy)	Applying essential oil on the affected skin twice a day (once in the morning and once in the evening)	Glycosides (picrocrocin)

TABLE 19.1 *(continued)*

Plant Species (Family)/ No. of Voucher Specimen	Common Name	Part Used **	Medicinal Use/ Properties	Forms of Use/Preparation/Administration	Active Constituents
Cucurbita pepo (Cucurbitaceae) A-S218	Pumpkin	S, F	Wounds and burns	Applying plant ashes on the affected skin	Isoprenoid
Cyclamen persicum Miller (Primulaceae) A-S219	Cyclamen	C	Skin ailments	Applying plant juice on the affected skin twice a day for one week	Glycoside, cyclamen, cyclamenoside
Ecballium elaterium (L.) Rick. (Cucurbitaceae) A-S220	Squirting cucumber	F	Skin ailments (warts)	Applying fruits juice on the affected skin	Elaterin, resin, lignin
Eruca sativa Milleu (Cruciferae) A-S221	Garden rocket	UGP	Skin ailments and hair loss	Applying in a mixture with olive oil, wheat bran, wheat grains, beans, and honey on the affected skin. For hair loss: applying in mixure with parsley, lupine, coconuts, fenugreek, aloe, and castor bean on the hair.	
Eucalyptus bicolor (Myrtaceae) A-S222	Eucalyp-tus	L	Allergy and pim-ples or pustules	Applying an infusion of the plant on the affected skin	Tannin, flavonoides, eucalyptine, es-sential oils
Inula viscosa (L.) Ait. (Compositae) A-S223	Inula	WP, L, ST, FL, F, R	Skin ailments and inflammatory burns	Applying an infusion of the plant on the affected skin	Sesquiterpenoids, essential oils (mainly thymol, and carvacrol), inulavosin, ψ-taraxasterol acetate, (narigenin, neoponcirin, flavonoid glycosides), monoterpenoids (terpenes, gamma-terpinene), phenols
Juglans regia L., (Juglandaceae) A-S5401	Walnut	LF, F, bark, leaves	Ringworm, psoriasis	Applying plant juice on affected skin	Inositol, juglane, carotein, vitamins A,B,C starch, proteins, hydrojuglone , essential oil, tannin, resin, volatile oil, juglone, fatty oil, sesquiterpenes, sesquiterpenoids, monoterpenoids (terpenes, gamma-terpinene), juglandin (naphthoquinones), pyrogallic acid
Laurus nobilis L. (Lauraceae) A-S224	Bay, lau-rel	Fruits, leaves	Skin ailments	Applying an infusion of the plant on the affected skin	Fatty oils, essential oils, bitter constitu-ents, tannine, alkaloids

Linum pubescens Banks and Sol. (Liniaceae) A-S225	Pink flax	Seeds	Skin ailments (psoriasis)	Applying the ground seeds mixed with olive oil on affected skin	Bitter compounds, essential oils, resin, tannins
Matricaria aurea (Loefl.) Sch. Bip. (Compositae) A-S226	Chamomile	Flowers and essential oils	Skin ailments	Applying a boiled and sun-dried mixture of the plant and oils of glycerin, fenugreek, paraffin, soya, anoline, black cumin, castor bean and vaseline on the affected skin	Essential oils, flavonoids
Mentha spicata L. (Labiatae) A-S227	Mint	Leaves	Eczema	Applying an infusion of the plant on the affected skin	Essential oils (e.g., menthol, menthone, jasmone), tannins, bitter compounds
Myrtus communis L. (Myrtaceae) A-S228	Common myrtle	Leaves	Skin ailments	Applying leaves' juice on the affected skin	Myrtol, tannin, resin
Nigella ciliaris DC. (Ranunculaceae) A-S229	Fennel flower	S, EO	Skin ailments (psoriasis)	Applying a boiled and sun-dried mixture of the plant and oils of glycerin, chamomile, fenugreek, paraffin, soya, anoline, black cumin, castor bean and vaseline on the affected skin	Saponoside (melanthine), bitter compound (e.g., nigelline), essential oil (nigellone), tannin
Olea europoea L. (Oleaceae) A-S230	Olives	EO	Skin ailments (psoriasis)	Applying a boiled and sun-dried mixture of the plant and oils of glycerin, chamomile, fenugreek, paraffin, soya, anoline, black cumin, castor bean and Vaseline on the affected skin	Olein, stearin, fixed oil, acids (palmitic and citric acids)
Opuntia ficus-indica L. (Cactaceae) A-S231	Prickly pear	ST	Wounds and hair loss	Applying the ground plant stem on the affected skin	Mucilage, glutamic, malic, and oxalic acids
Paronychia argentea L. (Caryophyllaceae) A-S232		WP	Skin ailments and ringworm	Applying the ground plant on the affected skin	
Petroselinum sativum Hoffm. (Umbelliferae) A-S233	Parsley	WP	Skin ailments and hair loss	Applying a mixture of the plant parsley, aloe, fenugreek, lupines, castor beans, coconuts and golden rocket on the hair for hair loss, Applying the ground plant on the affected skin	Essential oil (apiol, myristicin)
Phagnalon rupestre (L.) DC. (Compositae) A-S234	Phagnalon	L	Skin ailments	Applying an infusion of the plant on the affected skin	Thymol, carvacrol, quinones
Pinus halepensis L. (Pinaceae) A-S235	Aleppo pine	L	Skin ailments	Applying ground leaves on the affected skin	Urpentine, coniferin, tannic acid, resin, pinite, pinene, vitamin C
Pistacia lentiscus L. (Anacardiaceae) A-S5430	Mastic	L	Skin ailments (eczema)	Applying ground leaves on the affected skin	Masticadienoic acid, tucallol, resin, turpentines, thymol, triterpenoids, tannin, procyanidin polymer, essential oils (volatile oils)

TABLE 19.1 *(continued)*

Plant Species (Family)/ No. of Voucher Specimen	Common Name	Part Used **	Medicinal Use/ Properties	Forms of Use/Preparation/Administration	Active Constituents
Plumbago europea L. (Plumbaginaceae) A-S236	Leadwort	L, S	Skin ailments, warts, eczema, and ringworm	Applying the cooked ground plant on the affected skin	Plumbagin
Quercus calliprinos L. (Phagaceae) A-S237	Kermes oak	R, F	Skin ailments	Applying an infusion of the plant on the affected skin	Glycosides, tannin
Raphanus sativus L. (Cruciferae) A-S238	Mustard	L	Skin allergy and eczema	Applying ground leaves on the affected skin	Glycosides (glucoraphinine)
Ricinus communis L. (Euphorpiaceae) A-S239	Castor beans	EO	Skin ailments, psoriasis, and hair loss	Applying a boiled and sun-dried mixture of the plant and oils of glycerin, chamomile, fenugreek, paraffin, soya, anoline, black cumin, and Vaseline on the affected skin	Glycobrassicin, 4-methylthio-3-butenyl glucosinolate
Ruta chalapensis L. (Rutaceae) A-S240	Rue	UGP	Skin ailments and hair loss	Soaked in water with aloe and applied on hair. Boiled with olive oil and applied (massage) on the face	Alkaloids, rutin, caprinic acid, essential oil, saponins, sterols, triterpenes, coumarines, flavonoids (narigenin, neoponcirin, flavonoid glycosides), tannins, triterpenoids, quinones
Salix acmophylla L. Salicaceae A-S241	Willow	FL	Psoriasis and pustules or pimples	Applying as ointment on the affected skin	Glycosides, tannins
Salvia fruticosa L. (Labiatae) A-S242	White sage	L	Skin wounds and eczema	Applying as an infusion on the affected skin	Thymol, carvacrol, rosmarinic acid, saponins, essential oils, flavonoids (narigenin, neoponcirin, flavonoid glycosides) rosmarinic acid, monoterpenoids (terpenes, gamma-terpinene) 1-8-cineole, flavone aglycones

Sarcopoterium spinosum (L.) Sp. (Rosaceae) A-S243	Shrupy barnet	S	Skin ailments	Applying as an infusion on the affected skin	Tannin, triterpenoids, glycosides
Teucrium polium L. (Labiatae) A-S244	Cat thyme	UGP	Eczema and measles	Applying as infusion on the affected skin	Diterpenoids, sapogenin, hederagenin, alkanes, picropoline, β-sitosterol, stigmasterol, campesterol, brassicasterol, cholesterol, glucose, fructose, raffinose, rhamnose
Trigonella foenum- graecum (Papilionaceae) A-S245	Fenu-greek seed	EO	Skin ailments (psoriasis) and hair loss	Applying a boiled and sun-dried mixture of the plant and oils of glycerin, chamomile, paraffin, soya, anoline, black cumin, castor bean and Vaseline on the affected skin	Mucilage (galactomannanes), steroidal saponines, choline, essential oil, trigonilline
Triticum graminis (Gramineae) A-S246	Wheat	S	Skin ailments	Applying as an infusion on the affected skin	Glycosides
Urginea maritima (L.) Baker (Liliaceae) A-S247	Squill	P	Skin ailments (eczema) and hair loss	Applying the ground pulps mixed with salt on the affected skin	Uriginin, cardiotonic glycosides, scillaren, mucilage
Urtica pilulifera L. (Urticaceae) A-S248	Nettle	UGP	Skin ailments and allergy	Applying plant juice on the affected skin	Tannin, formic acid, glucoquinines
Verbascum sinuatum L. (Scrophulariaceae) A-S249	Mullein	WP, R	Skin ailments, including acnes, psoriasis, and wounds	Applying the ground plant on the affected skin several times	Verbascoside aucubin, mucilage, saponin, colourin matter
Vicia faba L. (Legumenosae) A-S250	Broad beans	WP	Skin ailments	Applying the ground plant on the affected skin several times	
Zea mays L. (Gramineae) A-S251	Corn	L	Hair loss	Applying as an infusion on the hair	Saponins, fatty oil, tannins, steroles, allantonine, alkaloides

REVIEW OF RECENT EXPERIMENTAL STUDIES ON ANTIMYCOTIC ACTIVITY OF SOME PLANTS USED FOR THE TREATMENT OF SKIN DISEASES IN PALESTINIAN FOLKLORIC MEDICINE

Achillea fragrantissima

Essential oil from *Achillea fragrantissima* was found to exert anti-candidal effect on *Candida albicans* (Barel et al., 1991). Two fractions which contained less polar compounds were active against *C. albicans* only. The fractions which contained more polar compounds inhibited the growth of all the microorganisms tested. One of these compounds was identified as terpinen-4-ol. Commercial terpinen-4-ol had a similar antimicrobial activity.

Allium cepa and Allium sativum

Aqueous extracts of garlic *(Allium sativum)* and onion *(Allium cepa)* were found to have significant growth inhibition against fungi including dermatophytes (Elnima et al., 1983; Venugopal and Venugopal, 1995; Damayanti et al., 1996; Singh and Rai, 2000). Allicin, one of the active principles of freshly crushed garlic homogenates, has a variety of antimicrobial activities. Allicin in its pure form was found to exhibit antifungal activity, particularly against dermatophytes (Al-Wareh et al., 1993). The main antimicrobial effect of allicin is due to its chemical reaction with thiol groups *Candida albicans* of various enzymes, e.g., alcohol dehydrogenase, thioredoxin reductase, and RNA polymerase. Antifungal activity and minimal fungicidal concentration (MFC) of extracts of seven *Allium* plants including garlic and onion against *Aspergillus niger, A. flavus,* and *A. fumigatus* were found to have antifungal activity, with garlic showing the lowest MFC (Yin and Tsao, 1999). Aqueous garlic extract and concentrated garlic oil along with various commercial garlic supplements and pharmaceutical prescriptions were found to have antifungal activity against *Aspergillus* species involved in otomycosis (Pai and Platt, 1995).

Artemisia *spp.*

The methanolic extracts of aerial parts of *Artemisia ludoviciana* and *A. tridentata* were found to exhibit the greatest fungal inhibition among 100 extracts screened, against nine fungi (Mc Cutcheon et al., 1994). Essential oil from *Artemisia nelagrica* showed strong antifungal activity against several dermatophytes as well as against *Aspergillus fumigatus* and *Cladosporium trichoides* (Kishore et al., 1993). This essential oil by formulation of ointment was able to cure experimental ringworm in guinea pigs within 7 to 12 days.

Carthamus tinctorius

Blaszczyk et al. (2000) screened extracts of 56 widely used, dried Chinese medicinal plants or their parts for their antimycotic properties against selected pathogenic fungi, and found that the highest antimycotic activity against *Aspergillus fumigatus* was shown by *Carthamus tinctorius* L. (Flos).

Eucalyptus globulus

Rai (1996) evaluated extracts of 17-medicinal plants against *Pestalotiopsis mangiferae,* a causative agent of leaf-spot disease of *Mangifera indica,* and found that the maximum activity was shown by *Eucalyptus globulus* (88 percent) followed by *Ricinus communis* (75 percent) and *Lawsonia inermis* (74.33 percent). Twelve methanolic plant extracts from botanical species used in traditional medicine in Morelos, Mexico to cure infectious diseases have been subjected to a screening study to detect potential antimicrobial activity against *Candida albicans.* The results showed that extracts from *Eucalyptus globolus* Labill, and *Artemisia mexicana* Wild., possess strong antimicrobial activity against *Candida albicans* (Navarro et al., 1996). The essential oil of *Eucalyptus pauciflora* showed strong antifungal activity against human pathogenic fungi, viz. *Epidermophyton floccosum, Microsporum audouinii, M. canis, M. gypseum, M. nanum, Trichophyton mentagrophytes, T. rubrum, T. tonsurans,* and *T. violaceum* (Shahi et al., 2000) and other several fungi (Pattnaik et al., 1996).

Inula viscosa

An antimycotic effect of an extract from *Inula viscosa* leaves was demonstrated on chitin synthesis in dermatophytes and *Candida albicans* (Maoz and Neeman, 2000). The best antimycotic effect was obtained with *Inula viscosa* flower extract. This activity was attributed to the presence of different flavonoids in the extract (Cafarchia et al., 1999). Aqueous extract of *Inula viscosa* was found to have strong antimycotic activity against dermatophytes including *Microsporum canis, Trichophyton mentagrophytes,* and *T. violaceum,* and *C. albicans* (Ali-Shtayeh et al., 1998; Ali-Shtayeh and Abu Ghdeib, 1999; Maoz and Neeman, 2000).

Juglans regia

The aqueous extract of *Juglans regia* was found to completely prevent the growth of dermatophytes and *Candida albicans* (Alkhawajah, 1997; Ali-Shtayeh and Abu Ghdeib, 1999).

Lawsonia inermis

The extracts of bark and leaves of *Lawsonia inermis* were found to exhibit strong fungitoxic activity when tested against dermatophytes and other fungi (Galal et al., 1965; Tripathi et al., 1978; Singh and Pandey, 1989).

Linum pubescens

A 25 kDa protein with a high degree of homology to other reported pathogenesis-related antifungal proteins was isolated and characterized from flax (Borgmeyer et al., 1992). The purified protein was found to exhibit antimycotic activity against *Alternaria solani,* the causative agent of tomato early blight and the human pathogen *Candida albicans.*

Mentha spicata

Essential oil of mint was found to have strong antimycotic activity against *Candida albicans* (Sow et al., 1995), *Aspergillus flavus,* and several other molds (Sarbhoy et al., 1978; Pattnaik et al., 1996), including *Ceratocystis paradoxa,* which causes soft rot of pineapples (Damayanti et al., 1996), and against several dermatophytes including *Aspergillus fumigatus* and *Cladosporium trichoides* (Kishore et al., 1993). This essential oil by formulation of ointment was able to cure experimental ringworm in guinea pigs within 7 to 12 days.

Pistacia lentiscus

The aqueous extract of *Pistacia lentiscus* was found to have strong antifungal activity against dermatophytes (Ali-Shtayeh and Abu Ghdeib, 1999). Alpha-pinene, myrcene, trans-caryophyllene, and germacrene D are the major components of *Pistacia lentiscus* shown to have antimycotic activity (Magiatis et al., 1999).

Ricinus communis

Rai (1996) evaluated extracts of 17 medicinal plants against *Pestalotiopsis mangiferae,* a causative agent of leaf-spot disease of *Mangifera indica,* and found *Ricinus communis* to show strong activity (75 percent).

Ruscus aculeatus and Salvia fruticosa

The aqueous extracts of these plants (15 micrograms ml^{-1} medium) were shown to exhibit strong antifungal activity (90-100 percent inhibition) against *T. violaceum* (Ali-Shatyeh and Abu Ghdeib, 1999).

Vitex agnus-castus

The hexanic extract from leaves of *Vitex agnus-castus* was found to completely inhibit the growth of the fungal plant pathogen *Fusarium* sp. (Hernandez et al., 1999).

Zea mays

Volatiles generated from corn silks of maize were found to exhibit antifungal activity against *Aspergillus flavus* (Zeringue, 2000). Wilson et al. (2000) isolated the antifungal protein zeamatin from the seeds of *Zea mays* and demonstrated its activity against *Candida albicans*. Extracts of maize kernels showed greater antifungal activity against *Aspergillus flavus* due to accumulation of aflatoxin (Guo et al., 1998). This result was attributed to the resistance of kernel proteins to *A. flavus* infection, and aflatoxin contamination in corn genotypes.

Ziziphus spina-christi

Extracts of different plants of the family Cruciferae were shown to have different active constituents including arachidonic and palmitic acids, nonadecane, cholesterol, stigmasterol, volatile constituents, and nonmethylated fatty acids and β-sitosterol, which possess strong antifungal activity against yeasts and fungi (Hashem and Saleh, 1999).

EXPERIMENTAL WORK ON ANTIMYCOTIC ACTIVITY OF SOME PLANTS FROM PALESTINE

Antifungal Activity of Plant Extracts

Plant Material: Collection and Extraction

Fifty-four Palestinian plant species, including most of the species reported in the ethnobotanical study (Table 19.1), used in folk medicine were selected (Table 19.3 and 19.4) for study of their antimycotic and anticandidal activity. Mature plants were collected from several sites in Nablus and Ramallah during May-June 1998. Collected plant material was used either fresh or dried in the shade, ground using a seed mill, and the powdered plant material was stored in plastic labeled bags at room temperature until used.

Crude extracts of plant parts were obtained using cold 95 percent ethanol (Kandil et al., 1994). A 500 gm portion of the powdered plant material was soaked in 2.5-3 L of 95 percent ethanol for 4-5 days at room temperature. The mixture was stirred daily for regular infusion. After a five-day period, the extract was filtered using Whatman filter paper no. 1, and the filtrate was then dried using a rotary evaporator at 60°C. The final dried extract was stored in labeled sterile glass bottles and kept at –20°C (Kandil et al., 1994).

Sterilization of Plant Extracts

One gram of the powdered dry extract was dissolved in 2.5 ml of solvent to obtain a final concentration of 400 mg/ml. Extracts were sterilized using sterile 0.45 mm membrane filter. Sterile filtrates were stored in sterile vials in the refrigerator at 5°C until use.

Test Microorganisms

One isolate of *Candida,* two isolates of dermatophytes, and five isolates of plant pathogenic fungi were employed in this study (Table 19.2).

Preparation of Inocula

Candidal Inocula

Part of an isolated *C. albicans* colony was transferred into a 5-ml Muller-Hinton broth tube and the tube was incubated for 4-18 hours at 37°C.

TABLE 19.2. Test microorganisms and source of collection

Microorganism	Number	Source/Researcher	Reference Antibiotic (Concentration)
Yeast			
Candida albicans (Robin) Berkhout	FCCAU* R10	Foot swab, patient/Mr. Suheil Abu-Ghdieb	Nystatin (10 μg/disc)
Dermatophytes			
Microsporum canis Bodin	FCCAU S14	Tinea capitis clinical specimens in Nablus area, patient/Mr. Abu-Ghdieb	Griseofulvin (0.6 μg/ml)
Microsporum gypseum (Bodin) Guiart and Grigorakis	FCCAU S15	Tinea capitis clinical specimens in Nablus area, patient/Mr. Abu-Ghdieb	Nystatin (0.6 μg/ml)
Phytopathogenic fungi			
Fusarium tricinctum (Corda) Sacc.	FCCAU M10	Water pool in Nablus/Mr. T. Khalid	Nystatin (5 μg/ml)
Pythium ultimum var *ultimum* Trow	FCCAU H5R3	Water and soil samples in Nablus area/Prof. Ali-Shtayeh	Hymexazol (25 μg/ml)
P. aphanidermatum (Edson) Fitzp.	FCCAU H739	Water and soil samples in Nablus area/Prof. Ali-Shtayeh	Hymexazol (25 μg/ml)
P. middletonii Sparrow	FCCAU PH 122	Water and soil samples in Nablus area/Prof. Ali-Shtayeh	Hymexazol (25 μg/ml)
Phytophthora citrophthora (R. E. Smith and E. H. Smith)	FCCAU AIE 2005	Water and soil samples in Nablus area/Dr. A-H Hamdan	Metalaxyl (10 μg/ml)

*FCCAU: Fungal Culture Collection of An-Najah University

The growth turbidity in the Muller-Hinton broth was adjusted by further incubation or dilution with sterile physiological saline, after comparison with that of a MacFarland nephelometer tube no. 0.5 (10^8 cfu/ml) using a spectrophotometer at 625 nm (optical density 0.08 - 0.1).

Mycelial Fungi Inocula

A 6 mm diameter agar plug with mycelium was aseptically cut out of an actively growing fungal culture on either SDA or CMA (corn meal agar) medium. The disc was then transferred onto the center of the test medium.

Anticandidal Activity Screening Test

Disk Diffusion Method

Using a sterile cotton applicator, 10^8 cfu/ml of *C. albicans* culture was swabbed on the surface of Muller-Hinton agar (MHA) plates (Murray et al., 1995). The specific reference antibiotic discs (Jawetz et al., 1995) were added on the agar plate beside the extract discs (Table 19.2). Three replicate plates were used for each test. The plates were incubated upside down at 37°C for 18 hours. The inhibition zone around each disc was then measured using a transparent ruler.

Screening for Antifungal Activity

Poisoned-Food Technique

In this method, all test isolates were inoculated onto SDA or CMA plates and incubated at 25°C for 1-4 days for *Pythium* and *Phytophthora,* and for 7-14 days for *Microsporum* and *Fusarium,* to obtain young, actively growing cultures consisting of mycelia and conidia. The required amount of the dried plant extract or reference antimycotic drug was dissolved in 2 ml sterile distilled water or 10 percent aqueous dimethylsulfoxide (DMSO), sterilized by filtration through a 0.45 mm membrane filter, and then mixed in requisite amount of presterilized SDA or CMA medium to give a final concentration of 15 mg/ml. A mycelial disc 6 mm in diameter, cut out from the periphery of 1-4 day-old cultures, was aseptically inoculated onto the medium. In controls, sterile DMSO or distilled water was used in place of plant extract as negative control and reference antibiotics as a positive control (Georgii and Korting, 1991; Mc Cutcheon et al., 1994). The inoculated plates were then incubated at 25°C and colony diameter was measured and recorded after seven days for dermatophytes and keratinophilic fungi, and after one day for *Pythium* and *Phytophthora* species. Percentage of mycelial inhibition was calculated as follows:

$$\% \text{ mycelial inhibition} = ((\text{dc-dt}) / \text{dc}) \times 100 = \frac{\text{dc - dt}}{\text{dc}} \times 100$$

where dc = colony diameter in control (-ve); dt = colony diameter in treatment. Three replicate plates were used for each treatment (Dikshit and Husain, 1984).

Statistical Analysis

Data were analyzed and treatments compared using analysis of variance with Duncan multiple range test (P < 0.05).

Diversity in Antimycotic Potential

Anticandidal and Antifungal Activity of Plant Extracts

Results of anticandidal and antifungal activity in vitro testing of extracts are presented in Tables 19.3 and 19.4.

Anticandidal Activity of Ethanolic Extracts

Among all the 37 plants studied, only *Coridothymus capitatus, Satureja thymbra, Quercus calliprinos, Anthemis tunictoria, Achillea fragrantissima, Sarcopoterium spinosum, Ajuga orientalis, Anthemis palestina, Ceratonia siliqua,* and *Papaver rhoeas* showed anticandidal activity with inhibition zone means ranging from 8.4 - 34.3 mm (Table 19.3). The plants differ significantly in their activity (F = 304.481, DF = 36, P < 0.01).

Antifungal Activity of Ethanolic Extracts

Results of antifungal activity in vitro testing of 48 ethanolic extracts of 48 plants against seven species of mycelial fungi are presented in Table 19.4. The extracts differed considerably in their antimycotic activity. About 8 percent (4/48)-58 percent (28/48) of these extracts had > 60 percent inhibition against one or the other of the test fungi.

Antifungal Activity of Ethanolic Extracts Against Human and Animal Pathogenic Dermatophytes (M. canis *and* M. gypseum)

The inhibitory effect of plants extracts against *M. canis* and *M. gypseum* varied (about 2-100 percent inhibition) significantly between plants (F = 34.457, DF = 47, P < 0.01). Extracts of *Coridothymus capitatus, Micromeria nervosa,* and *Satureja thymbra* completely prevented growth of *M. canis* and *M. gypseum* (Table 19.4).

Extracts of *Cichorium pumilum, Coridothymus capitatus, Micromeria nervosa, Pinus halepensis, Salvia fruticosa, Satureja thymbra,* and *Viscum cruciatum* were the most active (100 percent inhibition) against *M. canis.* Extracts of *Ceratonia siliqua, Alcea setosa, Gagea chloranth, Eryngium*

TABLE 19.3. Antimicrobial activity of ethanolic extracts against *Candida albicans*

	Means* of percent mycelial inhibition + SE
Plants	***C. albicans***
Achillea fragrantissima	12.9 ± 1.66 e**
Ajuga orientalis	11.6 ± 0.79 f
Alcea setosa	6 ± 0 i
Anthemis palestina	10.3 ± 0.19 g
Anthemis tunictoria	14.3 ± 0.50 d
Calycotome villosa	6 ± 0 i
Capparis spinosa	6 ± 0 i
Ceratonia siliqua	9.9 ± 0.11 g
Cichorium pumilum	6 ± 0 i
Companula rapunculus	6 ± 0 i
Coridothymus capitatus	34.3 ± 0.19 a
Crataegus aronia	6 ± 0 i
Cyclamen persicum	6 ± 0 i
Euphorbia hierosolymitana	6 ± 0 i
Foeniculum vulgare	6 ± 0 i
Gagea chloranth	6 ± 0 i
Lactuca serriola	6 ± 0 i
Lactuca tuberosa	6 ± 0 i
Linum pubescens	6 ± 0 i
Lupinus pilosus	6 ± 0 i
Lycium europeum	6 ± 0 i
Micromeria fruticosa	6 ± 0 i
Papaver rhoeas	8.4 ± 0.11 h
Paronychia argentea	6 ± 0 i
Pinus halepensis	6 ± 0 i
Pistacia lentiscus	6 ± 0 i
Quercus calliprinos	15.2 ± 0.40 d
Rhus coriaria	6 ± 0 i
Rubia tenuifolia	6 ± 0 i
Sarcopoterium spinosum	12.8 ± 0.11 e

	Means* of percent mycelial inhibition + SE
Plants	***C. albicans***
Satureja thymbra	6 ± 0 i
Scrophularia rubricaulis	6 ± 0 i
Teucrium polium	6 ± 0 i
Varthemia iphionoides	6 ± 0 i
Verbascum sinuatum	6 ± 0 i
Viscum cruciatum	6 ± 0 i
Vitex agnus-castus	6 ± 0 i
Control (+ve)	19.3 ± 0.19 c

*Means of three replicate plates for each isolate of each species.
**Values in the same column followed by the same letter were not significantly different based on Duncan's multiple-range test ($p < 0.05$). Disk diameter = 6 mm.

creticum, Capparis spinosa, Varthemia iphionoides, and *Crataegus aronia* were, on the other hand, the least active (< 20 percent inhibition).

Extracts of *Rhus coriaria, Satureja thymbra, Inula viscosa, Ruta chalepensis, Micromeria nervosa, Lawsonia inermis, Coridothymus capitatus,* and *Anthemis tunictoria* were the most active (80-100 percent inhibition) against *M. gypseum.* Extracts of *Juglans regia, Ceratonia siliqua, Alcea setosa,* and *Ziziphus spina-christi* were, on the other hand, the least active (< 20 percent inhibition).

Antifungal Activity of Ethanolic Extracts Against Phytopathogenic Pythium *Species* (P. ultimum, P. aphanidermatum, P. middletonii)

The inhibitory effect against the three fungi varied (about 2-100 percent inhibition) significantly between plants (F = 18.474, DF = 47, P < 0.01) (Table 19.4). Extracts of *Micromeria nervosa, Pinus halepensis,* and *Satureja thymbra* completely prevented growth of *P. ultimum, P. aphanidermatum,* and *P. middletonii.*

Extracts of *Achillea fragrantissima, Micromeria nervosa, Pinus halepensis, Quercus calliprinos, Satureja thymbra, Viscum cruciatum, Anthemis tunictoria,* and *Coridothymus capitatus* were the most active (80-100 percent inhibition) against *P. ultimum.* Extracts of *Ceratonia siliqua* and *Parietaria diffusa* were the least active (< 10 percent inhibition).

Extracts of *Calycotome villosa, Capparis spinosa, Micromeria nervosa, Paronychia argentea, Pinus halepensis, Salvia fruticosa, Sarcopoterium*

TABLE 19.4. Antifungal activity of ethanolic extracts of 48 plants

	Means* of percent mycelial inhibition + SE						
Plants	***M. canis***	***M. gypseum***	***P. ultimum***	***P. aphaniderm-atum***	***P. middletonii***	***F. tricinctum***	***Ph. citrophthora***
Achillea fragrantissima	36.4 ± 2.77 K**	48.6 ± 1.83 j	100 ± 0 a	81.1 ± 0.39 d	39.7 ± 0.85 h	30.4 ± 0.84 ghijk	37.3 ± 1.32 fg
Ajuga orientalis	33.6 ± 0.52 Kl	54.2 ± 0 ghi	45.5 ± 0.87 ij	69.3 ± 1.04 f	44.1 ± 1.69 gh	36.5 ± 0 f	33.3 ± 1.34 gh
Alcea setosa	10.3 ± 0.61 S	15.3 ± 0.69 u	61.1 ± 0.50 f	69.9 ± 0.69 f	16.7 ± 1.29 mnop	27.9 ± 1.66 ijklm	22 ± 1.16 kl
Anthemis palestina	23.6 + 1.04 nop	35.4 ± 0 mn	53 ± 0 h	73.9 ± 1.60 e	33.3 ± 1.29 i	28.8 ± 0 ijkl	52 ± 0 d
Anthemis tunictoria	61.8 + 1.04 ef	80.6 ± 0 b	85.9 ± 0 b	88.8 ± 0 b	81.4 ± 0.48 b	18.9 ± 0.31 qrs	ND
Asphodelin lutea	47.3 + 1.05 hi	74.3 ± 0.69 c	30.3 ± 0 o	88.8 ± 0 b	26.5 ± 0.85jk	32.7 ± 0 fghi	ND
Calycotome villosa	47.3 + 3.14 hi	54.9 ± 1.38 gh	54 ± 0.50 gh	100 ± 0 a	46.1 ± 6.02 g	18.7 ± 0.27 qrs	ND
Capparis spinosa	17.6 + 4.96 qr	38.4 ± 0.66 lm	77.8 ± 1 cd	100 ± 0 a	57.4 ± 0.85 def	10.9 ± 3.39 t	ND
Ceratonia siliqua	1.8 + 0 t	15.6 ± 0.60 u	5.3 ± 0.43 st	16.3 ± 1.04 r	12.5 ± 0.42 opqr	19.6 ± 0.64 pqr	50 ± 1.14 d
Cichorium pumilum	100 + 0 a	52.1 ± 0 hij	35.6 ± 0.43 mn	77.1 ± 1.39 e	16.9 ± 2.12 mno	27.4 ± 1.93 jklmn	ND
Clematis cirrhosa	28.5 + 2.18 lmn	20.8 ± 0 st	12.7 ± 0.48 qr	20.6 ± 1.39 qr	15.7 ± 4.01 mnopq	28.8 ± 0.55 hijkl	ND
Coridothymus capitatus	100 + 0 a	100 ± 0 a	81.3 ± 0.50 c	85.5 ± 0 bc	100 ± 0 a	51.6 ± 2.24 d	ND
Crataegus aronia	18.8 + 1.21 pqr	33.3 ± 1.20 no	20.7 ± 3.53 p	16.9 ± 1.39 r	18.4 ± 0.42 lmn	30.4 ± 2.62 ghijk	5.3 ± 1.34 n
Cyclamen persicum	56.4 + 5.25 g	59.7 ± 2.82 ef	46.2 ± 2.18 ij	52.6 ± 2.43 ijkl	43.1 ± 1.95 gh	51.4 ± 1.94 d	49.3 ± 2.66 d
Eryngium creticum	17.3 + 0.52 qr	35.4 ± 0 mn	13.5 ± 0.88 q	63.5 ± 0.39 gh	1.47 ± 0 u	30.1 ± 0.84 ghijkl	ND
Euphorbia hierosolymitana	20.6 + 0.61 pqr	21.5 ± 0.69 st	56.1 ± 0 gh	39.8 ± 1.38 p	23 ± 0.49 kl	30.1 ± 1.15 ghijkl	12. ± 0 m
Gagea chloranth	16.4 + 0 r	23.6 ± 0.69 rs	21.2 ± 0.87 p	55.4 ± 1.39 ij	16.9 ± 1.27 mno	22.4 ± 0.84 opq	12 ± 0 m
Inula viscosa	73.9 ± 0.60 bc	100 ± 0 a	69.2 ± 2.52 e	75.5 ± 2 e	58.8 ± 0 de	62.5 ± 0 c	ND

Juglans regia	26.7 ± 0.60 mno	19.5 ± 1.83 t	13.6 ± 0.87 q	51.2 ± 1.04 jkl	8.3 ± 0.98 rst	20.2 ± 0.96 pqr	ND
Lactuca serriola	27.3 ± 0 mno	27.1 ± 0 qr	44 ± 0 jk	76.3 ± 1.44 e	11.8 ± 1.47 opqrs	20.5 ± 1.39 pqr	28 ± 0 ij
Lactuca tuberosa	44.5 ± 2.62 i	41.7 ± 2.40 kl	60.6 ± 0.87 f	74.7 ± 2.08 e	46.6 ± 1.29 g	52.4 ± 1.38 d	5.3 ± 1.34 n
Lawsonia inermis	37.9 ± 5.61 jk	100 ± 0 a	31.7 ± 0.89 no	46.2 ± 3.99 mn	11.8 ± 0.84 opqrs	31.7 ± 1.46 fghij	ND
Linum pubescens	30.3 ± 0.61 lm	32.3 ± 0.60 nop	36.4 ± 0 lm	68.7 + 0.69 f	23 ± 1.29 kl	34 ± 0.64 fg	18 ± 1.16 l
Lupinus pilosus	42.7 ± 2.62 ij	50.7 ± 1.83 ij	58.1 ± 1.01 fg	61 ± 0.81 h	47 ± 2.24 g	44.6 ± 0.84 e	42 ± 0 e
Lycium europeum	56.4 ± 0 g	59.4 ± 0.60 ef	35.9 ± 0.50 mn	44.2 ± 1.06 no	31.6 ± 2.11 i	25.3 ± 0.84 lmno	30 ± 1.16 hij
Micromeria fruticosa	53.9 ± 1.21 g	55.2 ± 3.01 gh	40.3 ± 2.59 kl	44.2 ± 2.44 no	32.4 ± 3.06 i	29.3 ± 0.83 ghijkl	25.3 ± 1.32 jk
Micromeria nervosa	100 ± 0 a	100 ± 0 a	100 ± 0 a	100 ± 0 a	100 ± 0 a	67.6 ± 1.28 ab	ND
Papaver rhoeas	43.6 ± 1.04 i	40.6 ± 0.60 kl	20.5 ± 1.31 p	40.4 ± 1.73 op	14.7 ± 0.84 mnopq	14.9 ± 0.83 st	20 ± 0 l
Parietaria diffusa	33.9 ± 1.21 kl	27.1 ± 0 qr	2 ± 0.50 t	16.3 ± 1.04 r	5.9 ± 0.84 tu	26.2 ± 0.34 klmno	ND
Paronychia argentea	22.7 ± 1.57 opq	20.8 ± 0 st	35.6 ± 3.05 mn	100 ± 0 a	11 ± 0.48 pqrs	31.4 ± 2.62 ghij	26 ± 1.14 jk
Pinus halepensis	100 ± 0 a	75 ± 0 c	100 ± 0 a	100 ± 0 a	100 ± 0 a	44.2 ± 0.96 e	100 ± 0 a
Pistacia lentiscus	27.3 + 1.04 mno	30.2 ± 3.01 opq	48.5 ± 1.74 i	61.4 ± 1.38 h	40.4 ± 1.27 h	29.8 ± 1.66 ghijkl	6.6 ± 1.34 n
Quercus calliprinos	72.7 ± 1.04 c	59.4 ± 0.60 ef	100 ± 0 a	75.5 ± 1.60 e	63.7 ± 1.29 c	54.2 ± 1.95 d	62.8 ± 0 c
Retema raetam	58.2 ± 1.82 fg	32.6 ± 1.83 nop	35.4 ± 3.64 mn	49.4 ± 3.03 lm	19.1 ± 0.84 lm	32.7 ± 1.11 fghi	ND
Rhus coriaria	78.3 ± 0 b	100 ± 0 a	75.8 ± 0.87 d	81.9 ± 2.50 cd	56 ± 0.71 ef	54.5 ± 0.84 d	32 ± 0 hi
Rubia tenuifolia	30.9 ± 1.05 lm	30.6 ± 2.50 opq	20.5 ± 0.43 p	19.2 ± 0 qr	6.6 ± 1.27 st	17.6 ± 1.15 rs	ND
Ruscus acculeatus	63.6 ± 1.81 e	29.2 ± 1.20 pq	45 ± 1 ij	9.2 ± 0.39 s	22.5 ± 1.29 kl	23.1 ± 1.46 nopq	ND
Ruta chalepensis	67.3 ± 0 de	100 ± 0 a	43.9 ± 1.51 jk	54.1 ± 2.08 ijk	28.9 ± 1.76 ij	23.7 ± 1.94 mnop	ND
Salvia fruticosa	100 ± 0 a	61.5 ± 0.59 e	67.2 ± 2.52 e	100 ± 0 a	43.1 ± 2.73 gh	71.5 ± 2.73 a	ND
Sarcopoterium spinosum	71.8 ± 1.57 cd	56.3 ± 1.20 fg	49 ± 2.81 i	100 ± 0 a	52.9 ± 2.94 f	32.7 ± 1.11 fghi	40 ± 0 ef
Satureja thymbra	100 ± 0 a	100 ± 0 a	100 ± 0 a	100 ± 0 a	100 ± 0 a	71.2 ± 1.92 a	100 ± 0 ef
Solanum nigrum	50.9 ± 1.04 h	42.7 ± 0.59 k	9.1 ± 1.75 rs	50.6 ± 0 kl	26.5 ± 0 jk	26.9 ± 1.11 jklmno	ND

TABLE 19.4 *(continued)*

	Means* of percent mycelial inhibition + SE						
Plants	***M. canis***	***M. gypseum***	***P. ultimum***	***P. aphaniderm-atum***	***P. middletonii***	***F. tricinctum***	***Ph. citrophthora***
Teucrium polium	30.3 ± 2.18 lm	32 ± 1.38 nop	22.7 ± 0 p	55.4 ± 0 ij	12.5 ± 1.27 opqr	30.1 ± 1.78 ghijkl	18 ± 1.16 l
Varthemia iphionoides	18.8 ± 0.61 pqr	20.1 ± 0.69 st	54.5 ± 0.87 gh	64.5 ± 1.73 gh	22.5 ± 1.29 kl	42 ± 0.31 e	29.3 ± 1.34 hij
Verbascum sinuatum	41.8 ± 1.82 ij	38.5 ± 0.57 lm	33.2 ± 1.46 mno	23.3 ± 2 q	13.2 ± 0 nopqr	42.5 ± 0.23 e	ND
Viscum cruciatum	100 ± 0 a	65.6 ± 0.60 d	100 ± 0 a	100 ± 0 a	61.3 ± 1.29 cd	33.2 ± 0.27 fgh	42.7 ± 1.34 e
Vitex agnus-castus	37 ± 0.60 k	33.3 ± 2.08 no	56.1 ± 3.03 gh	65.9 ± 2.81 fg	15.2 ± 0.48 mnopq	44.2 ± 0 e	12 ± 0 m
Ziziphus spina-christi	21.8 ± 1.04 opqr	11.1 ± 2.77 v	31.8 ± mno	2.4 ± 0 t	11 ± 1.27 qrs	22.6 ± 3.36 opq	ND
+ ve control (reference antibiotic)***	63.6 ± 0 e	39.6 ± 0	60.6 ± 0 f	56.6 ± 0 i	25 ± 0 jk	64.4 ± 0 bc	71.9 ± 0 b*

*Means of three replicate plates for each isolate of each species.
**Values in the same column followed by the same letter were not significantly different based on Duncan's multiple-range test ($p < 0.05$).
***Type and (concentration) of reference antibiotics: *M. canis,* griseofulvin (0.6 µg/ml), for *Ph. citrophthora* metalaxyle (10 g/ml), *F. tricinctum* and *M. gypseum* nystatin (5 mg/ml).

spinosum, Satureja thymbra, Viscum cruciatum, Achillea fragrantissima, Anthemis tunictoria, Asphodelin lutea, Coridothymus capitatus, and *Rhus coriaria* were also most active (80-100 percent inhibition) against *P. aphanidermatum,* whereas extracts of *Ruscus acculeatus* and *Ziziphus spina-christi* were the least active (< 10 percent inhibition).

Extracts of *Coridothymus capitatus, Micromeria nervosa, Pinus halepensis, Satureja thymbra,* and *Anthemis tunictoria* were the most active (80-100 percent inhibition) against *P. middletonii.* Extracts of *Parietaria diffusa, Eryngium creticum, Rubia tenuifolia,* and *Juglans regia* were the least active (< 10 percent inhibition).

Antifungal Activity of Ethanolic Extracts Against Phytopathogenic Ph. citrophthora

The inhibitory effect against this fungus varied (5.3-100 percent inhibition) significantly between plants ($F = 256.138$, $DF = 26$, $P < 0.01$) (Table 19.4).

Extracts of *Pinus halepensis* and *Satureja thymbra* were the most active (100 percent inhibition). Extracts of *Pistacia lentiscus, Lactuca tuberosa,* and *Crataegus aronia* were the least active (< 10 percent inhibition).

Antifungal Activity of Ethanolic Extracts Against Phytopathogenic F. tricinctum

The inhibitory effect against these fungi varied (11-72 percent inhibition) significantly between plants ($F = 101.012$, $DF = 47$, $P < 0.01$) (Table 19.4).

Extracts of *Salvia fruticosa* and *Satureja thymbra* were the most active (> 70 percent inhibition). Extracts of *Capparis spinosa, Papaver rhoeas, Rubia tenuifolia, Calycotome villosa,* and *Anthemis tunictoria* were the least active (< 20 percent inhibition).

TRADITIONAL MEDICINE AS A SOURCE OF ANTIFUNGALS

Traditional medicine is an important source of potentially useful new compounds for the development of chemotherapeutic agents (Alonso paz et al., 1995). The first step toward achieving this goal is the screening of plants used in popular medicine. The search for new, safer, and more effective antimicrobial agents has grown with the increasing incidence of microbial infections (Larhsinin et al., 1996). Folkloric use suggests that there may be a scientific basis for their utility in traditional medicine for the treatment of deferent infections (Irobi, 1992).

The antimycotic effects of plant extracts of some plant species indicate the importance of many plant species as a natural source of antimycotic material (Table 19.3 and 19.4), (major active constituents present are indicated in Table 19.1). Antimycotic activity of medicinal plants, e.g., *Juglans* sp. and *Solanum* sp. extracts, against some dermatophytes including *M. canis* have also been reported previously (Ali-Shtayeh and Abu Ghdeib, 1999). In the present study, among 48 locally available plant species tested in vitro against dermatophytes, the ethanolic extracts of *C. capitatus, M. nervosa,* and *S. thymbra* were most active (100 percent inhibition). These results are therefore consistent with those of Vlietinck et al. (1995) who showed that, from 267 ethanolic crude extracts corresponding to 100 different plant species, 60 percent of the extracts had antidermatophytic activity including activity against *M. canis*. Also, Ali-Shtayeh and Abu Ghdeib (1999) found that among the 22 species they tested, 27 - 81 percent of the aqueous extracts showed high antimycotic activity against one or more of four dermatophytes including *M. canis.*

The majority of the plants that have been tested are an important source of antimycotic compounds that may provide renewable sources of useful antimycotic drugs against dermatophytic infections in humans. Among plant species tested, *Anthemis tunictoria, C. capitatus, M. nervosa,* and *S. thymbra* wre shown to have high antidermatophytic activity. This obviously justifies the use of many of these plants in traditional medicine to cure dermatophyte infections (Ali-Shtayeh and Abu Ghdeib, 1999).

The importance of natural products for plant disease control has been further investigated, and encouraging results on the subject have been reported (Akhtar and Munir, 1989; Al-Abed et al., 1993; Qasem and Abu-Blan, 1995). In vitro assessments of the potential of plant extracts as fungitoxicants have shown that plant species are different in their effects; some inhibit growth, others stimulate it or have no apparent effect (Al-Abed et al., 1993). In this study, results indicated the presence of active fungistatic materials in the extract of certain plant species. The most active extracts among 48 ethanolic plant extracts tested in vitro included *M. nervosa, P. halepensis,* and *S. thymbra* against *Pythium* sp.; extracts of *P. halepensis* and *S. thymbra* against *Ph. citropthora;* and extracts of *S. fruticosa* and *S. thymbra* against *F. tricinctum.* These results are consistent with those of Grosvenor et al. (1995) who showed that 20 percent of 114 plant species extracts they examined inhibited the growth of *Fusarium oxysporum.* Al-Abed et al. (1993) found that from 40 weeds from Jordan, aqueous extracts of *I. viscosa* and *Anagallis arvensis* were most active against the phytopathogenic fungi *Fusarium oxysporum* and *Helminthosporium sativum.*

Some of the ethanolic plant extracts (e.g., *C. capitatus* and *S. thymbra*) that were active against dermatophytes and plant pathogenic fungi, were also active against *C. albicans*. These results are in conformity with Ali-Shtayeh and Abu Ghdeib (1999) who also found some of the aqueous plant extracts to be highly active against dermatophytes and *C. albicans*.

On the other hand, many of the plant species, e.g., ethanolic extracts of *P. halepensis,* 87.5 percent, and *Alcea setosa* 8.7 percent, that showed high to low antimycotic activity against dermatophytes, were found to be inactive against *C. albicans*. These results are in conformity with Vlietinck et al. (1995), who also found that plants with high activity against dermatophytes were inactive against *C. albicans*. This indicates that anticandidal compounds are less frequently encountered in the ethanolic extracts of the test plants than other antimycotic compounds, and also indicates the difference in the mode of action of these compounds. Anticandidal compounds (e.g., nystatin, a polyene macrolide antibiotic) are characterized by having a large lactone ring with a number of conjugated double bonds. They are membrane-active agents that produce their effect by creating polar "pores" by insertion alongside the phospholipid and sterol molecules (Gale et al., 1981). On the other hand, antidermatophytic compounds such as griseofulvin, which affect only dermatophytes due to the presence of chitin in their walls, have no effect on yeastlike *C. albicans*. Their mode of action is binding strongly to the proteins associated with the microtubules so that they disrupt nuclear division in molecules (Gale et al., 1981).

CONCLUSIONS

This study demonstrates that plants are important sources of fungitoxic compounds, and that they may provide a renewable source of useful fungicides that can be utilized in integrated pest control programs. Further studies are therefore needed on these plants, especially in those that showed high antimycotic activity against the test fungi.

REFERENCES

Akhtar, M.S., Munir, M. (1989). Evaluation of the gastric antiulcerogenic effects of *Solanum nigrum, Brassica oleracea* and *Ocimum brasilicum* in rats. *Journal of Ethnopharmacology* 27: 163-176.

Al-Abed, A.S., Qasem, J.R., Abu-Blan, H.A. (1993). Antifungal effect of some common wild plant species on certain plant pathogenic fungi. *Dirasat (Pure and Applied Sciences)* 20: 149-158.

Ali-Shtayeh, M.S., Abu Ghdeib, S.I. (1999). Antifungal activity of plant extracts against dermatophytes. *Mycoses* 42: 665-672.

Ali-Shtayeh, M.S., Arda, H.M. (1986). A study of tinea capitis in Jordan (West Bank). *Journal of Tropical Medicine and Hygiene* 89: 137-141.

Ali-Shtayeh, M.S., Yaghmour, R.M., Faidi, Y.R., Salem, K., Al-Nuri, M.A. (1998). Antimicrobial activity of 20 plants used in folkloric medicine in the Palestinian area. *J Ethnopharmacol* 60: 265-271.

Ali-Shtayeh, M.S., Yaniv, Z., Mahajna, J. (2000). Ethnobotanical survey in the Palestinian area: A classification of the healing potential of medicinal plants. *Journal of Ethnopharmacology* 73: 221-232.

Alkhawajah, A.M. (1997). Studies on the antimicrobial activity of *Juglans regia*. *Am J Chin Med* 25: 175-80.

Alkofahi, A.S., Abdel Aziz, A., Mahmous, I., Abuirjie, M., Hunaiti, A., El-oqla, A. (1990). Cytotoxicity, mutagenicity and antimicrobial activity of forty Jordanian medicinal plants. *Int. J. Crude Dry Res.* 28: 139-144.

Alonso paz, E., Cerdeiras, M.P., Fernandez, J., Ferreira, F., Moyna, P., Soubes, M., Vazquez, A., Vero, S., Zunino, L. (1995). Screening of Uruguayan medicinal plants for antimicrobial activity. *Journal of Ethnopharmacology* 45: 67-70.

Al-Wareh, H.B., Al-Ghazal, R.K., Mouchanta, A.H. (1993). *Medicinal and Aromatic Plants.* Aleppo: Aleppo University Publications, Arabic, pp. 113-545.

Amoros, M., Fauconnier, B., Girre, R.L. (1987). *In vitro* antiviral activity of a saponin from *Anagallis arvensis,* Primulaceae, against herpes simplex virus. *Antiviral - Research* 8: 13-25.

Bagchi, G.D., Singh, A., Khanuja, S.P., Bansal, R.P., Singh, S.C., Kumar, S. (1999). Wide spectrum antibacterial and antifungal activities in the seeds of some coprophilous plants of north Indian plains. *Journal of Ethnopharmacology* 64: 69-77.

Barel, A., Segal, R., Yashphe, J. (1991). The antimicrobial activity of essential oil from *Achillea fragrantissima. Journal of Ethnopharmacology* 33: 187-191.

Blaszczyk, T., Krzyzanowska, J., Lamer-Zarawska, E. (2000). Screening for antimycotic properties of 56 traditional Chinese drugs. *Phytother Res* 14: 210-212.

Bokhari, M.A., Hussain, I., Jahangir, M., Haroon, T.S., Aman, S., Khurshid, K. (1999). Onychomycosis in Lahore, Pakistan. *Int J Dermatol.* 38: 591-595.

Borgmeyer, J.R., Smith, C.E., Huynh, Q.K. (1992). Isolation and characterization of a 25 kDa antifungal protein from flax seeds. *Biochem Biophys Res Commun* 187: 480-487.

Caceres, A., Jauregui, E., Herrera, D., Logemann, H. (1991). Plants used in Guatemala for the treatment of dermato-mucosal infections. 1. Screening of 38 Plant extracts for anticandidal activity. *Journal of Ethnopharmacology* 33: 277-283.

Caceres, A., Lopez, B.R., Giron, M.A., Logemann, H. (1991). Plants used in Guatemala for the treatment of dermatophytic infections. 1. Screening for Antimycotic activity of 44 plant extracts. *Journal of Ethnopharmacology* 31: 263-276.

Cafarchia, C., De Laurentis, N., Milillo, M.A., Losacco, V., Puccini, V. (1999). Research on antifungal activity of flowers and leaves of *Inula viscosa. Parassitologia* 41: 579-582.

Costa, T.R., Costa, M.R., da Silva, M.V., Rodrigues, A.B., Fernandes, O.D., Soares, A.J., Silva, M.D. (1999). The etiology and epidemiology of dermatophytoses in Goiania, GO, Brazil. *Rev Soc Bras Med Trop.* 32: 367-371.

Cowan, M.M. (1999). Plant products as antimicrobial agents. *Clinical Microbiology Reviews* 12: 564-582.

Damayanti, M., Susheela, K., Sharma, G.J. (1996). Effect of plant extracts and systemic fungicide on the pineapple fruit-rotting fungus, *Ceratocystis paradoxa. Cytobios* 86: 155-165.

Dikshit, A., and Husain, A. (1984). Antifungal action of some essential oils against animal pathogens. *Fitoterapia* 55: 171-176.

Elnima, E.I., Ahmed, S.A., Mekkawi, A.G., Mossa, J.S. (1983). The antimicrobial activity of garlic and onion extracts. *Pharmazie* 38: 747-748.

El-Sheikh, M., El-Hassan, G., El-Tayeb Abdel Hafeez, A., Abdalla, A., Antoun, M. (1982). Studies on Sudaneese medicinal plants III. *Planta Medica* 45: 116-119.

Fabry, W., Okemo, P., Ansorg, R. (1996). Fungistatic and fungicidal activity of East African medicinal plants. *Mycoses* 39: 67-70

Galal, E.E., Sabbagh, H., Ghali, B., Fawazi, G. (1965). Screening of antifungal activity of some Egyptian plant extracts; *"Lawsonia inermis." J Egypt Med Assoc* 48(Suppl): 153-160.

Gale, E.F., Cundcliffe, E., Reynolds, P.E., Richmond, M.H., Waring, M.J. (1981). *The Molecular Basis of Antibiotic Action* (Second Edition). London: Wiley.

Georgii, A., and Korting, H.C. (1991). Antifungal susceptibility testing with dermatophytes, *Mycoses* 34: 193-199.

Grayer, R.J., and Harborne, J.B. (1994). A survey of antifungal compounds from higher plants, 1982-1993. *Phytochemistry* 37: 19-42.

Grosvenor, P.W., Supriono, A., Gray, D.O. (1995). Medicinal plants from Riau Province, Sumatra, Indonesia. Part 2: Antibacterial and antifungal activity. *Journal of Ethnopharmacology* 45: 97-111.

Gruseck, E., Splanemann, V., Bleck, O., Ring, J., Abeck, D. (1996). Oral terbinafine in tinea capitis in children. *Mycoses* 39: 237-240.

Guo, B.Z., Brown, R.L., Lax, A.R., Cleveland, T.E., Russin, J.S., Widstrom, N.W. (1998). Protein profiles and antifungal activities of kernel extracts from corn genotypes resistant and susceptible to *Aspergillus flavus. J Food Prot* 61: 98-102.

Haiji, F., Fkih-Tetouani, S., Tantaoui-Elaraki, A. (1993). Antimicrobial activity of twenty-one *Eucalyptus* essential oils. *Fitoterapia* 64: 71-77.

Hanafy, M.S.M., and Hatem, M.E. (1991). Studies on the antimicrobial activity of *Nigella sativa* seed (black cumin). *Journal of Ethnopharmacology* 34: 275-278.

Hashem, F.A., and Saleh, M.M. (1999). Antimicrobial components of some Cruciferae plants. *Phytother. Res.* 13: 329-332.

Haslam, E., Lilley, T.H., Ya Cai Martin, R., Magnolato, D. (1989). Traditional herbal medicines—The role of polyphenols. *Planta Medica* 55: 1-8.

Irobi, O.N. (1992). Activities of *Chromolaena adorata* (Compositae) leaf extract against *Pseudomonas aeruginosa* and *Streptococcus faecalis. Journal of Ethnopharmacology* 37: 81-83.

Irobi, O.N., and Daramola, S.O. (1993). Antifungal activities of crude extracts of *Mitracarpus villosus* (Rubiaceae). *Journal of Ethnopharmacology* 40: 137-140.

Jawad, A.M., Jaffer, H.J., Al-naib, A., Saber, H., Razzak, A.A.W. (1988). Antimicrobial activity of some Iraq plants. *Fitoterapia* 59: 130-133.

Jawetz, E., Melnick, J.L., Adelberg, E.A., Brooks, G.F., Batel, J.S., Ornston, L.N. (1995). *Medical Microbiology* (Twentieth Edition). New York: Appleton and Lang.

Kandil, O., Radwan, N.M., Hassan, A.B., Amer, A.M.M., El-banna, H.A., Amer, W.M.M. (1994). Extracts and fractions of *Thymus capitatus* exhibit antimicrobial activities. *Journal of Ethnopharmacology* 45: 97-111.

Kishore, N., Mishra, A.K., Chansouria, J.P. (1993). Fungitoxicity of essential oils against dermatophytes. *Mycoses* 36: 211-215.

Larhsinin, M., Lazrek, H.B., Amarouch, H., Jana, M. (1996). Investigation of antifungal and analgesis activities of extracts from *Sium nodiflorum. Journal of Ethnopharmacology* 53: 105-110.

Magiatis, P., Melliou, E., Skaltsounis, A.L., Chinou, I.B., Mitaku, S. (1999). Chemical composition and antimicrobial activity of the essential oils of *Pistacia lentiscus* var. *chia. Planta Med* 65: 749-752.

Maoz, M., and Neeman, I. (2000). Effect of *Inula viscosa* extract on chitin synthesis in dermatophytes and *Candida albicans. J. Ethnopharmacol* 71: 479-482.

McCutcheon, A.R., Ellis, S.M., Hancock, R.E.W., Towers, G.H.N. (1994). Antifungal screening of medicinal plants of British Colombian native peoples. *Journal of Ethnopharmacology* 44: 157-169.

Mossa, J.S., Al-Yahya, M.A., Al-Meshal, I.A., Tariq, M. (1983). Phytochemical and biological screening of Saudi medicinal plants: Part 4. *Fitoterapia* 54: 75-80.

Murata, N., Sato, N., Takahashi, N. (1984). Very long chain saturated fatty acids in phosphatidyl serine from higher plant tissues. *Biochemica et Biophysica Acta* 795: 147-150.

Murray, P.R., Baron, E.J., Pfaller, M.A., Tenover, F.C., Yolke, R.H. (1995). *Manual of Clinical Microbiology,* Sixth Edition. Washington, DC: ASM Press, pp. 1327-1341.

Navarro, V., Villarreal, M.L., Rojas, G., Lozoya, X. (1996). Antimicrobial evaluation of some plants used in Mexican traditional medicine for the treatment of infectious diseases. *J Ethnopharmacol* 53: 143-147.

Oksuz, S. (1976). y-taraxasterol acetate from *Inula viscosa. Planta Medica* 29: 343-345.

Pai, S.T., and Platt, M.W. (1995). Antifungal effects of *Allium sativum* (garlic) extract against the *Aspergillus* species involved in otomycosis. *Lett Appl Microbiol* 20: 14-18.

Palevitch, P.D., and Yaniv, Z. (1991). *Medicinal Plants of Hollyland,* Volumes 1 and 2.Tel-Aviv: Tammuz Publisher Ltd., pp. 34-352.

Pattnaik, S., Subramanyam, V.R., Kole, C. (1996). Antibacterial and antifungal activity of ten essential oils *in vitro. Microbios* 86: 237-246.

Qasem, J.R., and Abu-Blan, H.A. (1995). Antifungal activity of aqueous extracts from some common weed species. *Annals of Applied Biology* 127: 215-219.

Rai, M.K. (1996). *In vitro* evaluation of medicinal plant extracts against *Pestalotiopsis mangiferae. Hindustan Antibiot Bull* 38: 53-56.

Sarbhoy, A.K., Varshney, J.L., Maheshwari, M.L., Saxena, D.B. (1978). Efficacy of some essential oils and their constituents on few ubiquitous molds. *Zentralbl Bakteriol [Naturwiss]* 133: 723-725.

Schmidt, M. (1933). *Planta (Berlin)* 20: 407.

Shahi, S.K., Shukla, A.C., Bajaj, A.K., Banerjee, U., Rimek, D., Midgely, G., Dikshit, A. (2000). Broad spectrum herbal therapy against superficial fungal infections. *Skin Pharmacol Appl Skin Physiol* 13: 60-64.

Singh, R., and Rai, B. (2000). Antifungal potential of some higher plants against *Fusarium udum* causing wilt disease of *Cajanus cajan*. *Microbios* 102: 165-173.

Singh, V.K., and Pandey, D.K.(1989). Fungitoxic studies on bark extract of *Lawsonia inermis* against ringworm fungi. *Hindustan Antibiot Bull* 31: 32-35.

Sow, A.I., Koyalta, D., Boye, C.S., Diedhiou/Badiane, D., Bassene, E. (1995). Antibacterial activity of essential oils from mint in Senegal. *Dakar Med* 40: 193-195.

Tanira, M.O.M., Bashir, A.K., Dib, R., Goodwine, C.S., Wasfi, I.A., Banna, N.R. (1994). Antimicrobial and phytochemical screening of medicinal plants of the United Arab Emirates. *Journal of Ethnopharmacology* 41: 201-205.

Tripathi, R.D., Srivastava, H.S., Dixit, S.N. (1978). A fungitoxic principle from the leaves of *Lawsonia inermis* lam. *Experientia* 34: 51-52.

Venugopal, P.V., and Venugopal, T.V. (1995). Antidermatophytic activity of garlic *(Allium sativum) in vitro*. *Int J Dermatol* 34: 278-279.

Vlietinck, A.J., Van Hoof, L., Totte, J., Lasure, A., Vanden Berghe, D., Rwangabo, P.C., Mvukiyumwami, J. (1995). Screening of hundred Rwandese medicinal plants for antimicrobial and antiviral properties. *Journal of Ethnopharmacology* 46: 31-47.

Wilson, S., Mahiou, B., Reiger, R., Tentler, S., Schimoler, R., Orndorff, S., Selitrennikoff, C.P. (2000). Pilot-scale purification of zeamatin, an antifungal protein from maize. *Biotechnol Prog* 16: 38-43.

Yegen, O., Berger, B., Heitefuss, R. (1992). Studies on the fungitoxic effects of extracts of six selected plants from Turkey on phytopathogenic fungi. *Zeitschrift-fur-P flanzenkrankheiten-und-Pflanzenschutz* 99: 349-359.

Yin, M.C., and Tsao, S.M. (1999). Inhibitory effect of seven *Allium* plants upon three *Aspergillus species*. *Int J Food Microbiol* 49: 49-56.

Yoshida, T., Moria, K., He, G.X. (1995). Inulavosin, A new thymol dimer with pesticidal activity from *Inula viscosa*. *Heterocycles* 41: 1923-1926.

Zeringue, H.J. Jr. (2000). Identification and effects of maize silk volatiles on cultures of *Aspergillus flavus*. *J Agric Food Chem* 48: 921-925.

Chapter 20

Native Use of Herbal Drugs for Treatment of Skin Diseases in Nepal

Narayan Prasad Manandhar

INTRODUCTION

Nepal is endowed with rich plant resources due to the variation in altitude, topography, rainfall, and climate. All of these factors have resulted in a diversity of vegetation types. The vegetation of Nepal ranges from tropical in the south (50-1200 m), temperate (1200-2900 m) in the midland, and alpine (2901-5000 m) in the north. The country has about 0.1 percent of the world's land surface but it claims over 2 percent of all flowering plants in its small land mass. The recorded list of medicinal plants in Nepal is over 800 species, which are mainly collected from the wild stocks. Of these, about 75 species are exported and as a result, are one of the sources of national as well as rural income.

Nepal is a mountainous country where about 77 percent of the total land area (147,181 km^2) is covered with mountains and rolling hills, and the rest is the flat lands of Terai in the south (CBS, 1998).

The population is about 20 million, and about 90 percent reside in rural areas where life-supporting facilities are lacking. It is estimated that there are more than 60 ethnic groups with their own traditions and culture. Most of the people in Nepal depend on plant resources for their livelihood because of the weak economy.

The adult literacy rate of the country is 40 percent. However, this percentage is even less in rural areas, where communication is through word-of-mouth. This is also the case with folklore medicine. Considering these facts, the goal of this study was to document folklore medicine used for the treatment of skin diseases.

The author wishes to thank the rural people for their cooperation and help during the field work.

COLLECTION OF ETHNOBOTANICAL INFORMATION

Information was collected during an ethnobotanical study in different parts of the country over the past two decades. The plant specimens were shown to the villagers to collect data on skin diseases. Information was obtained from the local healers and the experienced adults about the uses of plants for the treatment of diseases. In most cases, the villagers were observed applying drug plants for treatment of diseases.

Altogether, 85 informants were interviewed, including 22 local healers. These informants were between 40-65 years of age. Repeated queries were made at different times with the same informants and cross-checked at different places with other informants. The data were considered valid if at least five informants had similar replies about the medicinal properties of the plants.

ENUMERATION OF PLANTS

The plants studied are enumerated in alphabetical order by botanical name, family name (in parentheses), Nepali name, and, finally, usage of the plant parts. All plants were collected and identified with the help of books, photographs, and by comparing identified herbarium specimens preserved in the national herbarium (KATH), in Lalitpur, Nepal, where the present specimens have been deposited for future use.

- *Abrus precatories* L. (Leguminosae) 'Ratigeri'
  The juice of root, mixed with half the amount of the root juice of *Urena lobata* L. is applied on ringworm spots.
- *Acacia catechu* (L.F.) Willd. (Leguminosae) 'Khair'
  The leaf, mixed in equal amount with the leaf of *Abutilon indicum* (L.) Sweet is ground, and its paste is applied to treat boils and pimples.
- *Achyranthes aspera* L. (Amaranthaceae) 'Datiwan'
  The juice of root is applied to treat ringworm.
- *Aesandra butyracea* (Roxb.) Baehni (Sapotaceae) 'Chyuri'
  The oil-cake, mixed with half the amount of plant juice of *Adiantum capillus-veneris* L. and *Chenopodium ambrosoides* L. is applied in cases of boils and pimples.
- *Aesculus indica* (Colebr. ex Cambess.) Hook. (Hippocastanaceae) 'Ban khor' see (see Photo 20.1)
  The paste of seeds is applied to treat ringworm.

PHOTO 20.1. *Aesculus indica* plant used to treat ringworm infection

- *Ageratum conyzoides* L. (Composite) 'Ganne jhar'
  The juice of plant is applied on boils and pimples (Manandhar, 1989).
- *Allium sativum* L. (Amaryllidaceae) 'Lasun'
  The cloves, mixed with some salt, are pounded and the paste is applied to treat ringworm. After the application of the medicine for two to three times, a layer is formed which is removed and the treatment is continued for about two weeks.
- *Alternanthera sessilis* (L.) DC. (Amaranthaceae) 'Bisaune Jhar'
  The juice of plant, mixed in equal amount with the plant juice of *Eclipta prostata* (L.) L. is applied to treat ringworm.
- *Amaranthus spinosus* L. (Amranthaceae) 'Kamde linde'
  The juice of root, mixed with double the amount of root juice of *Paris polyphylla* Sm. is applied in case of itching.

- *Arctium lappa* L. (Compositae) 'Kurya'
  The juice of plant is applied on itching spots.
- *Argemone mexicana* L. (Papaveraceae) 'Sungure Kanda'
  The juice of seeds is applied in case of itching.
- *Artemisia indica* Willd. (Compositae) 'Titepati'
  The juice of leaf is applied to treat ringworm.
- *A. sieversinana* Willd. (Compositae) 'Guhya pati'
  The leaf, mixed with the root of *Parnassia nubicola* Wall. Ex. Royle and the resin of *Abies spectabilis* (D. Don) Mirb in equal amount is pounded with some water, and is applied for itching.
- *Arthomeris wallichiana* (Spreng.) Ching (Polypodiaceae) 'Harchur'
  The rhizome, mixed with some cumin seeds and water is pounded and this paste is applied to treat ringworm.
- *Artocarpus lakoocha* Wall. (Moraceae) 'Barahar'
  The juice of bark is warmed and is applied to treat boils.
- *Asclepias curassavica* L. (Asclepiadaceae) 'Machha Phul'
  The juice of root mixed with the latex of *Ficus religiosa* L. is warmed and applied to treat ringworm (see Photo 20.2).
- *Azadirachta indica* A. Juss. (Meliaceae) 'Nim'
  The juice of bark is applied in case of itching.
- *Bauhinia vahili* Eight & Arn. (Leguminosae) 'Bhorla'
  Chonemorpha fragrans (Moon) Alston and some yogurt is applied to treat ringworm.
- *Boehmeria ternifolia* D. Don (Urticeae) 'Kamle'
  The juice of leaf, mixed with half of the amount of root juice of *Urtica dioica* L. is used for itching.
- *Bupleurum falcatum* L. (Umbelliferae) 'Mariche Ghans'
  The juice of plant is applied to treat pimples (Manandhar, 1987).
- *Carica papaya* L. (Caricaceae) 'Mewa'
  The latex, mixed with the paste of *Eclipta prostrata* (L.) L. is applied is case of ringworm. The paste of seed is also applied to treat the same disease.
- *Chrysopogan aciculatus* (Retzius) Trinius (Gramineae) 'Ghore dubo'
  The juice of root, mixed in double amount with the root juice of *Imperata cylindrica* (L.) Beauv. and latex of *Calotropis gigantea* (L.) Dryand is applied to treat itching and other fungal infections.
- *Clerodendrum philippium* Schauer (Verbenaceae) 'Raj Beli'
  The juice of leaf is applied to treat pimples and fungal infections.
- *Coleogyne cristata* Lindley (Orchidaceae) 'Chandi Gabha'
  The paste of pseudo bulb is applied to treat boils and pimples.

PHOTO 20.2. *Asclepias curassavica* plant used to treat ringworm infection

- *Corydalis chaerophylla* DC. (Papaveraceae) 'Okhre Ghans'
  The root is pounded with some seeds of black pepper and is applied to treat boils and pimples.
- *Cynoglossum glochidiatum* Wall. Ex Benth. (Boraginaceae) 'Bhende kuro'
  The paste of plant is applied on ringworm infection.
- *Daphniphyllum himalense* (Benth.) Muell. -Arg. (Dahphniphyllaceae) 'Rakta Chandan'
  The paste of wood is applied to treat boils and fungal infections.
- *Debregesia longifolia* (Brum. F.) Wedd. (Urticaceae) 'Tusare'
  The juice of bark is applied in case of scabies and fungal infection.
- *Delphinium himalayai* Munz (Ranunclulaceae) 'Jaunde mulo'
  The paste of root is applied to treat boils.

- *Desmodium microphyllum* (Thunb.) DC. (Leguminosae) 'Bute kanike'
  The juice of root is applied to treat ringworm (dermatophytic infections).
- *Dioscorea deltoidea* Wallich ex Grisebach (Dioscoreaceae) 'Bhyakur'
  The juice of root is applied to treat itching caused by ringworm.
- *Entada phaseoloides* (L.) Merr. (Legumionsae) 'Pangra'
  The paste of cotyledon, mixed with half the amount of plant juice of *Crassocephalum crepidiodes* (Benth.) S. Moore is applied to treat boils and other skin infections.
- *Euphrasia himalayica* Wettst. (Scrophulariaceae) 'Hare'
  The juice of root is applied to treat ringworm infections.
- *Euphorbia royleana* Boiss (Euthorbiaceae) 'Syuri'
  The latex is applied to treat boils (Coburn, 1984).
- *Evolvulus nummularius* (L.) L. (Convolvulaceae) 'Shankhe phul'
  The juice of plant is applied to treat scabies (Manandhar, 1985).
- *Galium asperifolium* Wall. Ex Roxb. (Rubianceae) 'Kurkure jhar'
  The juice of the plant is applied on scabies and other skin problems.
- *Geniosporum coloratum* (D. Don.) Kuntze (Labiatae) 'Ban wari'
  The juice of plant is applied to treat itching.
- *Jatropha curcas* L. (Euphorbiaceae) 'Sajyun'
  Either the latex or the seed oil is applied to treat boils and pimples.
- *Launaea asplenifolia* (Willd.) Hook. f. (Compositae) 'Dhudhe jhar'
  The juice of plant is applied in the case of boils, pimples, and other skin diseases.
- *Leucas cephalotus* (Roth) Spreng. (Labiatae) 'Kanthe jhar'
  The paste of root is applied on pimples. The juice of plant, mixed in equal amounts with the plant juice of *Lablab purpureus* (L.) Sweet is applied to treat ringworm.
- *Lindenbergia grandiflora* (Buch. -Ham. ex D. Don) Benth. (Scrophulariaceae) 'Bhedi phul'
  The juice of plant is applied to relieve itching.
- *Lyonia ovalifolia* (Wall.) Drude (Ericaceae) 'Angeri'
  The juice of leaf is used for scabies (Manandhar, 1995; Shrestha, 1988).
- *Macaranga pustulata* King ex Hook f. (Euphorbiaceae) 'Mallato'
  The juice of bark is applied to treat itching.
- *Maclura cochinchinensis* (Lour.) Corner (Moraceae) 'Damaru'
  The juice of bark is applied on ringworm.
- *Maesa chisia* Buch.-Ham. Ex D. Don (Myrsinaceae) 'Bilauni'
  The paste of immature fruits is applied to treat itching.
- *Mallotus philippensis* (Lam.) Muell. Arg. (Euphorbiaceae) 'Sindure'
  The paste of unripe fruit is mixed with half the amount of plant juice of *Cynodon dactylon* (L.) Pers. and is applied to treat ringworm.

- *Milletia extensa* (Benth.) Baker (Leguminosae) 'Gaujo'
  The juice of bark is applied to treat scabies (Manandhar, 1998).
- *Mirabilis jalapa* L. (Nyctaginaceae) 'Lanka phul'
  The paste of leaf, mixed in equal amounts with leaf-paste of *Eclipta prostrata* (L.) L. is applied to treat boils and pimples.
- *Nepeta leucophylla* Benth. (Labiatae) 'Gumi'
  The juice of plant is used on itching spots.
- *Neolitsea cuipala* (Buch.-Ham ex D. Don) Kosterm. (Lauraceae) 'Kalchhe'
  The juice of seeds is applied to treat scabies.
- *Oxalis corniculta* L. (Oxalidaceae) 'Charimilo'
  The juice of plant mixed in equal amounts with the fruit paste of *Xanthium strumarium* L. and *Melia azadirach* L. is applied to treat boils and pimples.
- *Pedicularis longiflora* Rudolph (Scrophulariaceae) 'Mingi'
  The juice of plant is applied to treat ringworm.
- *Phyllanthus amarus* Schumacher & Thonn. (Euphorbiaceae) 'Bhuin amala'
  The juice of leaf is applied to treat pimples.
- *P. parvifolius* Buch-Ham ex. D. Don (Euphorbiaceae) 'Masino dudhi'
  The juice of bark is applied to treat boils and pimples (Manandhar, 1992).
- *Pieris formosa* (Wall.) D. Don (Ericaceae) 'Gineri'
  The juice of leaf is applied to treat scabies (Manandhar, 1993).
- *Plantago erosa* Wall. (Plantaginaceae) 'Bijauni phul'
  The root-paste is applied to treat boils and pimples (Manandhar, 1990).
- *Pouzolzia zeylanica* (L.) J. Bennett & Brown (Urticaceae) 'Barbere'
  The juice of leaf mixed with half the amount of paste of cotyledon of *Ricinus communis* L. is applied to treat ringworm.
- *Rabdosia coetsa* (Buch.-Ham D. Don) Hara (Labiatae) 'Mirre'
  The paste of plant is applied to treat ringworm.
- *Rabus rugosus* Smith (Rosaceae) 'Ban ainselu'
  The juice of bark, mixed in equal amounts with the fruit juice of *Smilax apera* L. is applied to treat scabies.
- *Scrophularia urticifolia* Wall. Ex. Benth. (Scrophulariaceae) 'Mokhi ghans'
  The juice of leaf is applied to treat ringworm.
- *Sida acuta* Brum. F. (Malvaceae) 'Kuro'
  The paste of root, mixed in equal amounts with the plant paste of *Tridax procumbens* L. is applied for itching.
- *Sorbus ursina* (Wenzig) Decne (Rosaceae) 'Ghuro'
  The paste of root is applied on boils.

- *Thalictrum foliolosum* DC. (Ranunculaceae) 'Ban suli'
  The juice of stem is applied on itching.
- *Thalictrum javanicum* Blume (Ranunculaceae) 'Guhya pati'
  The juice is applied in case of boils and pimples.
- *Trichilia connaroides* (Weight and Arn.) Bentvelzen (Meliaceae) 'Aank-hataruwa'
  The juice of seed is applied on scabies and other mycotic infections.
- *Triumfetta rhomboides* Jacq. (Tiliaceae) 'Bish khapre'
  The juice of plant is applied to treat boils.
- *Uraria lagapodioides* (L.) Desv. (Leguminosae) 'Sano Bhatte'
  The juice of plant is applied on itching.
- *Vernonia cineria* (L.) less. (Compositae) 'Phurke jhar'
  The juice of plant, mixed with the root juice of *Alternanthera sessilis* (L.) DC is applied to treat scabies.
- *Zizyphus mauritiana* Lam (Rhamnaceae) 'Bayar'
  The paste of stone is applied to treat ringworm.

DIVERSITY OF PLANTS USED IN THE TREATMENT OF SKIN INFECTIONS

The study revealed that 68 plants belonging to 35 families and 64 genera are used for the treatment of skin diseases. Of them, three species represent the family of monocotyledons and one species pteridophyte. Out of these, 22 species are used to treat boils and pimples, 16 for itches, 22 for ringworm, and 10 species for scabies. Twenty species are used in combination with other plants, including *Abies spectabilis* (D. Don) Mirb.; *Abutilon indicum* (L.) Sweet; *Adiantum capillus-veneris* L.; *Alternanthera sessilis* (L.) DC.; *Calotropis gigantea* (L.) Dryand; *Chenopodium ambrosoids* L.; *Chonemorpha fragrans* (Moon) Alston; *Crassocephalum crepidiodes* (Benth.) S. Moore; *Cynodon dactylon* (L.) Pers.; *Eclipta prostrata* (L.) L.; *Ficus religiosa* L.; *Imperata cylindrica* (L.) Beauv.; *Lablab purpureus* (L.) Sweet; *Melia azadirach* L.; *Paris polyphylla* Sm; *Parnassia nubicola* Wall. ex Royle; *Ricinus communis* L.; *Smilax aspera* L.; *Tridax procumbens* L.; *Urena lobata* L.; *Urtica dioica* L.; and *Xanthium strumarium* L. In cases of ringworm, scabies, and common itches, the affected part is first soaked in water for about five minutes, scratched slightly, and the medicine is applied so that it will deeply penetrate in skin. Local healers are consulted if help is needed. For Nepalese people, skin diseases do not hamper their normal activities. They are a common phenomenon.

Most of the information presented here is unrecorded (Sacherer, 1979; Shrestha, 1985; Dangol and Gurung, 1991). The plants with identical uses, which were recorded previously by different authors, are *Ageratum conyzoides, Bupleurum falcatum, Euphorbia royleana, Lyonia ovalifolia, Milletia extensa, Phyllanthus parvifolius, Pieris formosa,* and *Plantago erosa* (Coburn, 1984; Manandhar 1987, 1989, 1990, 1992, 1993, 1995, 1998; Shrestha 1988). In some cases, the treatment differs in the use of part of plant or in the disease (Toffin and Wart, 1985). Twenty species are used in combination with the parts of other plants, whereas the single plant of the rest is used to treat the diseases.

In course of the study, about 55 patients were also contacted to find out the efficiency of the medicine. Thirty-five patients gave positive replies, whereas the rest provided mixed replies: such as, some said the medicine had a negligible effect or had no effect at all. However, the efficiency of herbal remedies needs clinical trials, which may lead to determination of efficacy of native therapy and may also help to identify new therapeutic compounds. The medicinal plants, which are still untapped resources, should be documented before such indigenous knowledge is lost due to destruction of forestry and the acculturation of tribal communities.

CONCLUSIONS

Historically, humans have utilized plants as therapeutic aids to health. This knowledge has been acquired by trial and error. Further knowledge can be gained. The medical experience that has been acquired by native healers and priests has become a point of interest and importance (Vogel, 1970). An examination of herbal medicines used by tribal and native people may provide a good opportunity for discovering new drugs. The implication of advanced technology may determine, with a high degree of certainty, the efficacy of native medicine.

For most developing countries, herbal medicine is the main system of medical treatment. It is cheap, affordable, and offers a variety of new chemical constituents. These findings have proven useful for diseases such as cancer, AIDS, etc. The time has come to link rural technology with modern technology.

FUTURE PROSPECTS

Plant-based indigenous knowledge has an important role for human health. This knowledge is a characteristic element of sustainability, which deals with natural resource management. It may be an important source of information for policymakers, researchers, and doctors.

The main sources of medicinal plants are forests, which are depleting at a rate of 2.1 percent per year (Pudasaini, 1992). As a result, medicinal drug plants are being depleted and their sustainable exploitation is now a serious matter. There are various reasons for the depletion of medicinal plants: weak economy, lack of alternatives for livelihood, exploitation of florae, and deforestation. Excavation of sand and stones from natural habitats of plants, implementation of developmental activities, encroachment of forest-land for settlement and agriculture, and smuggling of forest products are also the causes of depletion of medicinal plant resources.

In order to continue the existence of natural resources, it is important to launch some applied programs and projects to bring a change in rural thinking about the forest. Experience has shown that effective conservation depends on direct involvement of local people in the management of natural resources. However, the protection of natural habitats, prevention of over-collection, trade control, effective management plans, and artificial propagation are some factors which will be useful in sustaining medicinal plants. Moreover, the result of ethnomedicinal data, after careful evaluation, should be implemented at the village level as far as possible. It is better to emphasize indigenous technology rather than impose sophisticated high technology.

REFERENCES

CBS (Central Bureau of Statistics) (1998). *Statistical Pocket Book*, Nepal: HMG of Nepal, Kathmandu.

Coburn, B. (1984). Some native medicinal plants of the Western Gurung. *Kailash* 11(1-2): 55-88.

Dangol, D. R. and Gurung, S. B. (1991). Ethnobotany of the Tharu of Chitwan District Nepal. *Int. J. Pharmacognosy* 29(3): 203-209.

Manandhar, N. P. (1985). Ethnobotanical notes on certain medicinal plants used by Tharus of Dang Deokhuri District. *Nepal Int. J. Crude Drug Res.* 23(4): 153-159.

Manandhar, N. P. (1987). An ethnobotanical profile of Manang valley. *J. Econ. Tax. Bot.* 10(1): 207-213.

Manandhar, N. P. (1989). Medicinal plants used by Chepang tribes of Makawanpur District, *Nepal. Fitoterapia* 60(1): 61-68.

Manandhar, N. P. (1990). Folklore medicine of Chitwan District, Nepal. *Ethnobotany* 2(1-2): 31-38.

Manandhar, N. P. (1992). Folklore medicine of Dhading District, Nepal. *Fitoterapia* 63(2): 163-177.

Manandhar, N. P. (1993). Herbal remedies of Surkhet District, Nepal. *Fitoterapia* 64(3): 265-272.

Manandhar, N. P. (1995). An inventory of some vegetable drug resources of Makawanpur District, Nepal. *Fitoterapia* 66(3): 231-238.

Manandhar, N. P. (1998). Native phytotherapy among the Raute tribes of Dadeldhura District Nepal. *Journal of Ethnopharmacology* 60: 199-206.

Pudasaini, S.P. (1992). "Environment crisis-integrated action required." *The Rising Nepal* (English Daily), August 26, p. 2.

Sacherer, J. (1979). The high altitude ethnobotany of Rolwaling Sherpa. *Contribution of Nepalese Studies* 6(2): 45-64.

Shrestha, P. (1985). Research note: Contribution to ethnobotany of the Palpa area. *Contribution to Nepalese Studies* 12(2): 63-74.

Shrestha, P. (1988). Contribution to ethnobotany of the Tamangs of Khatmandu valley. *Contribution to Nepalese Studies* 15(2): 247-266.

Toffin, G. and Wart, J. (1985). Recherches sur l' ethnobotanique des Tamangs du massif du Ganesh Himal (Nepal, central): Les plantes non cultivees. *J. d'Agric. Trad. et. de Bota. Appl* 32: 127-175.

Vogel, V.J. (1970). *American Indian Medicine*. Norman, OK: University Oklahoma Press, p. 583.

Chapter 21

The Antimycotic Potential of Fijian Plants

Kannoth Panicker Sreekumar
Subramaniam Sotheeswaran

INTRODUCTION

Fungi can cause annoying and sometimes dangerous infections in animals and humans (Bennett and Walzer, 1998). Most of these fungal infections are on the skin, particularly between the folds of the skin. Athlete's foot is a classic example of a fungal disease. The scalp, the groin, and the inside of the outer ear are other common sites of fungal attack. Some fungi penetrate more deeply into the body. The fungus known as *Monilia* often infects the mouth ("thrush") and the vagina, and *Histoplasma capsulatum* penetrates the lung, causing fungal infection. Fungal infections of the lungs are sometimes mistaken for tuberculosis.

There are several treatments for the ailments caused by fungi. Drying up the sites of infection is one way to cure fungal infections on the skin. This is not always easy, so long as the skin exudes perspiration. Fungus-killing drugs (antimycotics) that do not harm the skin are available and these are mainly organic acids. However, repeated applications of synthetic antimycotics can be harmful to the skin. For deep-seated fungus infections, antibiotics and sulfa drugs can be given either by mouth or as injections. In many instances, people are turning to medicinal plants for use as antimycotic agents. These are time-tested medicines used by rural healers and are becoming popular even in many urban cities in Fiji.

The authors wish to thank Ms. Mary Lou Quinn, managing director of the NAPRALERT database at the University of Illinois at Chicago, for permitting access to the database to one of the authors (Sotheeswaran) by courtesy of the World Health Organization's Western Pacific Regional Office in Manila, Philippines. The authors also wish to thank Professor Rai for the invitation to prepare this chapter for publication.

Most Fiji islanders still retain a faith in herbal medicines, even though they also utilize Western medicine for many health problems. In Fiji, traditional Fijian herbal medicine has been enriched by the introduction of the herbal system of medicine used by the Indians who came to Fiji more than 150 years ago, and who now comprise about 45 percent of the population. Many medicinal plants used in many tropical countries to treat for antimycotic activity are also recorded for such use in Fiji.

This chapter reviews the literature of plants (listed as follows) for which antimycotic activity is claimed by local healers, and some Fijian plants, which have been tested for antimycotic activity.

LIST OF PLANTS SHOWING ANTIMYCOTIC ACTIVITY

Ageratum conyzoides (Asteraceae)
Alphitonia viellardi (Rhamnaceae)
Annona muricata (Annonaceae)
Bischofia javanica (Euphorbiaceae)
Brachiaria mutica (Poaceae)
Cananaga odorata (Annonaceae)
Carica papaya (Caricaceae)
Cassia alata (Fabaceae)
Casuarina equisetifolia (Casuarinaceae)
Centella asiatica (Apiaceae)
Citrus aurantium (Rutaceae)
Citrus sinensis (Rutaceae)
Cocos nucifera (Arecaceae)
Codiaeum variegatum (Euphorbiaceae)
Curcuma longa (Zingiberaceae)
Dolicholobium latifolium (Rubiaceae)
Dysoxylum richii (Meliaceae)
Elaeocarpus strockii (Elaeocarpaceae)
Erythrina variegata (Fabaceae)
Flagellaria indica (Flagellariaceae)
Hibiscus rosa-sinensis (Malvaceae)
Lycopodium cernuum (Lycopodiaceae)
Manihot esculenta (Euphorbiaceae)
Messerschmidia argentea (Boraginaceae)
Mikania micrantha (Asteraceae)
Morinda citrifolia (Rubiaceae)
Piper methysticum (Piperaceae)

Plumbago zeylanica (Plumbaginaceae)
Plumeria rubra (Apocynaceae)
Psidium guajava (Myrtaceae)
Psychotria calycosa (Rubiaceae)
Psychotria confertiloba (Rubiaceae)
Rhizophora samoensis (Rhizophoraceae)
Ricinus communis (Euphorbiaceae)
Semecarpus vitiensis (Anacardiaceae)
Solanum torvum (Solanaceae)
Syzygium corynocarpum (Myrtaceae)
Terminalia catappa (Combretaceae)
Thespesia populnea (Malvaceae)
Vavae amricorum (Meliaceae)
Xylocarpus granatum (Meliaceae)

ANTIMYCOTIC ACTIVITY OF PLANTS

Ageratum conyzoides *L. (Asteraceae)*

Fijian name: Botebotekoro
Description: An herb growing up to 1 m tall. The leaves are simple, hairy, and the flowers are white to pale blue in color. The flowers are borne in small, sunflowerlike heads.
Habitat: Occurs along roadsides, forest trails, clearings, and is cultivated in mountainous areas.
Chemical constituents: Kaempferol, beta-sitosterol, stigmasterol, fumaric acid, pyrrolizidine alkanloids, coumarin, and alkanes
Antimycotic activity: Confirmed (WHO, 1998).
Traditional antifungal use: Use by Fijians not confirmed.

Alphitonia viellardi *Lenorm (Rhamnaceae)*

Fijian name: Doi, Doi damu
Description: A large shrub with elliptical to orbicular-shaped leaves with conspicuous veins. The flowers are fragrant with white to pale yellow calyx and white to greenish white petals. Fruits appear black on maturity. Fruits and flowers are seen throughout the year. This is endemic to Fiji.
Habitat: Occurs in dry or open forest, rocky places and open slopes. Occurs in elevation about 1,200 m above sea level.
Chemical constituents: Not available.

Antimycotic activity: Confirmed (Singh and Siwatibau, 1980).
Traditional antifungal use: Use by Fijians not confirmed.

Annona muricata ***L. (Annonaceae)***

Fijian name: Seremaia, Sarifa
Description: Small tree (7 m tall) with leaves alternate with leathery blades. The flowers are greenish with fleshy petals. Fruit is fleshy and covered with pseudospines and has black seeds. They flower and fruit throughout the year.
Habitat: Occurs in low-lying areas.
Antimycotic activity: Confirmed (WHO, 1998).
Chemical constituents: Annomonicin, anomurine, gigantetronenin, solamin, lipids, stigmasterol, and tannins.
Traditional antifungal use: Use by Fijians not confirmed.

Bischofia javanica ***Blume (Euphorbiaceae)***

Fijian name: Togo, Koko, Koka damu, Toga toga
Description: Normally 30 m tall, produces latex when bruised. Leaves alternate with toothed margins. Cream to yellowish colored, minute, unisexual flowers. Fruits are brown, globose, berry, and fleshy. They fruit throughout the year.
Habitat: Common in sea-level and mid-mountainous regions, forest edges, grassy slopes.
Antimycotic activity: Not confirmed (WHO, 1998).
Chemical constituents: Tannins, betulinic acid, glucoside, quercetin, beta-sitosterol, stigmasterol, and ursolic acid.
Traditional antifungal use: Fijians use the bark to treat athlete's foot (Cambie and Ash, 1994; Weiner, 1984; Whistler, 1992).

Brachiaria mutica ***(Forsk.) Stapf (Poaceae)***

Fijian name: Covatu
Description: A common grass, up to 2 m high with hairy sheath and nodes. The leaves are lanceolate and the inflorescence purple-tinged in color. It is also called para grass or Mauritius grass. It is a fodder grass.
Habitat: Occurs in wet zones, pastures, cultivated fields, along roadsides, on riverbanks, in swamps, in clearings, and along forest trails. It is also a weed of sugarcane fields and cultivation.
Chemical constituents: Hydrocyanic acid, gentisic acid, and benzenoid.

Antimycotic activity: Not confirmed.
Traditional antifungal use: The sap of the leaves is used for antifungal therapy (Singh and Siwatibau, 1980).

Cananaga odorata ***(Lam) Hook. F. and Thoms (Annonaceae)***

Fijian name: Mokosoi, Makosoi, Mokohoi, Makasui
Description: Tree grows up to 20 m tall. Leaves are alternate and the flowers are pale green to yellowish in color. Flowers appear in clusters. Fruits are oblong with brown seeds embedded in yellow, oily pulp. The flowers and fruits are seen throughout the year.
Habitat: Occurs in forests, slopes, or gullies from sea level to mid-montane.
Antimycotic activity: Confirmed (WHO, 1998).
Chemical constituents: Essential oils, sesquiterpenoides, isoquinoline, alkaloids, aromatic compounds, lipids, and cyanogenic materials.
Traditional antifungal use: Use by Fijians not confirmed.

Carica papaya ***L. (Caricaceae)***

Fijian name: Weleti, Wi, Maoli
Description: A palmlike, usually unbranched, soft tree up to 10 m high with milky latex. It bears fragrant flowers, white in color and moderate in size. The fruits are large, fleshy, yellow to orange in color with plenty of small black seeds. The leaves and skin of the fruit contain papain which digests protein and is used as a meat tenderizer. This plant is also known as papaya or pawpaw.
Habitat: Widely cultivated in plantations and naturalized dwellings and garden patches. Occurs from sea level to mid-mountain regions.
Chemical constituents: Papain, carpaine, nicotine, cryptoxanthin, citric acid, malic acid, glutaric acid, ascorbic acid, benzyl glucosinolate, benzyl isothiocyanate, avenasterol, and cycloartenol.
Antimycoytic activity: Confirmed (Singh and Siwatibau, 1980).
Traditional antifungal use: The white latex from green fruit is used for antifungal therapy (Singh, 1986).

Cassia alata ***L. (Fabaceae)***

Fijian name: Bai nicagi, Mbai ni thangi
Description: Ornamental shrub with large pinnate leaves. Flowers are bright yellow in spikes which have earned them local names such as "Ro-

man candle tree" and "Candlelabra tree." The fruit is a legume with two prominent, wavy wings extending along its full length.

Habitat: Naturalized in wet habitats from sea level to 250 m but also cultivated in gardens.

Chemical constituents: Kaempferol, luteolin, chrysophanic acid, sitosterol, lectin, daucosterol, and emodin.

Antimycotic activity: Confirmed (WHO, 1998).

Traditional antifungal use: The leaf and or seeds are used for antifungal therapy (Singh and Siwatibau, 1980).

Casuarina equisetifolia *L. (Casuarinaceae)*

Fijian name: Noko-noko, Qaro, Nakure

Description: The tall tree (25 m) has drooping branches with needlelike branchlets. Leaves are highly reduced and scalelike giving the branchlets a pine-needle-like appearance. Male flowers have spikes and female flowers are borne in globose heads. The fruit resembles a small pine cone with many small winged nuts. This is a flowering plant and is mistaken for a type of pine tree. They flower throughout the year.

Habitat: Occur along beaches, rocky coasts, limestone outcroppings, dry hillsides, in both wet and dry zones from sea level to mid-montane.

Chemical constituents: Beta-sitosterol, glycosides, trifolin, ellagic acid, cholesterol, stigmasterol, cyanidins, citrullin, gentisic acid, and hydroquinone.

Antimycotic activity: Confirmed (WHO, 1998).

Traditional antifungal use: Use by Fijians not confirmed.

Centella asiatica *(L.) Urban (Apiaceae)*

Fijian name: Totodro

Description: A small creeping aromatic herb with long runners and short rhizomes. The leaves are alternate and form rosettes. Flowers are small and inconspicuous. The orbicular or ellipsoidal fruit is yellowish brown in color and available throughout the year.

Habitat: Usually occurs in open spaces, lawns, pastures, shaded roads, and trailsides.

Chemical constituents: Brahmoside, asiaticoside, kaempferol, madecassic acid, vitamin C, linamarase, and triterpenoid trisaccharides.

Antimycotic activity: Confirmed (WHO, 1998).

Traditional antifungal use: The juice of the leaves is used for antifungal therapy (Cambie and Ash, 1994).

Citrus aurantium *L. (Rutaceae)*

Fijian name: Moli jamu, Moli kurukuru
Description: Ten-meter high tree with spines on younger branches. The leaves are alternate, oval-shaped with blunt-toothed edges that emit a strong citrus odor from copious oil glands. They bear white fragrant flowers. The fruit is sour and yellowish orange in color with numerous seeds. It flowers in the warmer months.
Chemical constituents: Carotenoids, coumarins, caffeine, isoquinoline alkaloides, several flavonoids, desmosterol, ergosterol, and stigmasterol.
Antimycotic activity: Confirmed (WHO, 1998).
Traditional antifungal use: Use by Fijians not confirmed.

Citrus sinensis *(L.) Osbeck (Rutaceae)*

Fijian name: Moli taiti
Description: The tree is small but has large spines on the younger branches. The leaves are alternate, oval-shaped blades with blunt-toothed edges and emit a strong citrus odor from copious oil glands. Flowers are white, very fragrant, and seen singly or in bunches. Fruits are thin-skinned and contain sweet pulp and several seeds and usually are green or orange in color. Flowers in warmer months.
Habitat: Cultivated and possibly naturalized in some locations.
Chemical constituents: Gibberellic acid, phytol, limonin, nomillin derivatives, flavonoids, carotenoids, castasterone, sitosterol, hydroquinone, sinapic acid, citrusins, coumarins, and stigmasterol.
Antimycotic activity: Confirmed (WHO, 1998).
Traditional antifungal use: Use by Fijians not confirmed.

Cocos nucifera *L. (Arecaceae)*

Fijian name: Niudamu, Niu, Niu dina
Description: Monopodial tree with long, narrow, often curved trunk. Leaves are confined to apex of the trunk. Flowers and fruit are borne in drooping clusters arising from between lower leaf petioles. Fruit (coconut) is large, green, brown, or reddish. The husk is fibrous and the mature single seed is large with both solid and liquid endosperm. This tree flowers and bears fruit throughout the year.
Habitat: Usually occurs along the seashore to moderate elevation in inland areas. It is usually cultivated in the tropics and most abundant near human settlements.

Chemical constituents: Saccharose, myoinositol, sorbitol, diphenylurea, alcohols, ketones, leucoanthocyanins, glycerol, sucrose, glucose, fatty acids, polyphenols, campesterol, squalene, stigmasterol, ligustrazine alkaloid, and 2,3,5-trimethylpyrazine alkaloid.
Antimycotic activity: Confirmed (WHO, 1998).
Traditional antifungal use: Use by Fijians not confirmed.

Codiaeum variegatum ***(L.) Blume (Euphorbiaceae)***

Fijian name: Sacasaca damu, Vasa damu
Description: This is an aromatic shrub with multicolored leaves. Flowers are small and unisexual. Some forms do not have fruit or flowers and some have fruit and flowers throughout the year.
Habitat: Usually cultivated from sea level to mid-montane regions.
Chemical constituents: Phenolic compounds, coumaric acid, ellagic acid, vanilla acid, and alkanes.
Antimycotic activity: Confirmed (WHO, 1998).
Traditional antifungal use: Use by Fijians not confirmed.

Curcuma longa ***L. (Zingiberaceae)***

Fijian name: Avea, Cago, Rerega
Description: An herb with fleshy rhizomes and leafy pseudostems. The leaves are oblong in shape with parallel veins. The flowers are yellow, three-parted, and borne in an erect spike. Generally, fruits are absent and propagation is through rhizomes. The rhizomes are aromatic and are the source for the spice—turmeric.
Habitat: Naturalized and cultivated in low-lying areas and also in lower mountain areas.
Chemical constituents: Atlatones, bisabolenes, curcumenes, germacran derivatives, turmerine, zingiberene, borneol, camphor, cineol, terpinene, caffeic acid, cinnamoyl derivaties, cholesterol, lignan, vanillic acid, stigmasterol, and turmerins,
Antimycotic activity: Confirmed (WHO, 1998).
Traditional antifungal use: Use by Fijians not confirmed.

Dolicholobium latifolium ***A.Gray (Rubiaceae)***

Fijian name: Mbuambua ni waitui, Buabua ni baravi
Description: Tree reaching a height from 2-20 m. The fragrant flowers are white in color and are seen at the branch tips. Fruits turn from green to

dull red or brownish at maturity, and after dehiscing yield plenty of pale brown seeds.
Habitat: Occurs near sea level and above sea level (1,150 m) in dense, dry, open, or dry forest. Often found along crests, ridges, or creeks.
Chemical constituents: Not available.
Antimycotic activity: Confirmed (Singh and Siwatibau, 1980).
Traditional antifungal use: Use by Fijians not confirmed.

Dysoxylum richii ***(A.Gray) C. DC. (Meliaceae)***

Fijian name: Sasavira,Tarawau-kei-rakaka
Description: Tree up to 50 m high. The compound leaf has 20 leaflets and the flowers are cream to greenish in color. The fruits are round green to rusty brown in color. The bruised parts emit a strong alliaceous odor similar to garlic or onion. The trees flower and fruit throughout the year and are endemic to Fiji.
Habitat: Found in abundance from sea level to an elevation of 1,000 m in different types of forests, in thickets, and edges of mangrove swamps.
Chemical constituents: Dysoxylin, triterpenes, richenol, ocotillone, eicherianic acid, and dysoxysulfone.
Antimycotic activity: Confirmed (Cambie and Ash, 1994).
Traditional antifungal use: Use by Fijians not confirmed.

Elaeocarpus storckii ***Seem. (Elaeocarpaceae)***

Fijian name: Ngaingai, Qaiqai
Description: A medium-sized tree, 9-25 m in height with thick, leathery leaves. The inflorescences are borne in branchlets below the leaves and the sepals are pink or reddish purple or yellowish with red spots. The flowers are yellowish proximally and pink toward the apices. The large fruits are deep purple in color at maturity. Flowers and fruits are seen in warm climates. This species is endemic to Fiji.
Habitat: Usually in wet primary and secondary forests and edges of forests.
Chemical constituents: Not available.
Antimycotic activity: Confirmed (Singh and Siwatibau, 1980).
Traditional antifungal use: Use by Fijians not confirmed.

Erythrina variegata ***L. (Fabaceae)***

Fiji name: Drala, Segi, Drala dina, Rara, Damu, Rarawai
Description: The tree has coarse spines on the trunk and branches. Leaves are variable in size with triangular-shaped leaflet blades. The flowers are

bright orange to deep red in color. The fruits, contained in pods (legumes), have red seeds. They usually flower and fruit in the cooler season.

Habitat: Occurs mainly in the lowland coastal areas.

Chemical constituents: Erythraline alkaloid, erysonine alkaloid, isoquinoline alkaloid, lectin, isoflavones, polyphenols, glycoside, campesterol, and sitosterol.

Antimycotic activity: Biological activity tested (WHO, 1998).

Traditional antifungal use: Use by Fijians not confirmed.

Flagellaria indica *L. (Flagellariaceae)*

Fijian name: Walaki, Qalo, Wasila

Description: A high climber with solid stems. It has parallel veins terminating in a tendril which it uses to climb over shrubs and smaller trees. The flowers are cream-colored and the fruits are small and fleshy. The stems are used for making baskets. Flowers in summer.

Habitat: Occurs as a climber on shrubs, forest trees, and in wet and dry places.

Chemical constituents: Kaempferol, cyanogenic glycoside, glucose, galactose, and arabinose.

Antimycotic activity: Confirmed (Singh and Siwatibau, 1980).

Traditional antifungal use: Use by Fijians not confirmed.

Hibiscus rosa-sinensis *L. (Malvaceae)*

Fijian name: Kauti, Loloru, Senitoa yaloyalo, Senicikobia

Description: This is a shrub with alternate leaves with conspicuous serrate margins. Flowers are large and available in variable forms. The petals are attractive and range from orange to yellow in color. The fruit is capsular in shape with small black seeds. They flower and fruit throughout the year. This is the national flower of Fiji.

Habitat: Introduced and widely distributed throughout the tropics. Occurs in plain and mountainous regions.

Chemical constituents: Taraxeryl acetate, sitosterol, stigmasterol, cholesterol, ergosterol, citric acid, tartaric acid, oxalic acid, fructose, glucose, sucrose, flavonoid glycosides, hibiscetin, and cyanin glycosides.

Antimycotic activity: Confirmed (WHO, 1998).

Traditional antifungal use: Use by Fijians not confirmed.

Lycopodium cernuum *L. (Lycopodiaceae)*

Fijian name: Lewanini, Suinigata, Alewa nini
Description: This is a land club moss with creeping rhizomes. The stem is erect, branched, and bears small leaves. The tip of the fruiting stem is club-shaped or cone-shaped.
Habitat: Usually appears in open spaces and sites.
Chemical constituents: Ceruine alkaloid, nicotine, lycopodine, triterpenes, sterols, esters, phenolic acids, and ketones.
Antifungal activity: Not confirmed (WHO, 1998).
Traditional antifungal use: The spores are used for antifungal therapy.

Manihot esculenta *Crantz (Euphorbiaceae)*

Fijian name: Tavioka, Yabia ni vavalagi, Kasaleka, Coci, Vula tolu, Belaselika, Noumea, Yabia, Merelestia, Manioke, Sokobale, Katafaga, Aikavitu, Yabia vula
Description: Shrub with raised leaf scars on stems. The root is elongated and tuberous. Leaves are alternate, palmately divided, and occur on the upper part of the stem. The flowers are unisexual and small.
Habitat: Cultivated throughout the plains and the mountainous regions.
Chemical constituents: Amentofalvone, hydrocyanic acid, hydrogen sulphide, linamarin, oxalic acid, quercetin alkaloids, and yucalexins.
Antimycotic activity: Confirmed (WHO,1998).
Traditional antifungal use: Use by Fijians not confirmed.

Messerschmidia argentea *(L. f) Johnston (Boraginaceae)*

Fijian name: Kauniyalewa, Roronibebe, Samuna kirakira
Description: A small shrub with large leaves. Young parts, leaves, and inflorescences densely pilose with greyish white, silvery hairs. The flowers are numerous, small, and white in color. The fruits are small, round, and greenish white on maturing. Flowers and fruits throughout the year.
Habitat: Commonly found in beach thickets.
Chemical constituents: Not available.
Antimycotic activity: Not confirmed.
Traditional antifungal use: The leaves are used for antifungal therapy (Singh and Siwatibau, 1980).

Mikania micrantha ***H.B.&K (Asteraceae)***

Fijian name: Wa butako
Description: A climbing vine with triangular leaves with a broad cordate base. Flowers are minute, white, and borne in small, densely packed heads resembling sunflowers. Fruits have white bristles to help with wind dispersal of the seeds. They flower and fruit throughout the year.
Habitat: A common weed of pastures, roadsides, fences, forest edges, and clearings.
Chemical constituents: Terpenoid compounds, taraxasterol, stigmasterol, and sitisterol.
Antimycotic activity: Confirmed (Singh and Siwatibau, 1980).
Traditional antifungal use: Use by Fijians not confirmed.

Morinda citrifolia ***L. (Rubiaceae)***

Fijian name: Kura
Description: Small tree or shrub with large, opposite, lanceolate leaves and square stems. The white flowers are borne on a globose syncarp. Fleshy fruits are greenish white on maturity, pungent, and tasteless. The roots yield a yellow dye, and the bark yields a red dye. Also known as "Indian mulberry." The flowers and fruits are available throughout the year.
Habitat: Occurs in wet areas, creeks, in beach tickets, rocky shores, and roadsides.
Chemical constituents: Alizarin, morindin, anthraquinones, flavonoids, sitosterol, caproic acid, and caprylic acid.
Antimycotic activity: Not confirmed (WHO, 1998).
Traditional antifungal use: The young shoots are used for antifungal therapy (Parham, 1943).

Piper methysticum Forst.f ***(Piperaceae)***

Fijian name: Yaqona
Description: Woody, aromatic shrub with green zig-zag stems. The leaves are heart-shaped. Flowers are minutely arranged in erect, greenish white spikes.
Habitat: Occurs in wet and damp areas especially near streams.
Chemical constituents: Alpha-pyrone derivatives such as kawain, yangonine, methysticin, etc., flavokawains, pipermethysticine, cepharadione A, campesterol, cholesterol, and stigmasterol.
Antimycotic activity: Confirmed (WHO, 1998).
Traditional antifungal use: Use by Fijians not confirmed.

Plumbago zeylanica ***L. (Plumbaginaceae)***

Fijian name: Natuna, Tutuna, Kenikeni
Description: An erect shrub with oblong leaves. The flowers have corollas that are white and/or at times faintly blue-tinged. The anthers are usually pale yellow and the fruits are green and sticky. Flowers in cooler climates.
Habitat: Occurs near sea level to an elevation of about 400 m in exposed thickets, cliff fences, talus slopes, along beaches and mangrove swamps.
Chemical constituents: Naphthaquinone plumbagin, elliptinone, sitosterol, and cyclopropenoid fatty acid.
Antimycotic activity: Confirmed (Ray and Majumdar, 1976).
Traditional antifungal use: The root is used for antifungal therapy (Parham, 1943).

Plumeria rubra Linn. ***(Apocynaceae)***

Fijian name: Bua, Bua ni vavalagi, Frangipani
Description: Freely branching tree with branches and latex. Flowers usually have five petals that are white, yellow, red, pink, or maroon in color with fragrance. Fruits contain numerous winged seeds. They flower throughout the year.
Habitat: Cultivated as an ornamental tree but occasionally naturalized.
Chemical constituents: Plumieride, monoterpenes, benzoquinonine derivatives, oleanolic acid, stigmasterol, benzoic acid, acetic acid, benzyl benzoate, para-coumaric acid, kaempferol, vallinic acid, and farsenol.
Antimycotic activity: Confirmed (WHO, 1998).
Traditional antifungal use: Use by Fijians not confirmed.

Psidium guajava ***L. (Myrtaceae)***

Fijian name: Quwawa
Description: This is a shrub or tree with smooth, patchy, peeling bark, and oval-shaped leaves. The flowers are whitish with numerous stamens. The fruit is fleshy yellow with edible pink mesocarp containing numerous small hard white seeds.
Habitat: Occurs in pastures, plantations, and other similar habitats.
Chemical constituents: Tannins, vitamins B and C, eugenol, gallic acid, asiatic acid, oleanolic acid, amyrins, cathechin, quaverin, daucosterol, leucocyanidin, and alkaloid zeatin.
Antimycotic activity: Confirmed (WHO, 1998).
Traditional antifungal use: Use by Fijians not confirmed.

Psychotria calycosa ***A.Gray (Rubiaceae)***

Fijian name: Not available
Description: This a freely branched tree about 2-5 m in height with oblong-lanceolate leaves having a breadth of 2.5 cm. The inflorescence is terminal, solitary, and flowers are pure white in color. The fruit is dull pink and usually occurs in a cool climate. Endemic to Fiji.
Habitat: Occurs in dense forests and at elevations up to 600 m above sea level.
Antimycotic activity: Confirmed (Singh and Siwatibau, 1980).
Traditional antifungal use: Use by Fijians not confirmed.

Psychotria confertiloba ***A.C. Smith (Rubiaceae)***

Fijian name: Tabulina, Kaimila, Matika, Nduralevu, Takala
Description: A tree occasionally noted as a shrub, erect, 2-10 m in height with few leaves that are glossy, and oblong to elliptical in shape. The flowers are white in color with fragrance. The fruits are spherical and range in color from red, to orange, to purple. They fruit throughout the year. Endemic to Fiji.
Habitat: Abundant in dense, open, or secondary forests from near sea level to an elevation of 1,200 m.
Chemical constituents: Not available.
Antimycotic activity: Confirmed (Singh and Siwatibau, 1980).
Traditional antifungal use: Use by Fijians not confirmed.

Rhizophora samoensis ***(R. mangle) L. (Rhizophoraceae)***

Fijian name: Tiriwai, Ndongo
Description: Tree appears about 9 m in height with many branches. The leaves are opposite and elliptical in shape. The bark is used for making dye. There are descending aerial roots from the branches. The flowers are white to pale yellow in color with orange or dull yellow anthers. Fruits and flowers throughout the year.
Habitat: Common in swampy coastal areas. Occurs in seaward edge of mangrove swamps.
Chemical constituents: Sitosterol, cholesterol, stigmasterol, betulin, oleanolic acid, galactose, arabinose, gentisic acid, and benzenoid.
Antimycotic activity: Confirmed (Singh and Siwatibau, 1980; Cacerese et al., 1993).
Traditional antifungal use: Use by Fijians not confirmed.

Ricinus communis *L. (Euphorbiaceae)*

Fijian name: Bele ni vavalagi, Toto ni vavalagi, Utouto
Description: Small tree or shrub with conspicuous, ringlike scars on the hollow stem. The leaves have serrate edges with red tinge. The terminal inflorescence is a narrow panicle. The fruit is spiny, splitting when mature, and has brown seeds. Flowers and fruits throughout the year.
Habitat: Occurs mainly in wastelands and in disturbed areas.
Chemical constituents: Beta-amyrin, brassicasterol, stigmasterol, quercetin, coumarin, haemaglutinin, glycoproteins, kaempferol glycoside, and vitamins B6 and B1.
Antimycotic activity: Confirmed (WHO, 1998).
Traditional antifungal use: Use by Fijians not confirmed.

Semecarpus vitiensis *(A. Gray) Engl. (Anacardiaceae)*

Fijian name: Kaukaro
Description: A large tree up to 30 m high with large, oblong leaves. The fruit is small, corky, with a fleshy base. The yellowish brown latex of the tree causes intense irritation to the skin. Hence, it is also called "itch wood."
Habitat: Usually occurs in dense or dry forests or forest edges. It can be seen near sea level to an altitude of 900 m above sea level.
Chemical constituents: Alkenylcatechols and dihydroisocoumarine 5 methylmellein
Antifungal activity: Not confirmed.
Traditional antifungal use: The acrid and burning sap is applied as an antifungal agent (Parham, 1943).

Solanum torvum *Swartz (Solanaceae)*

Fijian name: Soni, Kausoni, Kaisurisuri, Kauvotovotoa
Description: Freely branching spiny shrub 1-5 m high. The leaves are elliptical, lobed, and about 10-17 cm long with hairs on the lower surface. The flowers are white with yellow anthers. The berry fruit is yellow to orange-red at maturity with reddish seeds. It flowers throughout the year. Agriculturally, it is a serious noxious weed and is also known as "prickly solanum" or "devil's fig."
Habitat: Occurs near sea level to 900 m above sea level. Often forms dense thickets and is abundant in clearings, waste places, pastures, green fields, and along roadsides.

Chemical constituents: Spirostanes, glycoalkaloid solasine, sisalagenin, jurubine, caffeic acid, sapogenin, steroids, and lipids.
Antimycotic activity: Confirmed (Singh and Siwatibau, 1980, Ajaiyeoba, 1999).
Traditional antifungal use: Use by Fijians not confirmed.

Syzygium corynocarpum ***(A. Gray) C. Muell. (Myrtaceae)***

Fijian name: Yasi yasi
Description: Tree about 15 m tall with glossy leaves that are 15 cm long with veins inside the leaf margin. Fragrant flower is cream-colored with yellow stamens. The fruits are elongated, and are red or purple in color with hard seeds. Fruits occur in cooler climates.
Habitat: Occurs in dense, wet forests from sea level to an elevation of 1,000 m.
Chemical constituents: Calcium oxalate and essential oils.
Antimycotic activity: Confirmed (Singh and Siwatibau, 1980).
Traditional antifungal use: Use by Fijians not confirmed.

Terminalia catappa ***L. (Combretaceae)***

Fijian name: Tavola, Tavola lata and Tivi
Description: Large, tall tree (30 m) with obovate, bladed leaves that turn orange to red before falling. Flowers are small and white. The fruit is ovoid in shape with a fibrous outer layer, red in color and has only a single seed. They flower and fruit throughout the year.
Habitat: Occurs in lowland clearings, beaches, mangrove swamps, and rocky coasts.
Chemical constituents: Tannic acid, ellargic acid, palmitic acid, oleic acid, linoleic acid, myristic acid, flavonoids, and amino acids.
Antimycotic activity: Confirmed (WHO, 1998).
Traditional antifungal use: Use by Fijians not confirmed.

Thespesia populnea ***(L.) Soland. ex Correa (Malvaceae)***

Fijian name: Mulomulo, Wiriwiri
Description: This tree has dark green, glossy leaves with cordate blades. The flower has yellow petals with maroon-purple centers. The brown-colored fruit has a sticky yellow sap and the seeds are hairy. They flower and fruit throughout the year.
Habitat: Occurs along riverbanks, beaches, littoral forests.

Chemical constituents: Gossypol, populin, quercetin, rutin, kaempferol, beta-carotene, cyanidin glycoside, and sitosterol.
Antimycotic activity: Confirmed(WHO, 1998).
Traditional antifungal use: Use by Fijians not confirmed.

Vavae amicorum *Benth. (Meliaceae)*

Fijian name: Cevua, Tubua ni cevua, Wawaro, Mariko
Description: This is a small tree about 12 m in height. The flowers are white to pale yellow in color and are fragrant. Upon fully maturing, the fruit takes a purple to black color. It is also called false sandalwood. It bears fruits and flowers throughout the year.
Habitat: Appears in beach thickets, mangrove swamps, limestone cliffs, forests, and forest grassland transition areas. Also occurs in places 915 m above sea level.
Chemical description: Not available.
Antimycotic activity: Confirmed (Singh and Siwatibau, 1980).
Traditional antifungal use: Use by Fijians not confirmed.

Xylocarpus granatum *Koenig (Meliaceae)*

Fijian name: Dabi, Legilegi
Description: Tree (15 m high) has alternate leaves with oblong leaflets. The flowers have greenish white petals. The large fruits are green to light brown in color, 10-25 cm in diameter with numerous irregularly shaped seeds. Flowers in cooler periods.
Habitat: Occurs in lowland riverbanks, rocky coasts, mangrove swamps, littoral forests.
Chemical constituents: Xylomollin, fructose, gedunin, glucose, sucrose, tannins, and xylomollin.
Antimycotic activity: Confirmed (WHO, 1998).
Traditional antifungal use: Use by Fijians not confirmed.

CONCLUSIONS

Out of the forty-one species reviewed for their reported antimycotic potential, biological activity has been confirmed on thirty-three plants. Only ten of these plants are used traditionally as antifungal medicines. The University of the South Pacific has been in the forefront in the research of the chemistry and biological activity of natural products from terrestrial plants for the past twenty-five years. Plans have been made to study the plants with antimycotic potential with a view to isolating the biologically active components with confirmed antimycotic activity.

REFERENCES

Ajaiyeoba, E.O. (1999). Comparative phytochemical and antimicrobial studies of *Solanum macrocarpum* and *S. torvum* leaves. *Fitoterapia* 70(2): 184-186.

Bennett, J.E. and Walzer, P.D. (1998). Fungal infections. In *Harrison's principles of internal medicine,* Volume 1. Fauci, A.F., Braunwald, E., Isselbacher, K.J., Wilson, G.D., Martin, J.B., Kasper, D.L., Hauser, S.L. and Longo, D.L. (Eds.), New York: McGraw-Hill Health Profession Division, pp. 1148-1162.

Cacerese, A., Lopez B., Juarex, X., Del Aguila, J. and Garcia, S. (1993). Plants used in Guatemala for the treatment of dermatophyte infections: Evaluation of antifungal activity of the seven American plants. *J. Ethnopharmacol.* 40(3): 207-313.

Cambie, R.C. and Ash, J. (1994). *Fijian medicinal plants.* Melbourne, Australia: CSIRO.

Parham, H.B.R. (1943). *Fijian plants with their medicinal and other uses.* Auckland, New Zealand: The Polynesian Society.

Ray, P.G. and Majumdar, S.K. (1976). Antimicrobial activity of some Indian plants. *Econ. Bot.* 30: 317-320.

Singh, A. and Siwatibau, S. (1980). *Medicinal plants in Fiji and other South Pacific islands.* Research Report. Suva, Fiji: The University of the South Pacific, pp. 1-39.

Singh, Y.N. (1986). Traditional medicine in Fiji: Some herbal folk cures used by Fiji Indians. *J. Ethnopharmacol.* 15: 57-88.

Weiner, M.A. (1984). *Secrets of Fijian medicine.* Suva, Fiji: Government Printers.

Whistler, W.A. (1992). *Polynesian herbal medicine.* Hong Kong: Everbest.

World Health Organization (WHO) (1998). *Medicinal plants in the South Pacific.* Manila, Philippines: World Health Organization.

Chapter 22

Triterpenic Glycosides—Their Isolation Methods and Antifungal Activities

Süheyla Kirmizigül

INTRODUCTION

Glycosides are widely distributed in nature and are present in nearly 100 plant families, even primitive ones (Wagner and Horhammer, 1971). This group of natural products and their genins were studied many years ago. Due to improved techniques in isolation, structural determination, and biological and biochemical importance, studies were concentrated on these compounds (Agarwal and Rastogi, 1974). Many papers and reviews on the glycosides in plants have appeared in the literature. Their properties, especially biological activities, have been examined, such as membrane permeability, protein synthesis, and therapeutic and pharmaceutical activities (Mahato et al., 1988). Some glycosides have also been isolated which have shown similar genins, but which have different configurations and different types of carbohydrate units. Noteworthy progress has been witnessed in the isolation of pure glycosides and in their complete structure determination including the configuration of the carbohydrate moieties. Oligosaccharides, part of these compounds, are generally linear or branched and attached to a hydroxyl group. The sites of attachment may be monodesmosidic, bisdesmosidic, and tridesmosidic (Hiller and Adler, 1982; Adler and Hiller, 1985). In some cases, the glycosides from different parts of the same plant have been found to possess different properties. Differences in glycoside constituents were observed in the case of ginseng when it was cultivated in different soils and climatic conditions (Ju-Hyun et al., 1977).

Triterpenic glycosides have an important place in the study of glycosides depending on their genins (Connolly and Hill, 1996, 1997, 1999, 2000). The genins of triterpenic glycosides can be grouped into eight categories. These

The author is grateful to Dr. Sabire Karaçali (Molecular and Cell Biology Section, Biology Department, Faculty of Science, Ege University) for her help.

are: oleanane, squalen, lupane, ursane, fusidane-lanostane, dammarane-euphane, tetranortriterpenoids, and miscellaneous. Hederagenin glycosides are one of the most important groups of glycosides which have a genin of oleanane type. This chapter will explore the isolation, structural determination, and biological activities of triterpenic glycosides.

TECHNIQUES

The known methods of isolation and structural determination are available (Mahato et al., 1988), but sometimes these methods are not effective for the separation of mixtures of glycosides and their genins. Improved techniques have provided even more valuable means for isolation of pure glycosides.

Some chemical methods, such as the Liebermann-Burchard test, the colorimetric method, and some chemicals such as thionyl chloride, phosphomolybdic acid, silica tungstic acid, antimony pentachloride, and periodate-alkaline potassium permanganate have been used for detection of glycosides (Chandel and Rastogi, 1980). For isolation, polyamide charcoal-celite, DEAE-Sephadex, and different types of silica gel have been used. Droplet counter-current chromatography (DCCC), high performance liquid chromatography (HPLC), centrifugal liquid chromatography (CLC), rotation locular counter-current chromatography (RLCC), and flash chromatography are main techniques used for this purpose (Shany et al., 1972; Yahara et al., 1976; Chandel and Rastogi, 1980; Domon and Hostettmann, 1984; Mahato et al., 1988).

Air-dried and powdered plant material were extracted with 80 percent MeOH. After removing the solvent under vacuum at ~40°C, the waxy residue washed with hexane, $CHCl_3$, and Me_2CO to remove nonglycosidic substances. The remaining residue was dissolved in H_2O and extracts with n-BuOH. The BuOH layer evaporated under reduced pressure. The crude mixture fractioned chromatographs repetitively over Si-gel columns, generally using $CHCl_3$:MeOH:H_2O solvent systems with an increasing polarity. Fractions were purified by repetitive column and preparative chromatography. Generally, glycosides formed as amorphous powder.

Isolated glycosides hydrolyzed for yielding the aglycone and the sugar moieties, which are separately investigated. Sugar moieties of the compounds are identified by co-chromatography and GC methods. Different kinds of derivatives of pure glycosides serve as a source for finding the exact formulae. Recently, different spectroscopic methods, including ^{1}H-nmr, ^{13}C-nmr, 2D-nmr, mass spectroscopy, X ray, CD, and crystallography techniques have been used for structural determination of glycosides (Allerland and Doddrell, 1971; Neszmelyi et al., 1977; Yahara et al., 1977; Seo et al., 1978; Jeener et al., 1979; Kasai et al., 1979; Barber et al., 1981; Bax and Freeman, 1981; Kitagawa et al., 1981; Williams et al., 1981; Konoshima and Sawada, 1982; Mahato and Das, 1983; Fenselau, 1984; Patt, 1984; Van der Kaaden et al., 1984; Agarwal et al., 1985; Nishino et al., 1986; Waltho et al., 1986; Mahato et al., 1987).

One of the most prevalent and important classes of genins are oleananes (see Table 22.1).

Hederagenin glycosides have a prominent role among triterpenic glycosides. They show important biological activities (see Table 22.2).

GLYCOSIDES AS A PHARMACODYNAMIC GROUP

Because of the importance of the glycosides in pharmaceuticals, herbal drugs, and folk remedies (Tokuda et al., 1988), many studies have reported on the biological activities of these compounds. The use of glycosides as a source of crude drugs for treatment of chronic hepatitis, kidney problems, autoimmune diseases, and stomach problems have been reported in China (Sanchez-Contreras et al., 2000) and Japan (Yoshimitsu et al., 1999). Traditional usage has encouraged scientists to study the plants showing biological activity. A wide variety of biological effects of glycosides have been described on several levels. Some activities are harmful for organisms and some are beneficial (Agarwal and Rastogi, 1974; Chandel and Rastogi, 1980; Mahato et al, 1988).

Accumulated evidence suggests that the glycosides have antiprotective activities. The haemolytic properties of triterpenic glycosides have been known for a long time (Agarwal and Rastogi, 1974; Chandel and Rastogi, 1980; Mahato et al., 1988). *Hedera helix* saponins have different haemolytic activities. From these saponins, Compound IV is devoid of activity, while Compound V and VI have a potent haemolytic activity (Balansard and Bernard, 1966). According to Schloesser and Wulff (1969), optimum haemolytic activity depends on number, position, and acetyl derivatives of –OH groups in genins. Thalictosids I, II, III, IV, and XII were examined for immunosuppressive activity. The activity of S-isomer at C-21 was greater than R-isomer at the same carbon (Yoshimitsu et al., 1999). Some glycosides have a role as agglutination agents between phospholipid vesicles (Fukuda et al., 1985). This is not a beneficial activity when it occurs between the blood cells in the vascular system. In addition, triterpene glyco-

TABLE 22.1. Some important oleananes from different plants

Plant	Compounds	Formula No.*	References
Acanthothammus aphylus	2α-Hydroxypolpulnonic acid	1	Estrada et al. (1994)
	Friedela-2,4(23)-dien-29-oic acid	2	Estrada et al. (1994)
	29-Norfriedelane	3	Estrada et al. (1994)
Adina rubella	3β,23,24-trihydroxyolean-12-ene-28-oic acid	4	Fang et al. (1996)
	3β,6β,24-trihydroxyolean-12-ene-28-oic acid	5	Fang et al. (1996)
Alibertia edulis	3β,19α,23,24-Tetrahydroxyolean-12-en-28-oic acid	6	Brochini et al. (1994)
Alternanthera philoxeroides	Philoxeroic acid	7	He and Meng (1995)
Ardisia crenata	13β,28-Epoxyoleanane-3β,16α,30-triol	8	Jia et al. (1994)
Aster auriculatus	28-Noroleananes	9	Gao et al. (1994)
Atriplex stocksii	Atriplexinol	10	Siddiqui et al. (1994)
Bellis bernardii	Bellisonic acid	11	Schöpke et al. (1995)
Betula pubescens	27-Caffeoyloxy-3β-hydroxyolean-2-en-28-oic acid	12	Pan et al. (1994)
Bupleurum smithii var. *parvifolium*	Saikaogenin Q	13	Luo et al. (1995)
Caloncoba lophocarpa	Norfriedelanes lophocarpin	14	Tchuendem et al. (1996)
	21β-lophocarpin	15	Tchuendem et al. (1996)
Canarina canariensis	12α, 13α-epoxide of β-amyrin acetate	16	Gaydou et al. (1996)
Clinopodium chinensis	3β,23-Hydroxyoleana-12,21-dien-28-oic acid	17	Liu et al. (1995)
Combretum nigricans	Combregenin	18	Jossang et al. (1988)
Cuphea carthagensis	Friedel-7-en-3β-ol	19	González et al. (1994)
Dillenia papuana	Dillenic acid A, B and C	20,21,22	Nick et al. (1994)
	3-Oxooleana-1,12-dien-30-oic acid	23	Nick et al. (1994)
	2,3-Secooleanane dillenic acid D	24	Nick et al. (1995)
	Dillenic acid E	25	Nick et al. (1995)
	3-Oxoolean-12-en-30-oic acid	26	Nick et al. (1995)
Dolichos lablab	3β,22β,24-Trihydroxyolean-12-en-28-al	27	Yoshiki et al. (1995)
Euphorbia chamaesyce	3,4-Secooleana-4(23),18-dien-3-oic acid	28	Tanaka, Ida, et al. (1994)
Euphorbia cyparissias	2-Methylbutanoyl ester of glut-5-en-3α-ol	29	Öksüz et al. (1994)
Frullania species	3α-Hydroxyolean-18-en-28-oate	30	Tori et al. (1995)

Plant	Compounds	Formula No.*	References
Gentiana tibetica	1α,2α,3β,24-Tetrahydroxyolean-12-en-28-oic acid	31	Zhang and Yang (1994)
Glycyrrhiza pallidiflora	3β-Hydroxyoleana-11,13(18)-dien-30-oic acid	32	Kan et al. (1994)
Glycyrrhiza yunnanensis	Glyyunnansapogenin	33	Hu et al. (1995)
Gymnema sylvestre	Gymnemanol	34	Sahu et al. (1996)
Holoptelea integrifolia	Haloptelins A and B	35,36	Mondal et al. (1994)
Homalium longifolium	3β-hydroxyfriedelan-27,16α-olide acetate	37	Shaari and Waterman (1996)
	3β-hydroxyfriedelan-27,16α-olide benzoate	38	Shaari and Waterman (1996)
Hydrocotyle ranunculoides	17,22-Secooleanane	39	Della Greca et al. (1994)
	13b,28-Epoxyoleananes	40	Della Greca et al. (1994)
Ixeris chinensis	Ixerenol	41	Shiojima, Suzuki, Kodera, et al. (1995)
Lantana camara	Camarilic acid	42	Segum et al. (1995)
	Camaric acid	43	Siddiqui et al. (1995)
	25-hydroxy-3-oxoolean-12-en-28-oic acid	44	Fullas et al. (1996)
Laquidambar formosana	11α,12α-epoxy-3-oxooleanan-28,13β-olide	45	Sun and Sun (1996)
Ligularia veitchiana	16β,28-Epoxyolean-12-en-3β-ol	46	Zhao et al. (1995)
Limnophila heterophylla	Methyl 3-benzoylkatonoate	47	Mukherjee et al. (1995)
Lithospermum caroliniense	1α,3β,23-trihydroxyolean-12-ene-28-oic acid	48	Fullas et al. (1996)
Maesa lanceolata	13β,28-epoxyoleanane-3β,16a,21β,22α, 28S-pentol	49	Sindambiwe et al. (1996)
Maprounea africana	Maprounic acid derivatives A and B	50,51	Pengsuparp et al. (1994)
Maytenus blepharodes	7α-Hydroxyblepharodol	52	González et al. (1995)
Maytenus canariensis	6-oxoiguesterol	53	González, Alvarenga, Ravelo, Jiménez, et al. (1996)
Maytenus chuchuhuasca	6-oxopristimerol	54	Shirota et al. (1994)
	7',8'-dihydroxuxuarine Aβ	55	Shirota et al. (1995)
Maytenus ilicifolia	Norfriedelane 6-oxotingenol	56	Shirota et al. (1994)
Maytenus krukovii	Krukovines A and C	57,58	Shirota et al. (1996)
Maytenus magellanica	Magellanin	59	González et al. (2001)
Maytenus scutioides	Scutione	60	González, Alvarenga, Ravelo, Bazzocchi, et al. (1996)

TABLE 22.1 *(continued)*

Plant	Compounds	Formula No.*	References
	Scutionin αA	61	González, Alvarenga, Estévez-Braun, et al. (1996)
Melilotus messanensis	Soyasapogenol G	62	Macías et al. (1996)
	3β,27-dihydroxy-18α-oleanan-28,19β-olide	63	Macías et al. (1996)
Mimusops elengi	Mimusopgenone	64	Sen et al. (1995)
	Mimugenone	65	Sen et al. (1995)
Minquarita guianensis	13β,28-Epoxyolean-11-en-3β-ol	66	El-Seedi et al. (1994)
Myrianthus liberecus	3-Coumaroylarjunolic acid	67	Tapondjou et al. (1995)
Myrsine australis	13b,28-Epoxyoleanane-3β,16α,28 ξ-triol	68	Bloor and Qi (1994)
Nephthea albida	29,30-Dinoroleanane-13b,15b,16b,26-tetrol	69	Fu and Zhang (1995)
Penicillium simplissimum	7β,15α,,24-trihydroxyolean-12-ene-3,11,22-trione	70	Hayashi et al. (1996)
Pentaclethra eetveldeana	3β-palmitoyloxyoleanan-13β-ol	71	Byla and Herz (1996)
	23-palmitoyloxyoleanonic acid	72	Byla and Herz (1996)
Phyllanthus flexuosus	11β-Hydroxyfriedel-1-en-3-one	73	Tanaka, In, et al. (1994)
Phytolacca dodecandra	Dodecandral	74	Spengel (1996)
	Dodecandralol	75	Spengel (1996)
Picris hieracioides	Olean-12-ene-2β,3β,22α-triol	76	Shiojima, Suzuki, and Ageta (1995)
Polygala japonica	Polygalagenin	77	Zhang et al. (1995)
	3β,19α-dihydroxyolean-12-ene-24,28-dioic acid	78	Zhao et al. (1996)
Polypodiodes formosana	Preoleanatetraene	79	Arai et al. (1996)
Poraquciba guaianensis	Icacinik acid	80	Goulart et al. (1994)
Rathbunia alamosensis	Alamosenogenin	81	Takizawa et al. (1995)
Rhoiptelea chiliantha	Bis-caffeoyl ester	82	Jiang et al. (1995)
	Rhoipteleic acid A and B	83,84	Jiang et al. (1994)
Sabia schumanniana	11α-Hydroxyoleana-12-en-3-one	85	Yaun et al. (1994)
Sabia swinhoei	28-Hydroxyolean-13(18)-en-3-one	86	Liang et al. (1995)
Salacia beddomei	1β,15-α-dihydroxyfriedelan-3-one	87	Hisham et al. (1996)
Salacia reticulata	Salaciquinone	88	Tezuka et al. (1994)
	Isoiguesterinol	89	Dhanabalasingham et al. (1996)
	30-hydroxypristimerin	90	Dhanabalasingham et al. (1996)

Plant	Compounds	Formula No.*	References
Salvia nemorosa	Olean-12-ene-3β,11α,21α-triol	91	Ulubelen et al. (1994)
Salvia pomifera	23-Hydroxygermanicone	92	Topcu et al. (1994)
Salvia regla	Reglin	93	Ortega et al. (1995)
Saussurea japonica	11α,12α-epoxytaraxer-14-en-3-one	94	Kuo et al. (1996)
Serjania triquetra	11α-Hydroperoxyhederagenin	95	Chavez and Delgado (1994)
Staurogyne merguensis	Olean-12-ene-3β,21α,22β,29-tetrol	96	Hiura et al. (1996)
	Olean-12-ene-3β,21α,22β,23,29-pentol	97	Hiura et al. (1996)
Stephanotis lutchuensis var. *japonica*	2α,3β-Dihydroxy-23-oxoolean-12-en-28-oic acid	98	Maeda et al. (1994)
Thomandersia laurifolia	Olean-12-ene-2α,3α,19α-triol	99	Ngadjui et al. (1994)
Trichoceresus pachanoi	Bridgesigenin C	100	Kinoshita et al. (1995)
	Pachanol A, B and C	101,102,103	Kinoshita et al. (1995)
Trichosanthes kirilowii	3-Epikaounidiol	104	Akihisa et al. (1994)
	3-Epibryonol	105	Akihisa et al. (1994)
	3α,29-Dihydroxymultiflor-9(11)-en-7-one	106	Akihisa et al. (1994)
Tripterygium wilfordii	Wilforol A and B	107,108	Morota, Ogino, et al. (1995)
	Wilforol C and D	109,110	Morota, Sasaki, et al. (1995)
Vahlia capensis	3,4-Secooleana-4(23),18-dien-3-oic acid	111	Majinda et al. (1995)
Vicoa indica	Vicogenin	112	Balakrishna et al. (1995)

sides have potential as chemosterilants in fertility control (Chou et al., 1971; Uchendu et al., 2000). This effect would be of benefit for pest control.

A saponin from *Cyclamen ibericum* increased the permeability of dermal capillaries (Gladkikh, 1966). The effect of regulation on capillary permeability (Vogel et al., 1968) makes saponins very important in cases of inflammation. Inhibitory activity at the cellular level was observed with some triterpene glycosides from *Actinopiga agassizzi* in protein synthesis (Anisimov et al., 1971), with a glycoside from *Stichopus japonicus selenka* in both protein and nucleic acid synthesis (Elyakov et al., 1972), and another triterpene glycoside from *Hormonema* species in glucan synthesis (Schwartz et al., 2000). However, some contradictory effects have also been shown for protein synthesis (Oura et al., 1975). In addition, enzymes working in DNA synthesis, such as DNA polymerase beta (Ma et al., 1999; Sun et al., 1999; Deng et al., 2000) and HIV-1 reverse transcriptase activities (Pengsuparp et al., 1994) were inhibited by triterpenoids.

TABLE 22.2. Some important Hederagenin glycosides

Plant	Compounds	References
Akebia quinata	Saponin B and C	Higuchi et al. (1972)
	Saponin D, E, F, and G	Higuchi and Kawasaki (1972)
	Saponin PA, PC, PD, PF, PG, and PJ_2	Ryuichi and Toshio (1976)
	Saponin PK	Higuchi and Kawasaki (1976)
Caullophyllum robustum	Saponin I	Murakami et al. (1968)
	Caulosides A, B, C, and E	Strigina, Chetyrina, and Elyakov (1970); Strignia, Chetyrina, Lepshkina, et al. (1970)
		Chetyrina and Kalinovsky (1979)
	Cauloside D and F	Chetyrina et al. (1975)
	Cauloside G	Strigina et al. (1976)
Cephalaria gigantea	Giganteaside E1 and H1	Zviadadze et al. (1983)
Cephalaria kotchyi	Cephalaroside E	Aliev and Movsumov (1976)
Cephalaria transsylvanica	Transsylvanoside A	Kirmizigul and Anil (1994b)
	Transsylvanoside B	Kirmizigul and Anil (1994a)
	Cephalaria saponin A	Çaliskan et al. (1994)
	Transsylvanoside E and F	Kirmizigul et al. (1995)
	Cephalaria saponin B	Alankus-Çaliskan and Anil (1995)
	Transsylvanoside G and H	Kirmizigul, Anil, and Rose (1996)
	Transsylvanoside I, J, and K	Kirmizigul and Rose (1997)
Clematis chinensis	Prosapogenin CP^6 and CP^8	Kizu and Tomimori (1979)
	Prosapogenin CP^1 and CP^5	Kizu and Tomimori (1980a)
	Prosapogenin CP^0, $CP2^a$ and CP^{3a}	Kizu and Tomimori (1982b)
	Saponin CP^{3b} and CP^{10}	Kizu and Tomimori (1980b)
	Saponin CP^{8a} and CP^{10a}	Kizu and Tomimori (1982a)
Clematis songarica	Songaroside A	Krokhmalyuk et al. (1975)
Clematis vitalba	Vitalboside B	Chirva et al. (1975)
	Vitalboside D	Kintya et al. (1974b)
	Vitalboside G	Kintya et al. (1974a)
	Vitalboside H	Chirva et al. (1974)
Climacoptera transoxana	Copteroside B, C, and D	Annaev et al. (1983a, 1983c)
	Copteroside F	Annaev, Jsaev, et al. (1983)

Plant	Compounds	References
Dipsacus azureus	Dipsacoside A and B	Mukhamedziev et al. (1971), Mukhamedziev and Alimbaeva (1969)
Fatsia japonica	Fatsiaside B1 and D1	Kemoklidze et al. (1982)
	Fatsiaside D and F	Kemoklidze et al. (1984)
	Saponin XXXV and XXVII	Aoki et al. (1981)
	Saponin I	Aoki et al. (1976)
Ficaria ranunculoides	Saponin LXI	Texier et al. (1984)
Hedera helix	Saponin lactone	Balansard and Bernard (1966)
	Saponin LVI, LVII, LVIII, and LIX	Hostettmann (1980)
Hedera pastuchovii	Pastuchoside A	Dekanosidze et al. (1970)
	Pastuchoside B and C	Iskenderov (1971a,b)
Hedera rhombea	Kizuta saponin K8 and K11	Kizu et al. (1985)
Ladyginia bucharica	Ladyginoside B	Pakkhullaeva et al. (1972)
	Ladyginoside D and F	Pakkhullaeva et al. (1973)
Leontice eversmanni	Leontoside A,B,C, D, and E	Mzhel'skaya and Abubakirov (1967)
Lonicera nigra	Monodesmosidic saponin I and III	Domon and Hostettmann (1983)
Patrinia scabiosaefolia	Esculetin	Choi and Woo (1984)
	Saponin LXV	Choi and Woo (1984)
	Scabioside A and C	Bukharov et al. (1970)
Patrinia sibrica	Sibiroside A and C	Bukharov and Karlin (1970)
Phytolacca dodecandra	Saponin XXXXVII	Domon and Hostettmann (1984)

An antimitotic effect on inhibition of cell growth with saponin was observed (Balansard and Bernard, 1966; Anisimov et al., 1972). Some triterpene glycosides have very important roles for regulation of apoptosis (Arntzen et al., 1999) and as antitumor agents for cancer research (Arntzen et al., 1999; Abdel-Kader et al., 2000; Avilov et al., 2000; Lee et. al., 2000).

The pharmaceutical activities of saponins have been studied extensively. Ginseng saponin stimulates the central nervous system using a mechanism different from amphetamines (Lee, 1974). Thus, this saponin apparently reduces sleep and increases reclining and grooming in rats. *Astragalus* saponin I has an analgesic effect that is weaker and shorter than morphine, depending on the applied dose (Zhang et al., 1984). The effects of saponin on involuntary muscle contractions were shown on isolated rabbit ears (Gladkikh, 1966), on posterior parts of the intestine of frogs (Bellet and Cerede, 1967) and guinea pigs (Zhang et al., 1984). Two utero-contracting saponins were also isolated from *Ardisia crispa* (Jansakul et al., 1987).

Several protective effects of glycosides have been reported. Stimulation of humoral and cellular immune response by triterpene glycosides (Krivorutchenko et al., 1999) is one of the most beneficial protective effects. A significant hepatoprotective effect was suggested against D-galactosamine (a tumour necrosis factor-a) -induced cell death in primary cultured mouse hepatocytes (Adnyana et al., 2000). Glycosides protect the host by exhibiting antibacterial (Kirmizigul, Anil, Ucar, et al., 1996; Zhou, 2000) activities. Triterpene glycosides not only showed antibacterial activity in animals, but also against phytopathogenic bacteria in plants (Timbekova et al., 1996). Many studies on antiprotozoal activity have been recorded (Balansard and Bernard, 1966; Neal, 1967; Mshvildadze et al., 1998).

ANTIFUNGAL ACTIVITY

Important developmental and medical problems of plants and animals caused by fungi have caused some scientists to evaluate antifungal properties of triterpene glycosides in detail.

The knowledge about antifungal activity of triterpene glycosides from plants (Kitagawa et al., 1974; Bader et al., 1987; Kirmizigul, Anil, Ucar, et al., 1996; Srivastava et al., 2001), animals (Pivkin, 2000), and even fungi (Leet et al., 1996) has increased.

The level of activity against fungi differs from compound to compound and differs in applied conditions. Transsylvanoside A-C from *Cephalaria transsylvanica* showed high antifungal potency against *Aspergillus oryzae* and *Aspergillus flavus* (Kirmizigul, Anil, Ucar, et al., 1996), and some triterpene glycosides from *Holothurioidea* showed antifungal activity against *Cladosporium brevicompactum* and *Cladosporium sphaerospermum* (Pivkin, 2000). An important compound isolated from *Juniperus communis* has shown high antifungal activity in vitro and moderate activity in vivo in mouse model systems against *Candida* and *Aspergillus* (Pelaez et al., 2000). The effect of this compound on *Aspergillus fumigatus* is comparable to the effect of pneumocandin B(0), which is a glycan synthesis inhibitor. Some triterpene glycosides from the bark of *Pithecellobium racemosum* and some from *Hyalodendron* species have shown moderate activity (Khan et al., 1997; Bills et al., 2000). Three triterpenoid glycosides from the stem of *Anamirta cocculus* showed antifungal activities (Jayasinghe et al., 1993). Some synthetic glycosides also showed no antifungal effects (Takechi et al., 1996, 1997).

The influence of antifungal activity changes with bonding of carbohydrate units at C-3 (monodesmosidic glycosides) and at C-28 (bisdesmosidic glycosides) to aglycone (Bader et al., 2000). In addition, types of sugars and linkage points of sugars are also important to the antifungal activity of these glycosides. In some studies, (1–4) sugar-sugar linkages showed stronger activity than (1–6) sugar linkages (Takechi et al., 1997, 1998).

Several mechanisms for antifungal activities of these compounds occur at the cellular level. First of all, some triterpenic glycosides provide a chemical barrier that produces extracellular enzymes against fungi (Roldan et al., 1999). In another mechanism, compounds had a direct effect on cell membranes by modifying the structural-functional properties (Anisimov et al., 2000).

CONCLUSION

The significance of triterpene glycosides in pharmaceuticals has been reviewed according to their different activities. The potential activities of these natural products are essentially dependent on their structures and physical properties.

Solubility is the first factor. The water insoluble saponins from *Hedera helix* have haemolytic properties whereas water soluble saponins are devoid of this property (Agarwal and Rastogi, 1974).

Other factors include positions of –OH groups, acetylation of –OH groups, attachment points of sugars on aglycone, and number of sugar units and type of sugar linkages able to change polarity of these compounds. Polar groups in ring A and moderately polar groups in ring D or E in triterpene aglycone provide optimum activity. The substances with the highest haemolytic potency contain a 16α-OH or 16-keto group together with 3β-OH group (Schloesser and Wulff, 1969). Acetylation of hydroxyl groups resulted in loss of activity (Schloesser and Wulff, 1969; Tokuda et al., 1988). In connection with polarity, the cytotoxic activities of triterpene glycosides depend on the number of monosaccharide units attached to the hydroxyl group at C-3 of the aglycone. Four sugar units are more toxic than two sugar units. On the other hand, a saccharide chain with six units has little effect on activity (Agarwal and Rastogi, 1974). Number of sugar chains is also important in haemolytic action. Monodesmosidic glycosides are more active than bisdesmosidic glycosides. Activity of monodesmosidic glycosides decreases with long and branched sugar chains (Fukuda et al., 1985; Mahato et al., 1988; Tokuda et al., 1988). Anti-inflammatory activity becomes increased when tetra or pentasaccharite units are present at C-28 of aglycone (Pfander and Stoll, 1991). Linkaging points of sugars to aglycone and sugar-sugar linkages are also important for antifungal activity. According to the type of sugar linkages, (14) sugar-linked compounds have strong activity (Khan et al., 1997; Takechi et al., 1997, 1998).

In addition to these, many inactive precursors are present in plants, but they are converted into active antibiotics by enzymes that remove the sugars bound to the –COOH group. These active molecules may be harmful because of their high activity if the plant tissues are damaged by fungal infection (Kitagawa et al., 1976; Chandel and Rastogi, 1980). In addition to presence and amount of sugars, the configuration of glycosides have a high importance (Chandel and Rastogi, 1980), for example, S-isomer of a glycoside is active but R-isomer is not (Yoshimitsu et al., 1999).

The last and most important point is the applied dose. In one study, ginseng saponins in low doses (2,5-5 mg/kg) increased spontaneous activity in rats and mice but at high doses (100 mg/kg) showed a decrease in motor activity (Lee, 1974; Oura et al., 1975; Chandel and Rastogi, 1980). Another saponin from *Astragalus* species changes the analgesic effects depending on the applied dose (Zhang et al., 1984).

All of these explanations show that solubility, structure, polarity, number of sugars, type of sugars, linkaging point of sugars, attachment point of sugars, and applied dose influence the different activities of these compounds.

FUTURE DIRECTIONS

In the near future, significant functions of biologically active plant products will be explained in more detail. It is clear that triterpene glycosides have a number of biological effects on organisms. Modern analysis techniques will enable an understanding of their new functions. This is important in understanding many problems in the molecular biology of the single cell and the physiology of whole organs or organisms. The future of biology and biotechnology is in studies of cellular interactions and occurrence of pathological conditions at molecular levels. The critical roles of triterpenic glycosides in the integration of molecular physiology and pharmacology is promising.

APPENDIX: FORMULAS

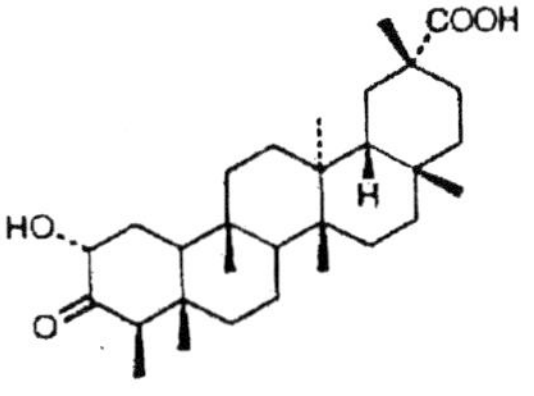

1

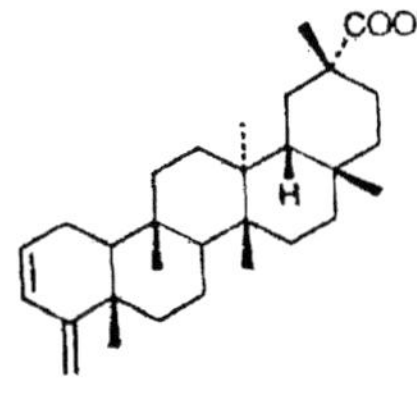

2

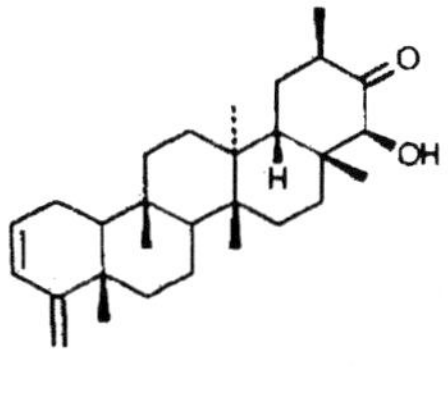

3

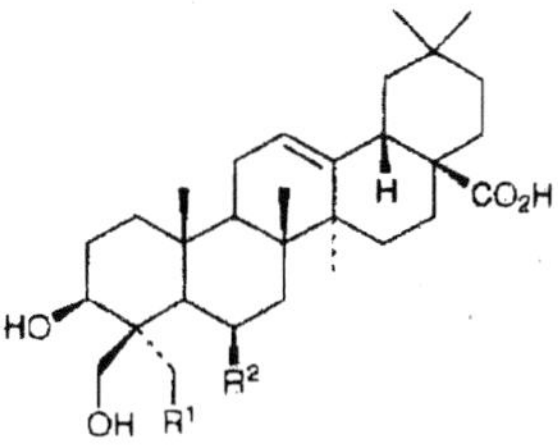

4 R^1 = OH ; R^2 = H
5 R^1 = H ; R^2 = OH

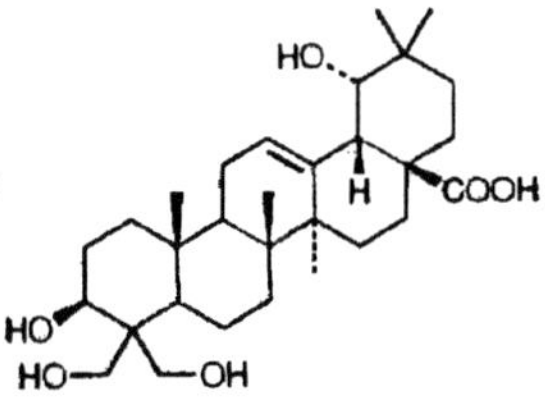

6

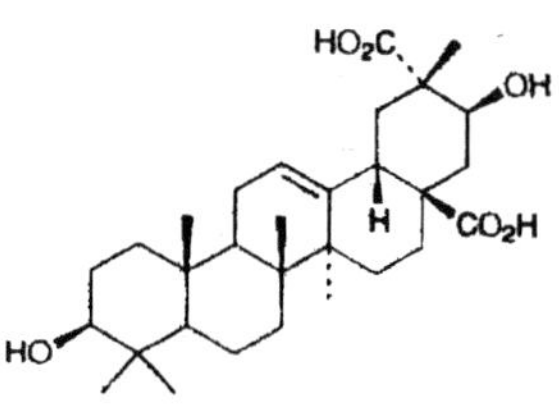

7

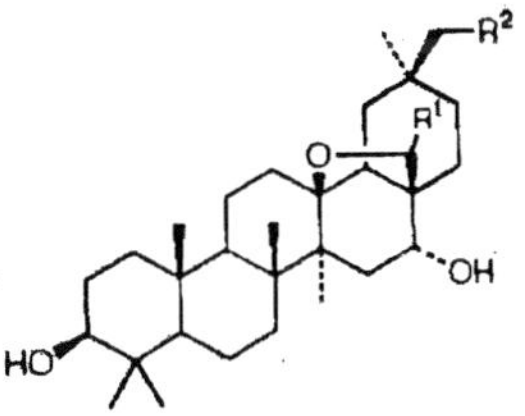

8 R^1 = H ; R^2 = OH

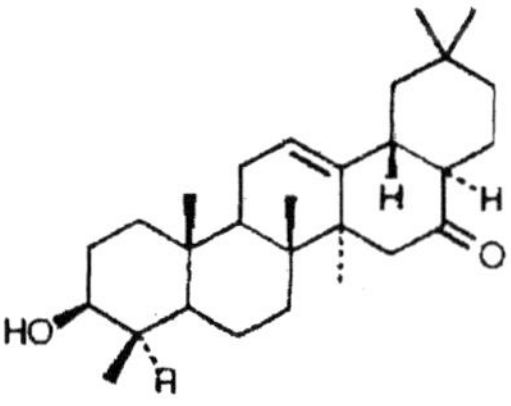

9

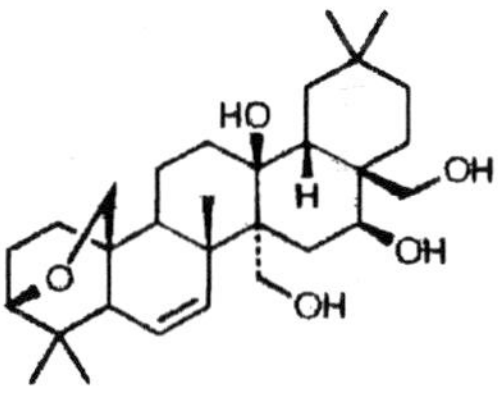

10

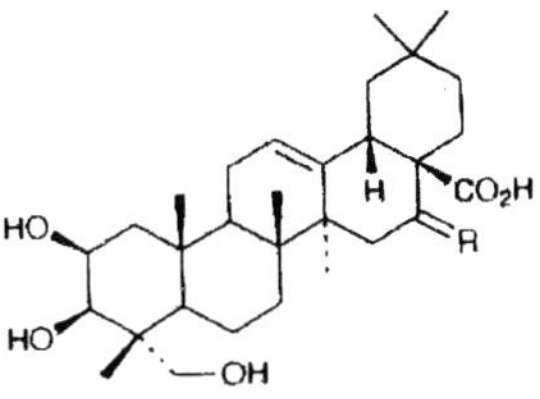

11 R = O

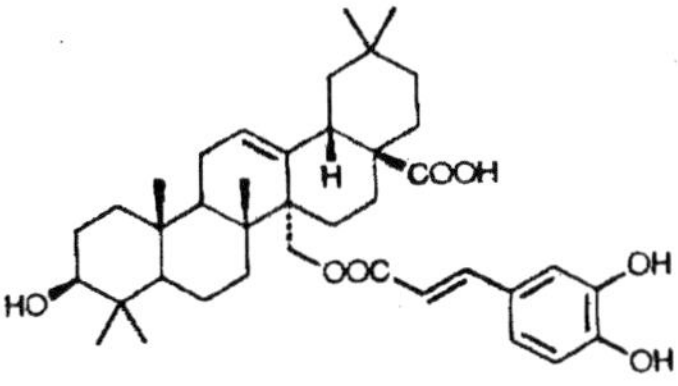

12

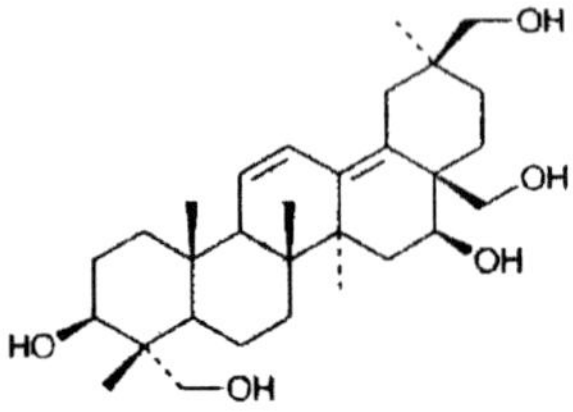

13

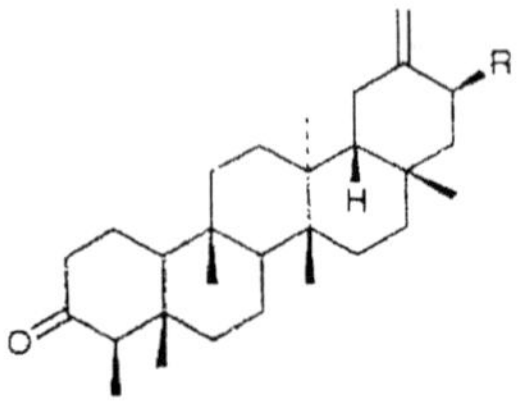

14 R = H
15 R = OH

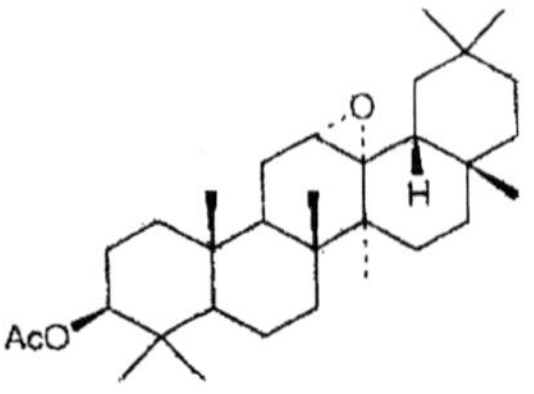

16

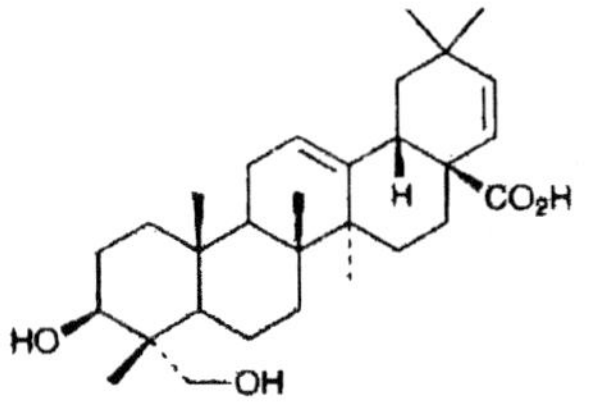

17

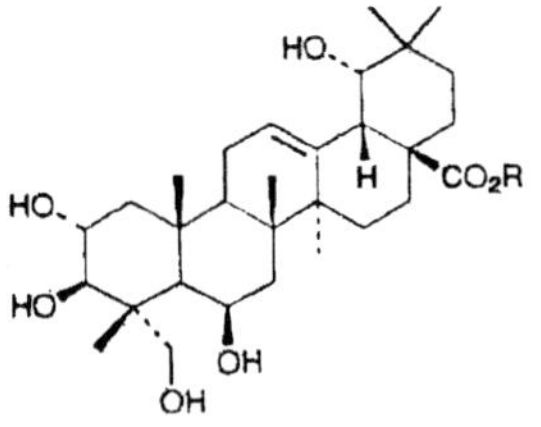

18 R = H

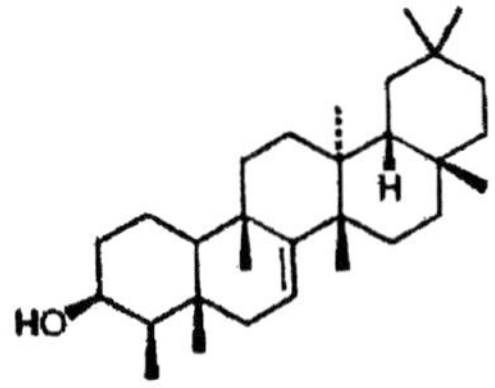

19

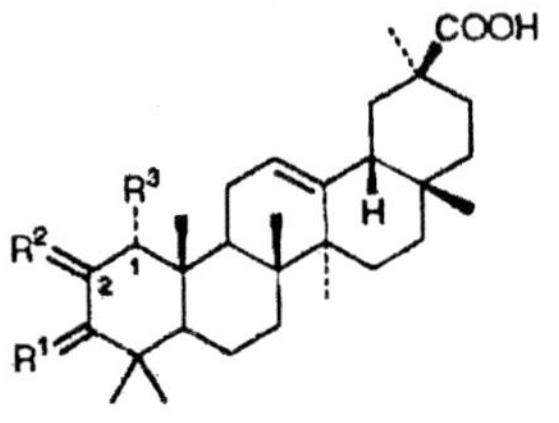

20 R^1=O; R^2 =α-OH,H; R^3 =H
21 R^1=β-OH,H; R^2 =O; R^3 =H
22 R^1=O; R^2 =H_2; R^3 =OH
23 R^1=O; R^2 =H_2; R^3 =H; 1,2-didehydro

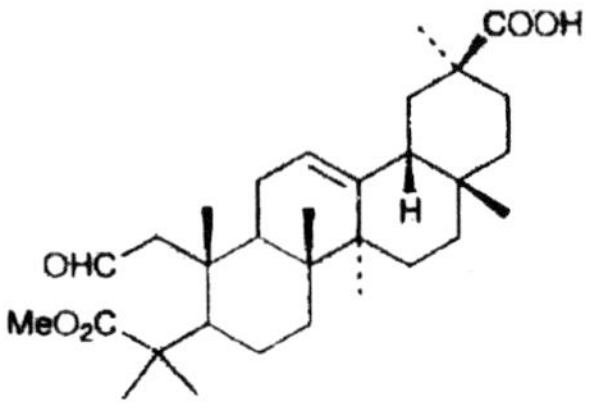

24

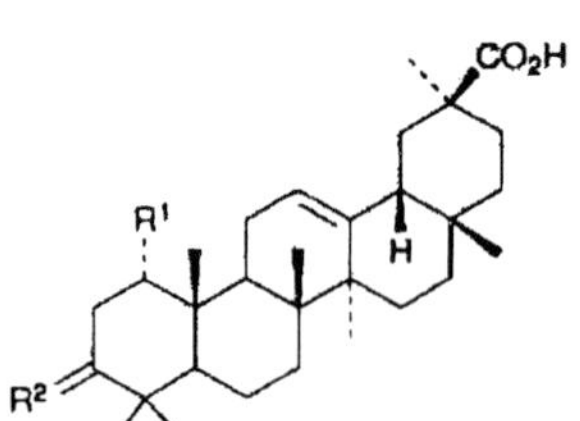

25 R^1=OH; R^2 =β-OH,H
26 R^1=H; R^2 =O

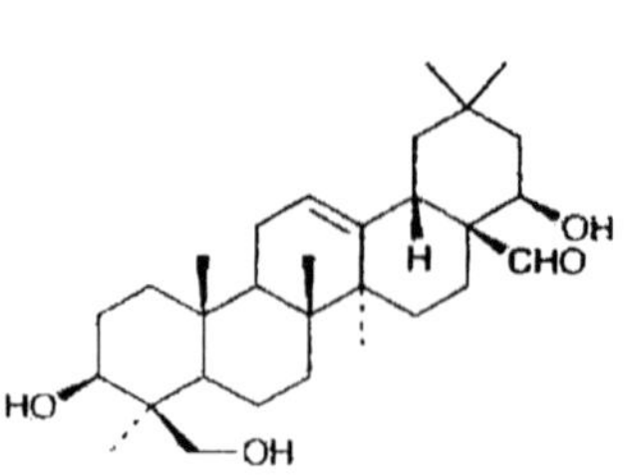

27

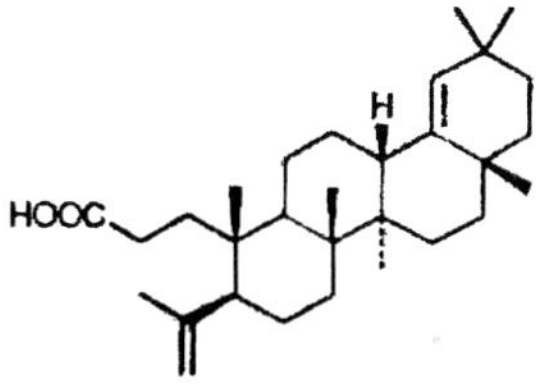

28

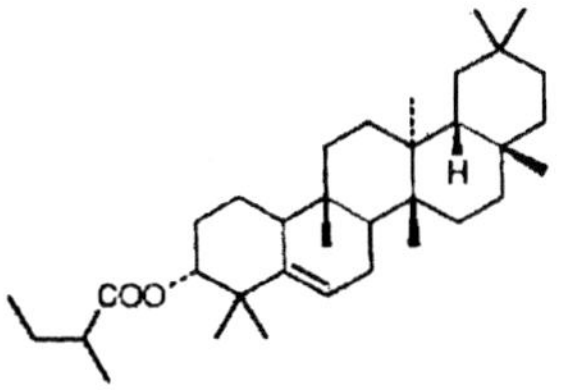

29

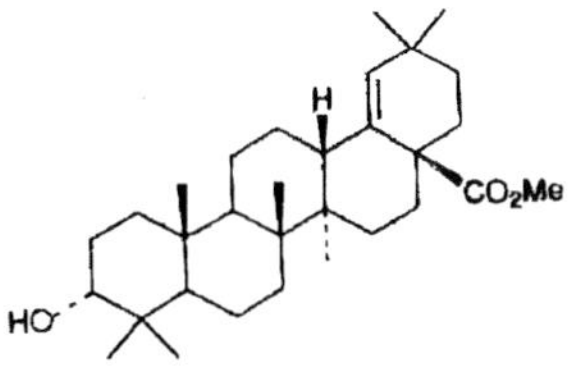

30

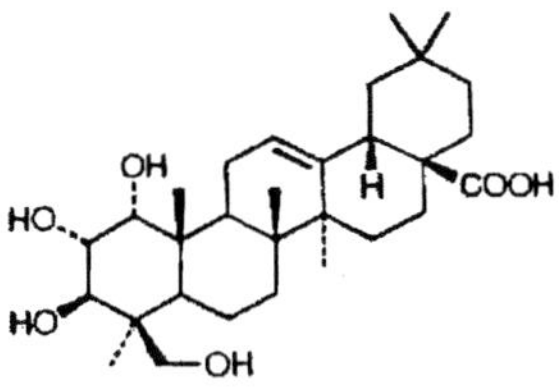

31

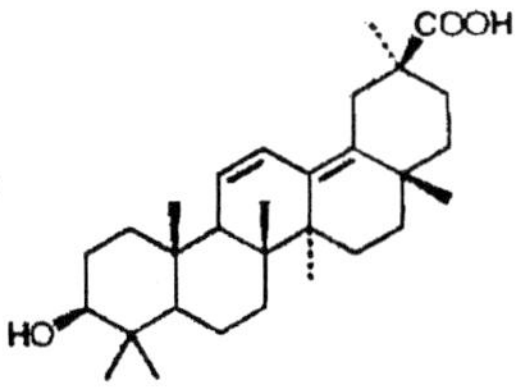

32

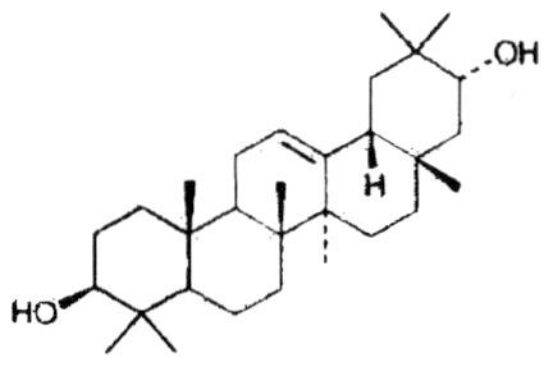

33

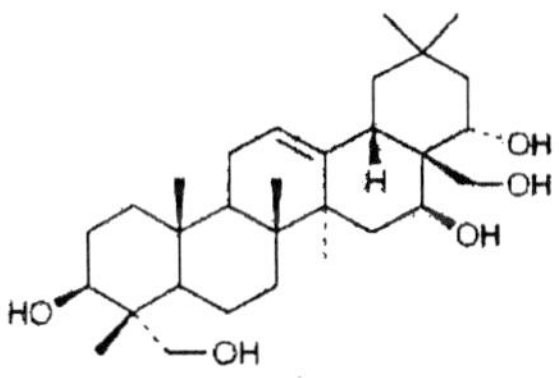

34

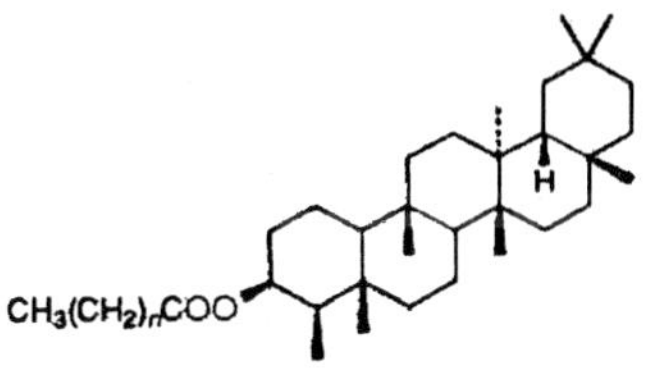

35 n = 14
36 n = 16

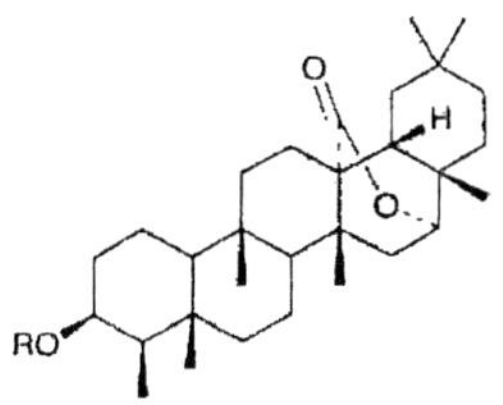

37 R = Ac
38 R = PhCO

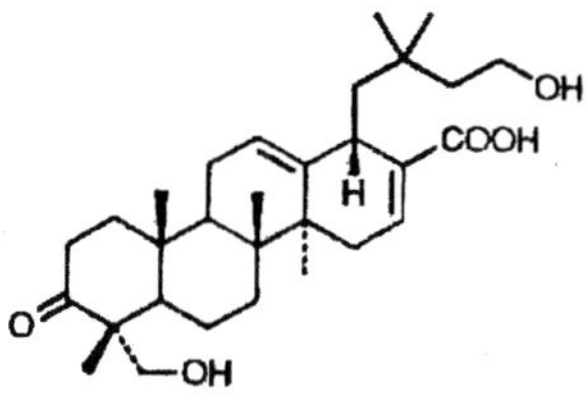

39

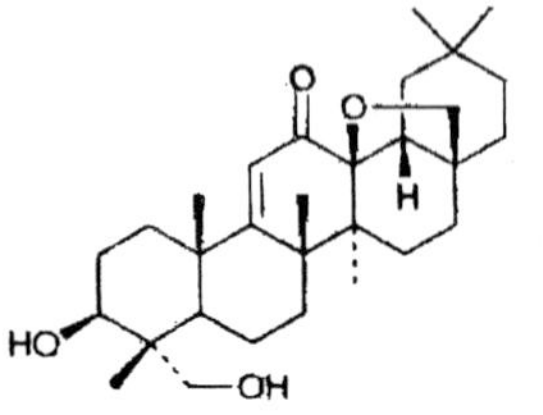

40

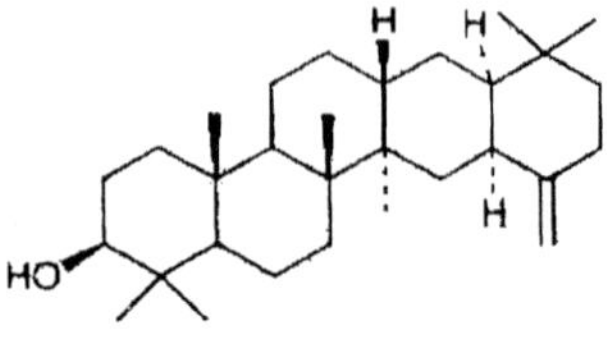

41

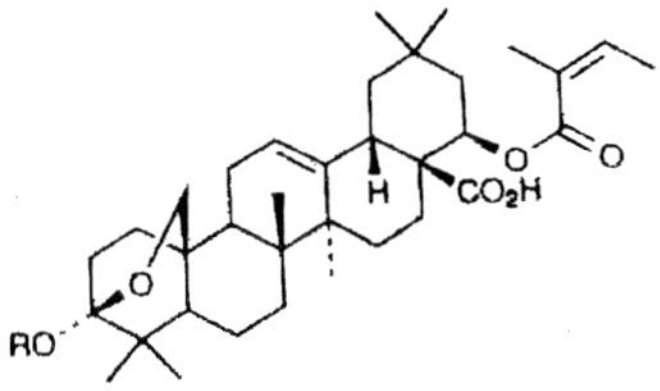

42 R = Me
43 R = H

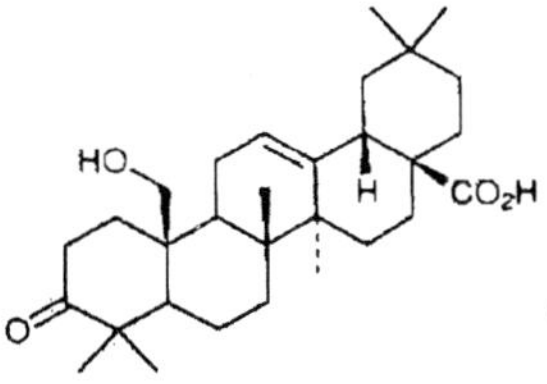

44

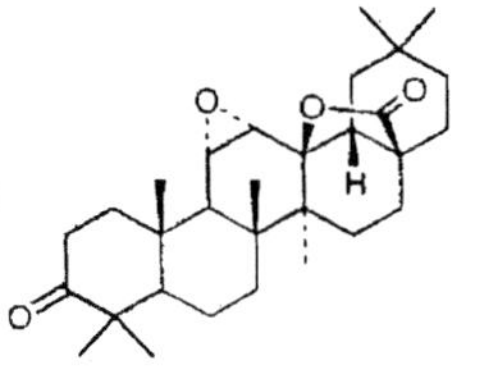

45

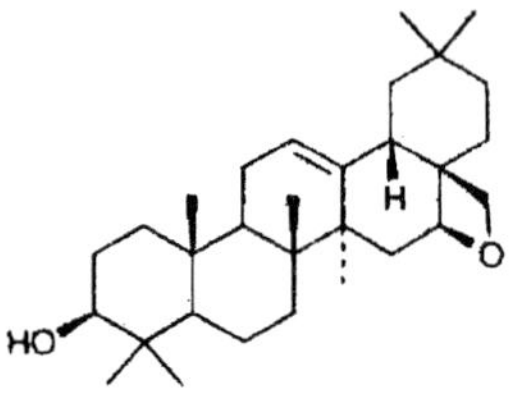

46

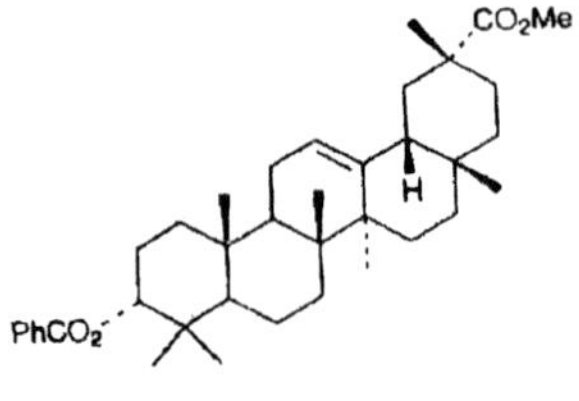

47

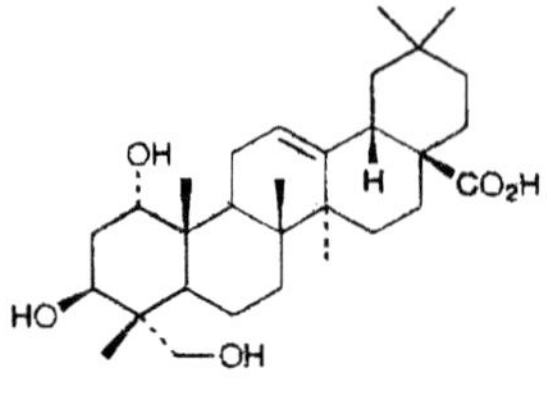

48

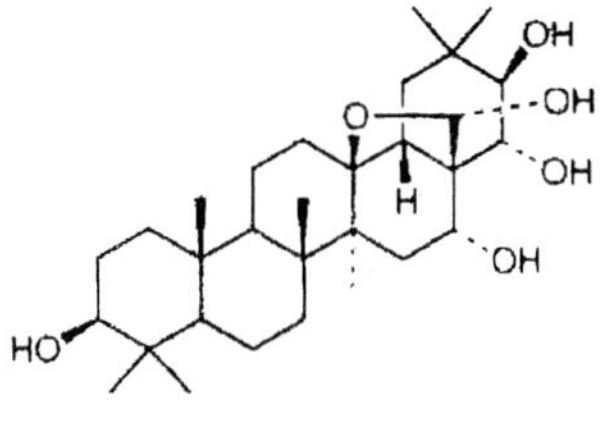

49

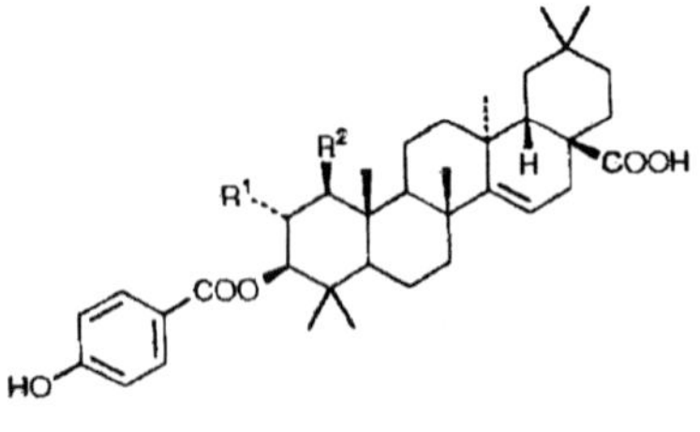

50 R^1 = 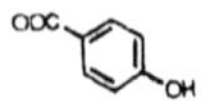 R^2 = H

51 R^1 = H; R^2 = OH

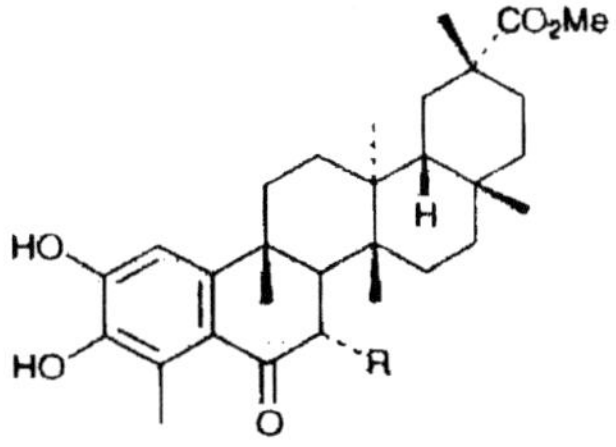

52 R = H

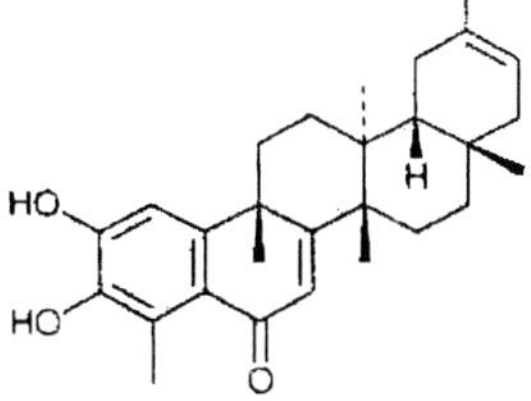

53

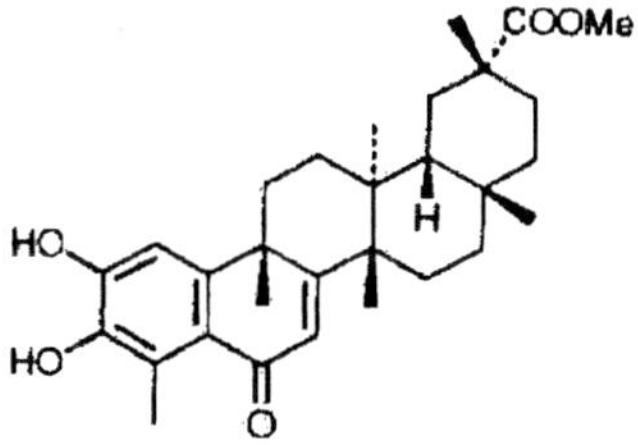

54

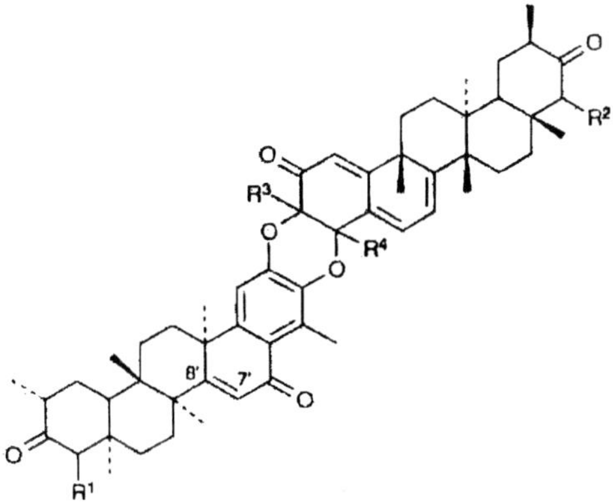

55 R[1]= R[2] = H; R[3] =α-OH; R[4] =α-CH_3; 7′,8′-dihydro

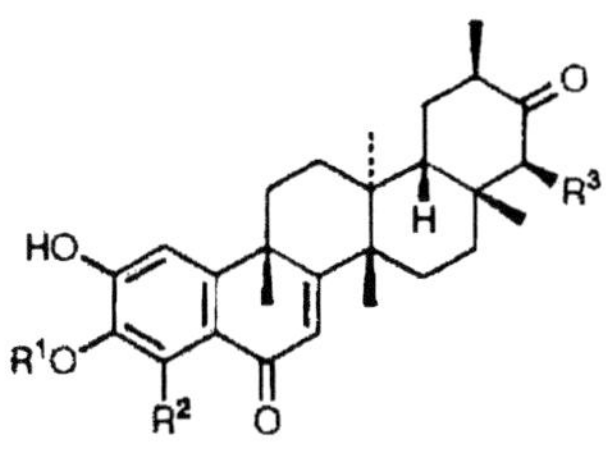

56 R[1] = R[3] = H; R[2] = Me

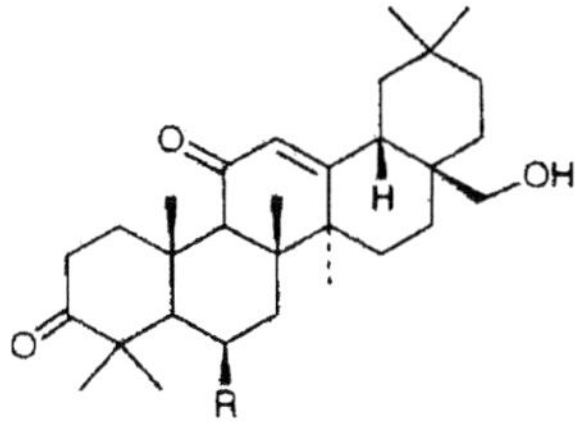

57 R = H
58 R = OH

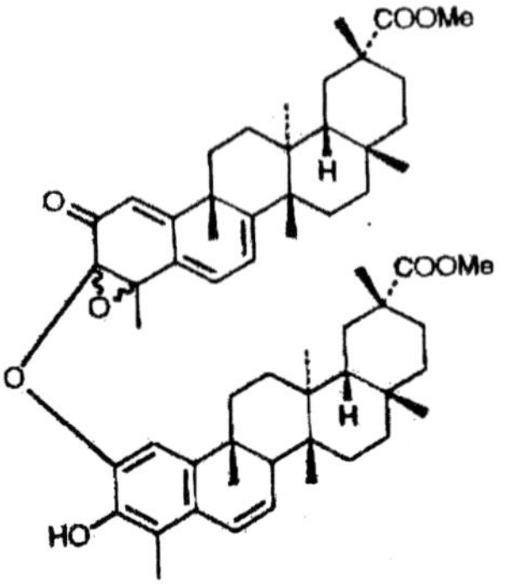

59

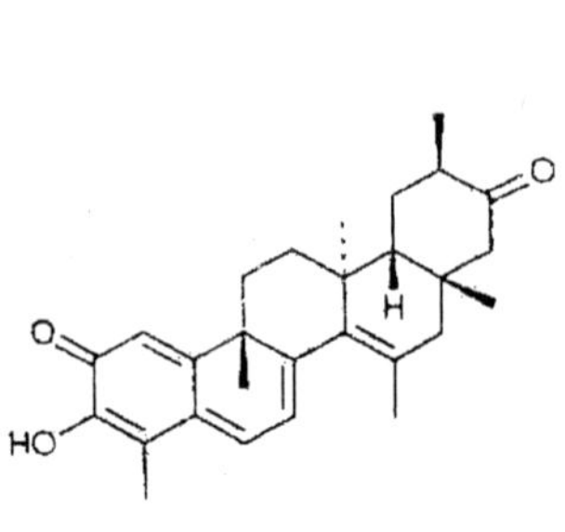

60

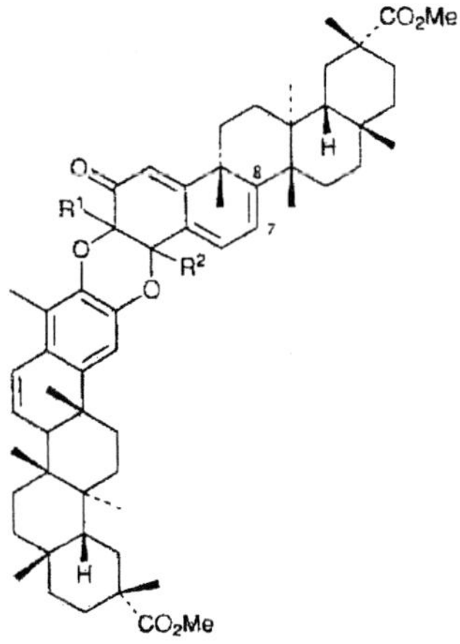

61 R[1]=β-OH; R[2] = β-Me

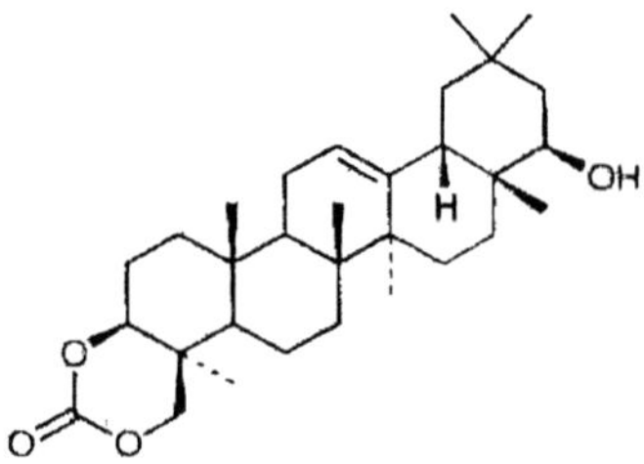

62

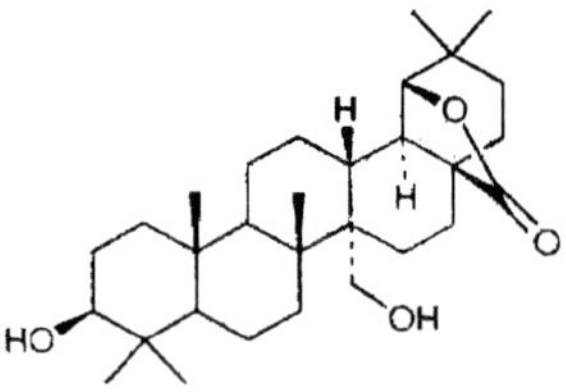

63

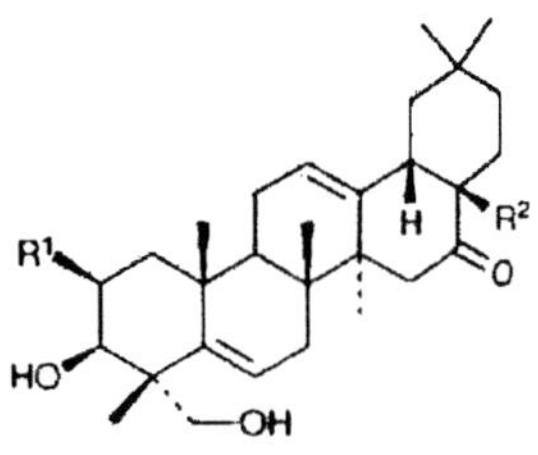

64 R[1] = OH ; R[2] = H
65 R[1] = H ; R[2] = CH_3

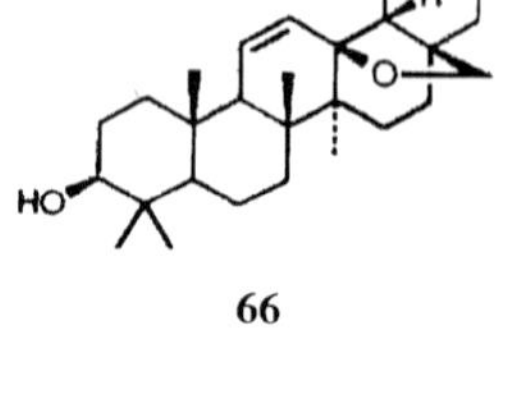

66

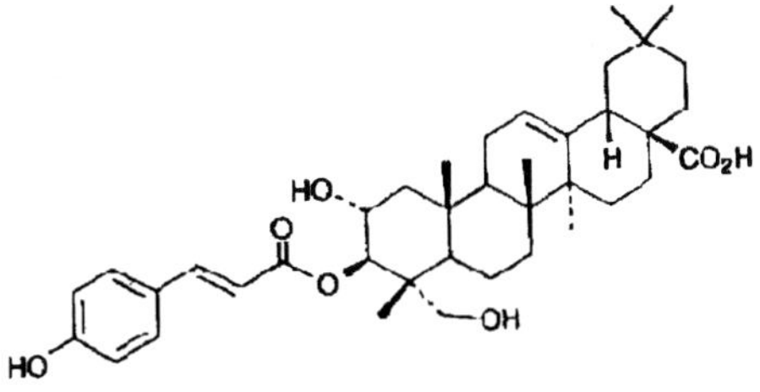

67

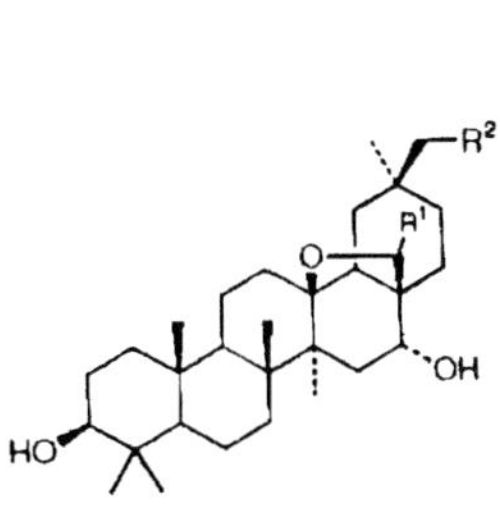

68

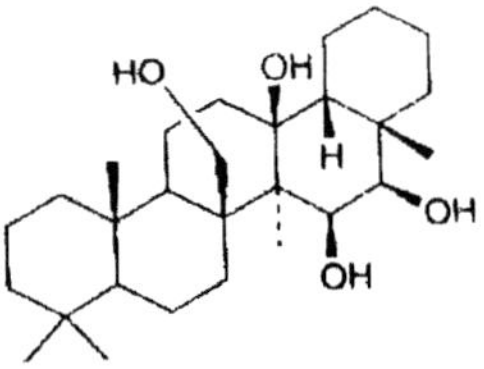

69

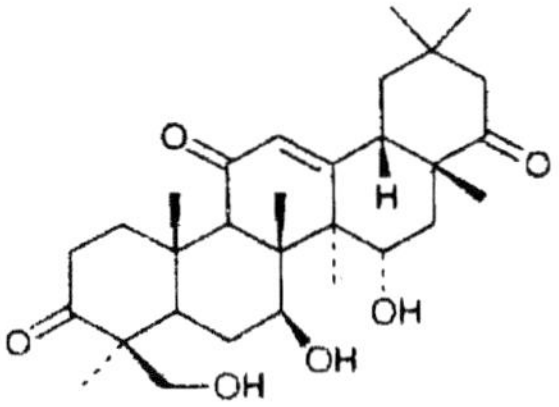

70

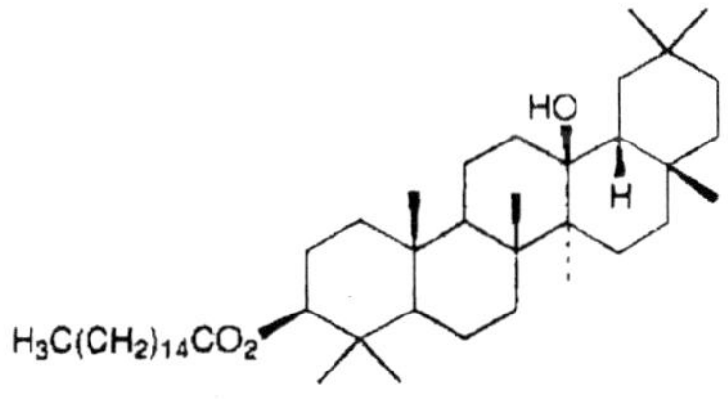

71

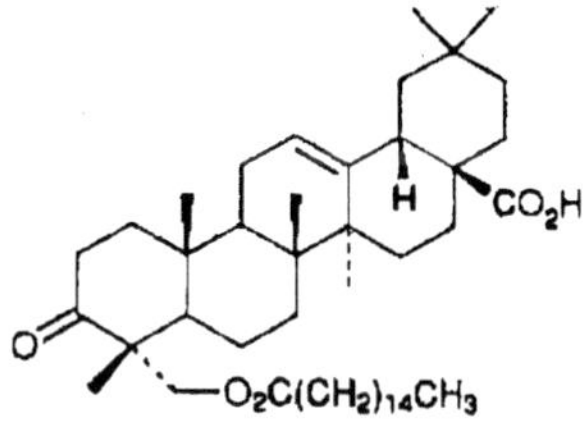

72

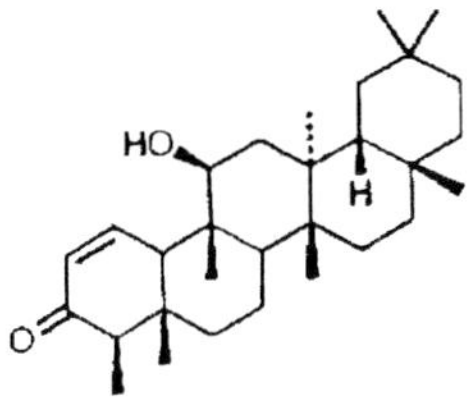

73

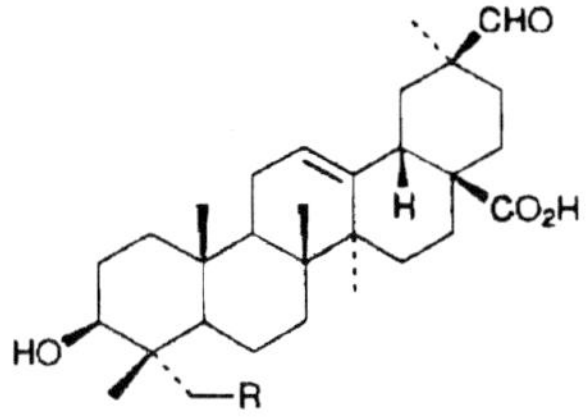

74 R = H
75 R = OH

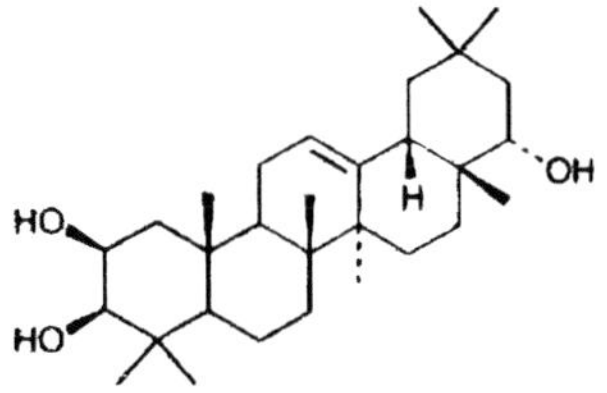

76

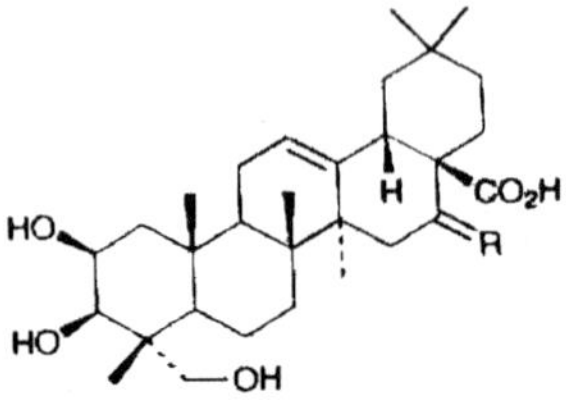

77 R = H_2

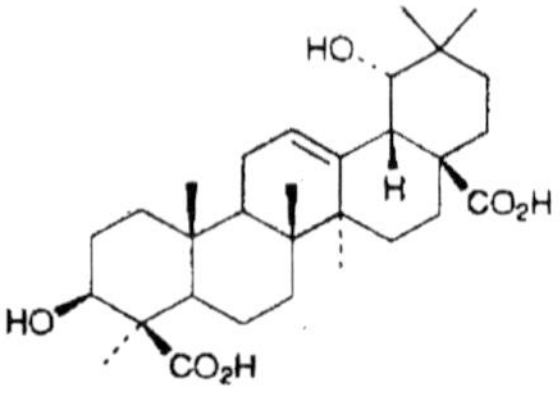

78

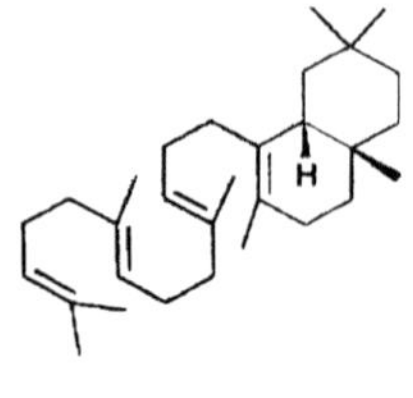

79

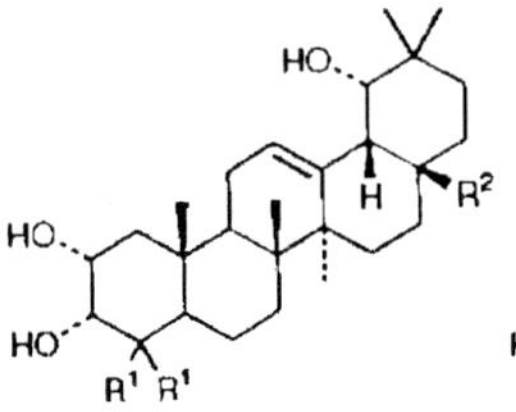

80 $R^1 = CH_2OH$; $R^2 = COOH$

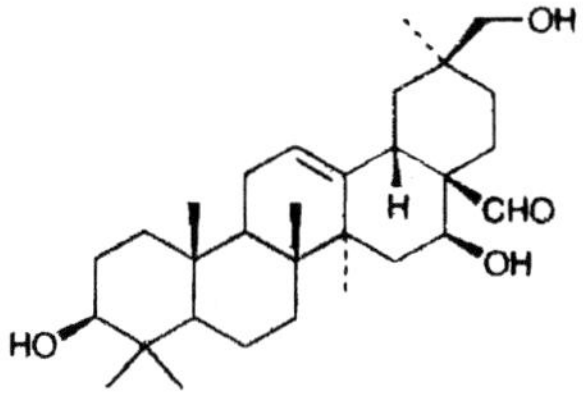

81

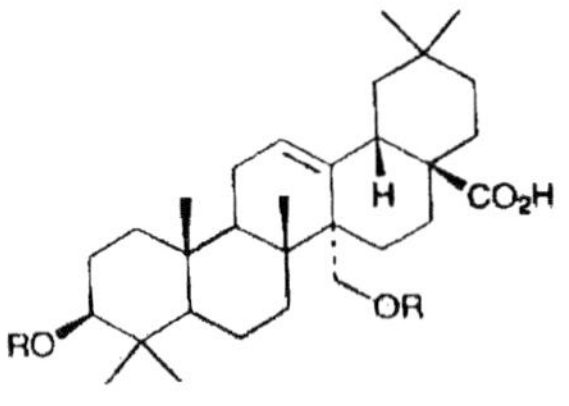

82 R = 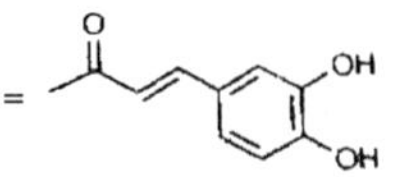

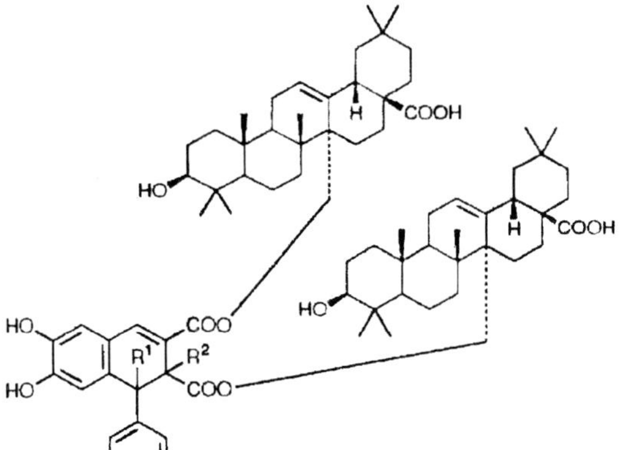

83 $R^1 = \alpha$-H; $R^2 = \beta$-H
84 $R^1 = \beta$-H; $R^2 = \alpha$-H

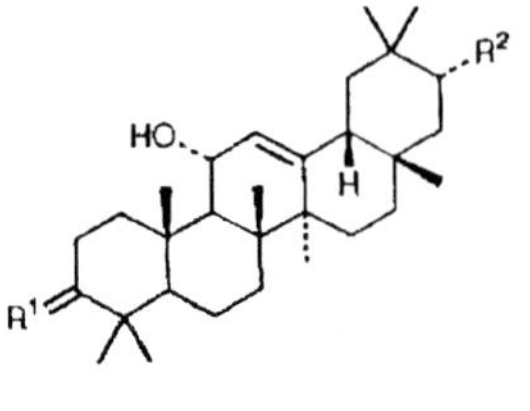

85 $R^1 = O$; $R^2 = H$

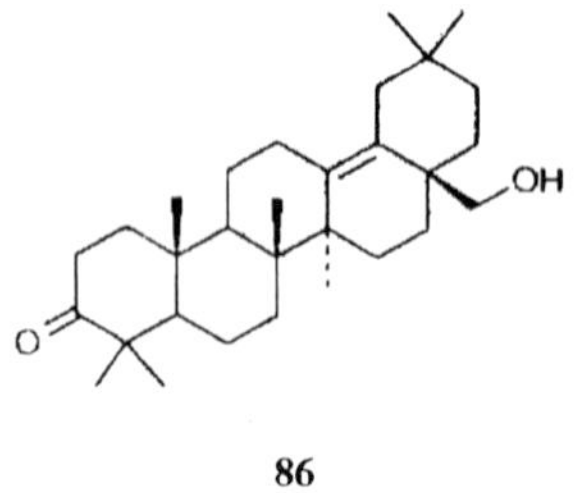

86

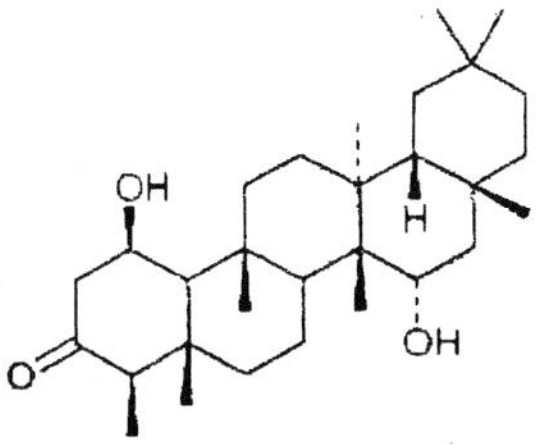

87

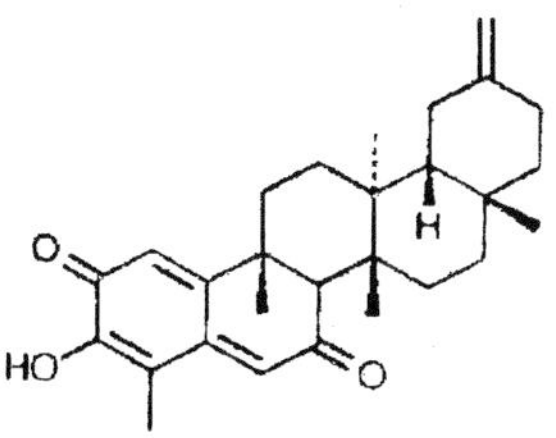

88

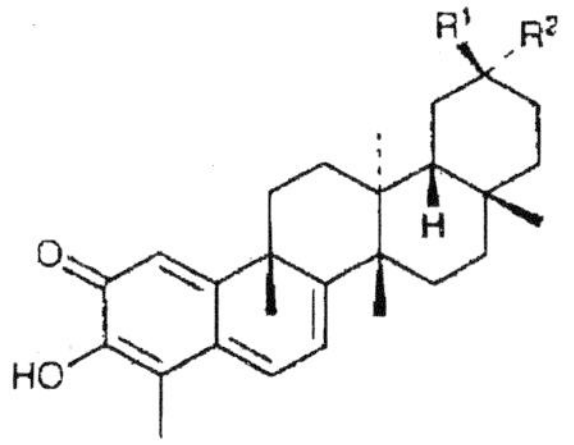

89 R^1= H; R^2 = CH_2OH
90 R^1= CH_2OH; R^2 = CO_2Me

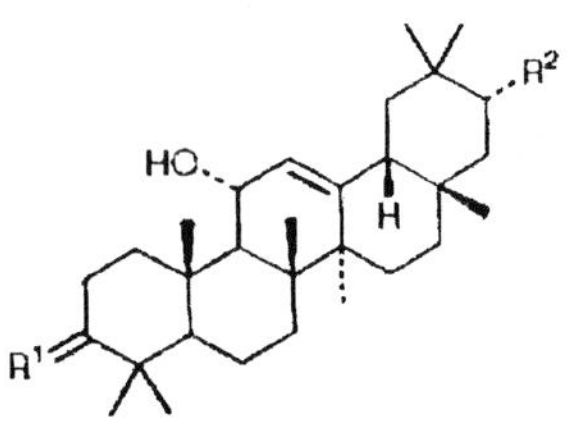

91 R^1= β-OH,H; R^2 = OH

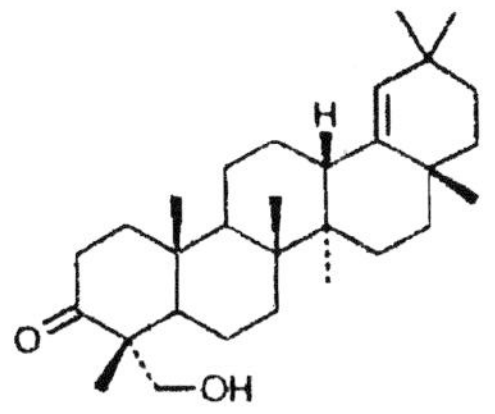

92

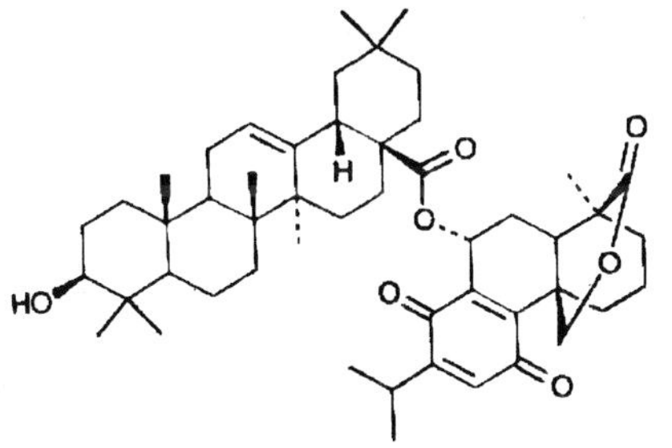

93

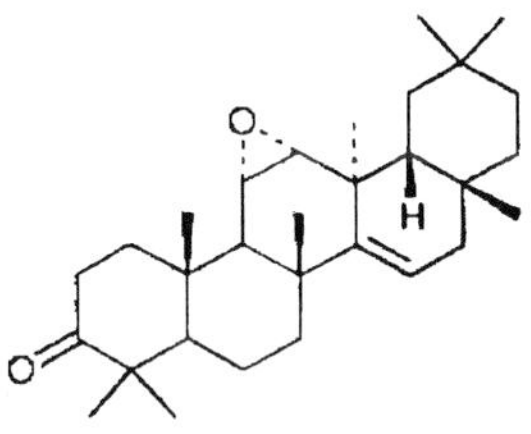

94

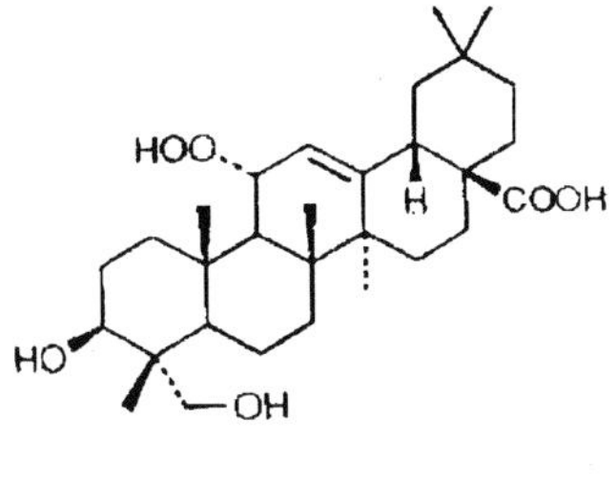

95

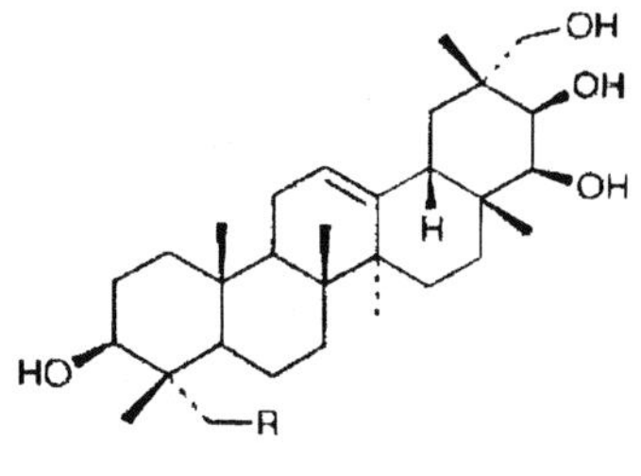

96 R = H
97 R = OH

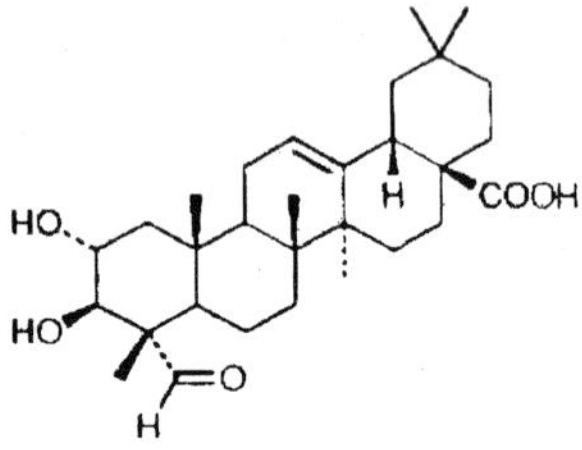

98

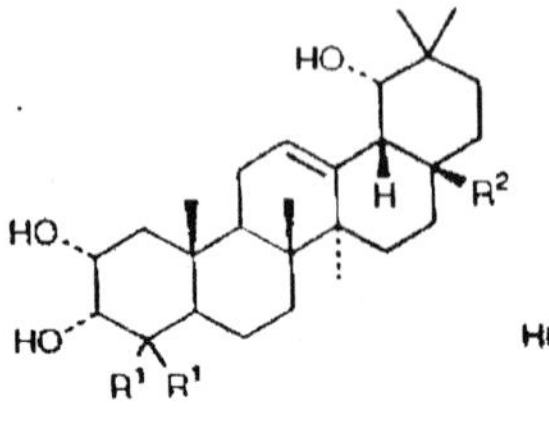

99 $R^1 = R^2 = Me$

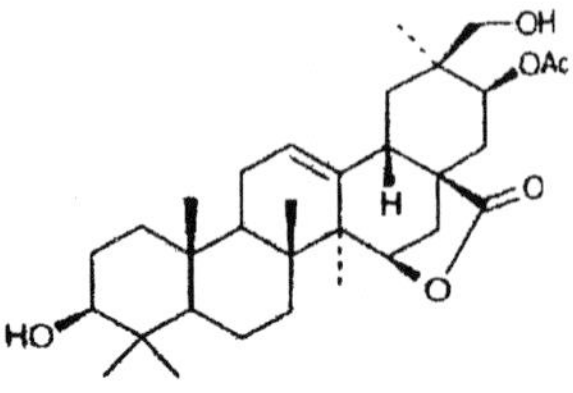

100

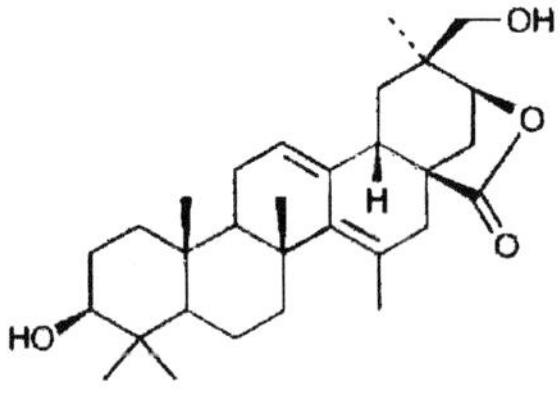

101

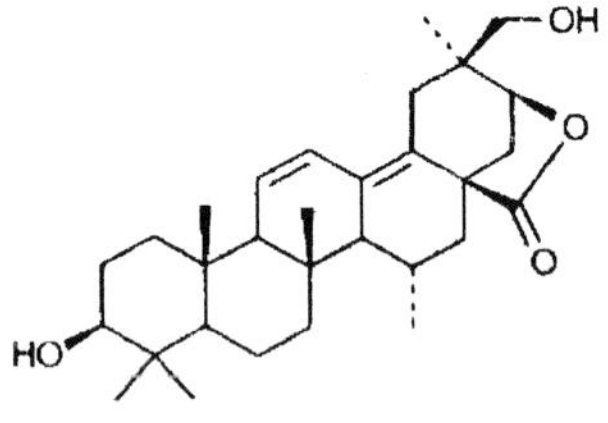

102

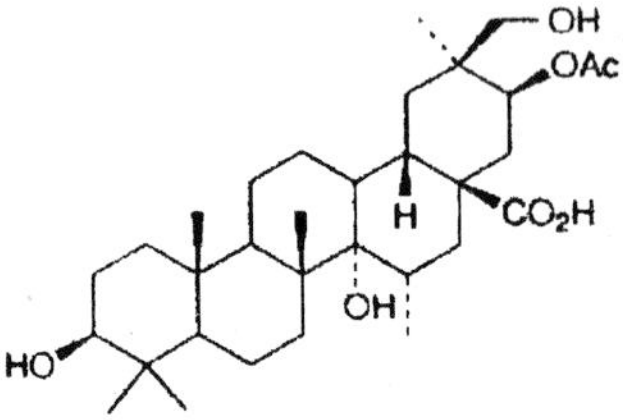

103

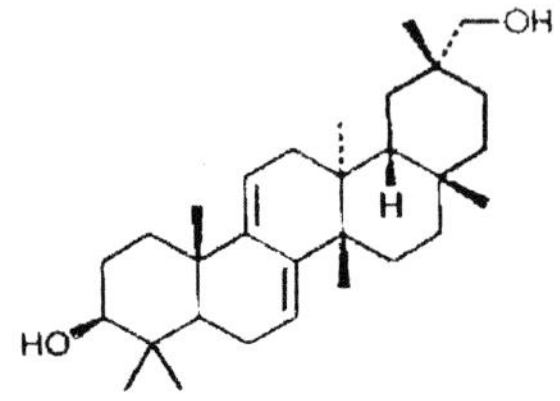

104

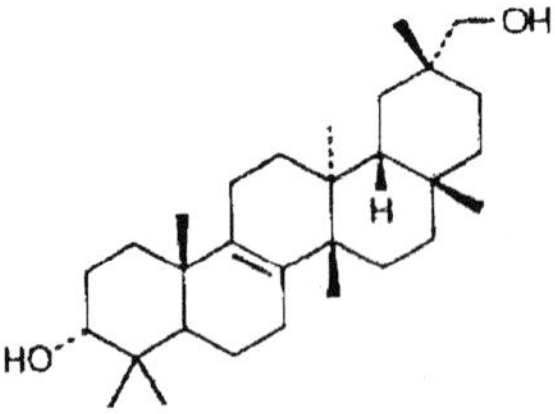

105

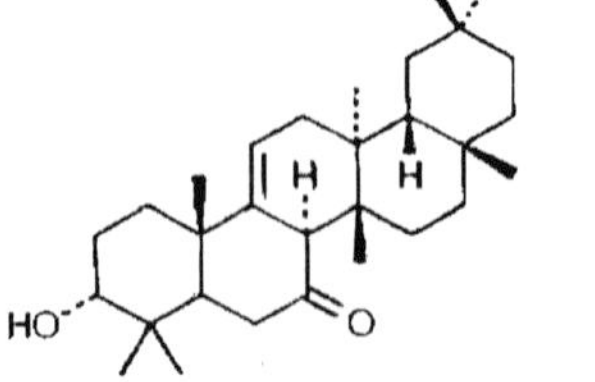

106

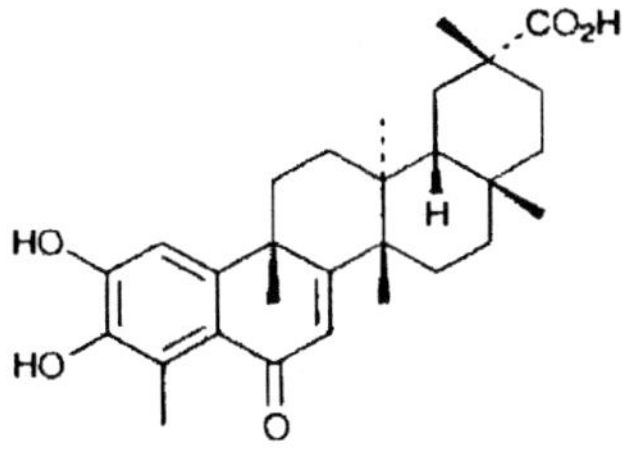

107

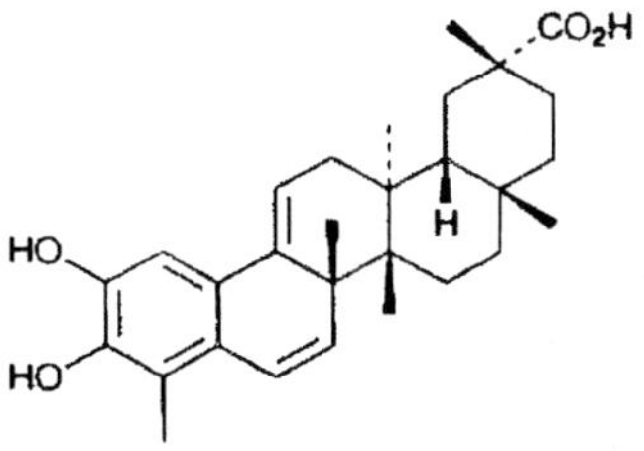

108

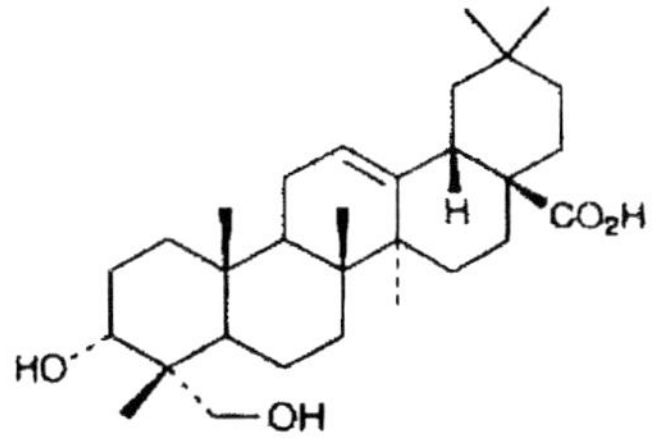

109

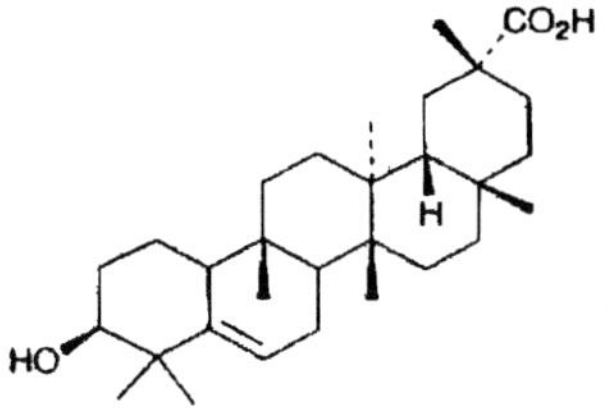

110

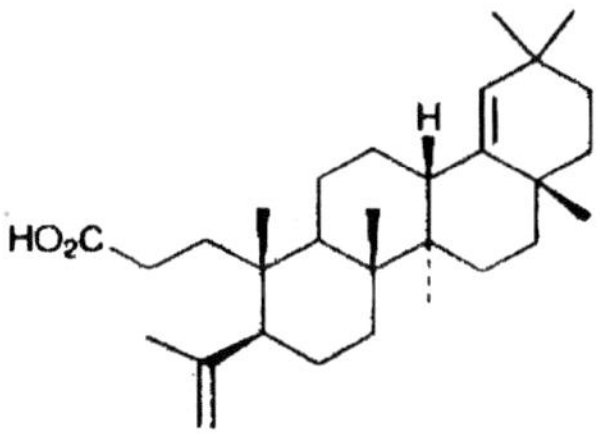

111

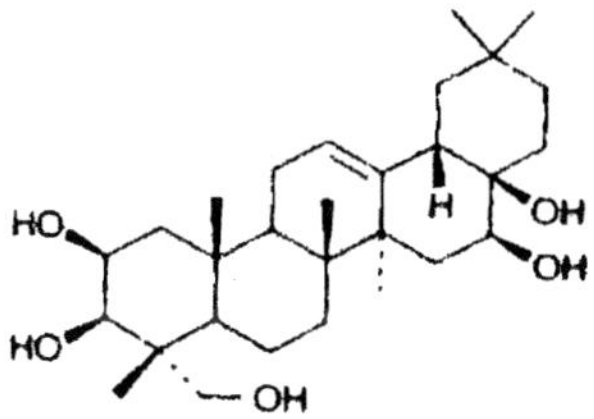

112

REFERENCES

Abdel-Kader, M.S., Bahler, B.D., Malone, S., Werkhoven, M.C.M., Wisse, J.H., Neddermann, K.M., Bursuker, I. and Kingston, D.G.I. (2000). Bioactive saponins from *Swartzia schomburgkii* from the Suriname rainforest. *J. Nat. Prod.* 63: 1461-1464.

Adler, C. and Hiller, K. (1985). Bisdesmosidic triterpene saponins. *Pharmazie* 40: 676-693.

Adnyana, I.K., Tezuka, Y., Banskota, A.H., Xiong, Q., Tran, K.Q. and Kadota, S. (2000). Quadranosides I-V, new triterpene glucosides from the seeds of *Combretum quadrangulare*. *J. Nat. Prod.* 63: 496-500.

Agarwal, P.K., Jain, D.C., Gupta, R.K. and Thakur, R. (1985). C-13 NMR spectral investigations. 10. C-13 NMR-Spectroscopy of steroidal sapogenins and steroidal saponins. *Phytochem.* 24: 2479-2496.

Agarwal, S.K. and Rastogi, R. P. (1974). Triterpenoid saponins and their genins. *Phytochem.* 13: 2623-2645.

Akihisa, T., Yasukawa, K., Kimura, Y., Takido, M., Kokke, W.C.M. and Twamura, T. (1994). 5-δ-C-Friedo-Oleanane triterpenes from the seeds of *Trichosanthes kirilowii* maxim and their anti-inflammatory effects. *Chem. Pharm. Bull.* 42: 1101-1105.

Alankus-Çaliskan, Ö. and Anil, H. (1995). A new bisdesmosidic triterpene saponin from *Cephalaria transsylvanica*. *Phytochem.* 38: 1493-1495.

Aliev, A.M. and Movsumov, I.S. (1976) Triterpene glycosides from *Cephalaria kotschyi* flowers. *Khim. Prir. Soedin.* 12: 264-265.

Allerland, A. and Doddrell, D. (1971). Strategies in the application of partially relaxed fourier transform nuclear magnetic resonance spectroscopy in assignments of Carbon-13 resonance of complex molecules. *J. Am. Chem. Soc.* 93: 2777-2779.

Anisimov, M.M., Kuznetsova, T.A., Shirokov, V.P., Prokof'eva, N.G. and Elyakov, G.B. (1972). Toxic effect of stichoposide from *Stichopus japonicus* on early embryogenesis of the sea urchin. *Toxicon.* 10: 187-188. [(1972) *Chem. Abstr.* 76, 136599d]

Anisimov, M.M., Prokof'eva, N.G., Kuznetsova, T.A. and Peretolchin, N.V. (1971). Effect of some triterpene glycosides on protein synthesis in tissue cultures of rat bone marrow. *Izv. Akad. Nauk SSSR, Ser. Biol.* (1), 137-140. [1971, *Chem. Abstr.* 74,73825k]

Anisimov, M.M., Strigina, L.I., Gorovoi, P.G., Aminin, D.L. and Agafonova, I.G. (2000). Chemical composition and medical biological properties of triterpene glycosides of the far eastern plant *Caulophyllum robustum* Maxim. (Berberidaceae). *Rastitel'nye Resursy* 36: 107-129.

Annaev, C., Isamuchamedova, M. and Abubakirov, N.K. (1983a). Triterpene glycosides of *Climacoptera transoxana*. 1. *Khim. Prir. Soedin.* 19: 202-205.

Annaev, C., Isamuchamedova, M. and Abubakirov, N.K. (1983b). Triterpene glycosides of *Salsola micranthera*. 1. The structure of Salsoloside-C and Salsoloside-D. *Khim. Prir. Soedin.* 19: 727-732.

Annaev, C., Isamuchamedova, M. and Abubakirov, N.K. (1983c). Triterpene glycosides of *Climacoptera transoxana*. 2. Structure of the Kopterosides-D. *Khim. Prir. Soedin.* 19: 453-456.

Annaev, C., Jsaev, M.I. and Abubakirov, N.K. (1983). Triterpene glycosides of *Climasoptera transoxana*. 3. Structure of the Kopterosides-E and Kopterosides-F. *Khim. Prir. Soedin.* 19: 596-601.

Aoki, T., Shido, K., Takahasi, Y. and Suga, T. (1981). Structures of 3,28-O-bisglycosidic triterpenoid saponins of *Fatsia japonica*. *Phytochem.* 20: 1681-1686.

Aoki, T., Tanio, Y. and Suga, T. (1976). Triterpenoid saponins from *Fatsia japonica*. *Phytochem.* 15: 781-784.

Arai, Y., Hirohara, M., Ogawa, R., Masuda, K., Shiojima, K., Ageta, H., Chang, H.C. and Chen, Y.P. (1996). Fern constituent: Preoleanatetraene, A novel bicyclic triterpenoid hydrocarbon from *Polypodiodes formosona*. *Tetrahedron Lett.* 37: 4381-4384.

Arntzen, C.J., Tracey, M.B., Gutterman, J.U., Hoffmann, J.J., Bailey, D.T. and Jayatilake, G.S. (1999). Triterpene glycosides of *Acacia victoriae* as antitumor agents. *PCT Int. Appl.* WO 9959578 A1, pp. 313.

Avilov, S.A., Antonov, A.S., Drozdova, O.A., Kalinin, V.I., Kalinovsky, A.I., Stonik, V.A., Riguera, R., Lenis, L.A. and Jimenez, C. (2000). Triterpene glycosides from the far-eastern sea cucumber *Pentamera calcigera*. 1. Monosulfated glycosides and cytotoxicity of their unsulfated derivatives. *J. Nat. Prod.* 63: 65-71.

Bader, G., Binder, K., Hiller, K. and Ziegler-Bohme, H. (1987). The antifungal action of triterpene saponins of *Solidago virgaurea* L. *Pharmazie* 42(2): 140-149.

Bader, G., Seibold, M., Tintelnot, K. and Hiller, K. (2000). Cytotoxicity of triterpenoid saponins: Part 2: Relationships between the structures of glycosides of polygalacic acid and their activities against pathogenic *Candida* species. *Pharmazie* 55: 72-74.

Balakrishna, K., Vasanth, S., Kundu, A.B., Rao, R.B. and Connolly, J.D. (1995). Vicogenin, a 28-nor-12-oleanenepentol from *Vicoa indica*. *Phytochem.* 40: 335-336.

Balansard, J.J.J. and Bernard, P. J. L. (1966). Extraction of saponins from plants. *Neth. Appl.* 6513.079 (Cl. A 61K), (1964) *Fr. Appl.* [(1967) *Chem. Abstr.* 66, 14034]

Barber, M., Bordoli, R.S., Sedgwick, R.D. and Tyler, A.N. (1981). Fast atom bombardment of solids (fab) - A new ion-source for mass-spectrometry. *J. Chem. Soc. Chem. Comm.* (7): 325-327.

Bax, A. and Freeman, R. (1981). Investigation of complex networks of Spin-Spin coupling by two-dimensional NMR. *J. Magn. Reson.* 44: 542-561.

Bellet, P. and Cerede, P. (1967). Preparation of new active substances from a species of *Araliaceae* and derived products. Roussel-UCLAF, Fr. 1,477,421 (Cl. C 07g), 3 pp.

Bills, G., Dombrowski, A., Morris, S.A., Hensens, O., Liesch, J.M., Zink, D.L., Onishi, J., Meinz, M.S., Rosenbach, M., Thompson, J.R., et al. (2000). Hyalodendrosides A and B, antifungal triterpenoid glycosides from a *Lignicolous hyphomycete*, *Hyalodendron* species. *J. Nat. Prod.* 63: 90-94.

Bloor, S.J. and Qi, L. (1994). Cytotoxic saponins from New Zealand *Myrsine* species. *J. Nat. Prod.* 57: 1354-1360.

Brochini, C. B., Martins, D., Rogue, N. F. and Bolzani, V. Da S. (1994). An oleanane acid from *Alibertia edulis*. *Phytochem.* 36: 1293-1295.

Bukharov, V.G. and Karlin, V.V. (1970). *Patrinia sibirica* glycosides I and II. *Khim. Prir. Soedin.* 6: 60-64. and 64-69. [(1970) *Chem. Abstr.* 73, 56361, 56359]

Bukharov, V.G., Karlin, V.V. and Sidorovich, T.N. (1970). Triterpene glycosides of *Patrinia scabiosofaelia* I. *Khim. Prir. Soedin.* 6: 60-74. [(1970) *Chem. Abstr.* 73, 99162w]

Byla, B. and Herz, W. (1996). Triterpenes and 1-(ω-hydroxyceratyl) glycerols from *Pentaclethra eetveldeana* root bark. *Phytochem.* 42: 501-504.

Çaliskan, Ö.A., Anil, H. and Stephenson G.R. (1994). Cephalaria saponin A, A new bidesmosidic triterpene saponin from *Cephalaria transsylvanica*. *J. Nat. Prod.* 57: 1001-1003.

Chandel, R. S. and Rastogi, R. P. (1980). Triterpenoid saponins and sapogenins 1973-1978. *Phytochem.* 19: 1889-1908.

Chavez, M.I. and Delgado, G. (1994). Isolation and relay synthesis of 11-α-hydroperoxy diacetyl hederagenin, a novel triterpenoid derivative from *Serjania triquetra* (Sapindaceae)—Biogenetic implications. *Tetrahedron* 50: 3869-3878.

Chetyrina, N.S. and Kalinovsky, A.I. (1979). Triterpene glycosides of *Caulophyllum robustum*—Structure of kaulosides-B and kaulosides-C. *Khim. Prir. Soedin.* 15: 174-176.

Chetyrina, N.S., Strigina, L.I., El'kin, Y.N. and Isakov, V.V. (1975). New triterpene glycosides of *Caulophyllum robustum* maxim: Cauloside E, F and G. *Tezisy Dokl.-Vses. Simp. Biorg. Khim.* 13. [(1976) *Chem. Abstr.* 85, 106644h]

Chirva, V.Y., Kintya, P.K. and Melnikov, V.N. (1974). Structure of triterpenic glycoside of *Clematis vitalba* L. with unusual sugar. *Dokl. Akad. Nauk. SSR* 217: 969-970. [(1974) *Chem. Abstr.* 81, 136449h]

Chirva, V.Y., Kintya, P.K. and Melnikov, V.N. (1975). New data on triterpene glycosides. Dokl. Soobshch.-Mendeleevsk. *S'ezed Obsch. Prikl. Chim.* (Rozinskaya, V. N., ed.) pp. 6, 109. 'Nauka' Moscow. [(1978) *Chem. Abstr.* 88, 148971e]

Chirva, V.Y., Kintya, P.K. and Sosnovskii, V.A. (1969). Triterpene glycosides of *Sapindus mukorossi*. *Khim. Prir. Soedin.* 5: 450-451. [(1970) *Chem. Abstr.* 72, 75669v]

Choi, J.S. and Woo, W.S. (1984). Coumarins and triterpenoid glycosides from the roots of *Patrinia scabioaefolia*. *Arch. Pharmacol. Res.* 7: 121-126.

Chou, S.C., Ramanathan, S., Matsui, A., Rogers, J. and Cutting, W.C. (1971). Isolation of saponins with antifertility activity from *Gleditschia horrida*. *Indian J. Exp. Biol.* 9: 503-504.

Connolly, J.D. and Hill, R.A. (1996). Triterpenoids. *Nat. Prod. Rep.* 13: 151-169.

Connolly, J.D. and Hill, R.A. (1997). Triterpenoids. *Nat. Prod. Rep.* 14: 661-679.

Connolly, J.D. and Hill, R.A. (1999). Triterpenoids. *Nat. Prod. Rep.* 16: 221-240.

Connolly, J.D. and Hill, R.A. (2000). Triterpenoids. *Nat. Prod. Rep.* 17: 463-482.

Dekanosidze, G.E., Pkheidze, T.A. and Kemertelidze, E.P. (1970). Sapogenins from *Hedera pastuchovii*. *Khim. Prir. Soedin.* 6: 375-376. [(1970) *Chem. Abstr.* 73, 127731b]

Della Greca, M., Fiorentino, A., Monaco, P. and Previtera, L. (1994a). Oleanane glycosides from *Hydrocotyle ranunculoides*. *Phytochem.* 36: 1479-1483.

Della Greca, M., Fiorentino, A., Monaco, P. and Previtera, L. (1994b). Polyoxygenated oleanane triterpenes from *Hydrocotyle ranunculoides*. *Phytochem.* 35: 201-204.

Deng, J.Z., Starck, S.R., Sun D.A., Sabat, M. and Hecht, S.M. (2000). A new 7,8-Euphadien type triterpenoid from *Brackenridgea nitida* and *Bleasdalea bleasdalei* that inhibits DNA polymerase beta. *J. Nat. Prod.* 63: 1356-1360.

Dhanabalasingham, B., Karunaratne V., Gunatilaka, A.A.L., Tezuka Y. and Kikuchi, T. (1996). Biogenetically important quinonemethides and other triterpenoid constituents of *Salacia reticulata*. *Phytochem.* 42: 1377-1385.

Domon, B. and Hostettmann, K. (1983). Saponins with molluscicidal properties from *Lonicera nigra* L. *Helv. Chim. Acta* 66: 422-428.

Domon, B. and Hostettmann, K. (1984). New saponins from *Phytolacca dodecandra* Lherit. *Helv. Chim. Acta* 67: 1310-1315.

El-Seedi, H.R., Hazell, A.C. and Torssell, K.B.G. (1994). Triterpenes, lichexanthone and an acetylenic acid from *Minquartia guianensis*. *Phytochem.* 35: 1297-1299.

Elyakov, G.B., Anisimov, M.M., Prokof'eva, N.G., Kuznetsova, T.A. and Fronert, E.B. (1972). Sensitivity of rat marrow cells in culture to the toxic effect of stichoposide A_1 from *Stichopus japonicus* selenka. *Toxicon.* 10: 299-300. [1972, *Chem. Abstr.* 77, 57344j]

Estrada, R., Cardenas, J., Esquivel, B. and Rodriguez-Hahn, L. (1994). D-A-friedo-oleanane triterpenes from the roots of *Acanthothamnus aphyllus*. *Phytochem.* 36: 747-751.

Fang, S.Y., He, Z.S., Fan, G.J., Wu, H.M. and Xu, J.F. (1996). Triterpenoids from *Adina rubella*. *J. Nat. Prod.* 59: 304-307.

Fenselau, C. (1984). Fast Atom Bombardment and Middle Molecule Mass-Spectrometry. *J. Nat. Prod.* 47: 215-225.

Fu, J. and Zhang, M. (1995). Studies on the chemical constituents of the chirese soft corals. *Huanan Ligong Daxue Xuebao, Ziran Kexueban,* 23: 120-124. [(1995) *Chem. Abstr.* 123, 139301a]

Fukuda, K., Utsumi, H., Shoji, J. and Hamada, A. (1985). Saponins can cause the agglutination of phospholipid vesicles. *Biochim. Biophys. Acta.* 820: 199-206.

Fullas, F., Wani, M.C., Wall, M.E., Tucker, J.C., Beecher, C.W.W. and Kinghorn, A.D. (1996). Triterpenes from the combined leaf and stem of *Lithospermum caroliniense*. *Phytochem.* 43: 1303-1305.

Gao, C.W., Chen, Y.S., Zhao, S.N., Lu, Y., Tian, Z.Y. and Zheng, Q.T. (1994). A new norpentacyclic triterpenoid auriculatone from *Aster auriculatus* Franch. *Gaodeng Xuexiao Huaxue Xuebao* 15: 521-523. [(1994) *Chem. Abstr.* 121, 297129p]

Gaydou, E.M., Faure, R. and Wollenweber, E. (1996). β-amyrin acetate epoxide from *Canarina canariensis*. *Phytochem.* 42: 1115-1118.

Gladkikh, A.S. (1966). Action of saponin from *Cyclamen ibericum* on some parameters of the condition of cardiovascular system of the animals. *Farmakol. Toksikol.* 29: 559-562.

González, A.G., Alvarenga, N.L., Estévez-Braun, A., Ravelo, A.G., Bazzocchi, I.L., and Moujir, L.M. (1996). Structure and absolute configuration of triterpene dimers from *Maytenus scutioides*. *Tetrahedron* 52: 9597-9608.

González, A.G., Alvarenga, N.L., Ravelo, A.G., Bazzocchi, I.L., Ferro, E., Navarro, A.G. and Moujir, L.M. (1996). Scutione, A new bioactive norquinonemethide triterpene from *Maytenus scutioides* (Celastraceae). *Bioorgan. Med. Chem.* 4: 815-820.

González, A.G., Alvarenga, N.L., Ravelo, A.G., Jiménez, I.A., Bazzocchi, I.L., Canela, N.J. and Moujir, L.M. (1996). Antibiotic phenol nor-triterpenes from *Maytenus canariensis*. *Phytochem.* 43: 129-132.

González, A.G., Alvarenga, N.L., Rodriguez, F., Ravelo, A.G., Jimenez, I.A., Bazzocchi, I.L. and Gupta, M. (1995). New phenolic and quinone-methide triterpenes from *Maytenus* species (Celastraceae). *Nat. Prod. Lett.* 7: 209-218.

González, A.G., Kennedy, M.L., Rodriguez F.M., Bazzocchi, I.L., Jimenez, I.A., Ravelo, A.G. and Moujir, L. (2001). Absolute configuration of triterpene dimers from *Maytenus* species (Celastraceae). *Tetrahedron* 57: 1283-1287.

González, A.G., Valencia, E., Exposito, T.S., Barrera, J.B. and Gupta, M.P. (1994). Chemical-components of *Cuphea* species-carthagenol-A new triterpene from *Cuphea carthagenensis*. *Planta Med.* 60: 592-593.

Goulart, M.O.F., Santana, A.E.G., Alves, R.J., De S., Filho, J.D., Maia, J.G.S., De Oliveira, G.G. and De Oliveira, A.B. (1994). Chemistry of Brazilian Icacinaceae. 1. Icacinic acid, a triterpenoid from *Poraqueiba guianensis*. *Phytochem.* 37: 1139-1142.

Hayashi, H., Asabu, Y., Naito, K., Nakayama, M., Nozaki, H. and Arai, M. (1996). New oleanane triterpene with three ketones produced by *Penicillium simplicissimum* ATCC 90288. *Biosci. Biotech. Biochem.* 60: 1732-1734.

He, L. and Meng, Z., (1995). Studies on chemical constituents of *Alternanthera philoxeroides* (Mort) griseb. *Zhongguo Yaoke Daxue Xuebao* 26: 263-267. [(1996) *Chem. Abstr.* 124, 170617c]

Higuchi, R. and Kawasaki, T. (1972). Seed saponins of *Akebia quinata* decne. 2. Hederagenin 3,28-O-bisglycosides. *Chem. Pharm. Bull. Jpn.* 20: 2143-2149.

Higuchi, R. and Kawasaki, T. (1976). Pericarp saponins of *Akebia quinata* decne. 2. Arjunolic and nor-arjunolic acids, and their glycosides. *Chem. Pharm. Bull. Jpn.* 24: 1314-1323.

Higuchi, R., Miyahara, K. and Kawasaki, T. (1972). Seed saponins of *Akebia quinata* decne. 1. Hederagenin 3-O-glycosides. *Chem. Pharm. Bull.* 20: 1935-1943.

Hiller, K. and Adler, C. (1982). New findings on triterpene saponins. A review. *Pharmazie* 37: 619-634.

Hisham, A., Jayakumar, G., Fujimoto, Y. and Hara, N. (1996). 1β,15α-dihydroxyfriedelan-3-one, a triterpene from *Salacia beddomei*. *Phytochem.* 43: 843-845.

Hiura, A., Akabane, T., Ohtani, K., Kasai, R., Yamasaki, K. and Kurihara, Y. (1996). Taste-modifying triterpene glycosides from *Staurogyne merguensis*. *Phytochem.* 43: 1023-1027.

Hostettmann, K. (1980). Saponins with molluscicidal activity from *Hedera helix* L. *Helv. Chim. Acta* 63: 606-609.

Hu, J.F., Ye, Z.L. and. Shen, F.J. (1995). New triterpenoidal sapogenins from the roots of *Glycyrrhiza yunnanensis*. *Yaoxue Xuebao* 30: 27-33. [(1995) *Chem. Abstr.* 123, 222884p]

Iskenderov, G.B. (1971a). Tritepenoid glycosides from *Hedera pastuchovii*. IV. Structure of pastuchoside B. *Khim. Prir. Soedin.* 7: 425-429. [(1971) *Chem. Abstr.* 75, 152041a]

Iskenderov, G.B. (1971b). Tritepenoid glycosides of *Hedera pastuchovii*. *Farmatsiya* 20: 27-30. [(1971) *Chem. Abstr.* 75, 106045]

Jansakul, C., Baumann, H., Kenne, L. and Samuelsson, G. (1987). Ardisiacrispin A and B, two utero-contracting saponins from *Ardisia crispa*. *Planta Med.* 53: 405-409.

Jayasinghe, L.G., Wannigama, P. and Macleod, J.K. (1993). Triterpenoids from *Anamirta cocculus*. *Phytochem.* 34: 1111-1116.

Jeener, J., Meier, B.H., Bachmann, P. and Ernst, R.R. (1979). Investigation of exchange processes by 2-dimensional Nmr spectroscopy. *J. Chem. Phys.* 71: 4546-4553.

Jia, Z., Koike, K., Ohmoto, T. and Ni, M. (1994). Triterpenoid saponins from *Ardisia crenata*. *Phytochem.* 37: 1389-1396.

Jiang, Z.H., Tanaka, T. and Kouno, I. (1994). Two diastereomeric triterpene-lignan esters having dimeric structure and their biosynthetically related triterpene caffeate from *Rhoiptelea chiliantha*. *Tetrahedron Lett.* 35: 2031-2034.

Jiang, Z.H., Tanaka, T. and Kouno, I. (1995). A lupane triterpene and two triterpene caffeates from *Rhoiptelea chiliantha*. *Phytochem.* 40: 1223-1226.

Jossang, A., Seuleiman, M., Maidou, E. and Bodo, B. (1988). Pentacyclic triterpenes from *Combretum nigricans*. *Phytochem.* 41: 591-594.

Ju-Hyun, J., Hai-Jung, K., Yu-Ryang, P. and Sung-Hi, N. (1977). Studies on the extraction of Korean ginseng component. Part 1. Differences of saponins in Korean ginseng by cultivation area and processing. *Han'guk Sikp'um Kwahakhoe Chi.* 9: 13. [(1978) *Chem. Abstr.* 89, 20319]

Kan, Y., Zhao, H., Liu, X., Zhu, Y., Xia, R., Zhu, Y. and Lin, M. (1994). Chemical constituents of prickly fruit licorice *(Glycyrrhiza pallidiflora)*. *Zhongcaoyao* 25: 3-6. [(1994) *Chem. Abstr.* 121, 91431x]

Kasai, R., Okihara, M., Asakawa, J., Mizutani, K. and Tanaka, O. (1979). ^{13}C-nmr study of α-anomeric and β-anomeric pairs of D-Mannopyranosides and L-Rhamnopyranosides. *Tetrahedron* 35: 1427-1432.

Kemoklidze, Z.S., Dekanosidze, G.E., Gikia, D.D., Vugalter, M.M. and Kemertelidze, E.P. (1982). Triterpene glycosides of fresh leaves *Fatsia japonica*. *Khim. Prir. Soedin.* 18: 788-789.

Kemoklidze, Z.S., Dekanosidze, G.E., Gikia, D.D., Vugalter, M.M. and Kemertelidze, E.P. (1984). Triterpenic glycosides from the leaves of *Fatsia japon-*

ica—Structure of fatsiaside-D, fatsiaside-E and fatsiaside-F. *Khim. Prir. Soedin.* 20: 252-253.

Khan, I.A., Clark, A.M. and McChesney, J.D. (1997). Antifungal activity of a new triterpenoid glycoside from *Pithecellobium racemosum*. *Pharm. Res.* 14: 358- 361.

Kimata, H., Nakashima, T., Kokuban, S., Nakayama, K., Mitoma, Y., Kitahara, T., Yata, N. and Tanaka, O. (1983). Saponins of pericarps of *Sapindus mukurossi* gaertin and solubilization of monodesmosides by bisdesmosides. *Chem. Pharm. Bull.* 31: 1998-2005.

Kinoshita, K., Takizawa, T., Koyama, K., Takahashi, K., Kondo, N. Yuasa, H. and Kawai, K.I. (1995). New triterpenes from *Trichocereus pachanoi*. *J. Nat. Prod.* 58: 1739-1744.

Kintya, P.K., Melnikov, V.N. and Chirva, V.Y. (1974a). Structure of vitalboside G from *Clematis vitalba*. *Khim. Prir. Soedin* 10: 802-803. [(1975) *Chem. Abstr.* 82, 167467n]

Kintya, P.K., Melnikov, V.N. and Chirva, V.Y. (1974b). Structure of vitalboside D from *Clematis vitalba*. *Khim. Prir. Soedin* 10: 803-804. [(1975) *Chem. Abstr.* 82, 167468p]

Kirmizigul, S. and Anil, H. (1994a). An acidic triterpene glycoside from *Cephalaria transsylvanica*. *Phytochem.* 36: 1555-1556.

Kirmizigul, S. and Anil, H. (1994b). Transsylvanoside A, an acidic triterpene glycoside from *Cephalaria transsylvanica*. *Phytochem.* 35: 1075-1076.

Kirmizigul, S., Anil, H. and Rose, M.E. (1995). Triterpenoid glycosides from *Cephalaria transsylvanica*. *Phytochem.* 39: 1171-1174.

Kirmizigul, S., Anil, H. and Rose, M.E. (1996). Triterpenoid saponins from *Cephalaria transsylvanica*. *J. Nat. Prod.* 59: 415-418.

Kirmizigul, S., Anil, H., Ucar, F. and Akdemir, K. (1996). Antimicrobial and antifungal activities of three new triterpenoid glycosides. *Phytother. Res.* 10: 274-276.

Kirmizigul, S. and Rose, M.E. (1997). Hederagenin glycosides from the flowers of *Cephalaria transsylvanica*. *Planta Med.* 63: 51-54.

Kitagawa, I., Nishino, T., Kobayshi, M. and Kyogoku, M. (1981). Marine natural products. 8. Bioactive triterpene oligoglycosides from the sea cucumber *Holothuria leucospilota* Brandt (2)—Structure of holothurin-A. *Chem. Pharm. Bull.* 29: 1951-1956.

Kitagawa, I., Sugawara, T. and Yosioka, I. (1974). Structure of Holotoxin-A, a major antifungal glycoside of *Stichopus japonicus* selenka. *Tetrahedron Lett.* 15: 4111-4114.

Kitagawa, I., Sugawara, T. and Yosioka, I. (1976). Saponin and sapogenol. 15. Antifungal glycosides from sea cucumber *Stichopus japonicus* selenka. 2. Structures of holotoxin-A and holotoxin-B. *Chem. Pharm. Bull. Jpn.* 24: 275-284.

Kizu, H., Hirabayashi, S., Suzuki, M. and Tomimori, T. (1985). Studies on the constituents of *Hedera rhombea* bean. 4. On the hederagenin glycosides. 2. *Chem. Pharm. Bull.* 33: 3473-3478.

Kizu, H. and Tomimori, T. (1979). Studies on the constituents of *Clematis* species. 1. Saponins of the root of *Clematis chinensis* osbeck. *Chem. Pharm. Bull.* 27: 2388-2393.

Kizu, H. and Tomimori, T. (1980a). Studies on the constituents of *Clematis* species. 2. On the saponins of the root of *Clematis chinensis* osbeck. 2. *Chem. Pharm. Bull.* 28: 2827-2830.

Kizu, H. and Tomimori, T. (1980b). Studies on the constituents of *Clematis* species. 3. On the saponins of the root of *Clematis chinensis* osbeck. 3. *Chem. Pharm. Bull.* 28: 3555-3560.

Kizu, H. and Tomimori, T. (1982a). Studies on the constituents of *Clematis* species. 4. On the saponins of the root of *Clematis chinensis* osbeck. 4. *Chem. Pharm. Bull.* 30: 859-865.

Kizu, H. and Tomimori, T. (1982b). Studies on the constituents of *Clematis* species. 5. On the saponins of the root of *Clematis chinensis* osbeck. *Chem. Pharm. Bull.* 30: 3340-3346.

Konoshima, T. and Sawada, T. (1982). Legume saponins of *Gleditsia japonica* miquel. 4. C-13-Nuclear Magnetic Resonance spectral studies for structure elucidation of gleditsia saponin-B and saponin-C. *Chem. Pharm. Bull.* 30: 2747-2760.

Krivorutchenko, Y.L., Grishkovets, V.I., Andronovskaya, I.B. and Krivoshein, Y.S. (1999). Triterpene glycosides stimulation of humoral and cellular immune response against target antigens. Russ. RU 2127605 C1 from *Izobreteniya* (8): 323.

Krokhmalyuk, V. V., Kintya, P. K., Chirva, V. Y. and Boshko, Z. I. (1975). Structure of the principle saponin from *Clematis songarica*. *Khim. Prir. Soedin.* 11: 470-474. [(1975) *Chem. Abstr.* 83, 175469a]

Kuo, Y.H., Way, S.T. and Wu, C.H. (1996). A new triterpene and a new lignan from *Saussurea japonica*. *J. Nat. Prod.* 59: 622-624.

Kurilenko, M.I. (1968). *Pulsatilla nigricans* root glycoside. *Farm. Zh.* 23: 75-80. [(1969) *Chem. Abstr.* 70, 99583x]

Lee, H.S., Seo, Y., Cho, K.W., Rho, J.R., Shin, J. and Paul, V.J. (2000). New triterpenoid saponins from the sponge *Melophlus isis*. *J. Nat. Prod.* 63: 915-919.

Lee, S.W. (1974). Effects of ginseng [saponin] on the central nervous system. *Taehan Yakrihak Chapchi* 10: 85-95. [(1975) *Chem. Abstr.* 83, 37829s]

Leet, J.E., Huang, S., Klohr, S.E. and McBrien, K.D. (1996). Ascosteroside, a new antifungal agent from *Ascotricha amphitricha*. II. Isolation and structure elucidation. *Journal of Antibiotics* 49: 553-559.

Liang, G.Y., Wu, K.Y., Xu, R.S., Chen, D.Y. and He, Z.Y. (1995). Constituents of *Sabia swinhoei* Hemsl. ex Forb. Et Hemsl. *Yaoxue Xuebao* 30: 367-371. [(1995) *Chem. Abstr.* 123, 107765p]

Liu, Z., Li, D., Owen, N.L., Grant, D.M., Cates, R.G. and Jia, Z.J. (1995). Olenaine triterpene saponins from the Chinese medicinal herb *Clinopodium chinensis*. *J. Nat. Prod.* 58: 1600-1604.

Luo, H.S., Zhao, Y.Y., Ma, L.B., Peng, J.R. and Zhang, R.Y. (1995). Isolation and identification of saikogenin Q and saikosaponin Q from *Bupleurum smithii wolff* var. *parvifolium*. *Yaoxue Xuebao* 30: 435-439. [(1995) *Chem. Abstr.* 123, 222788k]

Ma, J., Starck, S.R. and Hecht, S.M. (1999). DNA polymerase β inhibitors from *Tetracera boiviniana*. *J. Nat. Prod.* 62: 1660-1663.

Macías, F.A., Simonet, A.M., Esteban, M.D. and Galindo, J.C.G. (1996). Triterpenoids from *Melilotus messanensis;* Soyasapogenol G, the first natural carbonate derivative. *Phytochem.* 41: 1573-1577.

Maeda, C., Ohtani, K., Kasai, R., Yamasaki, K., Duc. N.M., Nham, N.T. and Cu, N.K.Q. (1994). Oleanane and ursane glycosides from *Schefflera octophylla. Phytochem.* 37: 1131-1137.

Mahato, S. B. and Das, M. C. (1983). Triterpenoids. *Phytochem.* 22: 1071-1095.

Mahato, S.B., Sahu, N.P., Luger, P. and Muller, E. (1987). Stereochemistry of a triterpenoid trisaccharide from *Centella asiatica*—X-ray determination of the structure of asiaticoside. *J. Chem. Soc. Perkin Trans. II* (10): 1509-1515.

Mahato, S.B., Sarkar, S.K. and Poddar, G. (1988). Triterpenoid saponins. *Phytochem.* 27: 3037-3067.

Majinda, R.R.T., Gray, A.I., Waigh, R.D. and Waterman, P.G., (1995). A seco-olean-18-ene triterpene acid from *Vahlia capensis. Phytochem.* 38: 461-463.

Mondal, D.N., Barik, B.R., Dey, A.K., Patra, A. and Kundu, A.B. (1994). Holoptelin A and B, two new triterpenoid fatty acid esters from *Holoptelea integrifolia. Indian Drugs* 31: 69-72. [(1994) *Chem. Abstr.* 121, 5164h]

Morota, T., Ogino, T., Katsuhara, T., Yoshino, T., Maruno, M., Yang, C.X., Qin, W.Z., Xu L.H., Miao K.L. and Yang B.H. (1995). D:A-friedo-24-noroleanane triterpenoids from *Tripterigium wilfordii. Phytochem.* 39: 1159-1163.

Morota, T., Sasaki, H., Sugama, K., Yoshino, T., Maruno, M., Yang, C.X., Qin, W.Z., Miao, K.L., Xu, L.H. and Yang, B.H. (1995). Triterpenes from *Tripterigium wilfordii. Phytochem.* 39: 1153-1157.

Mshvildadze, V., Elias, R., Delmas, F., Decanosidze, G., Kemertelidze, E. and Balansard, G. (1998) Antiprotozoal activity of triterpene glycosides from *Hedera* L. *Bull. Georgian Acad. Sci.* 158: 276-278.

Mukhamedziev, M.M. and Alimbaeva, P.K. (1969). Triterpene glycosides of *Dipsacus azureus. Khim. Prir. Soedin.* 5: 451-452. [(1970) *Chem. Abstr.* 72, 129407]

Mukhamedziev, M.M., Alimbaeva, P.K., Gorovits, T.T. and Abubakirov, N.K. (1971). Structure of the triterpenoid glycoside from *Dipsacus azureus,* dipsocoside B. *Khim. Prir. Soedin* 7: 153-158. [(1971) *Chem. Abstr.* 75, 494947]

Mukherjee, K.S., Brahmachari G. and Manna, T.K. (1995). Triterpene from *Limnophila heterophylla. Phytochem.* 38: 1273-1274.

Murakami, T., Nagasawa, M., Urayama, S. and Satake, T. (1968). New triterpenoid saponins in the rhizome and roots of *Caulophyllum robustum. Yakugaku Zasshi* 88: 321-324. [(1968) *Chem. Abstr.* 69, 67661]

Mzhel'skaya, L.G. and Abubakirov, N.K. (1967). Triterpenic glycosides of *Leontice eversmanni.* II. The structure of leontoside A and leontoside B. *Khim. Prir. Soedin.* 3: 101-105. [(1967) *Chem. Abstr.* 67, 100383d]

Neal, R.A. (1967). Antigenic vaccine with saponin adjuvant for immunizing vertebrates against protozoal diseases. Great Britain: Wellcome Foundation Ltd. 1,083,815 (Cl. A 61k), 8 pp.

Neszmelyi, A., Tori, K. and Lukacs, G. (1977). Use of C-13 spin-lattice relaxation times for sugar sequence determination in steroidal oligosaccharides. *J. Chem. Soc. Chem. Comm.* (17): 613-614.

Ngadjui, B.T., Donga, E., Ayafor, J.F. and Connolly, J.D. (1994). Thomandertriol, A triterpenoid from the twigs of *Thomandersa laurifolia*. *J. Nat. Prod.* 57: 161-163.

Nick, A., Wright, A.D. and Sticker, O. (1994). Antibacterial triterpenoid acids from *Dillenia papuana*. *J. Nat. Prod.* 57: 1245-1250.

Nick, A., Wright, A.D., Sticker, O. and Rali, T. (1995). Antibacterial triterpenoids from *Dillenia papuana* and their structure-activity relationships. *Phytochem.* 40: 1691-1695.

Nishino, C., Manabe, S., Enoki, N., Nagata, T., Tsushida, T. and Hamaya, E. (1986). The structure of the tetrasaccharide unit of Camellidins, saponins possessing antifungal activity. *J. Chem. Soc. Chem. Comm.* (9): 720-723.

Öksüz, S., Gil, R.R., Chai, H.B., Pezzuto, J.M., Cordell, G.A. and Ulubelen, A. (1994). Biologically active compounds from the *Euphorbiaceae*. 2. Two triterpenoids of *Euphorbia cyparissias*. *Planta Med.* 60: 594-596.

Ortega, A., Cárdenas, J., Maldonado, E. and Gage, D.A. (1995). Abietane and clerodane diterpenes from *Salvia regal*. *Phytochem.* 39: 931-933.

Oura, H., Hiai, S., Odaka, Y. and Yokozawa, T. (1975). Studies on the biochemical action of ginseng saponin. I. Purification from ginseng extract of the active component stimulating serum protein biosynthesis. *J. Biochem. Tokyo* 77: 1057-1065.

Pakkhullaeva, M., Mzhel'skaya, L.G. and Abubakirov, N.K. (1972). Tritepenoid glycosides of *Ladyginia bucharica*. II. Structure of ladyginosides A and B. *Khim. Prir. Soedin.* 8: 466-471. [(1975 *Chem. Abstr.* 78, 16445v]

Pakkhullaeva, M., Mzhel'skaya, L.G. and Abubakirov, N.K. (1973). Tritepenoid glycosides of *Ladyginia bucharica*. IV. Structure of ladyginosides D and F. *Khim. Prir. Soedin.* 9: 733-739. [(1975 *Chem. Abstr.* 82, 28554 a]

Pan, H., Lundgrea, L.N. and Andersson, R. (1994). Triterpene caffeates from bark of *Betula pubescens*. *Phytochem.* 37: 795-799.

Patt, S.L. (1984). 2-Dimensional NMR in Carbohydrate Structural Analysis. *J. Carbohyd. Chem.* 3: 493-511.

Pelaez, F., Cabello, A., Platas, G., Diez, M.T., Del Val, A.G., Basilio, A., Martan, I., Vicente, F., Bills, G.F., Giacobbe, R.A., et al. (2000). The discovery of enfumafungin, a novel antifungal compound produced by an endophytic *Hormonema* species biological activity and taxonomy of the producing organisms. *Systematic and Applied Microbiology* 23: 333-343.

Pengsuparp, T., Cai, L., Fong, H.H.S., Kinghorn, A.D., Pezzuto, J.M., Wani, M.C. and Wall, M.E. (1994). Pentacyclic triterpenes derived from *Maprounea africana* are potent inhibitors of HIV-1 reverse transcriptase. *J. Nat. Prod.* 57: 415-418.

Pfander, H. and Stoll, H. (1991). Terpenoid glycosides. *Nat. Prod. Rep.* 8: 69-95.

Pivkin, M.V. (2000). Filamentous fungi associated with *holothurians* from the Sea of Japan, off the Primorye Coast of Russia. *Biological Bulletin Woods Hole* 198: 101-109.

Roldan, A.T., Perez, E.A. and Ruiz, R.M. (1999). Tomatinase from *Fusarium oxysporum* f. sp. *lycopersici* defines a new class of saponinases. *Molecular Plant Microbe Interactions* 12: 852-861.

Ryuichi, H. and Toshio, K. (1976). Hederagenin Glycosides of *Akebia quinata*. *Chem. Pharm. Bull. Jpn.* 24: 1021-1033.

Sahu, N.P., Mahato, S.B., Sarkar, S.K. and Poddar, G. (1996). Triterpenoid saponins from *Gymnema sylvestre*. *Phytochem.* 41: 1181-1185.

Sanchez-Contreras, S., Diaz-Lanza, A.M. and Bernabe, M. (2000). Four new triterpenoid saponins from the roots of *Bupleurum rigidum*. *J. Nat. Prod.* 63: 1479-1482.

Schloesser, E. and Wulff, G. (1969). Structural specifity of saponin hemolysis. I. Triterpene saponins and aglycones. *Z. Naturforsch B.* 24: 1284-1290.

Schöpke, T., Thiele, H., Wray, V., Nimtz, M. and Hiller, K. (1995). Structure elucidation of a glycoside of 2β, 3β, 23-trihydroxy-16-oxoolean-12-en-28-oic acid from *Bellis bernardii* using mass spectrometry for the sugar sequence determination. *J. Nat. Prod.* 58: 152-155.

Schwartz, R.E., Smith, S.K., Onishi, J.C., Meinz, M., Krutz, M., Giacobbe, R.A., Wilson, K.E., Liesch, J., Zink, D., Horn, W., et al. (2000). Isolation and structural determination of enfumafungin, a triterpene glycoside antifungal agent that is a specific inhibitor of glucan synthesis. *J. Am. Chem. Soc.* 122: 4882-4886.

Segum, S., Raza, S.M., Siddiqui, B.S. and Siddiqui, S. (1995). Triterpenoids from the aerial parts of *Lantana camara*. *J. Nat. Prod.* 58: 1570-1574.

Sen, S., Sahu, N.P. and Mahato, S.B. (1995). Pentacyclic triterpenoids from *Mimusops elengi*. *Phytochem.* 38: 205-207.

Seo, S., Tomita, Y., Tori, K. and Yoshimura, Y. (1978). Determination of the absolute configuration of a secondary hydroxyl group in a chiral secondary alcohol using glycosidation shifts in carbon-13 nuclear magnetic resonance spectroscopy. *Am. Chem. Soc.* 100: 3331-3339.

Shaari, K. and Waterman, P.G. (1996). D:A-friedo-oleanane triterpenes from the stem of *Homalium longifolium*. *Phytochem.* 41: 867-869.

Shany, S., Gestetne, B., Birk, Y., Bondi, A. and Kirson, I. (1972). Alfalfa saponins. 8. Isolation of hederagenin and its saponin from alfalfa *(Medicago-sativa)*. *Isr. J. Chem.* 10: 881-886.

Shimizu, M., Shingyouchi, K., Morita, N., Kizu, H. and Tomimori, T. (1978). Triterpenoid saponins from *Pulsatilla cernua* spreng. 1. *Chem. Pharm. Bull. Jpn.* 26: 1666-1671.

Shiojima, K., Suzuki, H. and Ageta, H. (1995). Composite constituents—3 new triterpene triols isolated from fresh roots of *Picris hieracioides* Subsp. *japonica*. *Chem. Pharm. Bull.* 43: 1640-1642.

Shiojima, K., Suzuki, H., Kodera, N., Ageta, H., Chang, H.C. and Chen, Y.P. (1995). Composite constituent—Novel triterpenoid, ixerenol, from aerial parts of *Ixeris chinensis*. *Chem. Pharm. Bull.* 43: 180-182.

Shirota, O., Morita, H., Takeya, K. and Itokawa, H. (1995). Structures of xuxuarines, stereoisomeric triterpene dimers from *Maytenus chuchuhuasca*. *Tetrahedron* 51: 1107-1120.

Shirota, O., Morita, H., Takeya, K., Itokawa, H. and Iitka, Y. (1994). Cytotoxic aromatic triterpenes from *Maytenus ilicifolia* and *Maytenus chuchuhuasca*. *J. Nat. Prod.* 57: 1675-1681.

Shirota, O., Tamemura, T., Morita, H., Takeya, K. and Itokawa, H. (1996). Triterpenes from Brazilian medicinal plant 'chuchuhuasi' *(Maytenus krukovii). J. Nat. Prod.* 59: 1072-1075.

Siddiqui, B.S., Ahmed, S., Ghiasuddin, S. and Khan, M.A.U. (1994). Triterpenoids of *Atriplex stocksii. Phytochem.* 37: 1123-1125.

Siddiqui, B.S., Raza, S. M., Begum, S., Siddiqui, S. and Firdous, S. (1995). Pentacyclic triterpenoids from *Lantana camara. Phytochem.* 38: 681-685.

Sindambiwe, J.B., Baldé, A.M., De Bruyne, T., Pieters, L., Van den Heuvel, H., Claeys, M., Van Den Berghe, D.A., and Vlietinck, A.J. (1996). Triterpenoid saponins from *Maesa lanceolata. Phytochem.* 41: 269-277.

Spengel, S.M. (1996). Two pentacyclic triterpenes from *Phytolacca dodecandra* roots. *Phytochem.* 43: 179-182.

Srivastava, S.K., Srivastava, S.D. and Chouksey, B.K. (2001). New antifungal constituents from *Terminalia alata. Fitoterapia* 72: 106-112.

Strigina, L.I., Chetyrina, N.S., Elyakov, G.B. (1970). Triterpene glycosides of *Caulophyllum robustum. Khim. Prir. Soedin.* 6: 552-555. [(1971) *Chem. Abstr.* 74, 50520t]

Strigina, L.I., Chetyrina, N.S. and Isakov, V.V. (1976). Cauloside G, a new triterpene glycoside from *Caulophyllum robustum*—Identification of cauloside C. *Khim. Prir. Soedin.* 12: 619-623. [(1977) *Chem. Abstr.* 86, 152616u]

Strigina, L.I., Chetyrina, N.S., Lepshkina, M.K., Kol'chuk, E.M., Remennikova, T.M. and Elyakov, G.B. (1970). Triterpene glycoside of *Caulophyllum robustum. Aktual. Probl. Izuch. Efirnomaslich. Rest. Efirm.* 166, from *Ref. Zh. Khim.* (1971). Abstr. No. 1R515. [(1972) *Chem. Abstr.* 76, 83537]

Sun, D.A., Deng, J.Z., Starck, S.R. and Hecht, S.M. (1999). Mispyric acid, a new monocyclic triterpenoid with a novel skeleton from *Mischocarpus pyriformis* that inhibits DNA polymerase beta. *J. Am. Chem. Soc.* 121: 6120-6124.

Sun, Y.R. and Sun, Y.F. (1996). Structure elucidation of liquidambaric lactone. *Yaoxue Xuebao* 31: 437-440.

Takechi, M., Kousaka, Y., Uno, C. and Tanaka, Y. (1998). Biological activities of synthetic triterpenoid and steroid beta-D-glucopyranosyl-(1,2)-alpha-D-glucopyranosides. *Planta Med.* 64: 179-180.

Takechi, M., Uno, C. and Tanaka, Y. (1996). Structure-activity relationships of synthetic saponins. *Phytochem.* 41: 121-123.

Takechi, M., Uno, C. and Tanaka, Y. (1997). Biological activities of synthetic triterpenoid and steroid beta-D-xylopyranosyl-(1,6)-beta-D-glucopyranosides. *Phytochem.* 44: 299-303.

Takizawa, T., Kinoshita, K., Koyama, K., Takahashi, K., Kondo, N. and Yuasa, H. (1995). A new triterpene from *Rathbunia akamosensis. J. Nat. Prod.* 58: 1913-1914.

Tanaka, R., Ida, T., Takaoka, Y., Kita, S., Kamisako, W. and Matsunaga, S. (1994). 3,4-seco-oleana-4(23),18-dien-3-oic acid and other triterpenes from *Euphorbia chamaesyce. Phytochem.* 36: 129-132.

Tanaka, R., In, Y., Ishida, T. and Matsunaga, S. (1994). 11-β-hydroxy-D:A-friedoolean-1-en-3one from the stem bark of *Phyllanthus flexuosus. J. Nat. Prod.* 57: 1523-1528.

Tapondjou, A.L., Ngounou, N.F., Lontsi, D., Sondengam, B.L., Martin, M.T. and Bodo, B. (1995). Pentacyclic triterpenes from *Myrianthus liberecus*. *Phytochem.* 40: 1761-1764.

Tchuendem, M.H.K., Ayafor, J.F. and Connolly, J.D. (1996). Lophocarpin and 21-β-hydroxylophocarpin, two new 30-norfriedelane triterpenes from *Caloncoba lophocarpa*. *Nat. Prod. Lett.* 9: 27-32.

Texier, O., Ahond, A., Regerat, F. and Pourrat, H. (1984). A triterpenoid saponin from *Ficaria ranunculoides* tubers. *Phytochem.* 23: 2903-2905.

Tezuka, Y., Kikuchi, T., Dhanabalasingham, B., Karunaratne, V. and Gunatilaka, A.A.L. (1994). Studies on terpenoids and stereoids. 25. Complete ^{1}H-and ^{13}C-NMR spectral assignments of salaciquinane, a new 7-oxo-quinonemethide dinortriterpenoid. *J. Nat. Prod.* 57: 270-276.

Timbekova, A.E., Isaev, M.I. and Abubakirov, N.K. (1996). Chemistry and biological activity of triterpenoid glycosides from *Medicago sativa*. *Adv. Exp. Med. Biol.* 405: 171-182.

Tokuda, H., Konoshima, T., Kozuka, M. and Kimura, T. (1988). Inhibitory effects of 12-O-tetradecanoylphorbol-13-acetate and teleocidin B induced Epstein-Barr virus by saponin and its related compounds. *Cancer Lett.* 40: 309-317.

Topcu, G., Ulubelen, A. and Eri, C. (1994). Diterpenoids and triterpenoids of *Salvia pomifera*. *Phytochem.* 36: 743-745.

Tori, M., Aoki, M., Nakashima, K. and Asakawa, Y. (1995). Terpenoids from the liverworts *Symphyogyna brasiliensis* and unidentified *Frullania* species. *Phytochem.* 39: 99-103.

Uchendu, C.N., Kamalu, T.N. and Asuzu, I.U. (2000). A preliminary evaluation of antifertility activity of a triterpenoid glycoside (DSS) from *Dalbergia saxatilis* in female Wistar rats. *Pharmacol. Res.* 41: 521-525.

Ulubelen, A., Topcu, G., Sönmez, U. and Eri, C. (1994). Terpenoids from *Salvia nemorosa*. *Phytochem.* 35: 1065-1067.

Van der Kaaden, A., Boon, J.J. and Haverkamp, J. (1984). The analytical pyrolysis of carbohydrates. 2. Differentiation of homopolyhexoses according to their linkage type, by pyrolysis—Mass spectrometry and pyrolysis—Gas chromatography—Mass spectrometry. *Biomed. Mass Spectrom.* 11: 486-492.

Vogel, G., Marek, M.L. and Oertner, R. (1968). Pharmacology of the tea seed saponin, a saponin mixture from *Thea sinensis*. *Arzneim. Forsch.* 18: 1466-1467. [1969, *Chem. Abstr.* 70, 36279g]

Wagner, H. and Horhammer, L. (1971). *Pharmacognosy and Phytochemistry*. Berlin: Springer.

Waltho, J.P., Williams, D.H., Mahato, S.B., Pal, B.C. and Barna, J.C.J. (1986). Structure elucidation of 2-triterpenoid tetrasaccharides from *Androsace saxifragifolia*. *J. Chem. Soc. Perkin Trans. I* (8): 1527-1531.

Williams, D.H., Bradley, C., Bojesen, G., Santikarn, S. and Taylor, L.C.E. (1981). Fast atom bombardment mass-spectrometry—A powerful technique for the study of polar-molecules. *J. Am. Chem. Soc.* 103: 5700-5704.

Yahara, S., Kasai, R. and Tanaka, O. (1977). New dammarane type saponins of leaves of *Panax japonicus* Meyer, Ca. 1. Chikusetsusaponins-L5, chikusetsusaponins-L9A and chikusetsusaponins-L10. *Chem. Pharm. Bull. Jpn.* 25: 2041-2047.

Yahara, S., Tanaka, O. and Komori, T. (1976). Saponins of leaves of *Panax ginseng* Meyer, Ca. *Chem. Pharm. Bull. Jpn.* 24: 2204-2208.

Yaun, X., Wang, G. and Gong, F. (1994). Studies on triterpenoid constituents isolated from the roots of *Sabia schumanniana. Zhiwu Xuebao* 36: 153-158. [(1994) *Chem. Abstr.* 121, 226448f]

Yoshiki, Y., Kim, J. H., Okubo, K., Nagoya, I., Sakabe, T. and Tamura, N. (1995). A saponin conjugated with 2,3-dihydro-2,5-dihydroxy-6-methyl-4H-pyran-4-one from *Dolichos lablab. Phytochem.* 38: 229-231.

Yoshimitsu, H., Nishida, M. and Nohara, T. (1999). Studies on cycloartane glycosides from genus *Thalictrum* plants. Tennen Yuki Kagobutsu Toronkai Koen Yoshishu, 41st, *Nippon Kagakki Bulletin,* pp. 349-354.

Zhang, D., Miyase, T., Kuroyanagi, M., Umehara, K. and Ueno, A. (1995). Studies on the constituents of *Polygala japonica* Houtt. 2. Structures of polygalasaponins XI-XIX. *Chem. Pharm. Bull.* 43: 966-970.

Zhang, Y., Shen, J., Song, J., Wang, Y. and Li, D. (1984). Analgesic and sedative effects of *Astragalus* saponin I. *Nanjing Yixueyuan Xuebao* 4: 225-227. [(1986) *Chem. Abstr.* 104, 122973f]

Zhang, Y.J. and Yang, C.R. (1994). 2-triterpenoids from *Gentiana tibetica. Phytochem.* 36: 997-999.

Zhao, W., Xu, R., Qin, G., Vaisar, T. and Lee, M.S. (1996). Saponins from *Mussaenda pubescens. Phytochem.* 42: 1131-1134.

Zhao, Y., Tian, J., Jia, Z. and Sun, H. (1995). Liguveitoside A, a new triterpenoid saponin from *Ligularia veitchiana. Yunnan Zhiwu Yanöiu* 17: 356-358. [(1996) *Chem. Abstr.* 124, 25611s]

Zhou, J. H. (2000). Herbal sweetening and preservative composition comprising licorice extract and mogrosides obtained from plants belonging to *Cucurbitaceae* and/or *Momordica.* U.S. Patent Number US 6103240 A, pp. 6.

Zviadadze, L.D., Dekanosidze, G.E., Gikia, O.D. and Kemertelidze, E.P. (1983). Triterpene glycosides of *Cephalaria gigantea.* 3. The structure of glycoside E and glycoside H. *Khim. Prir. Soedin.* 19: 46-49.

Chapter 23

Structure, Function, and Biological Activity of Rice Phytoalexins and Elicitors

Jinichiro Koga

INTRODUCTION

When plants interact with certain pathogens, they protect themselves by generating various chemical and physical barriers called the hypersensitive response (Keen, 1990). During the hypersensitive response, recognition of a pathogen triggers the activation of a cell-death pathway that results in the formation of a zone of dead cells (hypersensitive cell death) around the site of infection. This is accompanied by the accumulation of antimicrobial compounds called phytoalexins, the induction of pathogenesis-related proteins (chitinases, β-1,3-glucanases and proteinase inhibitors), increased expression of defense-related genes, the formation of lignin, a burst of active oxygen, and increased resistance to subsequent infection by pathogens

---

The author wishes to thank Dr. Takashi Adachi (Meiji Seika Kaisha, Ltd.) for support during the writing of this chapter. Thanks are also due to: Dr. Masaru Shimura, Mr. Toyozo Yamauchi, Ms. Noriko Ogawa, Mrs. Kiyomi Kagami, and Dr. Nagahiro Ogasawara (Plant Biological Defense System Laboratories); Dr. Shuichi Gomi, Mr. Kenji Umemura, Mrs. Naoyo Takahashi, and Mr. Yoshihisa Tomada (Meiji Seika Kaisha, Ltd.); Mr. Tarozaemon Kera (Konokan High School); and Mr. Osamu Kawakami (Miigata Agricultural Experiment Station), for providing helpful advice and useful information. This article is dedicated to the memory of the author's grandmother, Mrs. Fuji Ozawa, who passed away on November 11, 1998.

Abbreviations: cerebroside A = (4*E*,8*E*)-*N*-D-2'-hydroxy-(*E*)-3'-hexadecenoyl-1-*O*-β-D-glucopyranosyl-9-methyl-4,8-sphingadienine; cerbroside B = (4*E*,8*E*)-*N*-D-2'-hydroxypalmitoyl-1-*O*-β-D-glucopyranosyl-9-methyl-4,8-sphingadienine; cerbroside C = (4*E*, 8*E*)-*N*-D-2'-hydroxy-(*E*)-3'-octadecenoyl-1-*O*-β-D-glucopyranosyl-9-methyl-4,8-sphingadienine; cerebroside D = (4*E*,8*E*)-*N*-D-2'-hydroxystearoyl-1-*O*-β-D-glucopyranosyl-9-methyl-4,8-sphingadienine

(Dixon and Harrison, 1990; Keen, 1992; Staskawicz et al., 1995). Among these defense responses, phytoalexin production is thought to play an important role in the defense system against pathogenic fungi in rice. Numerous phytoalexins, momilactones A and B, oryzalexins A-F and S, phytocassanes A-E, and sakuranetin, have been isolated and characterized in rice plants (Harborne, 1999). Momilactones, oryzalexins, and phytocassanes are classified as diterpenes, whereas sakuranetin is classified as a flavonoid. A notable characteristic of these rice phytoalexins is high antifungal activity due to their specific structures.

Phytoalexin synthesis in plants is not only induced by molecules called elicitors, derived from culture filtrates, cell walls, and membranes of pathogens, but is induced by chemical and physical stresses also (Keen, 1975; Côté and Hahn, 1994; Ebel and Cosio, 1994). In a strict sense, elicitors derived from pathogens or plants are called biotic elicitors, whereas chemical and physical stresses are called abiotic elicitors. In rice, examples of chemical and physical stresses that induce phytoalexins are ultraviolet irradiation and $CuCl_2$. Although elicitors derived from rice pathogens have been studied extensively, there have been no reports on the determination of their chemical structures. Recently, however, the chemical structures of two types of elicitors derived from pathogens have been determined.

Cerebrosides A and C were isolated as sphingolipid elicitors of hypersensitive cell death and phytoalexin accumulation in rice from the fungus *Magnaporthe grisea* (Koga, Yamaguchi, et al., 1998). In animals, sphingolipids are known to play important roles in the regulation of cell growth, differentiation, and apoptosis (Hakomori, 1990; De Maria et al., 1997). Furthermore, sphingolipid metabolic products, called ceramides, function as second messengers in the signal transduction pathway involved in apoptosis (Obeid et al., 1993; Hannun, 1996). In light of recent studies indicating that hypersensitive cell death is a type of apoptosis (Dangl et al., 1996; Greenberg, 1996), it seems plausible that hypersensitive cell death in plants may be mediated by the same mechanism that is responsible for the sphingolipid or ceramide-mediated apoptosis in animals.

β-glucooligosaccharides from β-glucan that act as elicitors of phytoalexin synthesis in suspension-cultured rice cells were isolated from *M. grisea,* and their structures were identified as reduced compounds, tetraglucosyl glucitols (Yamaguchi et al., 2000). Tetraglucosyl glucitols are structurally different from β-glucooligosaccharide elicitors in soybeans (Sharp, Valent, et al., 1984; Sharp, McNeil, et al., 1984) and have hardly any elicitor activity in soybean. This is the first study to show that the structure of β-glucooligosaccharide elicitors recognized by each plant is quite different.

Studies of rice phytoalexins and elicitors are exciting and are going to open new areas of research in the field of plant pathology. This chapter will describe the research on the structure, function, and biological activity of rice phytoalexins and elicitors.

PHYTOALEXINS

Momilactones

WL-28325 (2,2-dichloro-3,3-dimethyl-cyclopropane carboxylic acid) is known to induce resistance against rice blast disease caused by *M. grisea.* Cartwright et al. (1977) found that momilactones A and B, initially isolated as plant growth inhibitors from rice husks (Kato et al., 1973), are produced in *M. grisea*-infected rice leaves previously treated with WL-28325 (Cartwright et al., 1977, 1981). The structures of momilactones A and B, classified as pimarane-type diterpenes, are shown in Figure 23.1.

Momilactones A and B have high antifungal activity against the pathogenic fungus *M. grisea.* The ED_{50} values of momilactones A and B in prevention of *M. grisea* spore germination were found to be 15 and 3 μg/ml, respectively (see Figure 23.1). In light of the low antifungal activity of oryzalexins, which are structurally similar to momilactones, but lack the lactone skeleton, it is quite likely that the lactone skeleton of momilactones contributes to the high antifungal activity.

Since momilactone A, among the various rice phytoalexins, is the main phytoalexin produced in some cases, for example, in rice leaves infected with *M. grisea* (Kato et al., 1993; Kato et al., 1994; Dillon et al., 1997), in rice stems infected with *Rhizoctonia solani* (Koga et al., 1995), or in rice leaves treated with various elicitors (Koga, Yamaguchi, et al., 1998; Umemura et al., 2000), production of higher levels of momilactone A may be indicative of a higher defense response.

Oryzalexins

The production of rice phytoalexins is known to be triggered by not only *M. grisea* invasion but also by treatment with ultraviolet irradiation. In 1984, Akatsuka et al. (1985) isolated oryzalexins A, B, and C as a group of new phytoalexins from the foliar part of rice plants infected with *M. grisea* (Kono et al., 1984; Akatsuka et al., 1985; Kono et al., 1985), and, subsequently, oryzalexins D (Sekido et al., 1986), E (Kato et al., 1993), F (Kato et al., 1994), and S (Kodama, Li, et al., 1992; Tamogami et al., 1993) were

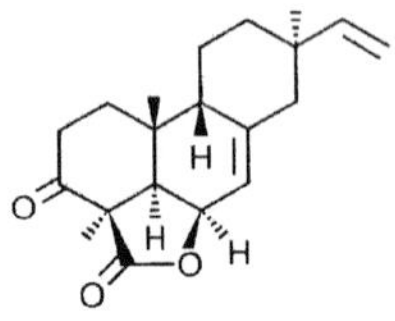

Momilactone A
(ED50=15 μg/ml)

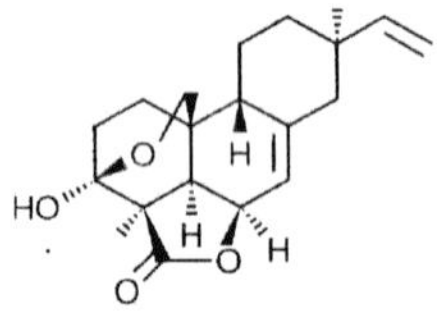

Momilactone B
(ED50=3 μg/ml)

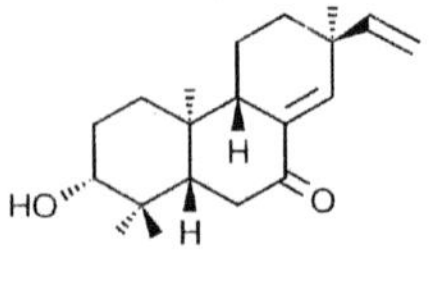

Oryzalexin A
(ED50=116 μg/ml)

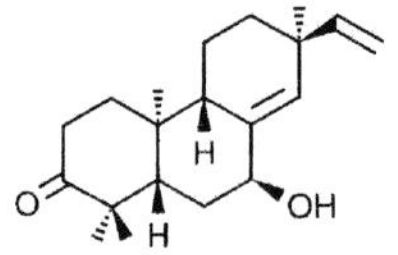

Oryzalexin B
(ED50=74 μg/ml)

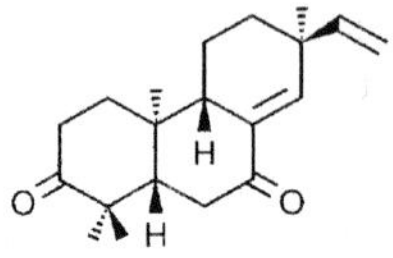

Oryzalexin C
(ED50=136 μg/ml)

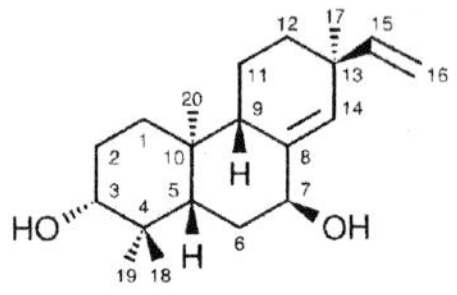

Oryzalexin D
(ED50=23 μg/ml)

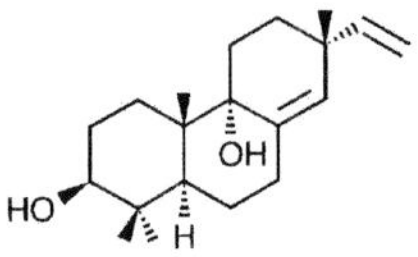

Oryzalexin E
(ED50=62.5 μg/ml)

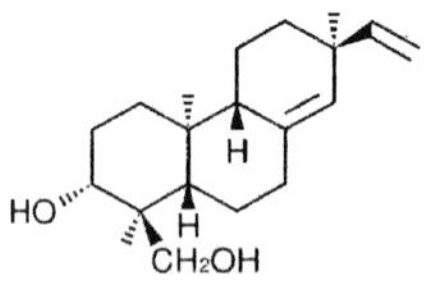

Oryzalexin F
(ED50=103 μg/ml)

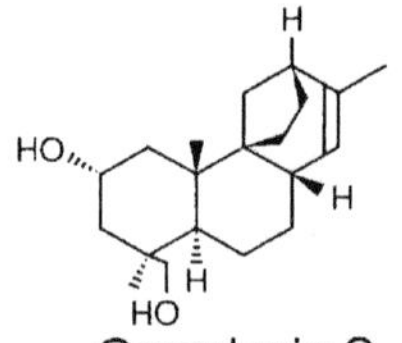

Oryzalexin S
(ED50=20 μg/ml)

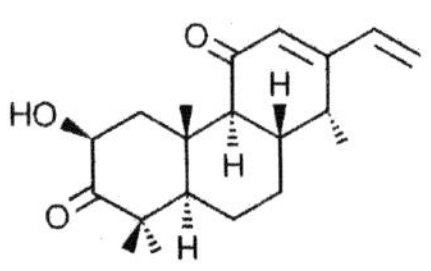

Phytocassane A
(ED50=20 μg/ml)

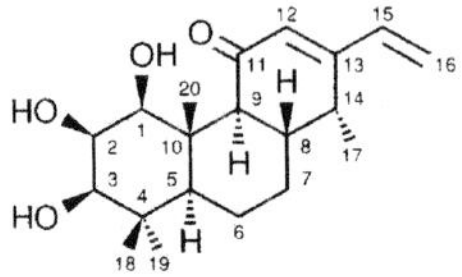

Phytocassane B
(ED50=4 μg/ml)

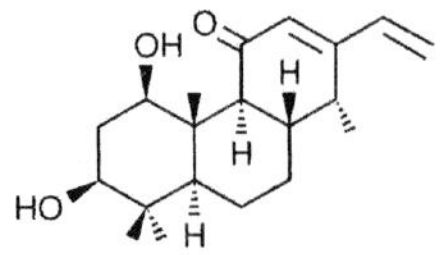

Phytocassane C
(ED50=7 μg/ml)

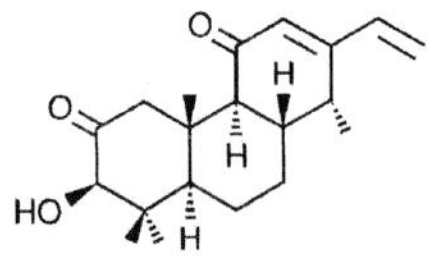

Phytocassane D
(ED50=25 μg/ml)

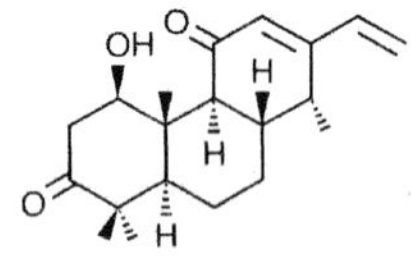

Phytocassane E
(ED50=6 μg/ml)

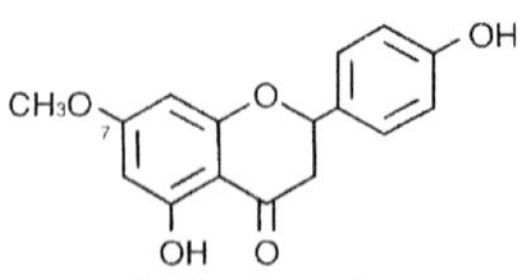

Sakuranetin
(ED50=15 μg/ml)

FIGURE 23.1. Structures of rice phytoalexins. The ED_{50} value of each phytoalexin in prevention of *M. grisea* spore germination is indicated under the name of the phytoalexin.

found in ultraviolet-irradiated rice leaves or in rice leaves infected with *M. grisea*. Oryzalexins A, B, C, D, E, and F are classified as pimarane-type diterpenes, whereas oryzalexin S is classified as a stemarane- type diterpene (see Figure 23.1).

The ED_{50} values of oryzalexins A, B, C, D, E, F, and S in prevention of *M. grisea* spore germination were found to be 116, 74, 136, 23, 62.5, 103, and 20 μg/ml, respectively (see Figure 23.1). Oryzalexins have relatively low antifungal activity compared to momilactones and phytocassanes. Oryzalexin D has the highest antifungal activity among the oryzalexins A-F, suggesting that the attachment of hydroxyl groups at the C-3 and C-7 positions and the number of hydroxyl groups are important for the antifungal activity of oryzalexins (see Figure 23.1, oryzalexin D). Furthermore, 7β-OH-oryzalexin A with an β-hydroxyl group at the C-7 position of oryzalexin A showed less activity than oryzalexin D with an α-hydroxyl group at the C-7 position (see Table 23.1, Sekido et al., 1987). Thus, it seems likely that the α-hydroxyl group at the C-7 position is indispensable for the high antifungal activity at oryzalexins.

As shown in Table 23.1, (–)-enantiomers of oryzalexins A, B, C, and D have less antifungal activity than the corresponding (+)-enantiomers, the natural configuration, suggesting that the natural configuration is important for the high antifungal activity of oryzalexins (Sekido et al., 1987).

TABLE 23.1. The ED_{50} values of oryzalexin enantiomers in prevention of *Magnaporthe grisea* spore germination and germ tube growth

	ED_{50} value (μg/ml)	
Enantiomer of Oryzalexins	**Spore Germination**	**Germ Tube Growth**
(+)-Oryzalexin A	116	31
(–)-Oryzalexin A	165	120
(+)-Oryzalexin B	74	16
(–)-Oryzalexin B	125	40
(+)-Oryzalexin C	136	44
(–)-Oryzalexin C	>200	>200
(+)-Oryzalexin D	23	9
(–)-Oryzalexin D	116	36
(+)-7β-OH-Oryzalexin A	100	29
(–)-7β-OH-Oryzalexin A	153	76

Source: Adapted from Sekido et al., 1987.

Sakuranetin

Kodama, Miyakawa, et al. (1992) identified sakuranetin as a new phytoalexin from ultraviolet-irradiated rice leaves. Since sakuranetin was also detected in rice leaves infected with *M. grisea,* this compound is categorized as a phytoalexin as well as a stress compound. The other phytoalexins, momilactones, oryzalexins, and phytocassanes, are classified as diterpenes, whereas sakuranetin is a structurally rare phytoalexin, classified as a flavonoid (see Figure 23.1).

The ED_{50} values of sakuranetin (5,4'-dihydroxy-7-methoxyflavanone) in prevention of *M. grisea* spore germination was found to be 15 µg/ml (see Figure 23.1). On the other hand, naringenin (5,7,4'-trihydroxyflavanone), a precursor of sakuranetin, scarcely showed any inhibition of spore germination at concentrations up to 50 µg/ml (Kodama, Miyakawa, et al., 1992), suggesting that methylation of the hydroxyl group at the C-7 position of naringenin is important for the high antifungal activity of sakuranetin (see Figure 23.1, sakuranetin). Recently, it has been reported that naringenin 7-*O*-methyltransferase, which catalyzes the methylation of naringenin to sakuranetin, is induced in ultraviolet-irradiated rice leaves, but is not found in healthy leaves (Rakwal, Hasegawa, et al., 1996). Furthermore, jasmonic acid (Rakwal, Tamogami, et al., 1996; Tamogami et al., 1997a) and CuCl2 (Rakwal, Tamogami, et al., 1996; Rakwal et al., 2000) cause the induction of naringenin 7-*O*-methyltransferase and subsequent accumulation of sakuranetin. These observations suggest that naringenin 7-*O*-methyltransferase is a key enzyme involved in the rice defense response.

Phytocassanes

To determine which phytoalexins contribute the most to rice defense systems against fungal pathogens, rice leaves that had been infected with *M. grisea* were used to screen for phytoalexins possessing high antifungal activity. As a result, Koga et al. (1995) identified phytocassanes A, B, C, and D as a group of new phytoalexins, and, subsequently, phytocassane E was identified from suspension-cultured rice cells after treatment with a mycelial extract of the potato pathogenic fungus *Phytophthora infestans* (Koga et al., 1997). Phytocassanes were also produced in rice stems infected with the rice pathogenic fungus *Rhizoctonia solani,* and had high antifungal activity against *R. solani* (Koga et at., 1995). The structures of phytocassanes A-E, classified as cassane-type diterpenes, are shown in Figure 23.1.

The ED_{50} values of phytocassanes A, B, C, D, and E in prevention of *M. grisea* spore germination were found to be 20, 4, 7, 25, and 6 µg/ml, respectively (see Figure 23.1). A notable characteristic of phytocassanes is high antifungal activity equivalent to the activity of momilactones. In particular, phytocassanes B, C, and E have ED_{50} values in the range of 4-7 g/ml, whereas phytocassanes A and D have ED_{50} values 4-5 times higher. The common structural feature of the three more effective forms is the presence of a 1β-hydroxyl group that can hydrogen bond to the carbonyl group at the C-11 position (see Figure 23.1, Koga et al., 1997). This structural feature seems to enhance the antifungal activity. However, since phytocassanes A and D, which lack this feature, are active, albeit less so, it is possible that the other hydroxyl groups (e.g., 2β- and 3β-hydroxyl groups) contribute to the antifungal activity.

Phytocassanes have two characteristic stereostructures. First, in phytocassanes B, C, and E, the hydroxyl group at the C-1 position forms an intramolecular hydrogen bond to the carbonyl group at the C-11 position, and this structural feature seems to contribute to the high antifungal activity (Koga et al., 1997). Second, the cyclohexane rings (C-l, C-2, C-3, C-4, C-5, C-10) in phytocassanes B, C, and D, including the other diterpene phytoalexins, are in a chair conformation, whereas the cyclohexane ring in phytocassane A is in a boat conformation (see Figure 23.2; Koga et al., 1995). Since phytocassane A is unstable in basic solutions and converts nonenzymatically to phytocassane D, it may be in a highly reactive configuration.

Comparisons of the Production Level, Antifungal Activity, and Methods of Analysis of Each Phytoalexin

The relative level of each phytoalexin produced is highly dependent on passage hours after fungal infection, types of elicitors, and rice tissue. For example, there was a higher relative production of momilactone A and phytocassane A in rice leaves or stems, whereas there was a higher relative production of phytocassanes C and E, and momilactone B in suspension-cultured rice cells (Koga et al., 1997). Seven days after infection of rice leaves with *M. grisea,* production levels of momilactone A and phytocassane A were higher than the other momilactones and phytocassanes (Koga et al., 1995), whereas four days after infection, the production of momilactone A was highest (Koga, Oshima, et al., 1998). Similarly, seven days after treatment with fungal elicitor, momilactone A and phytocassanes A and B were produced at a higher level than the other momilactones and phytocassanes (Umemura et al., 2000), whereas one to two days after treatment, only momilactone A was predominantly produced (Koga, Oshima, et al., 1998; Koga, Yamaguchi, et al., 1998). Since there are some reports that momi-

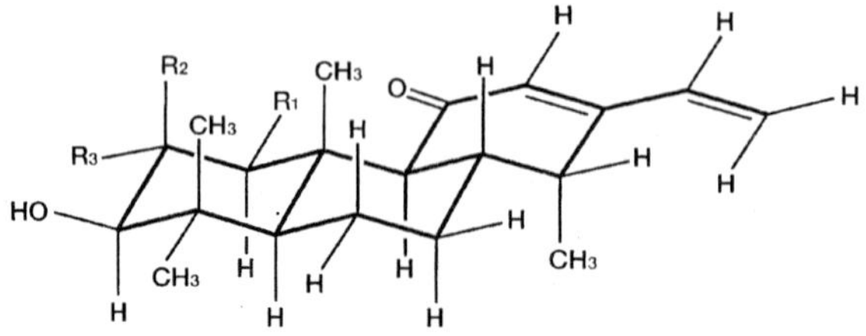

	R_1	R_2	R_3
Phytocassane B	OH	OH	H
Phytocassane C	OH	H	H
Phytocassane D	H	O	

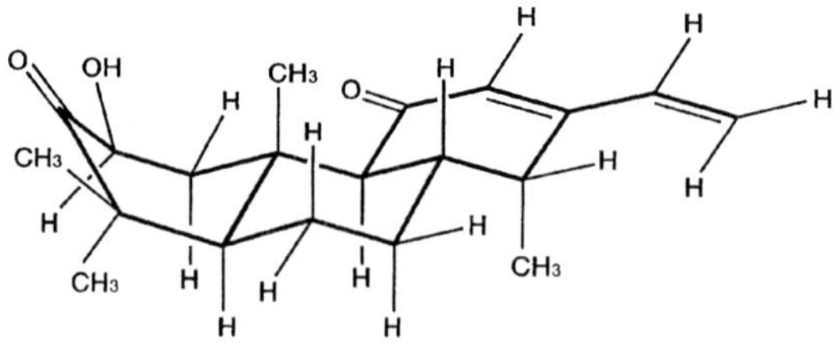

Phytocassane A

FIGURE 23.2. Stereostructures of phytocassanes A, B, C, D (*Source:* Adapted from Koga et al., 1995.)

lactone A is produced at a higher level than oryzalexins and sakuranetin in rice leaves infected with *M. grisea* (Kato et al., 1993; Kato et al., 1994; Dillon et al., 1997), it seems likely that the production level of momilactone A is highest among the phytoalexins induced in rice leaves in the early stage after *M. grisea* infection or elicitor treatment. Sakuranetin levels were 10-30 times higher than momilactone A levels in rice leaves treated with $CuCl_2$ or jasmonic acid (Rakwal, Tamogami, et al., 1996), but were about 30 times lower than momilactone A levels in rice leaves infected with *M. grisea* (Dillon et al., 1997) or treated with fungal elicitor (Koga et al., unpublished result). Based on these observations, sakuranetin seems to be a phytoalexin induced by chemical stress.

The relative level of antifungal activity of rice phytoalexins is generally as follows: momilactones = phytocassanes > sakuranetin > oryzalexins. In particular, momilactone B and phytocassanes B and C have notably high antifungal activity compared to the other phytoalexins (see Figure 23.1).

The specific structural features of rice phytoalexins contributing to such high antifungal activity are worth noting.

To measure elicitor activity by monitoring phytoalexin production, the amounts of phytoalexins produced can be quantified by high performance liquid chromatography (HPLC) or gas chromatography analysis. Since phytocassanes and sakuranetin have a characteristic UV absorption spectrum with a peak at about 280 nm, and momilactone A is accumulated to a high level, they can be detected and quantified by HPLC analysis. However, since momilactone B and oryzalexins have only an end UV absorption spectrum, they should be analyzed by gas chromatography with a high sensitivity.

ELICITORS

Cerebrosides

Sphingolipids are ubiquitous components of the membranes of all eukaryotic cells and are particularly abundant in plasma membranes. In animals, they modulate transmembrane signal transduction via their effects on protein kinases associated with growth factor receptors (Bremer et al., 1984; Hakomori, 1990) and on protein kinase C (Kreutter et al., 1987; Hakomori, 1990), thereby regulating cell proliferation (Hakomori, 1981; Hannun and Bell, 1989) and inducing cell differentiation (Obata et al., 1977; Roisen et al., 1981; Nojiri et al., 1986; Nojiri et al., 1988) and apoptosis (De Maria et al., 1997). In particular, sphingolipid metabolic products, ceramides, function as second messengers in the signal transduction pathway involved in apoptosis (Obeid et al., 1993; Hannun, 1996). In fungi, sphingolipids are known to function as inducers of cell differentiation. Fungal sphingolipids, such as cerebrosides B, C, and D (see Figure 23.3), induce cell differentiation in the fungus *Schizophyllum commune* (Kawai and Ikeda 1983; Kawai et al., 1985; Kawai et al., 1986; Kawai, 1989) and *Coprinus cinereus* (Mizushina et al., 1998), with resultant formation of the fruiting body. In animals and fungi, sphingolipids are known to have various biological functions as previously mentioned, whereas, in the case of plants, no reports have appeared demonstrating the involvement of sphingolipids in biological processes, such as the hypersensitive response, cell growth, differentiation, or apoptosis.

Koga, Yamaguchi, et al. (1998) succeeded in isolating fungal elicitors of hypersensitive cell death and phytoalexin accumulation in rice. These elicitors were isolated from the rice pathogenic fungus *M. grisea,* and their

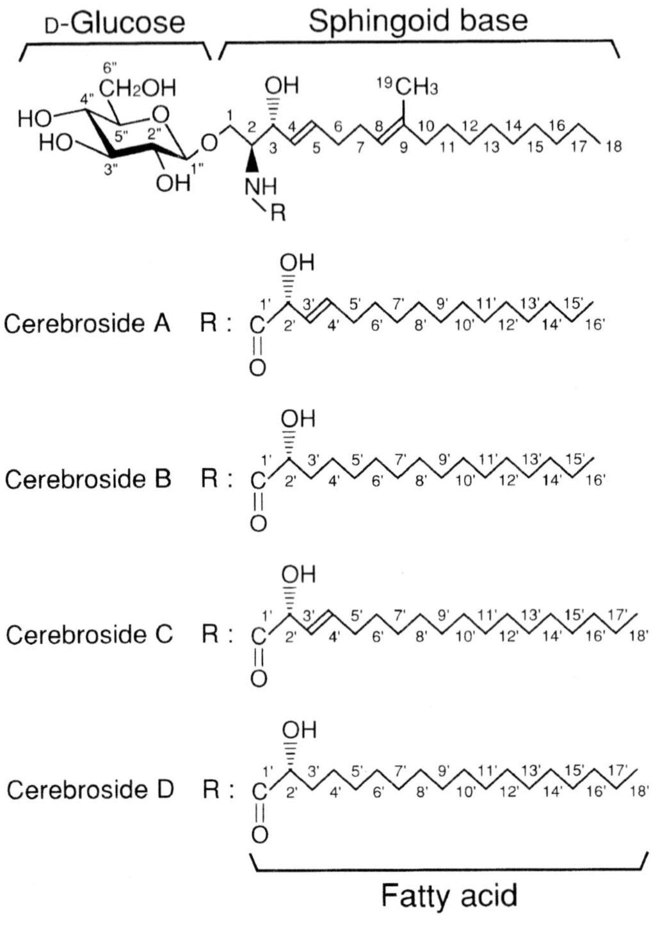

FIGURE 23.3. Structures of cerebrosides A, B, C, and D

structures were identified as cerebrosides A, B, and C (see Figure 23.3), sphingolipids that were isolated as inducers of cell differentiation in the fungus *S. commune* and *C. cinereus*. These cerebroside elicitors induced the accumulation of phytoalexins (momilactones A and B and phytocassanes A, B, C, D, and E), pathogenesis-related proteins (β-glucanase, chitinase, peroxidase, and *PBZl* protein) (Umemura et al., 2000), cell death (Koga, Yamaguchi, et al., 1998), and increased resistance to subsequent infection by compatible race of *M. grisea* (Koga, Yamaguchi, et al., 1998). Cerebrosides A, B, and C occur in a wide range of fungi, for example, *Rhizoctonia* sp., *Cochliobolus miyabeanus, Cercospora solani-melogenae, Mycosphaerella pinodes* (Umemura et al., 2000), *Pachybasium* sp. (Sitrin et al., 1988), *Penicillium funiculosum* (Kawai et al., 1985), *Schizophyllum commune* (Kawai and Ikeda, 1983), *Lentinus edodes* (Kawai, 1989), *Fusicoccum amygdali* Del. (Ballio et al., 1979), and *Ganoderma lucidum* (Mizushina et al., 1998), and their common structural feature is a methyl group at C-9 in the sphingoid base moiety (see Figure 23.3). Furthermore, treatment with cerebroside A protected rice plants from disease caused by C. *miyabeanus* as well as *M. grisea* (Umemura et al., 2000). These results indicate that cerebroside elicitors function as general elicitors in a wide variety of rice-pathogen interactions.

Comparison of the elicitor activity of cerebroside derivatives provided some important insight into the structural features required for elicitor activity in rice plants (Table 23.2, Koga, Yamaguchi, et al., 1998). Cerebroside is composed of D-glucose, fatty acid, and sphingoid base (see Figure 23.3). The fatty acid methyl ester and sphingoid base of cerebroside had no elicitor activity (see Table 23.2), whereas the ceramide, which lacks glucose, was active, albeit less so (Umemura et al., 2000). This observation indicates that the minimal component required for the elicitor activity is ceramide and that glucosylation of the ceramide enhances this activity. The cerebrosides from rice bran, which are structurally similar to cerebrosides A, B, and C, but lack the methyl group at C-9 (Fujino et al., 1985), showed hardly any activity (see Table 23.2). Furthermore, hydrogenation of the 4*E*-double bond in the sphingoid base moiety of cerebroside A led to a twelvefold decrease in activity. These observations indicate that the methyl group at C-9 in the sphingoid base moiety is essential for the elicitor activity and the 4*E*-double bond in the sphingoid base moiety enhances this activity (Koga, Yamaguchi, et al., 1998).

In the fungus *S. commune,* the methyl group at C-9 in the sphingoid base moiety is essential for inducing cell differentiation, and the 4*E*- and 8*E*-double bonds in the sphingoid base moiety enhance this activity (Kawai and Ikeda, 1983; Kawai et al., 1985; Kawai et al., 1986). In animals, the 4*E*-double bond in the sphingoid base moiety of ceramide is important for inducing apoptosis (Bielawska et al., 1993). The structural features required for elicitor activity in rice and for the induction of cell differentiation in *S. commune* are very similar. On the other hand, the structural feature required for elicitor activity in rice and for the induction of apoptosis in animals is similar in that the 4*E*-double bond is important for both types of activity. In both cases, it seems likely that the sphingoid base moiety is important for determining the recognition of specific sphingolipids.

TABLE 23.2. Elicitor activity of various sphingolipids and of the derivatives of cerebroside A

Compounds	Sugar	Fatty Acid	Sphingoid Base	Specific Elicitor Activity[a] (units/μg)
Derivatives of cerebroside A				
Cerebroside A	Glu[c]	16h:1[d]	d19:2(4*E*,8*E*,9Me)[e]	0.86
Cerebroside B	Glu	16h:0d	d19:2(4*E*,8*E*,9Me)	0.87
Cerebroside C	Glu	18h:1	d19:2(4*E*,8*E*,Me)	0.84
(4*E*) H2-Cerebroside A[b]	Glu	16h:1	d19:1(8*E*,9Me)[e]	0.072
(3*E*,4*E*,8*E*) H_5-Cerebroside A[b]	Glu	16h:0	d19:0(9Me)	0.071
Hexadecenoic acid methyl ester[b]	–	16h:1	–	<0.020
Methyl sphingadienine[b]	–	–	d19:2(4*E*,8*E*,9Me)	<0.020
Glucosyl methyl sphingadienine[b]	Glu	–	d19:2(4*E*,8*E*,9Me)	<0.020
Sphingolipids from animals				
Glucocerebrosides from Gaucher's spleen	Glu	14~24:0d 18~24:1	d18:1(4*E*)[e]	<0.020
Galactocerebrosides from bovine brain	Gal[c]	18~26h:0 24~26h:1	d18:1(4*E*)	<0.020
Sphingolipids from plants				
Cerebrosides from rice bran	Glu	14~26h:0[d] 14~26:0	d18:0~2(4*E*,8*E*or*Z*)[e] t18:0~1(8*E*or*Z*)[e]	<0.020

Source: Adapted from Koga, Yamaguchi, et al., 1998.

[a]One unit of elicitor activity is defined as the amount of the sample required to induce the half-maximal amount of momilactone A per leaf. The specific elicitor activity of the sample is defined as the number of units per mg dry weight of the sample.

[b](4*E*) H_2-cerebroside A [(8*E*)-*N*-D-2'-hydroxy-(*E*)-3'-hexadecenolyl-1-*O*-β-D-glucopyranosyl-9-methyl-8-sphingenine] and (3*E*,4*E*,8*E*) H_5-cerebroside A [*N*-D-2'-hydroxypalmitoyl-1-*O*-β-D-glucopyranosyl-9- methyl-phinganine] are the hydrogenation products of cerebroside A. Hexadecenoic acid methyl ester [2- hydroxy-(*E*)-3-hexadecoenoic acid methyl ester], methyl sphingadienine [(4*E*,8*E*)-9-methyl- 4,8- sphingadienine], and glucosyl methyl dphingadienine [(4*E*,8*E*)-D-glucopyranosyl- 9-methyl-4,8,-sphingadienine] are the hydrolysis products of cerebroside A.

[c]The abbreviations used are Glu = glucose; Gal = galactose.

[d]The abbreviations used are 14~26h:0, 2-hydroxy fatty acids having carbon chain length 14~26; 16:0, hexadecanoic acid; 16:1, (*E*)-3-hexadecenoic acid (hexadecanoic acid with 1 double bond); 16h:1,2-hydroxy-(*E*)-3-hexadecenoic acid.

[e]The abbreviations used are d18:0~2, sphinganine with 0~2 double bonds; d18:0, sphinganine, t18:0, 4-hydroxysphinganine; d18:1(4*E*), (*E*)-4-sphingenine; d19:0(9Me), 9-methylphinganine; d19:1(8*E*,9Me), (8*E*)-9-methyl-8-sphingenine, d19:2(4*E*,8*E*,9Me), (4*E*,8*E*)-9-methyl-4,8-sphingadienine.

β-*Glucooligosaccharides*

β-1,3-Glucanases [EC 3.2.1.39] and chitinases [EC 3.2.1.14] are widely distributed in higher plants and participate in defense systems against pathogen infection by directly attacking the cell walls of pathogenic fungi in which chitin and β-1,3-glucan are important constituents (Mauch and Staehelin, 1989; Collinge et al., 1993). Furthermore, oligosaccharides released from the fungal cell walls by β-1,3-glucanases and chitinases are thought to function as elicitors for further induction of these enzymes and for activation of other defense systems such as phytoalexin production (Sharp, Valent, et al., 1984; Sharp, McNeil, et al., 1984; Yamada et al., 1993).

Among β-glucooligosaccharide elicitors from pathogens, fragments of the β-glucan from the pathogenic fungus *Phytophthora sojae* are the best-characterized elicitor molecules, eliciting biosynthesis of a phytoalexin, glyceolin, in soybean cotyledon cells (Sharp, Valent, et al., 1984; Sharp, McNeil, et al., 1984). This elicitor was purified as a reduced compound, hexaglucosyl glucitol, consisting of a six-linked backbone and two three-linked, branched glucosyl residues (Figure 23.4, hexaglucosyl glucitol) (Sharp, McNeil, et al., 1984; Cheong et al., 1991). Since a mixture of fragments from β-glucan preparation also induced the defense response in several plant species other than soybean (Cline et al., 1978; Brady et al., 1993; Côté and Hahn, 1994), it was not clear whether this hexaglucosyl glucitol structure is generally important for recognition by the other plants or whether they recognize some other β-glucooligosaccharide structures.

Yamaguchi et al. (2000) succeeded in isolating β-glucooligosaccharide elicitors, which induce the synthesis of a phytoalexin, momilactone A, in suspension-cultured rice cells. These elicitors were isolated from the rice pathogenic fungus *M. grisea,* and their structures were identified as reduced compounds, tetraglucosyl glucitols, consisting of a three-linked backbone and a six-linked, branched glucosyl residue (see Figure 23.4, tetraglucosyl glucitols A, B, and C). Comparison of the elicitor activity of β-glucooligosaccharide derivatives provided some important information on the structural features required for elicitor activity in rice (see Figure 23.4, Yamaguchi et al., 2000). The three-linked backbone and the six-linked branched glucosyl residue were found to be essential for elicitor activity, and attachment of the six-linked branched glucosyl residue to the glucosyl residue adjacent to the reducing end of the β-glucooligosaccharide was required for maximum elicitor activity (see Figure 23.4, tetraglucosyl glucitol A). This tetraglucosyl glucitol elicitor was structurally different from the hexaglucosyl glucitol elicitor in soybean (see Figure 23.4, hexaglucosyl glucitol) and had hardly any elicitor activity in soybean. Conversely, the hexaglucosyl

Structures of β-glucooligosaccharide elicitors	Elicitor activity	
	in rice	in soybean
(Tetraglucosyl glucitol A) Glc β (1→3)-Glc β (1→3)-Glc β (1→3)-Glucitol Glc β (1→6)	High activity	No activity
(Tetraglucosyl glucitol B) Glc β (1→3)-Glc β (1→3)-Glc β (1→3)-Glucitol Glc β (1→6)	Low activity	No activity
(Tetraglucosyl glucitol C) Glc β (1→3)-Glc β (1→3)-Glc β (1→3)-Glucitol Glc β (1→6)	Low activity	No activity
(Hexaglucosyl glucitol) Glc β (1→6)-Glc β (1→6)-Glc β (1→6)-Glc β (1→6)-Glucitol Glc β (1→3) Glc β (1→3)	No activity	High activity

FIGURE 23.4. Structures of tetraglucosyl glucitol elicitors in rice and hexaglucosyl glucitol elicitor in soybean (Glc = glucopyranose)

glucitol elicitor in soybean had no elicitor activity in rice. These observations indicate that the β-glucooligosaccharide elicitors recognized by rice and soybean cells are quite different, probably reflecting differences in specificity of the receptors in these two plants (Yamaguchi et al., 2000). This is the first report to show that the structure of β-glucooligosaccharide elicitors recognized by each plant is quite different.

N-*Acetyl-chitooligosaccharides*

Although there is no report on the isolation of *N*-acetyl-chitooligosaccharide elicitors derived from rice pathogenic fungi, synthesized *N*-acetyl-chitooligosaccharides are known to function as elicitors effective in induction of phytoalexin (momilactone A), chitinase, and L-phenylalanine ammonia-lyase in suspension-cultured rice cells (Yamada et al., 1993; Inui et al., 1997). Also, *N*-acetyl-chitooligosaccharides induced many other cellular responses, such as changes of membrane potential (Kikuyama et al., 1997), ion flux (Kuchitsu et al., 1997), reactive oxygen generation (Kuchitsu et al., 1995), biosynthesis of jasmonic acid (Nojiri et al., 1996), and expression of several defense-related genes in rice cells (Minami et al., 1996). Furthermore, *N*-acetyl-chitooligosaccharides were larger than pentamers and were found to function as very potent phytoalexin elicitors (Yamada et al., 1993). Since *N*-acetyl-chitooligosaccharides have a linear structure and show little specific structural variation other than chain length, they may play a role in recognition of broader pathogens rather than recognition of specific pathogens.

Jasmonic Acid

Jasmonic acid (see Figure 23.5) is widely distributed in higher plants, and is shown to play an important role in the signal transduction pathway involved in plant-defense responses (Farmer and Ryan, 1992; Gundlach et al.,

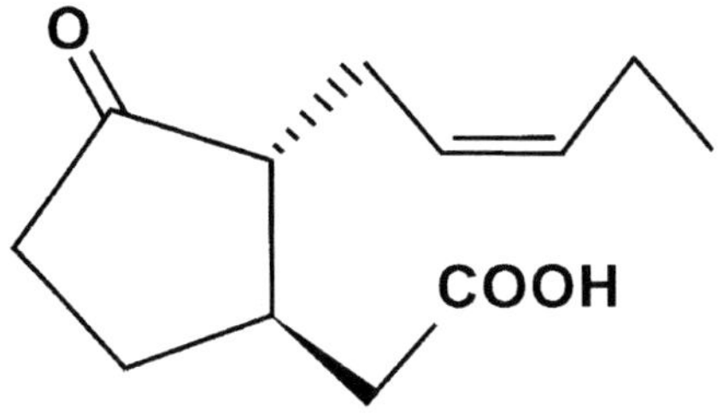

FIGURE 23.5. Structure of jasmonic acid

1992). In rice, jasmonic acid has also been shown to function as a signal transducer in the induction of phytoalexin production by treatment with *N*-acetyl- chitoheptaose (Nojiri et al., 1996) or $CuCl_2$, which is known as an abiotic elicitor (Rakwal, Tamogami, et al., 1996). Exogenously applied jasmonic acid induced the production of phytoalexin, such as momilactone A and sakuranetin. On the other hand, in rice treated with *N*-acetyl-chitoheptaose or $CuCl_2$, endogenous jasmonic acid was rapidly and transiently accumulated prior to accumulation of momilactone A and sakuranetin. Treatment with ibuprofen or salicylhydroxamic acid, inhibitors of jasmonic acid biosynthesis, reduced production of momilactone A and sakuranetin in rice treated with *N*-acetyl- chitoheptaose or $CuCl_2$, but treatment with ibuprofen or salicylhydroxamic acid did not alter production of momilactone A and sakuranetin in rice treated with jasmonic acid.

The induction of sakuranetin production by jasmonic acid resulted from the induction of naringenin 7-*O*-methyltransferase, a key enzyme in sakuranetin biosynthesis (Tamogami et al., 1997a), and the induction of naringenin 7-*O*-methyltransferase was inhibited by the plant hormones, kinetin and zeatin (Tamogami et al., 1997b).

Endo-β-l,4-xylanases

Xylans, major constituents of hemicellulose in higher plant cell walls, are composed of a linear backbone of β-1,4-linked xylose residues and branches of neutral or uronic monosaccharides and oligosaccharides. Endo-β-1,4-xylanases [EC 3.2.1.8] degrade the xylan backbone into small oligomers, and are known proteinaceous elicitors of defense response reactions in tobacco and tomato plants (Bailey et al., 1990; Avni et al., 1994). When applied to tobacco or tomato leaves, fungal endo-β-1,4-xylanases induced the accumulation of phytoalexins, pathogenesis-related proteins, ethylene, and hypersensitive cell death. Endo-β-1,4-xylanases from *Trichoderma viride* have been used as elicitors most extensively among the xylanases isolated from various plant pathogenic fungi. (Dean et al., 1989; Yano et al., 1998).

Recently, endo-β-1,4-xylanases from *T. viride* have been found to function as elicitors of phytoalexin accumulation in rice plants (Takahashi et al., unpublished results). Treatment with endo-β-1,4-xylanases from *T. viride* induced the accumulation of phytoalexins, momilactones A and B, and phytocassanes A, B, C, D, and E in rice plants, but the xylanase activity was not necessary for its elicitor activity (Takahashi et al., unpublished results) as is the case of tobacco and tomato plants (Enkerli et al., 1999; Furman-Matarasso et al., 1999).

CONCLUSIONS

Fifteen rice phytoalexins, momilactones A and B, oryzalexins A-F and S, phytocassane A-E, and sakuranetin, have been found and most of the structural features required for the high antifungal activity have been revealed. The lactone skeleton of momilactones, the hydroxyl groups of oryzalexins and phytocassanes, and the methoxyl group of sakuranetin were thought to contribute to their high antifungal activity. In consideration of the antifungal activity and the production level and stage, it seems likely that among these phytoalexins, momilactone A contributes the most to rice defense systems against pathogens.

N-Acetyl-chitooligosaccharides, β-glucooligosaccharides (tetraglucosyl glucitols), cerebrosides A-C, jasmonic acid, and endo-β-1,4-xylanases have been found to function as rice elicitors, however, elicitors isolated from pathogenic fungi are β-glucooligosaccharides and cerebrosides A-C only. The structural features of β-glucooligosaccharides required for elicitor activity in rice were found to be the three-linked backbone and the six- linked branched glucosyl residue, and the structures of β-glucooligosaccharide elicitors recognized by rice and soybean cells were found to be quite different. Considering the observations with respect to β-glucooligosaccharide and cerebroside elicitors, it seems quite probable that the structure of elicitors is specifically recognized by each plant, but the fungal sources of elicitors are unrelated to the plant-pathogen interaction.

The structural features of cerebrosides required for elicitor activity in rice and for the induction of cell differentiation in *S. commune* were found to be very similar in that the methyl group and the double bonds in the sphingoid base moiety are important for both types of activity. In light of the fact that sphingolipids modulate signal transduction pathways, thereby regulating cell proliferation, differentiation, and apoptosis in animals, studies of cerebroside elicitors are exciting and are going to open new areas of research in the field ofplant pathology.

FUTURE PERSPECTIVES

Comparison of the Contribution of Each Phytoalexin to Rice Defense Systems

To determine which phytoalexins contribute the most to rice defense systems against fungal pathogens, the antifungal activity and the levels or period of phytoalexin production after attack by pathogens must be compared.

With respect to antifungal activity, momilactones, phytocassanes, and sakuranetin have higher antifungal activity than oryzalexins (see Figure 23.1). Also, with respect to the production level after *M. grisea* infection or elicitor treatment, momilactone A and phytocassanes A and B have an advantage over the other phytoalexins. Furthermore, in the early stage after *M. grisea* infection or elicitor treatment, the production level of momilactone A is higher than phytocassanes A and B. From these observations, it seems likely that among the various phytoalexins, momilactone A contributes the most to rice defense systems against pathogens. Therefore, when measuring elicitor activity by monitoring phytoalexin production, the production of higher levels of momilactone A may be taken as an indicator of a higher defense response.

As described in the phytoalexins section, most of the structural features required for the high antifungal activity have been revealed. Furthermore, methods for the complete synthesis of momilactone A (Germain and Deslongchamps, 1999), oryzalexin S (Kaliappan and Subba Rao, 1996), and phytocassane A (Yajima and Mori, 2000) with high antifungal activity have been established, but the pathways of biosynthesis of these diterpene phytoalexins have not yet been unequivocally determined. If we are to understand the regulation of biosynthesis of each phytoalexin in the rice-defense response, further study of these phytoalexin biosynthetic pathways and the properties of the enzymes involved is necessary.

Relationship Between Cerebroside Elicitors and Plant Apoptosis

In animals, sphingolipids are known to play important roles in the regulation of cell growth, differentiation, and apoptosis (Hakomori, 1990; De Maria et al., 1997). Furthermore, it has been shown that sphingolipid metabolic products, ceramides, function as second messengers in the signal transduction pathway involved in apoptosis (programmed cell death) (Obeid et al., 1993; Hannun, 1996).

On the other hand, it is known that programmed cell death plays an important role in the development of plant organs and in the development of animal organs. Recent studies suggest that the hypersensitive response in plants is a type of programmed cell death (Dangl et al., 1996; Greenberg, 1996), since the appearance of the hypersensitive response is genetically controlled (Greenberg et al., 1994; Dietrich et al., 1994) and the hypersensitive response-associated cell death (hypersensitive cell death) is activated by certain elicitors (He et al., 1993; Levine et al., 1994). Furthermore, the induction of the hypersensitive response-associated cell death by some pathogens and elicitors has been shown to be mechanistically similar to apoptosis in animals, since apoptotic features in animals are present in some plant cells undergoing the hypersensitive response (Mittler et al., 1995; Levine et al., 1996; Wang et al., 1996; Ryerson and Heath, 1996). In light of the fact that the hypersensitive response is induced by sphingolipids or ceramide in rice plants (Koga, Yamaguchi, et al., 1998; Umemura et al., 2000), it seems plausible that the hypersensitive response in plants may be mediated by the same mechanism as that responsible for the sphingolipid- or ceramide-mediated apoptosis in animals.

Relationship Between Cerebroside Elicitors and Protein Kinases Involved in Plant Disease Resistance

In animal cells, sphingolipids modulate various protein kinases (Hakomori, 1990), suggesting that cerebroside elicitors may directly act on certain protein kinases in plants. Gene-for-gene interactions, in which a single disease resistance gene in the plant responds specifically to a single avirulence gene in the pathogen, have been described. Numerous resistance genes have been cloned from several plant species (Martin et al., 1993; Mindrinos et al., 1994; Whitham et al., 1994; Jones et al., 1994; Song et al., 1995; Staskawicz et al., 1995). The rice *Xa21* gene, which confers resistance to *Xanthomonas oryzae* pv. *oryzae* race 6 (Song et al., 1995), and the tomato *Pto* gene, which confers resistance to races of *Pseudomonas syringae* pv. *tomato* (Martin et al., 1993), encode serine-threonine protein kinases, suggesting that the protein kinases play an important role in plant disease resistance. If sphingolipids produced by pathogens can modulate these plant protein kinases in a manner similar to that observed in animal systems, the sphingolipid-mediated signaling pathway may function in plant disease resistance. If we are to understand the relationship between sphingolipid elicitors and plant disease resistance, further studies on the effects of these molecules on the signal transduction pathway are now essential.

Relationship Between Hypersensitive Cell Death in Plants and Cell Differentiation in Fungi

The structural features of cerebrosides required for the elicitor activity in rice and for the induction of cell differentiation in *S. commune* are very similar, probably reflecting the same substrate specificity of the receptor in rice and *S. commune.* Why do plants and fungi have receptors with the same substrate specificity? If fungi have diverged from plants during evolution, we could hypothesize that the gene for the cerebroside receptor in *S. commune* was originally derived from rice. From this point of view, the hypersensitive cell death in rice and cell differentiation in *S. commune* may be fundamentally the same phenomenon.

Recently, cerebrosides B and D (see Figure 23.3) have been isolated as eukaryotic DNA polymerase inhibitors from a basidiomycete, *G. lucidum* (Mizushina et al., 1998). These cerebrosides selectively inhibited the activity of replicative DNA polymerases, especially the α-type, from a wide range of eukaryotic species, whereas they had very little influence on the activity of repair-related DNA polymerase (DNA polymerase β), prokaryotic DNA polymerases, or RNA polymerase. Since these cerebrosides inhibited the activity of *C. cinereus* DNA polymerase α, it is deduced that the cell differentiation of *C. cinereus* caused by cerebrosides requires suppression of the activity of replicative DNA polymerase to stop mycelial cell proliferation (Mizushina et al., 1998). Since these cerebrosides also inhibited the activity of plant DNA polymerase I (α-like) (Mizushina et al., 1998), the hypersensitive cell death caused by cerebrosides may be mediated by the same mechanism as that responsible for the replicative DNA polymerase-mediated cell differentiation of *C. cinereus.*

Comparison of Cerebrosides and Probenazole As Agents Protecting Rice Plants Against Pathogens

Recently, because of environmental pollution caused by use of large amounts of fungicides, substances inducing plant disease resistance have attracted attention as alternative agents to fungicides. For example, probenazole (Oryzemate, 3-allyloxy-l,2-benzisothiazole-l,l-dioxide, Meiji Seika Kaisha Ltd.), an epoch-making agent, has been widely used in paddy fields to prevent infection by the blast fungus *M. grisea* through inducing rice defense systems (Watanabe et al., 1977; Sekizawa and Mase, 1980; Shimura et al., 1983; Midoh and Iwata, 1996). With respect to induction of rice defense systems, cerebrosides are very similar to probenazole. For example, pathogenesis-related proteins such as β-glucanase, chitinase, peroxidase, and PBZl protein are induced by probenazole and cerebrosides (Umemura et al., 2000), suggesting that disease resistance induced by these compounds has common features. However, there are significant differences between the functional mechanisms of the two substances. Cerebrosides induce phytoalexins and hypersensitive cell death (Koga, Yamaguchi, et al., 1998), but probenazole does not induce such defense responses (Umemura et al., 2000). From this point of view, cerebroside elicitors may be ideal new agents, being a different type of substance inducing plant disease resistance.

In most cases, isolated elicitors have been shown to be oligo- or polysaccharides derived from pathogens, and these elicitors exhibit hardly any elicitor activity when sprayed on whole plants due to their hydrophilic structures. In contrast, cerebroside A induced the synthesis of phytoalexins and pathogenesis-related proteins even when sprayed on whole rice plants (Umemura et al., 2000). This is because cerebroside elicitors may be effectively absorbed in the hydrophobic surface of rice leaves due to their hydrophobic structures. In fact, cerebroside A was found to protect rice plants against disease by *M. grisea* even in the field (Umemura et al., 2000). In the future, cerebroside elicitors may be used as powerful agents for the protection of plants against pathogens, in place of fungicides.

Specificity of Plant Recognition and Fungal Source of β-Glucooligosaccharide Elicitors

Elicitors have been studied extensively and, in most cases, have been shown to be oligo- or polysaccharides derived from β-glucan from mycelia of pathogenic fungi. Since the mixture of oligo- or polysaccharides from β-glucan of a pathogen was found to induce defense responses even in several plant species not infected by this pathogen, it has been thought that oligo- or polysaccharides derived from β-glucan are nonspecifically recognized by several plants. However, Yamaguchi et al. (2000) showed that the structures of β-glucooligosaccharide elicitors recognized by rice and soybean cells are quite different. This indicates that β-glucooligosaccharide elicitors are specifically recognized by each plant, probably reflecting differences in specificity of the receptors. On the other hand, a mycelial extract of the potato pathogenic fungus *P. infestans,* which is thought to contain a mixture of oligo- or polysaccharides derived from β-glucan, has been reported to induce momilactones and phytocassanes in suspension-cultured rice cells (Koga et al., 1997). This indicates the possibility that the β-glucooligosaccharide elicitors from *M. grisea* are also found in the mixture of oligo- or polysaccharides from the β-glucan of *P. infestans.* Therefore, β-glucooligosaccharide elicitors may exist not only in a specific pathogen related to a certain plant, but also in the nonpathogenic fungi, as in the case of cerebroside elicitors. Considering all of these observations with respect to β-glucooligosaceharide and cerebroside elicitors, it seems quite probable that the structure of elicitors is specifically recognized by each plant, but the fungal sources of elicitors are unrelated to the plant-pathogen interaction.

REFERENCES

Akatsuka, T., Kodama, O., Sekido, H., Kono, Y., and Takeuchi, S. (1985). Novel phytoalexins (oryzalexins A, B and C) isolated from rice blast leaves infected with *Pyricuiaria oryzae*. Part I: Isolation, characterization and biological activities of oryzalexins. *Agric. Bioi. Chem.* 49: 1689-1694.

Avni, A., Avidan, N., Eshed, Y., Zamir, D., Bailey, B.A., Stommel, J.R., and Anderson, J.D. (1994). The response of *Lycopersicon esculentum* to fungal xylanase is controlled by a single dominant gene (abstract no.872). *Plant Physiol.* 105: S-158.

Bailey, B.A., Dean, J.F.D., and Anderson, J.D. (1990). An ethylene biosynthesis-inducing endoxylanase elicits electrolyte leakage and necrosis in *Nicotiana tabacum* cv Xanthi leaves. *Plant Physiol.* 94: 1849-1854.

Ballio, A., Casinovi, C.G., Framondino, M., Marino, G., Nota, G., and Santurbano, B. (1979). A new cerebroside from *Fusicoccum amygdali* del. *Biochim Biophys. Acta* 573: 51-60.

Bielawska, A., Crane, H.M., Liotta, D., Obeid, L.M., and Hannun, Y.A. (1993). Selectivity of ceramide-mediated biology. *J. Bioi. Chem.* 268: 26226-26232.

Brady, K.P., Darvill, A.G., and Albersheim, P. (1993). Activation of a tobacco glycine-rich protein gene by a fungal glucan preparation. *Plant J.* 4: 517-524.

Bremer, E.G., Hakomori, S., Bowen-Pope, D.F., Raines, E., and Ross, R. (1984). Ganglioside-mediated modulation of cell growth, growth factor binding, and receptor phosphorylation. *J. Biol. Chem.* 259: 6818-6825.

Cartwright, D., Langcake, P., Pryce, R.J., Leworthy, D.P., and Ride, J.P. (1977). Chemical activation of host defence mechanisms as a basis for crop protection. *Nature* 267: 511-513.

Cartwright, D.W., Langcake, P., Pryce, R.J., Leworthy, D.P., and Ride, J.P. (1981). Isolation and characterization of two phytoalexins from rice as momilactones A and B. *Phytochemistry* 20: 535-537.

Cheong, J.J., Birberg, W., Fügedi, P., Pilotti, A., Garegg, P.J., Hong, N., Ogawa, T., and Hahn, M.G. (1991). Structure-activity relationships of oligo-β-glucoside elicitors of phytoalexin accumulation in soybean. *Plant Cell* 3: 127-136.

Cline, K., Wade, M., and Albersheim, P. (1978). Host-pathogen interactions. XV. Fungal glucans which elicit phytoalexin accumulation in soybean also elicit the accumulation of phytoalexins in other plants. *Plant Physiol.* 62: 918-921.

Collinge, D.B., Kragh, K.M., Mikkelsen, J.D., Nielsen, K.K., Rasmussen, U., and Vad, K. (1993). Plant chitinases. *Plant J.* 3: 31-40.

Côté, F. and Hahn, M.G. (1994). Oligosaccharins: Structures and signal transduction. *Plant Mol. Biol.* 26: 1379-1411.

Dangl, J.L., Dietrich, R.A., and Richberg, M.H. (1996). Death don't have no mercy. Cell death programs in plant-microbe interactions. *Plant Cell* 8: 1793-1807.

Dean, J.F.D, Gamble, H.R., and Anderson, J.D. (1989). The ethylene biosynthesis-inducing xylanase: Its induction in *Trichoderma viride* and certain plant pathogens. *Phytopathology* 79: 1071-1078.

De Maria, R., Lenti, L., Malisan, F., d'Agostino, F., Tomassini, B., Zeuner, A., Rippo, M.R., and Testi, R. (1997). Requirement for GD3 ganglioside in CD95- and ceramide-induced apoptosis. *Science* 277: 1652-1655.

Dietrich, R.A., Delaney, T.P., Uknes, S.J., Ward, E.R., Ryals, J.A., and Dangl, J.L. (1994). Arabidopsis mutants simulating disease resistance response. *Cell* 77: 565-577.

Dillon, V.M., Overton, J., Grayer, R.J., and Harborne, J.B. (1997). Differences in phytoalexin response among rice cultivars of different resistance to blast. *Phytochemistry* 44: 599-603.

Dixon, R.A. and Harrison, M.J. (1990). Activation, structure, and organization of genes involved in microbial defense in plants. *Adv. Genet.* 28: 165-234.

Ebel, J. and Cosio, E.G. (1994). Elicitor of plant defense responses. *Int. Rev. Cytol.* 148: 1-36.

Enkerli, J., Felix, G., and Boller, T. (1999). The enzymatic activity of fungal xylanase is not necessary for its elicitor activity. *Plant Physiol.* 121: 391-397.

Farmer, E.E. and Ryan, C.A. (1992). Octadecanoid precursors of jasmonic acid activate the synthesis of wound-inducible proteinase inhibitors. *Plant Cell* 4: 129-134.

Fujino, Y., Ohnishi, M., and Ito, S. (1985). Molecular species of ceramide and mono-, di-, tri- and tetraglycosylceramide in bran and endosperm of rice grains. *Agric. Bioi. Chem.* 49: 2753-2762.

Furman-Matarasso, N., Cohen, E., Du, Q., Chejanovsky, N., Hanania, U., and Avni, A. (1999). A point mutation in the ethylene-inducing xylanase elicitor inhibits the β-1-4- endoxylanase activity but not the elicitation activity. *Plant Physiol.* 121: 345-351.

Germain, J. and Deslongchamps, P. (1999). Transannular diels-alder approach to the synthesis of momilactone A. *Tetrahedron Lett.* 40: 4051-4054.

Greenberg, J.T. (1996). Programmed cell death: A way of life for plants. *Proc. Natl. Acad. Sci. U. S. A.* 93: 12094-12097.

Greenberg, J.T., Guo, A., Klessig, D.F., and Ausubel, P.M. (1994). Programmed cell death in plants: A pathogen-triggered response activated coordinately with multiple defense functions. *Cell* 77: 551-563.

Gundlach, H., Müller, M.J., Kutchan, T.M., and Zenk, M.H. (1992). Jasmonic acid is a signal transducer in elicitor-induced plant cell cultures. *Proc. Natl. Acad Sci. U. S. A.* 89: 2389-2393.

Hakomori, S. (1981). Glycosphingolipids in cellular interaction, differentiation, and oncogenesis. *Ann. Rev. Biochem.* 50: 733-764.

Hakomori, S. (1990). Bifunctional role of glycosphingolipids. *J. Biol. Chem.* 265: 18713-18716.

Hannun, Y.A. (1996). Functions of ceramide in coordinating cellular responses to stress. *Science* 274: 1855-1859.

Hannun, Y.A. and Bell, R.M. (1989). Functions of sphingolipids and sphingolipid breakdown products in cellular regulation. *Science* 243: 500-507.

Harborne, J.B. (1999). The comparative biochemistry of phytoalexin induction in plants. *Biochem. Syst. Ecol.* 27: 335-367.

He, S.Y., Huang, H.C., and Collmer, A. (1993). *Pseudomonas syringae* pv. *syringae* Harpinr$_{Pss}$: A protein that is secreted via the Hrp pathway and elicits the hypersensitive response in plants. *Cell* 73: 1255-1266.

Inui, H., Yamaguchi, Y., and Hirano, S. (1997). Elicitor actions of *N*-acetylchitooligosaccharides and laminarioligosaccharides for chitinase and L-phenylalanine ammonia-lyase induction in rice suspension culture. *Biosci. Biotech. Biochem.* 61: 975-978.

Jones, D.A., Thomas, C.M., Hammond-Kosack, K.E., Balint-Kurti, P.J., and Jones, J.D.G. (1994). Isolation of the tomato *Cf-9* gene for resistance to *Cladosporium fulvum* by transposon tagging. *Science* 266: 789-793.

Kaliappan, K. and Subba Rao, G.S.R. (1996). An expedient route to the preparation of key intermediates for the total synthesis of aphidacolin, stemodin and oryzalexin S. *Telrahedron Lett.* 37: 8429-8430.

Kato, H., Kodama, O., and Akatsuka, T. (1993). Oryzalexin E, a diterpene phytoalexin from UV-irradiated rice leaves. *Phylochemistry* 33: 79-81.

Kato, H., Kodama, O., and Akatsuka, T. (1994). Oryzalexin F, a diterpene phytoalexin from UV-irradiated rice leaves. *Phylochemistry* 36: 299-301.

Kato, T., Kabuto, C., Sasaki, N., Tsunagawa, M., Aizawa, H., Fujita, K., Kato, Y., and Kitahara, Y. (1973). Momilactones, growth inhibitors from rice, *Oryza saliva* L. *Teltahedron Lett.* 39: 3861-3864.

Kawai, G. (1989). Molecular species of cerebrosides in fruiting bodies of *Lentinus edodes* and their biological activity. *Biochim. Biophys. Acta* 1001: 185-190.

Kawai, G. and Ikeda, Y. (1983). Chemistry and functional moiety of a fruiting-inducing cerebroside in *Schizophyllum commune. Biochim. Biophys. Acta* 754: 243-248.

Kawai, G., Ikeda, Y., and Tubaki, K. (1985). Fruiting of *Schizophyllum commune* induced by certain ceramides and cerebrosides from *Penicillium .funiculosum. Agric. Biol. Chem.* 49: 2137-2146.

Kawai, G., Ohnishi, M., Fujino, Y., and Ikeda, Y. (1986). Stimulatory effect of certain plant sphingolipids on fruiting of *Schizophyllum commune. J. Biol. Chem.* 261: 779- 784.

Keen, N.T. (1975). Specific elicitors of plant phytoalexin production: Determinants of race specificity in pathogens? *Science* 187: 74-75.

Keen, N.T. (1990). Gene-for-gene complementarity in plant pathogen interactions. *Annu. Rev. Genet.* 24: 447-463.

Keen, N.T. (1992) The molecular biology of disease resistance. *Plant Mol. Biol.* 19: 109-122.

Kikuyama, M., Kuchitsu, K., and Shibuya, N. (1997). Membrane depolarization induced by *N*-acetylchitooligosaccharide elicitor in suspension-cultured rice cells. *Plant Cell Physiol.* 38: 902-909.

Kodama, O., Li, W.X, Tamogami, S., and Akatsuka, T. (1992). Oryzalexin S, a novel stemarane-type diterpene rice phytoalexin *Biosci. Biotech. Biochem.* 56: 1002-1003.

Kodama, O., Miyakawa, J., Akatsuka, T., and Kiyosawa, S. (1992). Sakuranetin, a flavanone phytoalexin from ultraviolet-irradiated rice leaves. *Phytochemistry* 31: 3807-3809.

Koga, J., Ogawa, N., Yamaguchi, T., Kikuchi, M., Ogasawara, N., and Shimura, M. (1997). Functional moiety for the antifungal activity of phytocassane E, a diterpene phytoalexin from rice. *Phytochemistry* 44: 249-253.

Koga, J., Oshima, K., Ogawa, N., Ogasawara, N., and Shimura, M. (1998). A new bioassay for measuring elicitor activity in rice leaves. *Ann. Phytopathol. Soc. Jpn.* 64: 97-101.

Koga, J., Shimura, M., Oshima, K., Ogawa, N., Yamaguchi, T., and Ogasawara, N. (1995). Phytocassanes A, B, C and D, novel diterpene phytoalexins from rice, *Oryza sativa* L. *Tetrahedron* 51: 7907-7918.

Koga, J., Yamaguchi, T., Shimura, M., Ogawa, N., Oshima, K., Umemura, K., Kikuchi, M., and Ogasawara, N. (1998). Cerebrosides A and C, sphingolipid elicitors of hypersensitive cell death and phytoalexin accumulation in rice plants. *J. Bioi. Chem.* 273: 31985-31991.

Kono, Y., Takeuchi, S., Kodama, O., and Akatsuka, T. (1984). Absolute configuration of oryzalexin A and structures of its related phytoalexins isolated from rice blast leaves infected with *Pyricularia oryzae. Agric. Biol. Chem.* 48: 253-255.

Kono, Y., Takeuchi, S., Kodama, O., Sekido, H., and Akatsuka, T. (1985). Novel phytoalexins (oryzalexins A, B and C) isolated from rice blast leaves infected with *Pyriclaria oryzae*. Part II: Structural studies of oryzalexins. *Agric. Biol. Chem.* 49: 1695- 1701.

Kreutter, D., Kim, J.Y.H., Goldenring, J.R., Rasmussen, H., Ukomadu, C., DeLorenzo, R.J., and Vu, R.K. (1987). Regulation of protein kinase C activity by gangliosides. *J. Biol. Chem.* 262: 1633-1637.

Kuchitsu, K., Kosaka, T., Shiga, T., and Shibuya, N. (1995). EPR evidence for generation of hydroxyl radical triggered by *N*-acetylchitooligosaccharide elicitor and a protein phosphatase inhibitor in suspension-cultured rice cells. *Protoplasma* 188: 138-142.

Kuchitsu, K., Yazaki, Y., Sakano, K., and Shibuya, N. (1997). Transient cytoplasmic pH change and ion fluxes through the plasma membrane in suspension-cultured rice cells triggered by *N*-acetylchitooligosaccharide elicitor. *Plant Cell Physiol.* 38: 1012-1 018.

Levine, A., Pennell, R.I., Alvarez, M.E., Palmer, R., and Lamb, C. (1996). Calcium-mediated apoptosis in a plant hypersensitive disease resistance response. *Curr. Biol.* 6: 427-437.

Levine, A., Tenhaken, R., Dixon, R., and Lamb, C. (1994) H_20_2 from the oxidative burst orchestrates the plant hypersensitive disease resistance response. *Cell* 79: 583-593.

Martin, G.B., Brommonschenkel, S.H., Chunwongse, J., Frary, A., Ganal, M.W., Spivey, R., Wu, T., Earle, E.D., and Tanksley, S.D. (1993). Map-based cloning of a protein kinase gene conferring disease resistance in tomato. *Science* 262: 1432-1436.

Mauch, F. and Staehelin, L.A. (1989). Functional implications of the subcellular localization of ethylene-induced chitinase and β-1,3-glucanase in bean leaves. *Plant Cell:* 447-457.

Midoh, N. and Iwata, M. (1996). Cloning and characterization of a probenazole-inducible gene for an intracellular pathogenesis-related protein in rice. *Plant Cell Physiol.* 37: 9-18.

Minami, E., Kuchitsu, K., He, D.Y., Kouchi, H., Midoh, N., Ohtsuki, Y. and Shibuya, N. (1996). Two novel genes rapidly and transiently activated in suspension-cultured rice cells by treatment with *N*-acetylchitoheptaose, a biotic elicitor for phytoalexin production. *Plant Cell Physiol.* 37: 563-567.

Mindrinos, M., Katagiri, F., Yu, G.-L., and Ausubel, F.M. (1994). The A. thaliana disease resistance gene *RPS2* encodes a protein containing a nucleotide-binding site and leucine- rich repeats. *Cell* 78: 1089-1099.

Mittler, R., Shulaev, V., and Lam, E. (1995). Coordinated activation of programmed cell death and defense mechanisms in transgenic tobacco plants expressing a bacterial proton pump. *Plant Cell* 7: 29-42.

Mizushina, Y., Hanashima, L., Yamaguchi, T., Takemura, M., Sugawara, F., Saneyoshi, M., Matsukage, A., Yoshida, S., and Sakaguchi, K. (1998). A mushroom fruiting body-inducing substance inhibits activities of replicative DNA polymerases. *Biochem. Biophys. Res. Commun.* 249: 17-22.

Nojiri, H., Kitagawa, S., Nakamura, M., Kirito, K., Enomoto, V., and Saito, M. (1988). Neolacto-series gangliosides induce granulocytic differentiation of human promyelocytic leukemia cell line HL-60. *J. Biol. Chem.* 263: 7443-7446.

Nojiri, H., Sugimori, M., Yamane, H., Nishimura, Y., Yamada, A., Shibuya, N. Kodama, O., Murofushi, N., and Omori, T. (1996) Involvement of jasmonic acid in elicitor-induced phytoalexin production in suspension-cultured rice cells. *Plant Physiol.* 110: 387-392.

Nojiri, H., Takaku, F., Terui, Y., Miura, Y., and Saito, M. (1986). Ganglioside GM3: An acidic membrane component that increases during macrophage-like cell differentiation can induce monocytic differentiation of human myeloid and monocytoid leukemic cell lines HL-60 and U937. *Proc. Natl. Acad. Sci. U.S.A.* 83: 782-786.

Obata, K., Oide, M., and Handa, S. (1977). Effects of glycolipids on *in vitro* development of neuromuscular junction. *Nature* 266: 369-371.

Obeid, L.M., Linardic, C.M., Karolak, L.A., and Hannun, Y.A. (1993). Programmed cell death induced by ceramide. *Science* 259: 1769-1771.

Rakwal, R., Agrawal, G.K., Yonekura, M., and Kodama, O. (2000). Naringenin 7-*O*-methyltransferase involved in the biosynthesis of the flavanone phytoalexin sakuranetin from rice *(Oryza sativa* L.). *Plant Sci.* 155: 213-221.

Rakwal, R., Hasegawa, M., and Kodama, O. (1996). A methyltransferase for the synthesis of the flavanone phytoalexin sakuranetin in rice leaves. *Biochem. Biophys. Res. Commun.* 222: 732-735.

Rakwal, R., Tamogami, S., and Kodama, O. (1996). Role of jasmonic acid as a signaling molecule in copper chloride-elicited rice phytoalexin production. *Biosci. Biotech. Biochem.* 60: 1046-1048.

Roisen, F.J., Bartfeld, H., Nagele, R., and Yorke, G. (1981). Ganglioside stimulation of axonal sprouting *in vitro. Science* 214: 577-578.

Ryerson, D.E. and Heath, M.C. (1996). Cleavage of nuclear DNA into oligonucleosomal fragments during cell death induced by fungal infection or by abiotic treatments. *Plant Cell* 8: 393-402.

Sekido, H., Endo, T., Suga, R., Kodama, O., Akatsuka, T., Kono, Y., and Takeuchi, S. (1986). Oryzalexin D (3,7-Dihydroxy-(+)-sandaracopimaradiene), a new phytoalexin isolated from blast-infected rice leaves. *J. Pesticide. Sci.* 11: 369-372.

Sekido, H., Kamada, K., Kodama, O., and Akatsuka, T. (1987). Antifungal activity of enantiomers of oryzalexins against *Pyricularia aryzae*. *Agric. Biol. Chem.* 51: 2017-2018.

Sekizawa, Y. and Mase, S. (1980). Recent progress in studies on non-fungicidal controlling agent, probenazole, with reference to the induced resistance mechanism of rice plant. *Rev. Plant Protec. Res.* 13: 114-121.

Sharp, J.K., McNeil, M., and Albersheim, P. (1984). The primary structures of one elicitor- active and seven elicitor-inactive hexa (β-D-glucopyranosyl)-D-glucitols isolated from the mycelial walls of *Phytophthora megasperma* f sp. *glycinea*. *J. Biol. Chem.* 259: 11321-11336.

Sharp, J.K., Valent, B., and Albersheim, P. (1984). Purification and partial characterization of a β-glucan fragment that elicits phytoalexin accumulation in soybean. *J. Biol. Chem.* 259: 11312-11320.

Shimura, M., Mase, S., Iwata, M., Suzuki, A., Watanabe, T., Sekizawa, Y., Sasaki, T., Furihata, K., Seto, H., and Otake, N. (1983). Anti-conidial germination factors induced in the presence of probenazole in infected host leaves. III. Structural elucidation of substances A and C. *Agric. Bioi. Chem.* 47: 1983-1989.

Sitrin, R.D., Chan, G., Dingerdissen, J., DeBrosse, C., Mehta, R., Roberts, G., Rottschaefer, S., Staiger, D., Valenta, J., Snader, K.M., et al. (1988). Isolation and structure determination of *Pachybasium* cerebrosides which potentiate the antifungal activity ofaculeacin. *J. Antibiot.* 41: 469-480.

Song, W.-Y., Wang, G.-L., Chen, L.-L., Kim, H.-S., Pi, L.-Y., Holsten, T., Gardner, J., Wang, B., Zhai, W.-X., Zhu, L.-H., et al. (1995). A receptor kinase-like protein encoded by the rice disease resistance gene, *Xa21*. *Science* 270: 1804-1806.

Staskawicz, B.J., Ausubel, F.M., Baker, B.J., Ellis, J.G., and Jones, J.D.G. (1995). Molecular genetics of plant disease resistance. *Science* 268: 661-667.

Tamogami, S., Mitani, M., Kodama, O., and Akatsuka, T. (1993). Oryzalexin S structure: A new stemarane-type rice plant phytoalexin and its biogenesis. *Tetrahedron* 49: 2025-2032.

Tamogami, S., Rakwal, R., and Kodama, O. (1997a). Phytoalexin production by amino acid conjugates of jasmonic acid through induction of naringenin-7-*O*-methyltransferase, a key enzyme on phytoalexin biosynthesis in rice (*Oryza sativa* L.). *FEBS Lett.* 401: 239-242.

Tamogami, S., Rakwal, R., and Kodama, O. (1997b). Phytoalexin production elicited by exogenously applied jasmonic acid in rice leaves (*Oryza sativa* L.) is under the control of cytokinins and ascorbic acid. *FEBS Lett.* 412: 61-64.

Umemura, K., Ogawa, N., Yamaguchi, T., Iwata, M., Shimura, M., and Koga, J. (2000). Cerebroside elicitors found in diverse phytopathogens activate defense responses in rice plants *Plant Cell Physiol.* 41: 676-683.

Wang, H., Li, J., Bostock, R.M., and Gilchrist, D.G. (1996) Apoptosis: A functional paradigm for programmed plant cell death induced by a host-selective phytotoxin and invoked during development. *Plant Cell* 8: 375-391.

Watanabe, T., Igarashi, H., Matsumoto, K., Seki, S., Mase, S., and Sekizawa, Y. (1977). The characteristics of probenazole (Oryzemate®) for the control of rice blast. *J. Pesticide. Sci.* 2: 291-296.

Whitham, S., Dinesh-Kumar, S.P., Choi, D., Hehl, R., Corr, C., and Baker, B. (1994). The product of the tobacco mosaic virus resistance gene *N:* Similarity to Toll and the interleukin-1 receptor. *Cell* 78: 1101-1115.

Yajima, A. and Mori, K. (2000). Absolute configuration of phytocassanes as proposed on the basis of the CD spectrum of synthetic (+)-2-deoxyphytocassane A. *Tetrahedron Lett.* 41: 351-354.

Yamada, A., Shibuya, N., Kodama, O., and Akatsuka, T. (1993). Induction of phytoalexin formation in suspension-cultured rice cells by *N*-acetyl-chitooligosaccharides. *Biosci. Biotech. Biochem.* 57: 405-409.

Yamaguchi, T., Yamada, A., Hong, N., Ogawa, T., Ishii, T., and Shibuya, N. (2000). Differences in the recognition of glucan elicitor signals between rice and soybean: β-glucan fragments from the rice blast disease fungus *Pyricularia oryzae* that elicit phytoalexin biosynthesis in suspension-cultured rice cells. *Plant Cell* 12: 817-826.

Yano, A., Suzuki, K., Uchimiya, H., and Shinshi, H. (1998). Induction of hypersensitive cell death by a fungal protein in cultures of tobacco cells. *Mol. Plant-Microbe Interact.* 11: 115-123.

Chapter 24

Antifungal Agents from Traditional Chinese Medicines Against Rice Blast Fungus *Pyricularia oryzae* Cavara

Ke Hu
Aijun Dong
Hisayoshi Kobayashi
Shigeo Iwasaki
Xinsheng Yao

INTRODUCTION

Rice Blast Fungus **Pyricularia oryzae**

Rice blast is an important worldwide issue and has been a devastating problem since ancient times. It is the most serious and damaging disease of rice in the world due to its distribution and destructiveness, and causes dramatic yield losses on susceptible varieties, altering what a farmer can or cannot grow, and costing large amounts in breeding for resistance and other control options. The blast does not develop every year but it is very destructive when it occurs. Moreover, the occurrence of blast is unpredictable from year to year. In the United States, rice blast is responsible for the loss of millions of dollars annually and often occurs in many states including Arkansas, Alaska, Arizona, Connecticut, Florida, Hawaii, Kentucky, Louisiana,

---

The research for this chapter was financially supported by Northeast Pharmaceutical Co. Ltd. The authors are grateful to Dr. Gang Shao, Mrs. Gexia Qu, Mrs. Yueping Pei, Mrs. Shuhua Dong, Mr. Zhenghuan He, and Mr. Yuan Su, Department of Natural Products Chemistry, Shenyang Pharmaceutical University, China, for preparing some ethanol and aqueous extracts. Many thanks to professors Qishi Sun and Zerong Jiang for identification of plant materials, and to Mr. Jincai Lu, Department of Pharmacognosy, Shenyang Pharmaceutical University, China, for collection of parts of medicinal plants from Yunnan, Sichuan, and Liaoning Provinces of China.

Mississippi, Missouri, New Jersey, South Carolina, and Texas, and in Asian rice-producing countries, including China, Japan, Korea, and Vietnam.

Rice blast is caused by a fungus known as *Pyricularia oryzae* Carava (Moniliaceae), which is a member of Deuteromycotina. *Pyricularia oryzae* (perfect stage, *Magnaporthe grisea*) is a filamentous heterothallic Ascomycete with uninucleate and septate hyphae. The fungus commonly produces three-celled, bowling pin-shaped microscopic conidia, which are the reproductive structures. The conidia can be disseminated by wind and blowing rain over a great distance to new rice plants during the growing season, and thus, initiate new infections. Therefore, *P. oryzae* is seed-borne and seeds from the infected plants can carry and spread if they are used in the next planting season.

Pyricularia cause a variety of plant diseases in members of Gramineae, such as, Oryzoideae and Bambusoideae (Itoi et al., 1978). As far as it is known, *P. oryzae* attacks only rice and has no other plant host. Closely related forms of the blast fungus infect wild grasses and some turfgrasses but do not infect rice.

Pyricularia oryzae overwinters in rice seeds and infects rice stubble. Rice blast symptoms can occur on leaves, leaf sheaths, nodes, and panicles. Leaf spots are typically elliptical (football-shaped), with gray-white centers and brown to red-brown margins. Fully developed leaf lesions are approximately 1.0 to 1.8 cm long and 0.2 to 0.5 cm wide. Variation in shape and color depends on the environment, age of the lesion, and rice variety.

Blast damages plants and reduces yield in a number of ways. Leaf spots or lesions reduce the effective leaf area. Lesions from nodes of the stem and panicle, panicle branches, and the small panicle branches that support individual grains, cause girdling that results in complete grain fill or total grain failure. Losses in severely affected fields may exceed 50 percent. The most serious damage occurs when *P. oryzae* attacks nodes just below the head because the stems often break at the disease node. This stage of the disease is referred to as "rotten neck." Disease in the node prevents the flow of water and nutrients to the kernels which stops development. Heads of plants damaged in this way may be completely blank to nearly normal, depending on the stage of head development when infection occurs. The poorly developed grain usually breaks up badly in milling, reducing quality.

The Approaches to Control Rice Blast

Breeding for resistance is an economical way to control disease within the integrated disease management strategy. However, *P. oryzae* is capable of undergoing rapid genetic mutation and variation under favorable conditions, such as weather with high humidity, thus producing a new strain. Therefore, new virulent strains could cause the resistance of the existing rice varieties to break down and become easily infected. As a result, the useful life span of many cultivars is only one or a few years in disease-conducive environments due to the breakdown of resistance in the face of high pathogenic variability of *P. oryzae* (Atkins, 1962; Padmanabhan et al., 1970; Kiyosawa, 1982; Yamada, 1985; Bonman et al., 1986). Since the efforts to control rice blast by breeding resistant rice cultivars have been frustrated by the frequent emergence of new pathogenic races (Parsons et al., 1987; Leung and Taga, 1988), the breeding for more durably resistant cultivars has become a priority in rice improvement.

The principal approach to controlling rice blast is the use of fungicides against *P. oryzae*. Based on classical tests of minimum inhibitory concentration (MIC), many antibiotics, such as blasticidin S (Takeuchi et al., 1958; Otake et al., 1966), kasugamycin (Hamada et al., 1965; Ishiyama et al., 1965; Umezawa et al., 1965), bramycin (Sakagami et al., 1966), miharamycin (Shomura et al., 1967; Tsuruoka et al., 1967), and aabomycin A (Aizawa et al., 1969), were found to inhibit the growth of *P. oryzae*. However, the development of fungicides against *P. oryzae* was not limited to MIC tests alone. More efficient, target- and mechanism-based designs as well as mechanism elucidation of antifungal agents against *P. oryzae*, have been developed.

Blasticidin S, which is produced by *Streptomyces griseochromogenes* (Takeuchi et al., 1958), has been widely used to control rice blast and was proved to exert significant inhibition of protein synthesis in *P. oryzae* (Huang et al., 1964).

Chitin, a β-1,4-linked polymer of *N*-acetylglucosamine, is one of the most important structural components of the cell wall of fungi (Bartnicki-Garcia, 1968). Although abundant in arthropods, most fungi, and other eukaryotes, chitin is absent from plant and mammalian species. Thus, biosynthesis of chitin is an attractive target in the design of antifungal drugs (Cohen, 1987; Gow et al., 1994; Park et al., 1999).

Edifenphos was marketed by Bayer in 1968 to control *P. oryzae* (Scheinpflug and Jung 1968). The related organophosphorus fungicide, iprobenfos, inhibited incorporation of [^{14}C]-glucosamine into the cell wall of *P. oryzae*, suggesting that the fungicide inhibited chitin synthesis either directly, by inhibiting chitin synthase, or indirectly, by altering membrane permeability (Maeda et al., 1970). Further studies have revealed that *P. oryzae* is very sensitive to edifenphos because of inhibition of phosphatidylcholine biosynthesis (Binks et al., 1993).

Carpropamid was developed as a potent control agent against rice blast (Kagabu and Kurahashi, 1998). Recent biochemical studies have revealed that carpropamid strongly inhibits scytalone dehydratase (Motoyama et al., 1998; Nakasako et al., 1998), a member of the group of enzymes involved in melanin biosynthesis.

Antifungal Agents from TCM

Antifungal agents for use against *P. oryzae* can be obtained from either synthetic or natural products. As far as natural products are concerned, many antibiotics have been isolated from the secondary metabolites of microorganisms (Harada and Kishi, 1977; Okeke et al., 1994, 1995; Tamamura and Sato, 1999; Zhang et al., 1999; Hwang, Yun, Kim, Kwon, Kim, Lee, Bae, et al., 2000; Hwang, Yun, Kim, Kwon, Kim, Lee, Jeong, et al., 2000; Li et al., 2000; Suzuki et al., 2000). Very few studies have been done on antifungal agents from herbal medicine, and then only by MIC tests (Satoh et al., 1996; Flamini et al., 1999). This is also true of traditional Chinese medicines (TCM), which has a long legacy, and of novel bioassay systems.

Morphologically deformed mycelia and/or conidia of fungi, such as curling, swelling, hyperdivergence, bead formation, and inhibition of germination, are often observed in the presence of bioactive substances (Brain, 1949; Isono et al., 1968; Richmond, 1975; Gunji et al., 1983). A bioassay detecting morphological deformation of mycelia and conidia of *P. oryzae* was developed to detect antifungal compounds (Kobayashi et al., 1996). This bioassay was efficiently used in the screening of antifungal agents from fungus metabolites such as rhizoxin (Tsuruo et al., 1986) and fusarielin A (Kobayashi et al., 1995). However, this assay was never tried in herbal medicines. The next section summarizes a first-time application of the assay in the discovery of antifungal compounds from TCM.

ACTIVE TCM EXTRACTS AGAINST P. ORYZAE

Table 24.1 shows different organs of 247 species of herbal TCMs from eighty-three families. They were selected on the basis of their historic treatment of cancer, fungal, or viral infections. The crude materials of herbal medicines were collected from various provinces in China and given a series number. Each crude material was extracted by water and 75 percent ethanol to yield the aqueous and ethanol extracts, respectively, which were then tested for activity against *P. oryzae* with morphological deformation and growth inhibition of mycelia and conidia of *P. oryzae* as activity index. It was assumed that the minimum activity concentration of active extracts should be less than 500 μg/ml.

TABLE 24.1. The activity of ethanol and aqueous extracts of TCM against *Pyricularia oryzae*

No.	TCM	Family	Part	E[a]	W[b]
1	*Andrographis paniculata* (Burm. f.) Nees.	Acanthaceae	whole plant	–[c]	–
2	*Baphicacanthuis cusia* (Nees.) Brem.	Acanthaceae	whole plant	–	–
3	*Dicliptera roxburghiana* Nees.	Acanthaceae	stem	–	–
4	*Strobilanthus auriculates* Bremek.	Acanthaceae	whole plant	–	–
5	*Actinidia arguta* Planch.	Actinicliaceae	root	–	–
6	*Alangium chinense* (Lour.) Harms	Alangiaceae	leaf	125[d]	–
7	*Alangium kurzii* Craib	Alangiaceae	stem	–	–
8	*Curculigo capitulata* Kunth	Amaryllidaceae	stem	–	–
9	*Angiopteris latemarginata* Ching	Angipteridaceae	root	–	–
10	*Fissistigma kwangsiense* Tsiang et P. T. Li	Annonaceae	stem	–	–
11	*Fissistigma minuticalyx* (Mcgr. et W. W. Sm.) Chatterjee	Annonaceae	stem	–	–
12	*Fissistigma oldhamii* (Hemsl.) Merr.	Annonaceae	stem	500	125
13	*Goniothalamus cheliensis* Hu	Annonaceae	bark	125	–
14	*Goniothalamus griffithii* Hook. f. et Thoms.	Annonaceae	bark	125	–
15	*Polyalthia cheliensis* Hu	Annonaceae	bark	125	–
16	*Epigynum auritum* (Scheid.) Tsiang et P. T. Li	Apocynaceae	stem	–	–
17	*Rauvolfia tetraphylla* L.	Apocynaceae	leaf	–	–
18	*Trachelospermum jasminoides* (Lindl.) Lem.	Apocynaceae	stem	–	–
19	*Winchia calophylla* A. DC.	Apocynaceae	stem	–	–
20	*Acorus gramineus* Soland.	Araceae	root	–	–
21	*Arisaema cousanguineum* Schott	Araceae	rhizome	–	–
22	*Rhaphidophora lancifolia* Schott	Araceae	whole plant	–	–
23	*Heteropanax fragrans* (Roxb.) Seem	Araliaceae	stem	–	–
24	*Arca inflata* Reeve.	Arcidae	shell	–	–
25	*Aristolochia kaempferi* Willd.	Aristolochiaceae	root	–	–
26	*Aristolochia mollissima* Hance	Aristolochiaceae	whole plant	–	–
27	*Cynanchum otophyllum* Schneid.	Asclepiadaceae	root	–	–
28	*Dregea volubilis* (L. f.) Benth. ex Hook. f.	Asclepiadaceae	stem	–	–
29	*Basella rubra* L.	Basellaceae	whole plant	–	–
30	*Boussingaultia gracilis* Miers var. *pesudobaselloides* Bailey	Basellaceae	stem	250	–
31	*Mayodendron igneum* (Kurz) Kurz	Bignoniaceae	stem	–	–
32	*Oroxylum indicum* (L.) Vent.	Bignoniaceae	stem	–	–
33	*Bombyx mori* L.	Bombycidae	whole worm	–	–
34	*Arnebia euchroma* (Royle) Johnst.	Boraginaceae	root	–	–

TABLE 24.1 *(continued)*

No.	TCM	Family	Part	E[a]	W[b]
35	*Commiphora myrrha* Engl.	Burseraceae	root	–	–
36	*Garuga floribunda* Decne. var. *gamblai* (King et Smith) Kalkm.	Burseraceae	stem	–	–
37	*Buthus martensi* Karsch	Buthidae	whole worm	–	–
38	*Codonopsis convolvulacea* Kurz var. *forrestii* (Diels) Ballard	Campanulaceae	root	–	–
39	*Codonopsis lanceolata* (Sieb. et Zucc.) Trautv.	Campanulaceae	root	–	–
40	*Lobelia chinensis* Lour.	Campanulaceae	whole plant	–	–
41	*Sambucus chinensis* Lindl.	Caprifoliaceae	whole plant	–	–
42	*Dianthus superbus* L.	Caryophyllaceae	whole plant	–	–
43	*Vaccaria segetalis* (Neck.) Garcke	Caryophyllaceae	seed	–	–
44	*Maytenus hookeri* Loes.	Celastraceae	leaf	–	–
45	*Tripterygium hypoglaucum* (Levl.) Levl. et Hutch.	Celastraceae	root	125[d]	–
46	*Cephalotaxus fortunei* Hook. f.	Cephalotaxaceae	leaf	–	–
47	*Sarcandra glabra* (Thunb.) Nakai	Chloranthaceae	whole plant	–	–
48	*Combretum yunnanensis* Exell	Combretaceae	trig	–	–
49	*Terminalia chebula* Retz.	Combretaceae	fruit	–	–
50	*Arctium lappa* L.	Compositae	fruit	–	–
51	*Artemisia argyii* Levi. et Van.	Compositae	leaf	–	–
52	*Artemisia scoparia* Waldst. et Kitaib.	Compositae	stem	–	–
53	*Atractylodes macrocephala* Koidz.	Compositae	root	–	–
54	*Bidens bipinnata* L.	Compositae	whole plant	–	–
55	*Blumea balsamifera* (L.) DC.	Compositae	stem	–	–
56	*Carthamus tinctorius* L.	Compositae	flower	–	–
57	*Cirsium segetum* Bge.	Compositae	root	–	–
58	*Chrysanthemum indicum* L.	Compositae	flower	–	–
59	*Doellingeria scabra* (Thunb.) Nees.	Compositae	root	–	–
60	*Eclipta prostrata* L.	Compositae	whole plant	–	–
61	*Erigeron breviscapus* (Van.) Hand.-Mazz.	Compositae	root	–	–
62	*Gynura segetum* (Lour.) Merr.	Compositae	leaf	–	–
63	*Laggera pterodonta* (DC.) Benth.	Compositae	whole plant	–	–
64	*Senecio scandens* Buch.-Ham. ex D. Don	Compositae	whole plant	–	–
65	*Siegesbeckia orientalis* L.	Compositae	whole plant	–	–
66	*Taraxacum mongolicum* Hand-Mazz.	Compositae	whole plant	–	–
67	*Isatis tinctoria* L.	Cruciferae	root	–	–
68	*Bolbostemma paniculatum* (Maxim.) Franq.	Cucurbitaceae	rhizome	–	–

No.	TCM	Family	Part	E[a]	W[b]
69	*Gymnopetalum chinensis* (Lour.) Merr.	Cucurbitaceae	stem	–	–
70	*Gynostemma pentaphylla* (Thunb.) Makino	Cucurbitaceae	whole plant	–	–
71	*Melothria indica* Lour.	Cucurbitaceae	whole plant	–	–
72	*Trichosanthes kirilowii* Maxim.	Cucurbitaceae	root	–	–
73	*Cyperus iria* L.	Cyperaceae	whole plant	–	–
74	*Cyperus rotundus* L.	Cyperaceae	root	–	–
75	*Dioscorea bulbifera* L.	Dioscoreaceae	rhizome	250	500
76	*Dioscorea collettii* Hook. f. var. *hypoglauca* Palib.	Dioscoreaceae	rhizome	62	62
77	*Dryopteris crassirhizoma* Nakai	Dryopteridaceae	root	–	–
78	*Elaeocarpus apiculatus* Mast.	Elaeocarpaceae	bark	–	–
79	*Equus asinus* L.	Equidae	bark	–	–
80	*Eucommia ulmoides* Oliv.	Eucommiaceae	bark	–	–
81	*Anemone raddeana* Regel.	Euphorbiaceae	root	–	–
82	*Bischofia javanica* Bl.	Euphorbiaceae	stem	125	–
83	*Croton argyratus* Bl.	Euphorbiaceae	stem	–	–
84	*Euphorbia humifusa* Willd.	Euphorbiaceae	whole plant	–	–
85	*Euphorbia kansui* Liou.	Euphorbiaceae	root	–	–
86	*Euphorbia lathyris* L.	Euphorbiaceae	seed	–	–
87	*Euphorbia royleana* Boiss.	Euphorbiaceae	stem	–	–
88	*Glochidion sphaerogynum* Kurz	Euphorbiaceae	root	–	–
89	*Homonoia riparia* Lour.	Euphorbiaceae	stem	–	–
90	*Knoxia valerianoides* Thorel.	Euphorbiaceae	root	–	–
91	*Mallotus philippinensis* (Lam.) Muell.-Arg.	Euphorbiaceae	stem	–	–
92	*Phyllanthus emblica* L.	Euphorbiaceae	whole plant	–	–
93	*Sauropus androgynus* (L.) Merr.	Euphorbiaceae	whole plant	–	–
94	*Securinega virosa* (Roxb. et Willd.) Baill.	Euphorbiaceae	root	125	–
95	*Geranium nepalense* Sweet	Geraniaceae	whole plant	–	–
96	*Geranium wilfordii* Maxim.	Geraniaceae	whole plant	–	–
97	*Ornithoboea henryi* Craib	Gesneriaceae	whole plant	–	–
98	*Gnetum montanum* Makgr.	Gnetaceae	stem	250	–
99	*Coix lachryma-jobi* L.	Gramineae	seed	–	–
100	*Imperata cylindrica* (L.) P. Beauv. var. Major (Nees) C. E. Hubb.	Gramineae	root	–	–
101	*Setaria viridis* (L.) Beauv.	Gramineae	whole plant	–	–
102	*Hypericum japonicum* Thunb. ex Murray	Guttiferae	whole plant	–	–
103	*Belamcanda chinensis* (L.) DC.	Iridaceae	root	–	–
104	*Juglans cathayensis* Dode	Juglandaceae	leaf	–	–
105	*Pterocarya tonkinensis* (Fr.) Dode	Juglandaceae	bark	250	–

TABLE 24.1 *(continued)*

No.	TCM	Family	Part	E[a]	W[b]
107	*Colebrookea oppositifolia* Sm.	Labiatae	leaf	–	–
108	*Elsholtzia winitiana* Craib	Labiatae	whole plant	–	–
109	*Leucas ciliata* Benth.	Labiatae	whole plant	–	–
110	*Lycopus lucidus* Turcz.	Labiatae	stem	250	–
111	*Ocimum gratissimum* L.	Labiatae	whole plant	–	–
112	*Plectranthus excisus* Maxim.	Labiatae	whole plant	125	–
113	*Prunella vulgaris* L.	Labiatae	whole plant	125	–
114	*Rabdosia lophanthoides* Hara	Labiatae	leaf	125	–
115	*Rabdosia rosthornii* Hara	Labiatae	whole plant	250	–
116	*Salvia chinensis* Benth.	Labiatae	whole plant	4	–
117	*Salvia digitaloides* Diels	Labiatae	root	–	–
118	*Salvia miltiorrhiza* Bge.	Labiatae	root	–	–
119	*Scutellaria baicalensis* Georgi	Labiatae	root	–	–
120	*Scutellaria barbata* D. Don	Labiatae	whole plant	–	–
121	*Scutellaria likiangensis* Diels	Labiatae	root	–	–
122	*Laminaria japonica* Aresch.	Laminariaceae	thallus	–	–
123	*Cassytha filiformis* L.	Lauraceae	stem	–	–
124	*Litsea cubea (Lour.)* Pers.	Lauraceae	stem	–	
125	*Astragalus membranaceus* (Fisch.) Bge.	Leguminosae	root	–	–
126	*Caesalpinia sappan* L.	Leguminosae	stem	–	–
127	*Cassia laevigata* Willd.	Leguminosae	leaf	125	–
128	*Crotalaria calycina* Schrank	Leguminosae	leaf	–	–
129	*Crotalaria dubia* Grah. ex Benth.	Leguminosae	root	–	–
130	*Crotalaria ferruginea* Grah.	Leguminosae	whole plant	–	–
131	*Crotalaria pallida* Ait.	Leguminosae	whole plant	–	–
132	*Desmodium pulchellum* L.	Leguminosae	whole plant	–	–
133	*Erythrina lithosperma* Bl. ex Miq.	Leguminosae	stem	–	–
134	*Gleditsia sinensis* Lam	Leguminosae	trig	–	–
135	*Glycyrrhiza uralensis* Fisch.	Leguminosae	root	–	–
136	*Gueldenstaedtia yunnanesis* Franch.	Leguminosae	root	–	–
137	*Lespedeza juncea* (L. f.) Pers.	Leguminosae	whole plant	–	–
138	*Millettia dielsiana* Harms	Leguminosae	stem	125	–
139	*Millettia leptobotrya* Dunn	Leguminosae	root	–	–
140	*Phyllodium longipes* (Craib) Schindl.	Leguminosae	whole plant	–	–
141	*Psoralea corylifolia* L.	Leguminosae	fruit	–	250
142	*Sophora flavescens* Ait.	Leguminosae	root	–	–
143	*Sophora subprostrata* Chun et T. Chen	Leguminosae	root	–	–

No.	TCM	Family	Part	E[a]	W[b]
144	*Asparagus cochinchinensis* (Lour.) Merr.	Liliceae	root	–	500
145	*Paris polyphylla var. chinensis* (Franch.) Hara	Liliceae	rhizome	125	–
146	*Polygonatum kingianum* Coll. et Hemsl.	Liliceae	whole plant	–	–
147	*Smilax china* L.	Liliceae	root	250	–
148	*Buddleja davidii* Franch.	Loganiaceae	whole plant	31	–
149	*Buddleja officinalis* Maxim.	Loganiaceae	flower	–	–
150	*Scurrula parasitica* L.	Loranthaceae	stem	–	–
151	*Hibiscus mutabilis* L.	Malvaceae	stem	–	–
152	*Malvastrum coromandelianum* (L.) Gurcke	Malvaceae	root	–	–
153	*Sida alnifolia* L.	Malvaceae	whole plant	–	–
154	*Sida rhombifolia* L.	Malvaceae	whole plant	–	–
155	*Melastoma polyanthum* H. L. Li	Melastomaceae	whole plant	–	–
156	*Cipadessa cinerascens* (Pell.) Hand.-Mazz.	Meliaceae	stem	250	–
157	*Melia toosanden* Sieb. et Zucc.	Meliaceae	fruit	–	–
158	*Menispermum dauricum* DC.	Menispermaceae	root	–	–
159	*Stephania delavayi* Diels	Menispermaceae	stem	–	–
160	*Stephania hernandifolia* (Willd.) Walp.	Menispermaceae	stem	–	–
161	*Stephania tetrandra* S. Moore	Menispermaceae	root	–	–
162	*Tinomiscium tonkinense* Gagnep.	Menispermaceae	stem	–	–
163	*Antiaris toxicaria* (Pers.) Lesch.	Moraceae	stem	–	–
164	*Cudrania tricuspidata* (Carr.) Bur.	Moraceae	fruit	–	–
165	*Ficus tinctoria* Forst. f. subsp. *gibbosa* (Bl.) Corner	Moraceae	bark	–	–
166	*Streblus indica* (Bur.) Corner	Moraceae	stem	125	–
167	*Psidium guajava* L.	Myrtaceae	leaf	–	–
168	*Syzygium szemaoense* Merr. et Perry	Myrtaceae	stem	–	–
169	*Jasminum laurifolium* Roxb.	Oleaceae	stem	–	–
170	*Ligustrum lucidum* Ait.	Oleaceae	fruit	–	–
171	*Ludwigia hyssopifolia* (G. Don) Exell.	Onagraceae	whole plant	125	–
172	*Caryota ochlandra* Hance	Palmae	stem	–	–
173	*Chelidonium majus* L.	Papaveraceae	whole plant	–	–
174	*Gallus gallus domesticus* Brisson.	Phasianidae	membrane	–	–
175	*Piper betle* L.	Piperaceae	whole plant	–	–
176	*Piper fasciata* Fr.	Piperaceae	whole plant	–	–
177	*Piper longum* L.	Piperaceae	stem	–	–
178	*Piper sarmentosum* Roxb. ex Hunter	Piperaceae	whole plant	–	–
179	*Plantago asiatica* L.	Plantaginaceae	whole plant	–	–
180	*Antenoron neofiliforme* (Nakai) Hara	Polygonaceae	whole plant	–	–

TABLE 24.1 *(continued)*

No.	TCM	Family	Part	E[a]	W[b]
181	*Polygonum cuspidatum* Sieb. et Zucc.	Polygonaceae	root	–	–
182	*Polygonum multiflorum* Thunb.	Polygonaceae	root	–	–
183	*Polygonum tinctorium* Ait.	Polygonaceae	leaf	–	–
184	*Polyporus umbellatus* (Pers.) Fr.	Polyporaceae	sclerotium	–	–
185	*Poria cocos* (Fries) Wolf.	Polyporaceae	sclerotium	–	–
186	*Baliospermum effusum* Pax et Hoffm.	Ranunculaceae	stem	–	–
187	*Clematis armandii* Franch.	Ranunculaceae	stem	–	–
188	*Clematis rubifolia* C. H. Wright	Ranunculaceae	stem	–	–
189	*Coptis chinensis* Franch.	Ranunculaceae	aerial part	125	125
190	*Coptis chinensis* Franch.	Ranunculaceae	root	–	–
191	*Paeonia lactiflora* Pall.	Ranunculaceae	root	–	–
192	*Agrimonia pilosa* Ledeb.	Rosaceae	whole plant	–	–
193	*Duchesnea indica* (Andr.) Forke	Rosaceae	whole plant	–	–
194	*Prunus armeniaca* L.	Rosaceae	seed	–	–
195	*Prunus mume* (Sieb.) Sieb. et Zucc.	Rosaceae	fruit	–	–
196	*Prunus persica* (L.) Batsch.	Rosaceae	seed	–	–
197	*Rubus multibracteatus* Levl. et Vent.	Rosaceae	stem	–	–
198	*Sanguisorba officinalis* L.	Rosaceae	root	–	–
199	*Anthocephalus chinensis* (L.) Rich. ex Walp.	Rubiaceae	bark	–	–
200	*Hedyotis diffusa* Willd.	Rubiaceae	whole plant	125	–
201	*Mussaenda hossei* Craib	Rubiaceae	stem	–	–
202	*Mussaenda mollissima* C. Y. Wu ex Hsue	Rubiaceae	stem	–	–
203	*Mussaenda pubescens* Ait. f.	Rubiaceae	stem	125	–
204	*Oldenlandia corymbosa* L.	Rubiaceae	whole plant	–	–
205	*Oldenlandia diffusa* (Willd.) Roxb.	Rubiaceae	whole plant	–	–
206	*Psychotria calocarpa* Kurz.	Rubiaceae	stem	–	–
207	*Rubia cordifolia* L.	Rubiaceae	root	500	500
208	*Citrus aurantium* L.	Rutaceae	leaf	–	–
209	*Evodia austrosinensis* Hand.-Mazz.	Rutaceae	trig	–	–
210	*Glycosmis pentaphylla* (Retz.) Correa	Rutaceae	stem	–	–
211	*Phellodendron chinensis* Schneid.	Rutaceae	bark	125	–
212	*Zanthoxylum planispinum* Sieb. et Zucc.	Rutaceae	stem	125	–
213	*Arytera litoralis* Bl.	Sapindaceae	stem	–	–
214	*Houttuynia cordata* Thunb.	Saururaceae	whole plant	–	–
215	*Schisandra chinensis* (Turcz.) Baill.	Schisandraceae	fruit	–	–
216	*Scolopendra subspinipes mutilans* L.	Scolopendridae	whole worm	–	–

No.	TCM	Family	Part	E[a]	W[b]
217	*Scrophularia ningpoensis* Hemsl.	Scrophulariaceae	root	–	–
218	*Brucea javanica* (L.) Merr.	Simaroubaceae	fruit	–	–
219	*Lycium chinensis* Mill.	Solanaceae	fruit	–	–
220	*Solanum indicum* L.	Solanaceae	root	–	–
221	*Solanum lyratum* Thunb.	Solanaceae	whole plant	<u>250</u>	<u>500</u>
222	*Solanum nigrum* L.	Solanaceae	whole plant	<u>125</u>	<u>250</u>
223	*Solanum spirale* Roxb.	Solanaceae	stem	–	–
224	*Solanum verbacifolium* L.	Solanaceae	root	–	–
225	*Sparganium stoloniferum* Buch.-Ham.	Sparganiaceae	rhizome	–	–
226	*Stemona tuberosa* Lour.	Stemonaceae	root	–	–
227	*Sterculia lanceolata* Cav.	Sterculiaceae	trig	–	–
228	*Wikstroemia indica* (L.) C. A. Mey.	Thymelaeaceae	root	<u>250</u>	<u>250</u>
229	*Bupleurum chinense* DC.	Umbelliferae	root	–	–
230	*Glehnia littoralis* F. Schmidt ex Miq.	Umbelliferae	root	–	–
231	*Heracleum rapula* Franch.	Umbelliferae	root	–	–
232	*Ligusticum chuanxiong* Hort.	Umbelliferae	root	–	–
233	*Boehmeria longispica* Steud.	Urticaceae	whole plant	–	–
234	*Pellionia tsoongii* (Merr.) Merr.	Urticaceae	whole plant	–	–
235	*Urtica mairei* Levl.	Urticaceae	whole plant	–	–
236	*Callicarpa arborea* Roxb.	Verbenaceae	whole plant	–	–
237	*Clorodendrum philippinum* Schauer var. *simplex* Moldenke	Verbenaceae	whole plant	–	–
238	*Vitex trifolia* L.	Verbenaceae	fruit	250	–
239	*Leea hispida* Gagnep.	Vitaceae	stem	–	–
240	*Leea macrophylla* Roxb. ex Hornem.	Vitaceae	leaf	<u>125</u>	–
241	*Tetrastigma crucistum* Craib et Gagnep.	Vitaceae	stem	–	–
242	*Caulokaempferia yunnanensis* (Gagn.) R. M. Sm.	Zingiberaceae	rhizome	–	–
243	*Costus speciosus* (Koenig) Sm.	Zingiberaceae	rhizome	–	–
244	*Curcuma aromatica* Salisb.	Zingiberaceae	root	–	–
245	*Curcuma longa* L.	Zingiberaceae	root	250	–
246	*Curcuma zedoaria* (Berg.) Rosc.	Zingiberaceae	root	–	–
247	*Hedychium coronarium* Koenig	Zingiberaceae	root	125	–
248	*Kaempferia galanga* L.	Zingiberaceae	root	–	–

Note: The data underlined indicate the extracts causing morphological deformation ("3+") otherwise inhibition of conidia germination and mycelium growth (×).
[a] Ethanol extract
[b] Aqueous extract
[c] Inactive
[d] The minimum active concentration values in the fifth and sixth columns are μg/ml.

Ten ethanol extracts from seven families caused morphological deformation of mycelia and conidia of *P. oryzae,* including *Tripterygium hypoglaucum* root (Celastraceae) (No. 45, 125 μg/ml), *Dioscorea bulbifera* rhizome (Dioscoreaceae) (No. 75, 250 μg/ml), *Dioscorea collettii* var. *hypoglauca* rhizome (Dioscoreaceae) (No. 76, 62 μg/ml), *Cassia laevigata* leaf (Leguminosae) (No. 127, 125 μg/ml), *Millettia dielsiana* stem (Leguminosae) (No. 138, 125 μg/ml), *Paris polyphylla* var. *chinensis* rhizome (Liliceae) (No. 145, 125 μg/ml), *Solanum lyratum* whole plant (Solanaceae) (No. 220, 250 μg/ml), *Solanum nigrum* whole plant (Solanaceae) (No. 221, 125 μg/ml), *Wikstroemia indica* root (Thymelaeaceae) (No. 227, 250 μg/ml) and *Leea macrophylla* leaf (Vitaecae) (No. 239, 125 μg/ml). Five aqueous extracts from three families, *Dioscorea bulbifera* rhizome (Dioscoreaceae) (No. 75, 500 μg/ml), *Dioscorea collettii* var. *hypoglauca* rhizome (Dioscoreaceae) (No. 76, 62 μg/ml), *Solanum lyratum* whole plant (Solanaceae) (No. 220, 500 μg/ml), *Solanum nigrum* whole plant (Solanaceae) (No. 221, 250 μg/ml) and *Wikstroemia indica* root (Thymelaeaceae) (No. 227, 250 μg/ml) also induced morphological deformation of mycelia and conidia of *P. oryzae.* Among these extracts, the aqueous and ethanol extracts of *D. collettii* var. *hypoglauca* rhizome (Dioscoreaceae) (No. 76, 62 μg/ml) were the most active.

Thirty ethanolic extracts from seventeen families showed inhibition against conidia germination and mycelium growth of *P. oryzae,* including *Alangium chinense* leaf (Alangiaceae) (No. 6, 125 μg/ml), *Fissistigma oldhamii* stem (Annonaceae) (No. 12, 500 μg/ml), *Goniothalamus cheliensis* bark (Annonaceae) (No. 13, 125 μg/ml), *Goniothalamus griffithii* bark (Annonaceae) (No. 14, 125 μg/ml), *Polyalthia cheliensis* bark (Annonaceae) (No. 15, 125 μg/ml), *Boussingaultia gracilis* stem (Basellaceae) (No. 30, 250 μg/ml), *Bischofia javanica* stem (Euphorbiaceae) (No. 82, 125 μg/ml), *Securinega virosa* root (Euphorbiaceae) (No. 94, 125 μg/ml), *Gnetum montanum* stem (Gnetaceae) (No. 98, 250 μg/ml), *Pterocarya tonkinensis* bark (Juglandaceae) (No. 105, 250 μg/ml), *Lycopus lucidus* stem (Labiatae) (No. 110, 250 μg/ml), *Plectranthus excisus* whole plant (Labiatae) (No. 112, 125 μg/ml), *Prunella vulgaris* whole plant (Labiatae) (No. 113, 125 μg/ml), *Rabdosia lophanthoides* leaf (Labiatae) (No. 114, 125 μg/ml), *Rabdosia rosthornii* whole plant (Labiatae) (No. 115, 250 μg/ml), *Salvia chinensis* whole plant (Labiatae) (No. 116, 4 μg/ml), *Smilax china* root (Liliceae) (No. 147, 250 μg/ml), *Buddleja officinalis* whole plant (Loganiaceae) (No. 148, 31 μg/ml), *Cipadessa cinerascens* stem (Meliaceae) (No. 156, 250 μg/ml), *Streblus indica* stem (Moraceae) (No. 166, 125 μg/ml), *Ludwigia hyssopifolia* whole plant (Onagraceae) (No. 171, 125 μg/ml), *Coptis chinensis* aerial part (Ranunculaceae) (No. 188, 125 μg/ml), *Hedyotis diffusa* whole plant (Rubiaceae) (No. 199, 125 μg/ml), *Mussaenda pubescens* stem (Rubiaceae) (No. 202, 125 μg/ml), *Rubia cordifolia* root (Rubiaceae) (No. 206, 500 μg/ml), *Phellodendron chinensis* bark (Rutaceae) (No. 210, 125 μg/ml), *Zanthoxylum planispinum* stem (Rutaceae) (No. 211, 125 μg/ml), *Vitex trifolia* fruit (Verbenaceae) (No. 237, 250 μg/ml), *Curcuma longa* root (Zingiberaceae) (No. 244, 250 μg/ml), and *Hedychium coronarium* root (Zingiberaceae) (No. 246, 125 μg/ml). Activity was mainly observed in the families of Labiatae, Annonaceae, and Rubiaceae. The ethanol extract of *S. chinensis* whole plant (Labiatae) (No. 116, 4 μg/ml) exhibited the strongest activity. Five aqueous extracts from five families also showed inhibition against conidia germination and mycelium growth of *P. oryzae,* including *Fissistigma oldhamii* stem (Annonaceae) (No. 12, 125 μg/ml), *Psoralea corylifola* fruit (Leguminosae) (No. 141, 250 μg/ml), *Asparagus cochinchinensis* root (Liliceae) (No. 144, 500 μg/ml), *Coptis chinensis* aerial part (Ranunculaceae) (No. 188, 125 μg/ml), and *Rubia cordifolia* root (Rubiaceae) (No. 206, 500 μg/ml).

This kind of morphological deformation has never been reported before. Therefore, novel mechanism(s) were probably involved. Active extracts inducing morphological deformation of mycelia and conidia of *P. oryzae* were of particular interest. Further studies were conducted to identify chemical constituents from the extracts that were responsible for the activity against *P. oryzae*. Further bioactivity-guided fractionation with *P. oryzae* assay resulted in the isolation of fourteen, three, and six antifungal compounds from *D. collettii* var. *hypoglauca* rhizome (Dioscoreaceae), *S. nigrum* whole plant (Solanaceae), and *W. indica* root (Thymelaeaceae), respectively, which are discussed in the following sections.

ANTIFUNGAL COMPOUNDS 1-14 FROM D. COLLETTII *VAR.* HYPOGLAUCA *AGAINST* P. ORYZAE

The rhizome of *D. collettii* var. *hypoglauca* (Dioscoreaceae) has been used for the treatment of fungus, cervical carcinoma, carcinoma of the bladder, and renal tumor. The plant is widely distributed in southeast China. It was included in the 1985, 1990, and 1995 editions of *The Pharmacopoeia of People's Republic of China* based on its traditional application in the treatment for fungus and cancer. However, very few studies have been carried out, except for three phytochemical reports (Lou et al., 1984; Tang and Pang, 1984; Tang et al., 1986).

In this study, bioactivity-guided fractionation with *P. oryzae* assay led to the isolation of fourteen steroidal saponins from the ethanol extract of *D.*

collettii var. *hypoglauca* rhizome (see Figure 24.1), including three spirostanol saponins, i.e., prosapogenin A of dioscin (1), dioscin (2), and gracillin (3) (Hu et al., 1996); nine furostanol saponins, i.e., protoneodioscin (4), protodioscin (5), protoneogracillin (6), protogracillin (7), methyl protoneodioscin (8), methyl protodioscin (9), methyl protoneogracillin (10), methyl protogracillin (11), (Hu et al., 1997a) and hypoglaucin F (12) (Hu et al., 1997b); and two prenane glycosides, i.e., hypoglaucin G (13) and hypoglaucin H (14) (Hu, Kobayshi, Dong, et al., 1998; Hu, Don, et al., 1999).

As shown in Figure 24.1, spirostanol saponins 1-3 have the same six-ring aglycone diosgenin but are different in C-3 sugar chains. Furostanol saponins 4-12 have five-ring aglycones as well as C-3 and C-26 sugar chains. The C-26 sugar chains of 4-12 are the same, i.e. β-D-glucose. In the structures of 4 and 5, 6 and 7, 8 and 9, 10 and 11 are four pairs of R/S stereoisomers at C-25. Pregnane glycosides 13 and 14 have four-ring aglycones and the same C-3 sugar chains. It is quite interesting that there are only three types of C-3 sugar chains among 1-14, i.e., type I: one inner β-D-glucose attached to C-3 of the aglycone, and one terminal α-L-rhamnose attached to C-2 of the inner glucose; type II: one inner β-D-glucose attached to C-3 of the aglycone, and two terminal α-L-rhamnose attached to C-2 and C-4 of the inner β-D-glucose; type III: one inner β-D-glucose attached to C-3 of the aglycone, one terminal α-L-rhamnose attached to C-2 of the inner β-D-glucose, and one terminal β-D-glucose attached to C-3 of the inner β-D-glucose. The moiety of β-D-Glc^2-α-L-Rha is the common point of these three types. Structure 1 has type I sugar chain; 2, 4, 5, 8, 9, 12, 13, and 14 have type II sugar chain; 3, 6, 7, 10, and 11 have type III sugar chain. In the viewpoint of biosynthesis, furostanol saponins 5 and 8 are the precursors of spirostanol saponin 2; furostanol saponins 7 and 11 are the precursors of spirostanol saponin 3; pregnane glycoside 14 is the precursor of pregnane glycoside 13.

As listed in Table 24.2, compounds1-14 induced morphological deformation of mycelia and conidia of *P. oryzae* with MMDC (minimum morphological deformation concentration) 5.5, 2.3, 9.0, 95.4, 95.4, 94.0, 94.0, 15.1, 15.1, 14.8, 14.8, 324, 135, and 236 μM, respectively. Compund 2 was the most active among 1-14. The positive controls were antifungal drugs, i.e., fusarielin A, griseofulvin, econazole, nocodazole, and thiabendazole. All of their MMDCs were 50 μM except fusarielin A (MMDC 15 μM). As far as the chemical structure is concerned, spirostanol saponins 1-3 were much stronger than furostanol saponins 4-12 and pregnane glycosides 13 and 14, suggesting the critical role of F close ring of 1-3. The 22-methoxyl furostanol saponins 8-11 were more active than 22-hydroxyl furostanol saponins 4-7, suggesting the critical role of methoxyl group attached to C-22.

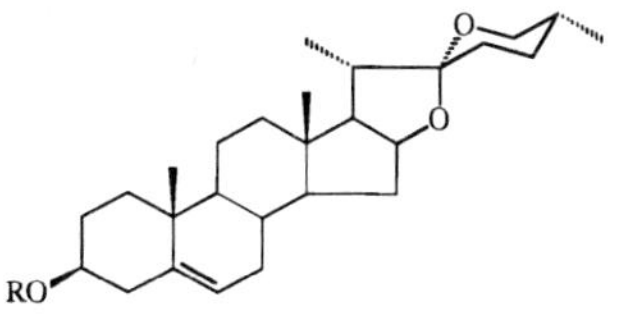

1 R = -β-D-Glc ^{2}-α-L-Rha

2 R = -β-D-Glc < 2 α-L-Rha; 4 α-L-Rha

3 R = -β-D-Glc < 2 α-L-Rha; 3 β-D-Glc

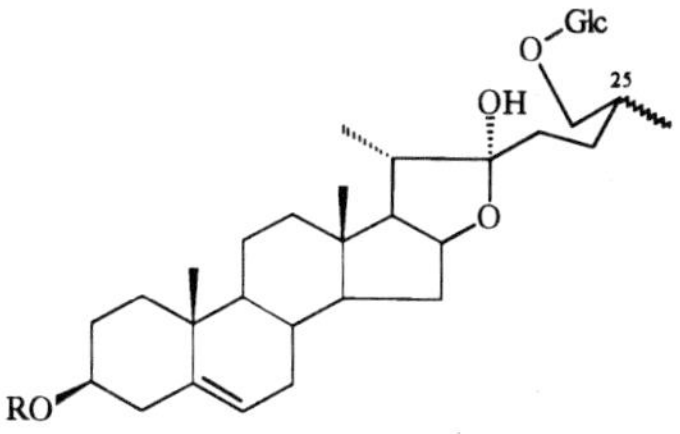

4 R = -β-D-Glc < 2 α-L-Rha; 4 α-L-Rha 25S

5 R = -β-D-Glc < 2 α-L-Rha; 4 α-L-Rha 25R

6 R = -β-D-Glc < 2 α-L-Rha; 3 β-D-Glc 25S

7 R = -β-D-Glc < 2 α-L-Rha; 3 β-D-Glc 25R

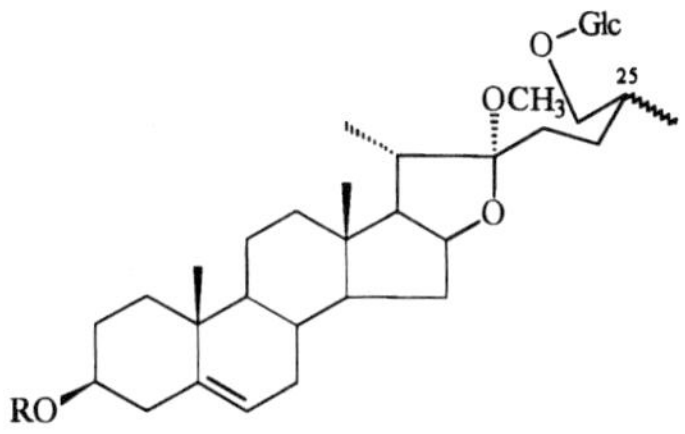

8 R = -β-D-Glc < 2 α-L-Rha; 4 α-L-Rha 25S

9 R = -β-D-Glc < 2 α-L-Rha; 4 α-L-Rha 25R

10 R = -β-D-Glc < 2 α-L-Rha; 3 β-D-Glc 25S

11 R = -β-D-Glc < 2 α-L-Rha; 3 β-D-Glc 25

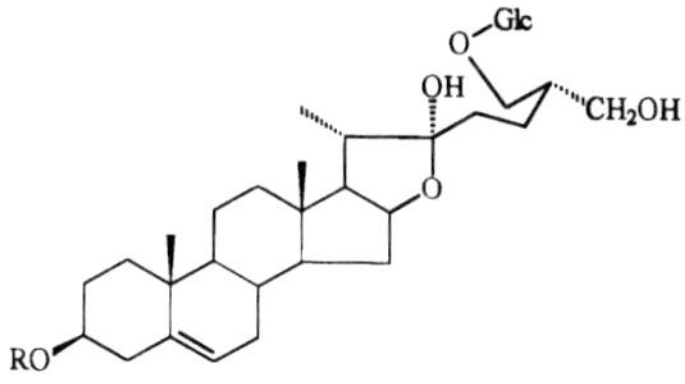

12 R = -β-D-Glc < 2 α-L-Rha; 4 α-L-Rha

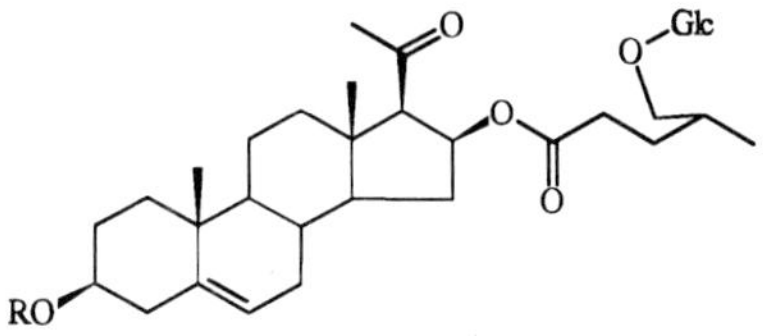

13 R = -β-D-Glc < 2 α-L-Rha; 4 α-L-Rha

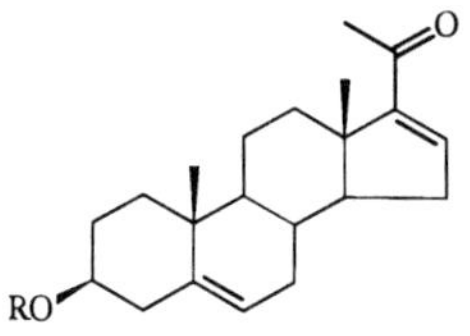

14 R = -β-D-Glc < 2 α-L-Rha; 4 α-L-Rha

FIGURE 24.1. The structures of compounds 1–14 from *Dioscorea collettii* var. *hypoglauca*

TABLE 24.2. The activity of compounds 1-23 against *Pyricularia oryzae*

Plant Sources	No.	Compounds	MMDC (μM)
Dioscorea collettii var. *hypoglauca*	1	prosapogenin A of dioscin	5.5
	2	dioscin	2.3
	3	gracillin	9.0
	4	protoneodioscin	95.4
	5	protodioscin	95.4
	6	protoneogracillin	94.0
	7	protogracillin	94.0
	8	methyl protoneodioscin	15.1
	9	methyl protodioscin	15.1
	10	methyl protoneogracillin	14.8
	11	methyl protogracillin	14.8
	12	hypoglaucin F	324
	13	hypoglaucin G	135
	14	hypoglaucin H	236
Solanum nigrum	15	β_2–solamargine	63.0
	16	solamargine	38.5
	17	degalactotigonin	97.2
Wikstroemia indica	18	daphnoretin	68.4
	19	(+)-nortrachalogenin	31.3
	20	genkwanol A	45.8
	21	wikstrol A	70.1
	22	wikstrol B	52.3
	23	daphnodorin B	73.7
Positive control		fusarielin A	15
		griseofulvin	15
		econazole	50
		nocodazole	50
		thibendazole	50

The 25S furostanol saponins 4, 6, 8, and 10 showed the same activity as their respective 25R furostanol saponins 5, 7, 9, and 11, suggesting that the R/S configuration at C-25 did not have an impact on the activity. Furostanol saponin 12 with a hydroxyl group attached to C-27 showed the weakest activity, suggesting the critical role of the methyl group attached to C-25. Spirostal saponins 1-3 and furostanol saponins 4-11 were much stronger than pregnane glycosides 13 and 14, suggesting the critical roles of E and F rings.

ANTIFUNGAL COMPOUNDS 15-17 FROM S. NIGRUM *AGAINST* P. ORYZAE

Solanum nigrum (Solanaceae) is widely distributed in most regions of China. It has been traditionally used for the treatment of fungus and cancer. Previous studies on *S. nigrum* mainly focused on the isolation of chemical constituents including flavonols, steroidal glycoalkaloids, and steroidal saponins (Varshney and Dube, 1970; Saijo et al., 1982; Doepke et al., 1987, 1988; Nawwar et al., 1989; Ridout et al., 1989). A biological study on *S. nigrum* only reported the cytotoxicity of solamargine, solasonine, and degalactotigonin against human JTC-26 ovarian cancer cell line in vitro (Saijo et al., 1982).

In this study, bioactivity-guided fractionation with *P. oryzae* assay led to the isolation of two alkaloidal glycosides, β_2-solamargine (15) and solamargine (16); and one steroidal saponin, degalactotigonin (17) (see Figure 24.2) (Hu, Kobayashi, Jing, et al., 1998; Hu, Kobayashi, et al., 1999). Compounds 15-17 induced morphological deformation of mycelia and conidia of *P. oryzae* with MMDC 63.0, 38.5, and 97.2 μM, respectively (see Table 24.2).

Compounds 15 and 16 have the same aglycone solasodine but differ in C-3 sugar chains. The sugar chain of 15 contains one inner β-D-glucose attached to C-3 of the aglycone, and one terminal α-L-rhamnose attached to C-4 of the inner β-D-glucose. The C-3 sugar chain of 16 contains one inner β-D-glucose attached to C-3 of the aglycone, and two terminal α-L-rhamnose attached to C-2 and C-4 of the inner β-D-glucose, respectively. Actually, the sugar chain of 16 is type II, which is the same as those of 2, 4, 5, 8, 9, 12, 13, and 14. Therefore, 16 is different from 2 only in the hetero-atoms of F ring. However, this minor structural difference did lead to 17-fold activity of 2 more than that of 16, suggesting the critical role of oxygen in F ring. It is the same reason why the activity of 1 is 11-fold more than that of 15.

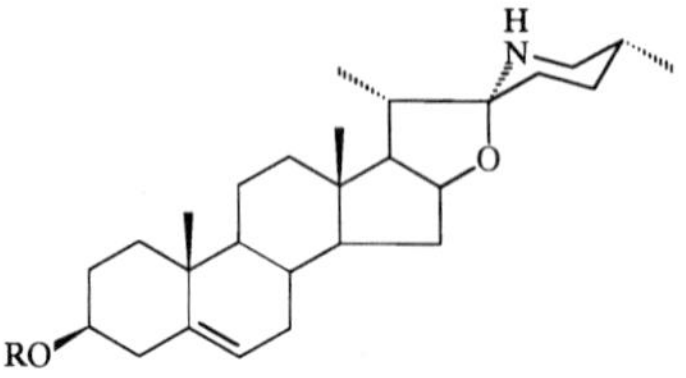

15 R = -β-D-Glc^{2}-α-L-Rha

16 R = -β-D-Glc$<^{2}_{4}$ α-L-Rha / α-L-Rha

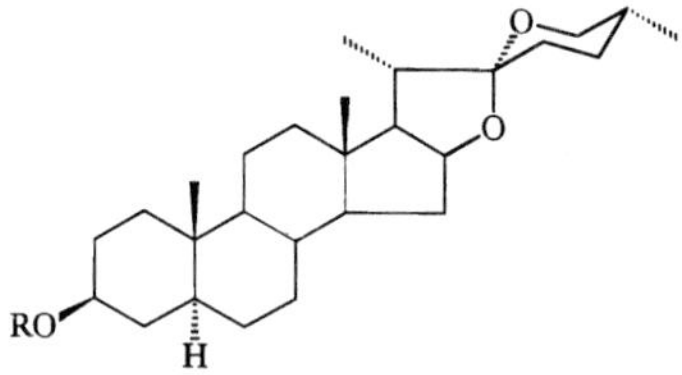

17 R = -β-D-Gal^{4}-β-D-Glc$<^{2}_{3}$ β-D-Glc / β-D-Xyl

FIGURE 24.2. The structures of compounds 15-17 from *Solanum nigrum*

ANTIFUNGAL COMPOUNDS 18-23 FROM W. INDICA *AGAINST* P. ORYZAE

Wikstroemia indica (Thymelaeaceae) is widely distributed in southern China, and its root has been used as an herbal remedy for the treatment of cancer and syphilis. Previous studies have reported on the isolation of flavanols, biscoumarins, lignans, and polysaccharides from root of *W. indica* (Tandon and Rastogi, 1976; Xie and Yang, 1978; Suzuki et al., 1982; Wang et al., 1985; Geng et al., 1988). Bioactivity studies regarding central nervous system (CNS) depression of (+)-nortrachelogenin (Kato et al., 1979), cytotoxicity of tricin, kaempferol 3-*O*-β-D-glucoside and (+)-nortrachelogenin against P-388 lymphocytic leukemia cell line in vivo (Lee et al., 1981), and the mechanism studies on daphnoretin as a protine kinase C (PKC) activator (Ko et al., 1993; Wang et al., 1995; Chen et al., 1996) have been reported.

In this study, bioactivity-guided fractionation with *P. oryzae* assay led to the isolation of one biscoumarin daphnoretin (18), one lignan (+)-nortrachelogenin (19), four biflavonoids genkwanol A (20), wikstrol A (21), wikstrol B (22) and daphnodorin B (23) from root of *W. indica* (see Figure 24.3) (Hu et al., 2000). Compounds 18-23 induced morphological deformations of mycelia and conidia of *P. oryzae* with similar MMDCs 68.4, 31.3, 45.8, 70.1, 52.3, and 73.7 μM, respectively (see Table 24.2).

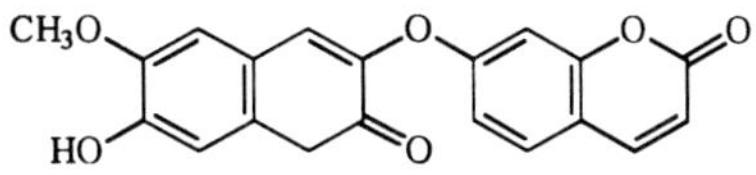

18

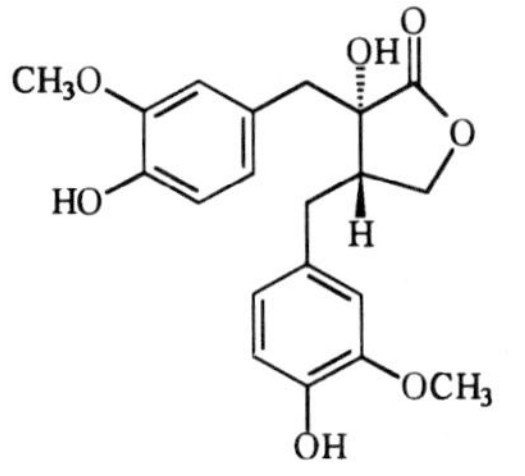

19

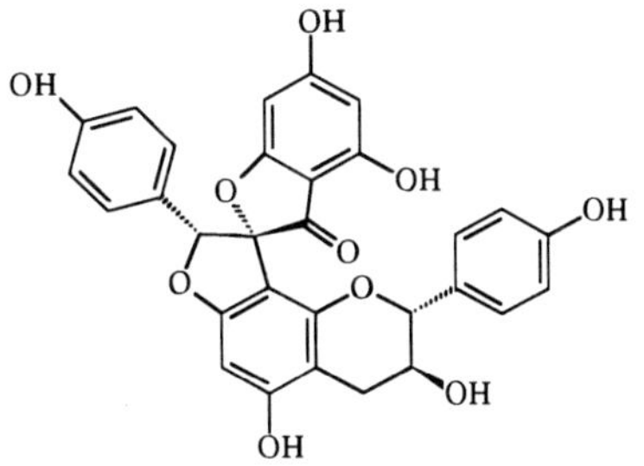

20

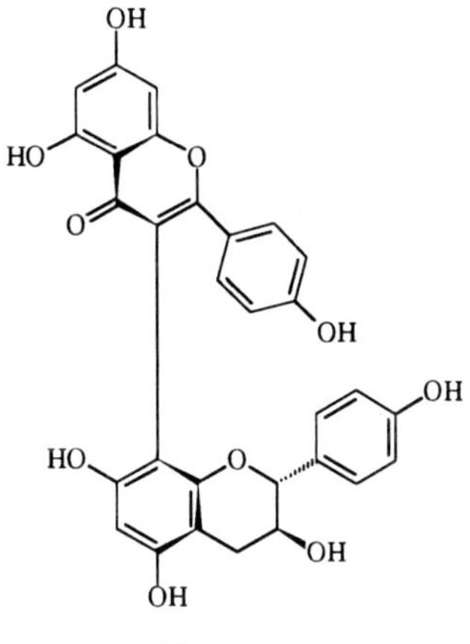

21

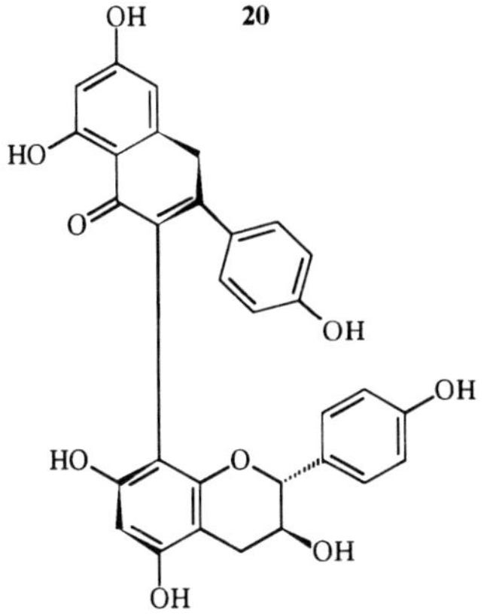

22

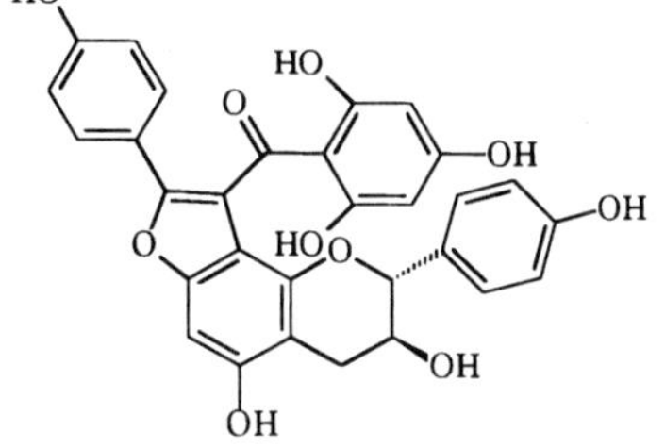

23

FIGURE 24.3. The structures of compounds 18-23 from *Wikstroemia indica*

CONCLUSION

The study of novel, biologically active agents from natural sources contributes to the development of effective screening systems. This is also true for screening antifungal compounds against rice blast fungus, *P. oryzae*. Although different from classical MIC tests and other enzyme-based bioassays, the bioassay-detecting morphological deformation of mycelia and conidia of *P. oryzae* has been developed (Kobayashi et al., 1996), though the mechanism has not yet been figured out. The application of this bioassay in TCM has been successfully proven in screening bioactive agents against *P. oryzae* through a five-year investigation in 247 species of TCM. Typically, active TCM extracts which induced morphological deformation of mycelia and conidia of *P. oryzae* were studied. These targeted herbal TCMs might have antifungal compounds with mechanism(s) of action. If so, the findings would probably open up new areas in antimycotic research, not only for the control of phytopathogenic fungi, such as, *P. oryzae,* but also for the treatment of human pathogenic fungi.

With *P. oryzae* bioassay, fourteen (1-14), three (15-17), and six (18-23) antifungal compounds from *D. collettii* var. *hypoglauca* rhizome (Dioscoreaceae) (see Figure 24.1), *S. nigrum* whole plant (Solanaceae) (see Figure 24.2), and *W. indica* root (Thymelaeaceae) (see Figure 24.3) were isolated respectively, by bioactivity-guided fractionation (Hu et al., 1996, 1997a,b; Hu, Kobayashi, Dong, et al., 1998; Hu, Kobayashi, Jing, et al., 1998; Dong et al., 1999; Hu et al., 2000). The compounds 1-23 have diversified chemical structures, including steroidal saponins, alkaloidal glycosides, coumarin, lignan, and biflavonoids. The structure-activity relationship has been analyzed. However, antifungal mechanisms against *P. oryzae* have not yet been known and are under investigation. Other active extracts of TCM are in study as well.

REFERENCES

Aizawa, S., Nakamura, Y., Shirato, S., Taguchi, R., Yamaguchi, I., and Misato, T. (1969). Aabomycin A, a new antibiotic. I. Production, isolation and properties of aabomycin A. *J. Antibiotics* 22: 457-462.

Atkins, J. G. (1962). Prevalence and distribution of pathogenic races of *Pyricularia oryzae* in the US. *Phytopathol.* 52: 2.

Bartnicki-Garcia, S. (1968). Cell wall chemistry, morphogenesis, and taxonomy of fungi. *Annu. Rev. Microbiol.* 22: 87-108.

Binks, P. R., Robson, G. D., Goosey, M. W., and Trinci, A. P. (1993). Inhibition of phosphatidylcholine and chitin biosynthesis in *Pyricularia oryzae, Botrytis fabae* and *Fusarium graminearum*. *J. Gen. Micro.* 139: 1371-1377.

Bonman, J. M., Vergei de Dios, T. I., and Khin, M. M. (1986). Physiologic specialization of *Pyricularia oryzae* in the Philippines. *Plant Dis.* 70: 767-769.

Brain, P. W. (1949). Studies on the biological activity of griseofulvin. *Ann. Botany* 13: 59-77.

Chen, H. C., Chou, C. K., Kuo, Y. H., and Yeh, S. F. (1996). Identification of a protein kinase C (PKC) activator, daphnoretin, that suppresses hepatitis B virus gene expression in human hepatoma cells. *Biochem. Pharm.* 52: 1025-1032.

Cohen, E. (1987). Chitin biochemistry: Synthesis and inhibition. *Annu. Rev. Entomol.* 32: 71-93.

Doepke, W., Duday, S., and Matos, N. (1987). Alkaloids and sapogenins from *Solanum nigrum*. *Z. Chem.* 27: 64.

Doepke, W., Duday, S., and Matos, N. (1988). 23-O-acetyl-12β-hydroxysolasodine, a new alkaloid from *Solanum nigrum*. *Z. Chem.* 28: 185-186.

Flamini, G., Cioni, P. L., Cheong, C. S., and Chung, B. Y. (1999). Antimicrobial activity of the essential oil of *Calamintha nepeta* and its constituent pulegone against bacterial and fungi. *Phytother. Res.* 13: 349-351.

Geng, J. X., Wang, L. X., Xu, Y. S., Du, Y. X., and Chen, S. R. (1988). Isolation and identification of the polysaccharide of Indian stingbush *(Wikstroemia indica)*. *Chin. Trad. Herbal Drugs* 19: 6-8.

Gow, N. A. R., Robbin, P. W., Lester, J. W., Brown, A. J. P., Fonzi, W. A., Chapman, T., and Kinsmon, O. S. (1994). A hyphal-specific chitin synthase gene (CHS2) is not essential for growth, dimorphism, or virulence of *Candida albicans*. *Proc. Natl. Acad. Sci. USA* 91: 6216-6220.

Gunji, S., Arima, K., and Beppu, T. (1983). Screening of antifungal antibiotics according to activities inducing morphological abnormalities. *Agric. Biol. Chem.* 47: 2061-2069.

Hamada, M., Hashimoto, T., Takahashi, T., Yokoyama, S., Miyake, M., Takeuchi, T., Okami, Y., and Umezawa, H. (1965). Antimicrobial activity of kasugamycin. *J. Antibiotics Ser. A* 18: 104-106.

Harada, S., and Kishi, T. (1977). Isolation and characterization of a new nucleoside antibiotic, amipurimycin. *J. Antibiotics* 30: 11-16.

Hu, K., Dong, A. J., Kobayashi, H., Jing, Y. K., Iwasaki, S., and Yao, X. S. (1999). A new pregnane glycoside from rhizomes of *Dioscorea collettii* var. *hypoglauca*. *J. Nat. Prod.* 62: 299-301.

Hu, K., Dong, A. J., Yao, X. S., Kobayashi, H., and Iwasaki, S. (1996). Antineoplastic agents; I. Three spirostanol glycosides from rhizomes of *Dioscorea collettii* var. *hypoglauca*. *Planta Med.* 62: 573-575.

Hu, K., Dong, A. J., Yao, X. S., Kobayashi, H., and Iwasaki, S. (1997a). Antineoplastic agents; II. Four furostanol glycosides from rhizomes of *Dioscorea collettii* var. *hypoglauca*. *Planta Med.* 63: 161-165.

Hu, K., Dong, A. J., Yao, X.S., Kobayashi, H., and Iwasaki, S. (1997b). A furostanol glycoside from rhizomes of *Dioscorea collettii* var. *hypoglauca*. *Phytochemistry* 44: 1339-1342.

Hu, K., Kobayshi, H., Dong, A. J., Iwasaki, S., and Yao, X. S. (1998). A new pregnane glycoside from *Dioscorea collettii* var. *hypoglauca*. *Chin. Chem. Lett.* 9: 739-742.

Hu, K., Kobayashi, H., Dong, A. J., Iwasaki, S., and Yao, X. S. (2000). Anti-fungal, anti-mitotic and anti-HIV agents from roots of *Wikstroemia indica*. *Planta Med.* 66: 564-567.

Hu, K., Kobayashi, H., Dong, A. J., Jing, Y. K., Iwasaki, S., and Yao, X. S. (1999). Antineoplastic agents III: Three steroidal glycosides from *Solanum nigrum*. *Planta Med.* 65: 35-38.

Hu, K., Kobayashi, H., Jing, Y. K., Dong, A. J., Yao, X. S., and Iwasaki, S. (1998). Cytotoxicity of steroidal glycosides from *Solanum nigrum*. *Ann. Oncol.* 9: 159.

Huang, K. T., Misato, T., and Asuyama, H. (1964). Effect of blasticidin S on protein synthesis of *Pyricularia oryzae*. *J. Antibiot. Ser. A* 17: 65-70.

Hwang, E. I., Yun, B. S., Kim, Y. K., Kwon, B. M., Kim, H. G., Lee, H. B., Bae, K. S., and Kim, S. U. (2000). Chaetoatrosin A, a novel chitin synthase II inhibitor produced by *Chaetomium atrobrunneum* F449. *J. Antibiotics* 53: 248-255.

Hwang, E. I., Yun, B. S., Kim, Y. K., Kwon, B. M., Kim, H. G., Lee, H. B., Jeong, W. J., and Kim, S. U. (2000). Phellinsin A, a novel chitin synthase inhibitor produced by *Phellinus* sp. PL3. *J. Antibiotics* 53: 903-911.

Ishiyama, T., Hara, I., Matsuoka, M., Sato, K., Shimada, S., Izawa, R., Hashimoto, T., Hamada, M., Okami, Y., Takeuchi, T., et al. (1965). Studies on the preventive effect of kasugamycin on rice blast. *J. Antibiotics Ser. A* 18: 115-119.

Isono, K., Kobinata, K., and Suzuki, S. (1968). Isolation and characterization of polyoxins J, K and L, new components of polyoxin complex. *Agric. Biol. Chem.* 32: 792-793.

Itoi, S., Nozu M., Satou, F., Yamamoto, J., Noda, T., and Utida, T. (1978). *Pyricularia* sp. hosting bamboo and bamboo-grass. *Jpn. J. Phytopathol.* 44: 209-213.

Kagabu S., and Kurahashi Y. (1998). Carpropamid, a new fungicide for rice blast control: Its selection based on stereoisomer-activity relationships. *J. Pestic. Sci.* 23: 145-147.

Kato, A., Hashimoto, Y., and Kidokoro, M. (1979). (+)-Nortrachelogenin, a new pharmacologically active lignan from *Wikstroemia indica*. *J. Nat. Prod.* 42: 159-162.

Kiyosawa, S. (1982). Genetic and epidemiological modeling of break-down of plant disease resistance. *Annu. Rev. Phytopathol.* 20: 93-117.

Ko, F. K., Chang, Y. L., Kuo, Y. H., Lin, Y. L., and Teng, C. M. (1993). Daphnoretin, a new protein kinase C activator isolated from *Wikstroemia indica*. *Biochem. J.* 295: 321-327.

Kobayashi, H., Namikoshi, M., Yoshimoto, T., and Yokochi, T. (1996). A screening model for antimitotic and antifungal substances using conidia of *Pyricularia oryzae*, modification and application to tropical marine fungi. *J. Antibiotics* 49: 873-879.

Kobayashi, H., Sunaga, R., Furihata, K., Morisaki, N., and Iwasaki, S. (1995). Isolation and structures of an antifungal antibiotic, fusarielin A, and related compounds produced by a *Fusarium* sp. *J. Antibiotics* 48: 42-52.

Lee, K. H., Tagahara, K., Suzuki, H., Wu, R. Y., Haruna, M., and Hall, I. H. (1981). Antitumor agents. 49. Tricin, kaempferol-3-O-beta-D-glucopyranosie and (+)-nortrachelogenin, antileukemic principles from *Wikstroemia indica*. *J. Nat. Prod.* 44: 530-535.

Leung, H., and Taga, M. (1988). *Magnaporthe grisea* (*Pyricularia oryzae* species), the blast fungus. *Adv. Plant Pathol.* 6: 175-188.

Li, J. Y., Strobel, G., Harper, J., Lobkovsky, E., and Clardy, J. (2000). Cryptocin, a potent tetramic acid antimycotic from the endophytic fungus *Cryptosporiopsis* cf. *quercina*. *Org. Lett.* 2: 767-770.

Lou, W., Yang, Y. Q., and Chen, Y. Y. (1984). Isolation and identification of steroidal sapogenins from rhizome of *Dioscorea collettii* var. *hypoglauca*. *Acta Botanica Yunnanica* 6: 461-462.

Maeda, T., Abe, H., Kakiki, K., and Misato, T. (1970). Studies on the mode of action of organophosphorous, Kitazin. Part II. Accumulation of an amino sugar derivative from Kitazin-treated mycelia of *Pyricularia oryzae*. *Agric. Biol. Chem.* 34: 700-709.

Motoyama T., Imanishi K., Kinbara T., Kurahashi Y., and Yamaguchi I. (1998). Inhibition of scytalone dehydratase in melanin biosynthesis by carpropamid, a novel rice blast controlling agent. *J. Pestic. Sci.* 23: 58-61.

Nakasako, M., Motoyama, T., Kurahashi, Y., and Yamaguchi, I. (1998). Cryogenic x-ray crystal structure analysis for the complex of scytalone dehydratase of a rice blast fungus and its tight-binding inhibitor, carpropamid: The structural basis of tight binding inhibition. *Biochemistry* 37: 9931-9939.

Nawwar, M. A. M., El-Mousallamy, A. M. D., and Barakat, H. H. (1989). Quercetin 3-glycosides from the leaves of *Solanum nigrum*. *Phytochemistry* 28: 1755-1757.

Okeke, B., Seigle-Murandi, F., and Steiman, R. (1995). Biological activity of a heptaketide metabolite from *Pleiochaeta setosa*. *Biosci. Biotechnol. Biochem.* 59: 173-175.

Okeke, B., Seigle-Murandi, F., Steiman, R., and Buarque de Gusmao, N. (1994). Fungal metabolite extracts active against phytopathogens. *Sci. Total Environ.* 155: 125-130.

Otake, N., Takeuchi, S., Endo T., and Yonehara, H. (1966). Chemical studies on blasticidin S. III. The structure of blasticidin S. *Agrc. Biol. Chem.* 30: 132-141.

Padmanabhan, S. Y., Chakrabartl, N. K., Mathur, S. C., and Veeraaghovon, J. (1970). Identification of pathogenic races of *Pyricularia oryzae*. *Phytopathol.* 60: 1574-1577.

Park, I. C., Horiuchi, H., Hwang, C. W., Yeh, W. H., Ohta, A., Ryu, J. C., and Takagi, M. (1999). Isolation of *csm1* encoding a class V chitin synthase with a myosin motor-like domain from the rice blast fungus, *Pyricularia oryzae*. *FEMS Micro. Lett.* 170: 131-139.

Parsons, K. A., Chumley, F. G., and Valent, B. (1987). Genetic transformation of the fungal pathogen responsible for rice blast disease. *Proc. Natl. Acad. Sci. USA* 84: 4161-4165.

Richmond, D. V. (1975). Effects of toxicants on the morphology and fine structures of fungi. *Adv. Appl. Microbiol.* 19: 289-319.

Ridout, C. L., Price, K. R., Coxon, D. T., and Fenwick, G. R. (1989). Glycoalkaloids from *Solanum nigrum* L., α-solamargine and α-solasonine. *Pharmazie* 44: 732-733.

Saijo, R., Murakami, K., Nohara, T., Tomimatsu, T., Sato, A., and Matsuoka, K. (1982). Studies on the constituents of *Solanum nigrum*. II. On the constituents of the immature berries of *Solanum nigrum* L. *Yakugaku Zasshi* 102: 300-306.

Sakagami, Y., Sekine, H., Yamabayashi, S., Kitaura, Y., Ueda, A., and Kosaka, Y. (1966). The studies on bramycin, a new antifungal antibiotic. I. Isolation, purification and properties of bramycin. *J. Antibiotics Ser. A* 19: 99-103.

Satoh, A., Utamura, H., Ishizuka, M., Endo, N., Tsuji, M., and Nishimura, H. (1996). Antimicrobial benzopyrans from the receptacle of sunflower. *Biosci. Biotechnol. Biochem.* 60: 664-665.

Scheinpflug, H., and Jung, H. F. (1968). Organophosphates for the control of fungal disease of crops. *Pflanzenschutz-Nachrichten* 21: 79-91.

Shomura, T., Hamamoto, K., Ohashi, T., Amano, S., Yoshida, J., Moriyama, C., and Niida, T. (1967). New antibiotics, miharamycins A and B. II. Some biological characteristics of miharamycin. *Sci. Rep. Meiji Seika Kaisha* 9: 5-10.

Suzuki, H., Lee, K. H., Haruna, M., Iida, T., Ito, K., and Huang, H. C. (1982). (+)-Arctigenin, a lignan from *Wikstroemia indica*. *Phytochemistry* 21: 1824-1825.

Suzuki, Y., Takahashi, H., Esumi, Y., Aric, T., Morita, T., Koshino, H., Uzawa, J., Uramoto, M., and Yamaguchi, I. (2000). Haematocin, a new antifungal diketopiperazine produced by *Nectria haematococca* Berk. Et Br. (880701a-1) causing nectria blight disease on ornamental plants. *J. Antibiotics* 53: 45-49.

Takeuchi, S., Hirayama, K., Ueda, K., Saki, H., and Yonehara, H. (1958). Blasticidin S, a new antibiotic. *J. Antibiotics Ser. A* 11: 1-5.

Tamamura, T., and Sato, K. (1999). Comparative studies on *in vitro* activities of kasugamycin and clinically-used aminoglycoside antibiotics. *Jpn. J. Antibiotics* 52: 57-67.

Tandon, S., and Rastogi, R. P. (1976). Wikstromol, a new lignan from *Wikstroemia viridiflora*. *Phytochemistry* 15: 1789-1791.

Tang, S. R., Jiang, Z. D., and Pang, Z. J. (1986). Isolation and identification of steroid saponins of *Dioscorea collettii* Hook f. var. *hypoglauca* (Palibin) Pei et Ting (II). *Acta Botanica Sinica* 28: 453-456.

Tang, S. R., and Pang, Z. J. (1984). Isolation and identification of steroid saponins of *Dioscorea collettii* Hook f. var. *hypoglauca* (Palibin) Pei et Ting. *Acta Botanica Sinica* 26: 419-424.

Tsuruo, T., Oh-hara, T., Iida, H., Tsukagoshi, S., Sato, Z., Matsuda, I., Iwasaki, S., Okuda, S., Shimizu, F., Sasagawa, K., et al. (1986). Rhizoxin, a macrocyclic lactone antibiotic, as a new antitumor agent against human and murine tumor cells and their vincristine-resistant sublines. *Cancer Res.* 46: 381-385.

Tsuruoka, T., Yumoto, H., Ezaki, N., and Niida, T. (1967). New antibiotics, miharamycins A and B. I. Isolation and characterization of miharamycins A and B. *Sci. Rep. Meiji Seika Kaisha* 9: 1-4.

Umezawa, H., Okami, Y., Hashimoto, T., Suhara, Y., Hamada, M., and Takeuchi, T. (1965). A new antibiotic, kasugamycin. *J. Antibiotics Ser. A* 18: 101-103.

Varshney, I. P., and Dube, N. K. (1970). Chemical investigation of *Solanum nigrum* berries. *J. Indian Chem. Soc.* 47: 717-718.

Wang, J. P., Ruang, S. L., Kuo, Y. H., and Teng, C. M. (1995). Daphnoretin-induced respiratory burst in rat neutrophils is, probably, mainly through protein kinase C activation. *Euro. J. Pharm.* 288: 341-348.

Wang, Z. D., Yan, P., and Yu, D. Q. (1985). Studies on the constituents of *Wikstroemia indica*. *Chin. Trad. Herbal Drugs* 16: 39-40.

Xie, P. S., and Yang, J. P. (1978). Studies on the constituents of *Wikstroemia indica:* Isolation and identification of daphnoretin. *Chin. Trad. Herbal Drugs* 9: 97-101.

Yamada, M. (1985). Pathogenic specialization of rice blast fungus in Japan. *TARO* 19: 178-183.

Zhang, H., Tomoda, H., Tabata, N., Oohori, M., Shinose, M., Takahashi, Y., and Omura, S. (1999). Zelkovamycin, a new cyclic peptide antibiotic from *Streptomyces* sp. K96-0670. I. Production, isolation and biological properties. *J. Antibiotics* 52: 29-33.

Index

Page numbers followed by the letter (t) indicate tables.